CMOS
Analog
Circuit
Design

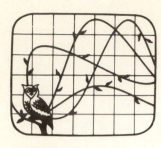

HRW
Series in
Electrical and
Computer Engineering

M. E. Van Valkenburg, Series Editor Electrical Engineering
Michael R. Lightner, Series Editor Computer Engineering

CMOS
Analog
Circuit
Design

PHILLIP E. ALLEN
Georgia Institute of Technology

DOUGLAS R. HOLBERG
Crystal Semiconductor, Inc.

HOLT, RINEHART AND WINSTON, INC.
New York   Chicago   San Francisco   Philadelphia
Montreal   Toronto   London   Sydney   Tokyo
Mexico City   Rio de Janeiro   Madrid

To Margaret and Candy
And in memory of Edward R. Klinkovsky

Acquisitions Editor	Deborah Moore
Publisher	Ted Buchholz
Senior Project Manager	Chuck Wahrhaftig
Production Manager	Paul Nardi
Design Supervisor	Bob Kopelman

Library of Congress Cataloging-in-Publication Data
Allen, P. E. (Phillip E.)
  CMOS analog circuit design.

  Bibliography: p.
  Includes index.
  1. Linear integrated circuits.   2. Metal oxide
semiconductors, Complementary.   I. Holberg, Douglas R.
II. Title.
TK7874.A428 1987   621.381′73   86-14275

ISBN 0-03-006587-9

7 8 9   038   9 8 7 6 5 4 3 2

Holt, Rinehart and Winston
The Dryden Press
Saunders College Publishing

preface

The objective of this book is to present the principles and techniques of the design of analog circuits that are to be implemented in a CMOS technology. It has been developed with two goals in mind. The first is to teach the methodology of analog integrated circuit design by using a hierarchically-oriented approach to the subject. The second goal is to mix the academic and practical viewpoints in a treatment that is neither superficial nor overly detailed. Many of the circuits, techniques, and principles presented are derived from the industrial experience and knowledge of the authors and the many other individuals who have contributed to this text both directly and indirectly.

Prior to the mid-1970s, analog circuits were not widely implemented by integrated circuit technology, particularly MOS technology. This was because most analog circuits require a precise definition of a resistor, a capacitor, or of resistor-capacitor products. A precision resistor or precision capacitor is not within the capability of normal integrated-circuit technology. Analog circuits that were integrated did not require these elements internally. Examples of such circuits include the op amp and voltage regulators.

In the 1970s a technique emerged, called *switched-capacitor circuits,* that allowed resistor-capacitor products to be related to capacitor ratios. The result was analog circuits that were compatible with MOS technology. The relative accuracy of capacitors (and therefore the RC product) in MOS technology is better than 1%. As a consequence, simple and complex analog circuits and systems have been implemented in MOS technology and are now familiar products in the integrated-circuit marketplace.

This book was developed from several sources. One was notes used in a graduate course at Texas A&M University on analog circuit design. These notes were used to supplement the first edition of *Analysis and Design of*

v

Analog Integrated Circuits by Gray and Meyer [1], which covers the subject of analog bipolar integrated circuits. The second source was a 300-page set of notes titled "CMOS Analog Circuit Design" developed as part of a General Electric contract in 1981 to describe and identify the principles and techniques of CMOS analog integrated-circuit design. Early versions of this text were used in short courses taught at the Royal Institute of Technology in Stockholm and the Lockheed Research Laboratory at Palo Alto, California. More recently, these notes have been used as the text in a short course on CMOS Analog Integrated-Circuit Design at the Berlin Continuing Engineering Education Program, Schlumberger Well Services in Houston, Texas and in graduate courses at the Georgia Institute of Technology, Texas A&M University, the University of Nebraska, and Southern Methodist University.

The most difficult decision of this project was to omit bipolar analog circuits. The reasons for doing so are very good texts on bipolar analog design (there are no texts on CMOS analog design with the exception of the second edition of *Analysis and Design of Analog Integrated Circuits* [2]), CMOS has become the dominant technology for analog integrated circuit design, and to reduce the size of this book. Consequently, the niche that this book fulfills is the treatment of the design of analog circuits from a CMOS viewpoint. While there are many similarities between CMOS and BJT analog circuits, there are also some important differences. One of the more important of these differences is that the small signal performance of CMOS circuits depends strongly on the device geometry and the dc variables whereas bipolar depends only on the dc variables. This means the design of CMOS circuits has twice the degree of freedom of bipolar circuits. While CMOS technology has become a dominant technology for analog integrated circuits, it is probable that in the future some combination of bipolar and CMOS might become dominant as technology matures.

CMOS Analog Circuit Design should be useful both as a reference book in industry and as a textbook in senior or graduate courses on the subject. The authors assume that the reader is familiar with a basic course on electronics—including biasing, modeling, and analysis of electronic circuits, and has some familiarity with frequency response. The total picture of design—including modeling, simulation, and testing—is presented, so that after finishing this text, the reader should be able to undertake the design of an analog circuit that can be implemented by CMOS technology. Additionally, many pitfalls in the path of the beginning designer have been identified. The engineer in industry will find the book useful as both an introduction and aid to designing CMOS analog circuits. The expert will find the organization and emphasis on design methodology helpful in putting past design experience into perspective.

The organization of this text is illustrated in Table 1.1-2. Material throughout the text is related back to this schema, helping the reader to gain the hierarchical perspective of this study. Chapter 1 is an introductory chapter that gives an overview of the subject, defines notation and convention, makes a brief survey of analog signal processing and gives an example of analog CMOS design. Each chapter has a set of problems, most of which have been given as quiz or exam problems, that are designed to reinforce and further develop the student's understanding of the concepts of that chapter.

Chapters 2, 3, and 4 are written at the device level of the hierarchy shown in Table 1.1-2. Chapter 2 reviews CMOS technology as applied to MOS devices, passive components, and miscellaneous components such as a vertical bipolar device and a Zener diode. Design rules and their implications are also considered in this chapter. Chapter 3 introduces the key subject of modeling, which is used throughout the remainder of the text. Various levels of models for the active device are presented along with models for the passive elements, and the use of these models in simulation is considered. Chapter 4 is unusual in that it emphasizes the characterization of MOS devices, showing the designer methods and techniques that can be used to characterize the parameters of models presented in Chapter 3.

The next chapters, 5 and 6, move from the device level to the simple circuit level of the design hierarchy. These chapters present the basic circuits that will be used as components of the more complex circuits. Subjects covered include the MOS switch, active resistors, current sources, current mirrors, current amplifiers, the inverting amplifier, the differential amplifier, the cascode amplifier, and output amplifiers.

Chapters 7, 8, and 9 cover more complex circuits. These chapters are the heart of this text and best illustrate the design process. Comparators are covered first because they are basically an op amp without compensation. Thus dc characteristics are dealt with first. In Chapter 8, unbuffered op amps are considered. The general design approach to op amps is followed by the subject of compensation, and simple, two-stage op amp design is considered in detail followed by cascode op amps. This chapter also includes a section on the simulation and measurement of op amp performance. Then chapter 9 considers more sophisticated op amp designs having higher performance than the two-stage op amp. This includes op amps with improved performance in temperature independence, power supply independence, power dissipation, noise, and dynamic range. This chapter is an excellent example of how the performance of a circuit can be optimized.

Chapter 10 covers CMOS digital-analog and analog-digital converters. These circuits are first characterized and then studied from a performance viewpoint. Digital-analog converters are divided into voltage and current scaling, and serial types. Analog-digital converters include serial, medium speed, and high performance converters.

Chapter 11 covers analog circuits and systems. This chapter treats bandgap voltage references, multipliers, waveshaping circuits, and oscillators and timers. It also examines analog circuits and systems. Circuits from previous chapters are treated as blocks to permit the focus on higher level circuits and systems.

Finally, the appendices include information supplementing the text material and provide simulation techniques and information. They also present a brief review of circuit analysis for CMOS analog circuits, a calculator program for analyzing CMOS circuits, and the time-frequency domain relationships for a second-order system.

The material is more than sufficient for a 10- or 15-week course. A typical 10-week course would cover most of Chapters 2 through 8 with parts of 9 and 10 as appropriate and time permitting. Chapters 5 and 6 are key to the remainder and should be covered in as much detail as possible. A 15-week course could expand the material accordingly. If the reader has a good background in bipolar integrated-circuit design, then much of the material can be skipped by taking advantage of the similarity between the two technologies. However, the much stronger dependence of the performance of MOS circuits on their geometry should not be neglected.

The background required for this text is a good understanding of basic electronics. A good example of this background is found in *Microelectronics*, by Sedra and Smith [3]. Topics of importance include large-signal models, biasing, small-signal models, frequency response, feedback, and op amps. It is also helpful to have a background in semiconductor devices and their conduction mechanisms, integrated-circuit processing, simulation, and modeling. With this background, the reader could start at Chapter 4 or Chapter 5.

The authors gratefully acknowledge the fabrication of many of the circuits included in this text by Texas Instruments, Inc., and Harris Semiconductor, Inc. The support and encouragement of Dr. Demetrius T. Paris, Director of the School of Electrical Engineering at the Georgia Institute of Technology and the engineering management of Crystal Semiconductor, Inc., is also gratefully acknowledged. This text has been strongly influenced by the research support of the Semiconductor Research Corporation in the area of analog integrated-circuit design methodology.

The authors would also like to express their appreciation and gratitude to the many individuals who have contributed to the development of this text. These include graduate students and their instructors who used early versions of this text in classes at the Georgia Institute of Technology, Texas A&M University, the University of Nebraska, and Southern Methodist University. Graduate students who have made a special contribution include Marius Breevoort and Jesse Still at Georgia Tech. Professors J.A. Connelly, Randall Geiger, Mohammed Ismail, and John Lowell have been most helpful in using the text in their classes. We would also like to acknowledge the efforts of Mike and Myle Buchanan in proofreading the final manuscript. A special thanks must go to the many individuals in industry who have contributed examples and suggestions to this text. Finally, as usual in a project of this magnitude, there is one person who has contributed far above and beyond what could be reasonably expected. Both authors especially thank Sherri Brenner for her selfless commitment to this project. Lastly, we are happy to acknowledge that without the understanding and support of our families this text would not have been possible.

REFERENCES

1. P.R. Gray and R.G. Meyer, *Analysis and Design of Analog Integrated Circuits,* First Edition John Wiley & Sons, New York: NY, 1977.
2. Ibid, Second Edition, 1984.
3. A.S. Sedra and K.C. Smith, *Microelectronic Circuits,* Holt, Rinehart and Winston, New York: NY, 1982.

contents

CMOS
Analog
Circuit
Design

chapter 1

Introduction and Background

The development of VLSI technology, coupled with the demand for more signal processing integrated on a single chip, has resulted in a tremendous potential for the design of analog integrated circuits [1]. It is clear that analog circuits and systems have an important role in the implementation and application of VLSI technology. For example, most VLSI systems require analog circuits or systems such as reconstruction filters, digital-analog and analog-digital converters, voltage comparators, automatic gain-control circuits, and so on.

Unfortunately, the design techniques used for discrete analog circuits are not useful in designing the more complex analog circuits and systems implemented by VLSI technology. It has become necessary to examine closely the design process of analog circuits and to identify those principles that will increase design productivity and the designer's chances for suc-

1

cess. Thus, this book attempts a hierarchical organization of the subject of analog integrated-circuit design and identification of its general principles.

The objective of this chapter is to introduce the subject of analog integrated-circuit design and to lay the groundwork for what follows. It deals with the general subject of analog integrated-circuit design; the notation, symbology, and terminology used in this book is presented next; the next section discusses the general considerations of an analog signal-processing system; and the last section gives an example of analog CMOS circuit design. The reader may wish to review other topics pertinent to this study before continuing on to Chapter 2. Such topics include electronic modeling, computer simulation techniques, Laplace and z-transform theory, and semiconductor device theory.

1.1   *Analog Integrated Circuit Design*

The objective of analog circuit design is to transform specifications into circuits that satisfy those specifications. It is a challenging activity because the problem has many variables and many decisions must be made to achieve a successful design. Depending on their individual experience and background, different designers may use different approaches to implement the same set of specifications. Until recently, the really creative analog circuit designs were relegated to the domain of a few "gifted" designers. However, as the complexity of analog circuit design has grown, it has become necessary to use the concepts of regularity, partitioning, and design hierarchy. These concepts, supported by cleverly designed computer tools, have begun to open the design of analog circuits to a much wider group. The objective of this book is to prepare that wider group of designers for their task.

Integrated-circuit design is separated into two major categories: analog and digital. To characterize these two design methods we must first define analog and digital signals. A *signal* will be considered to be any detectable value of voltage, current, or charge. A signal should convey information about the state or behavior of a physical system. An *analog signal* is a signal that is defined over a continuous range of time and a continuous range of amplitudes. Fig. 1.1-1(a) illustrates an analog signal. A *digital signal* is a signal that is defined only at discrete values of time and amplitude. Typically, the digital signal is a binary weighted sum of signals having only two defined values of amplitude as illustrated in Fig. 1.1-1(b) and shown in Eq. (1).

$$D = b_1 2^{-1} + b_2 2^{-2} + b_3 2^{-3} + \cdots + b_N 2^{-N} = \sum_{i=1}^{N} b_i 2^{-i} \qquad \textbf{(1)}$$

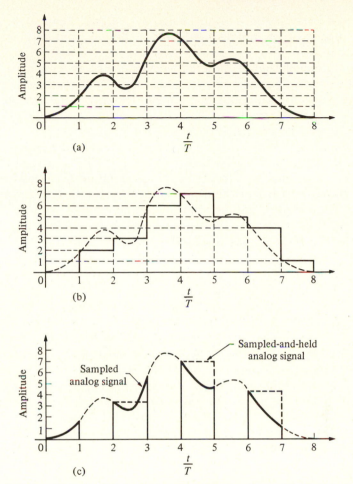

Figure 1.1-1   Signals. (a) Analog or continuous time. (b) Digital. (c) Analog sampled data or discrete time. T is the period of the digital or sampled signals.

The individual binary numbers, b_i, have a value of either zero or one. Consequently, it is possible to implement digital circuits using circuits with only two defined states. This leads to a great deal of regularity and to an algebra that can be used to describe the function of the circuit. As a result, digital circuit designers have been able to adapt readily to the design of more complex integrated circuits.

Another type of signal encountered in analog integrated-circuit design is an analog sampled-data signal. An *analog sampled-data signal* is a signal that is defined over a continuous range of amplitudes but only at discrete values of time. Switched-capacitor techniques result in this type of signal. Often the sampled analog signal is held at the value of the beginning of the period, resulting in a sampled-and-held signal. Figure 1.1-1(c) illustrates an analog sampled-data signal and a sampled-and-held analog sampled-data signal.

Circuit design is the creative process of developing a solution to a problem. Design can be better understood by comparing it to analysis. The *analysis* of a circuit, illustrated in Fig. 1.1-2(a), is the process by which one starts with the circuit and finds its properties. An important characteristic of the analysis process is that the solution or properties are unique. On the other hand, the synthesis or *design* of a circuit is the process by which one starts with a desired set of properties and finds a circuit which satisfies them. In a design problem the solution is not unique. This lets the designer be creative. Consider the design or synthesis of a 1.5 ohm resistance as a simple example. This resistance could be realized as the cascade of three 0.5 ohm resistors, the combination of a 1 ohm resistor cascaded with two 1 ohm resistors in parallel, and so on. All would satisfy the requirement of 1.5 ohms resistance although some might exhibit other properties that would favor their use. Fig. 1.1-2 illustrates the difference between synthesis, or design, and analysis.

Integrated-circuit design is clearly a technology-driven activity. As integrated-circuit technology has become smaller, more reliable, and better understood, the challenge to utilize technology fully has caused a radical change in circuit design. The designer must now understand the process deeply enough to be able to model, lay out, and test his circuits, for the design of an integrated circuit can rarely be separated from the technology used to build it. The implications of projected advances in technology are overwhelming to the future designer. For instance, Fig. 1.1-3 shows the trends in feature size (minimum device dimension) and number of components per chip as a function of calendar years [2]. This is an excellent

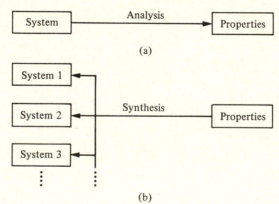

Figure 1.1-2   (a) The process of analysis. (b) The design process.

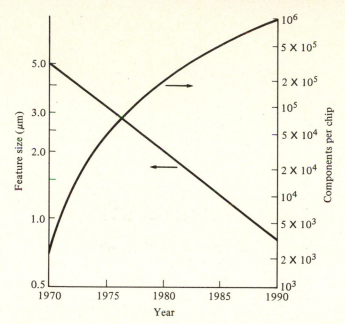

Figure 1.1-3 The trends in integrated-circuit feature size and number of components per chip over two decades.
[J. Cornell, "On VLSI Applications," NSF Workshop, Aug. 1982]

illustration of how technology influences circuit design. As technology improves, the minimum feature size decreases, making it possible to fabricate larger chips.

As a consequence of increased technological capability, the time needed to define and design an integrated circuit is also rapidly increasing. Fig. 1.1-4 shows that the time required to define and design an integrated circuit is doubling every 32 months [3]. While there may be some deviation from these trends, it is clear that the time required to define, create, and market a design must be drastically reduced in order to take advantage of the technological capability that presently exists.

The differences between integrated and discrete analog circuit design are important. Discrete circuits use active and passive components that are not on the same substrate, whereas all of the components of an integrated circuit are on the same substrate. The most obvious difference between the two design methods is that the geometry of the active devices and passive components in integrated circuit design are under the control of the designer. This gives the designer an entirely new degree of freedom in the design process. A second difference is that it is not feasible to breadboard the integrated-circuit design. Consequently, the designer must turn to sim-

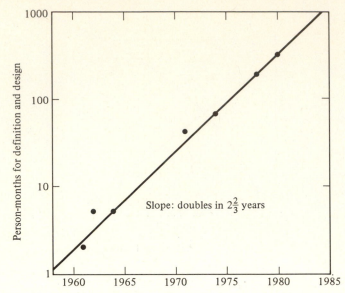

Figure 1.1-4 Time in person-months required to define and design integrated circuits as a function of calendar time.
(© 1979 IEEE)

ulation methods to confirm the design's performance. Another difference is that the integrated circuit designer is restricted to a limited class of components which are compatible with integrated-circuit technology.

The task of designing an analog integrated circuit includes many steps. Fig. 1.1-5 illustrates the general approach to the design of an integrated circuit. The major steps in the design process are:

1. Definition
2. Synthesis or implementation
3. Simulation or modeling
4. Geometrical description
5. Simulation including the geometrical parasitics
6. Fabrication
7. Testing and verification

The designer is responsible for all of these steps except fabrication. The first major task is to define and synthesize the design. This step is crucial since it determines the performance capability of the design. When this task is completed, the designer must be able to confirm his (or her) design before it is fabricated. This leads to the second major task—using simulation methods to predict the performance of the circuit. At this point, the designer may iterate using the simulation results to improve the circuit's

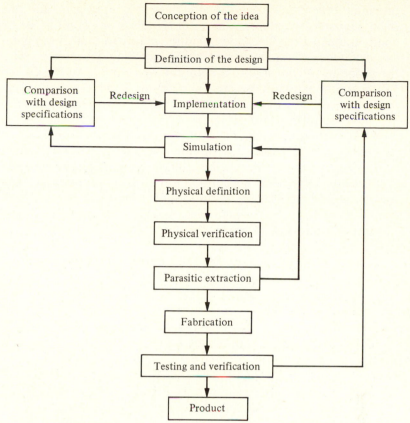

Figure 1.1-5   The design process for analog integrated circuits.

performance. Once satisfied with this performance, the designer can attack the third major task—a geometrical description (layout) of the circuit. This geometrical description typically consists of a computer database of variously shaped rectangles or polygons (in the x-y plane) at different levels (in the z-direction); it is intimately connected with the electrical performance of the circuit. Once the layout is finished, it is necessary to include the geometrical effects in a second simulation. If the results are satisfactory, the circuit is ready for fabrication. Then the designer is faced with the last major task—determining whether the fabricated circuit meets the design specifications. If the designer has not carefully considered this step in the overall design process, it is often impossible to test the circuit and determine whether or not the specifications have been met.

One distinction between discrete and integrated analog-circuit design is that it may be impractical to breadboard the integrated circuit. Computer simulation techniques have been developed that have several advantages, providing the models are adequate. These advantages include: elimination of breadboards, the ability to monitor signals at any point in the circuit, the ability to open a feedback loop, and the ability to easily modify the circuit. Disadvantages of computer simulation include: poor models, the failure of the program to converge to a solution, and the use of the computer as a substitute for thinking. Because simulation is closely associated with the design process, it will be included in the text where appropriate.

In accomplishing the design steps described above, the designer works with three different types of description formats. These include: the design description, the physical description, and the model/simulation description. The format of the design description is the means by which the circuit is specified; the physical description format is the geometrical definition of the circuit; the model/simulation format is the means by which the circuit can be simulated. The designer must be able to describe his design in each of these formats. For example, the first steps of analog integrated-circuit design could be carried out in the design description format. The geometrical description obviously uses the geometrical format. The simulation steps would use the model/simulation format.

Analog integrated-circuit design can also be characterized from the viewpoint of hierarchy. Table 1.1-1 shows a vertical hierarchy consisting of *devices, circuits,* and *systems,* and horizontal description formats consisting of *design, physical,* and *model.* The device level is the lowest level of design. It is expressed in terms of device specifications, geometry, or model parameters for the design, physical, and model description formats, respectively. The circuit level is the next higher level of design and can be expressed in terms of devices. The design, physical, and model description formats typically used for the circuit level include voltage and current relationships, parameterized layouts, and macromodels. The highest level of design is the systems level—expressed in terms of circuits. The design, physical, and model description formats for the systems level include mathematical or graphical descriptions, a chip floorplan, and a behavioral model.

This book on CMOS analog-circuit design has been organized as to emphasize the hierarchical viewpoint of integrated-circuit design, as illus-

Table 1.1-1

Hierarchy and Description of the Analog Circuit Design Process.

Hierarchy	Design	Physical	Model
Systems	System Specifications	Floorplan	Behavioral Model
Circuits	Circuit Specifications	Parameterized Blocks/Cells	Macromodels
Devices	Device Specifications	Geometrical Description	Device Models

Table 1.1-2

Relationship of the Book Chapters to Analog Circuit Design.

Design Level	MOS Technology		
Systems	Chapter 10 D/A and A/D Converters		Chapter 11 Analog Circuits and Systems
Complex Circuits	Chapter 7 Comparators	Chapter 8 Unbuffered Op Amps	Chapter 9 High-Performance Op Amps
Simple Circuits	Chapter 5 Analog Subcircuits		Chapter 6 Amplifiers
Devices	Chapter 2 Technology	Chapter 3 Modeling	Chapter 4 Characterization

trated in Table 1.1-2. At the device level, Chapters 2 through 4 deal with CMOS technology, models, and characterization. In order to design CMOS analog integrated circuits the designer must understand the technology, so Chapter 2 gives an overview of CMOS technology, along with the design rules that result from technological considerations. This information is important for the designer's appreciation of the constraints and limits of the technology. Before starting a design, one must have access to the process and electrical parameters of the device model. Modeling is a key aspect of both the synthesis and simulation steps and is covered in Chapter 3. The designer must also be able to characterize the actual model parameters in order to confirm the assumed model parameters. Ideally, the designer has access to a test chip from which these parameters can be measured. Finally, the measurement of the model parameters after fabrication can be used in testing the completed circuit. Characterization techniques are covered in Chapter 4.

Chapters 5 and 6 discuss circuits consisting of two or more devices that are classified as simple circuits. These simple circuits are used to design more complex circuits, which are covered in Chapters 7 through 9. Finally, the circuits presented in Chapters 7 through 9 are used in Chapters 10 and 11 to implement analog systems. Some of the dividing lines between the various levels will at times be unclear. However, the general relationship is valid and should leave the reader with an organized viewpoint of analog integrated circuit design.

1.2 *Notation, Symbology, and Terminology*

To help the reader have a clear understanding of the material presented in this book, this section dealing with notation, symbology, and terminology is included. The conventions chosen are consistent with those used in undergraduate electronic texts and with the standards proposed by technical societies. The International System of Units has been used throughout. Every effort has been made in the remainder of this book to use the conventions here described.

The first item of importance is the notation (the symbols) for currents and voltages. Signals will generally be designated as a quantity with a subscript. The quantity and the subscript will be either upper or lower case according to the convention illustrated in Table 1.2-1. Figure 1.2-1 shows how the definitions in Table 1.2-1 would be applied to a periodic signal superimposed upon a dc value.

This notation will be of help when modeling the MOS device. For example, consider the portion of the MOS model that relates the drain-source current to the various terminal voltages. This model will be developed in terms of the total instantaneous variables (i_D). For biasing purposes, the dc variables (I_D) will be used; for small signal analysis, the ac variables (i_d); and finally, the small signal frequency discussion will use the complex variable (I_d).

The second item to be discussed here is what symbols are used for the various components. (Most of these symbols will already be familiar to the reader. However, inconsistencies exist about the MOS symbol shown in Fig. 1.2-2.) The symbols shown in Figs. 1.2-2(a) through 1.2-2(d) are used for enhancement- and depletion-mode MOS transistors when the substrate or bulk (*B*) is connected to the source. Although the transistor operation will be explained later, the terminals are called *drain* (*D*), *gate* (*G*), and *source* (*S*). The advantage of these symbols is that they are very similar to the widely-used BJT symbols. If the bulk is not connected to the source, then the symbols shown in Fig. 1.2-2(e) through Fig. 1.2-2(h) are used for the enhancement- and depletion-mode MOS transistors. It will be important to know where the bulk of the MOS transistor is connected when it is used in circuits.

Fig. 1.2-3 shows another set of symbols that should be defined. Fig. 1.2-3(a) represents a differential-input op amp while Fig. 1.2-3(b) repre-

Table 1.2-1

Definition of the Symbols for Various Signals.

Signal Definition	Quantity	Subscript	Example
Total instantaneous value of the signal	Lowercase	Uppercase	q_A
dc value of the signal	Uppercase	Uppercase	Q_A
ac value of the signal	Lowercase	Lowercase	q_a
Complex variable, phasor, or rms value of the signal	Uppercase	Lowercase	Q_a

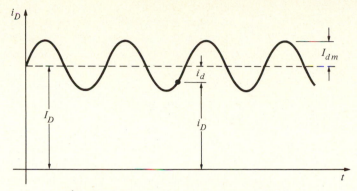

Figure 1.2-1   The notation for signals.

sents a finite-gain, differential amplifier. We will also use Fig. 1.2-3(b) to represent a comparator which may have a gain approaching that of the op amp. Figs. 1.2-3(c) and (d) represent an independent voltage and current source. Often, the battery symbol is used instead of Fig. 1.2-3(c). Finally, Figs. 1.2-3(e) through (h) represent the four types of ideal controlled sources. Fig. 1.2-3(e) is a voltage-controlled, voltage source (VCVS), Fig. 1.2-3(f) is a voltage-controlled, current source (VCCS), Fig. 1.2-3(g) is a current-controlled, voltage source (CCVS), and Fig. 1.2-3(h) is a current-controlled, current source (CCCS). The gains of each of these controlled sources are given by the symbols A_v, G_m, R_m, and A_i (for the VCVS, VCCS, CCVS, and CCCS, respectively).

The last subject of this section is the introduction of a subscript notation to help describe some characteristic or property of three-terminal devices. Fig. 1.2-4 shows an arbitrary three-terminal device. The general form of the notation is

$$Q_{ABC} \tag{1}$$

The quantity Q is some characteristic or property of the three-terminal device which is associated with terminals A and B. Examples of the quantity, Q, are voltage, current, and capacitance. The subscripts A and B refer to two of the terminals associated with the three-terminal device while the subscript C designates the condition of the third terminal with respect to terminal B. Normally, terminal A is the device terminal with the larger magnitude of dc potential and terminal B is the device terminal with the smaller magnitude of dc potential (typically ground or zero). The condition of terminal C with respect to terminal B is indicated by the following convention:

$C = O$ [There is infinite resistance between terminals B and C]

$C = S$ [There is zero resistance between terminals B and C]

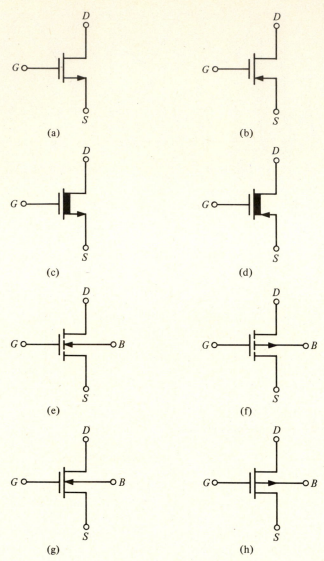

Figure 1.2-2 Examples of MOS device symbols. (a) Enhancement NMOS with $V_{BS} = 0$. (b) Enhancement PMOS with $V_{BS} = 0$. (c) Depletion NMOS with $V_{BS} = 0$. (d) Depletion PMOS with $V_{BS} = 0$. (e) through (h) same as (a) through (d) except $V_{BS} \neq 0$.

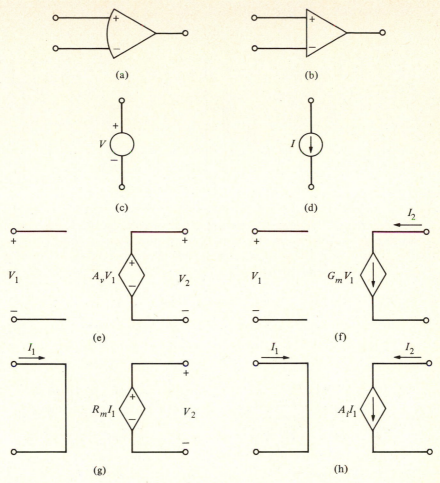

Figure 1.2-3 (a) Symbol for an op amp. (b) Symbol for a finite-gain, differential input, amplifier or comparator. (c) Independent voltage source. (d) Independent current source. (e) Voltage-controlled, voltage source (VCVS). (f) Voltage-controlled, current source (VCCS). (g) Current-controlled, voltage source (CCVS). (h) Current-controlled, current source (CCCS).

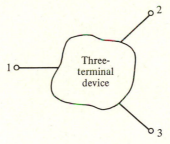

Figure 1.2-4 Arbitrary three-terminal device.

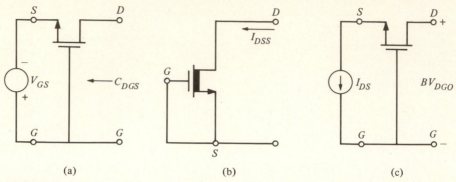

(a)   (b)   (c)

Figure 1.2-5   Examples of three-terminal notation. (a) Capacitance from drain to gate with the gate shorted to source. (b) Drain-source current when gate is shorted to source. (c) Breakdown voltage from drain to gate when the gate is open-circuited with respect to the source.

$C = R$ [There is a finite resistance between terminals B and C]

$C = X$ [An independent voltage source in series with a resistance is connected between terminals B and C in such a manner as to reverse bias a pn junction]

Fig. 1.2-5 shows several examples of how this notation is used.

Unfortunately, manufacturers are not always consistent with this notation. A good example is C_{CBO} for bipolar junction transistors. Quite often this capacitor is listed as C_{OBO}, which is loosely interpreted as "the output capacitor seen from the collector to base with the base grounded and an open circuit between the emitter and base." The designer must exercise flexibility when applying this three-terminal convention. Other terminology that does not follow the above convention will be adopted in specific applications where it has become standard. In Chapter 3, for example, C_{BS0} will be used to mean the value of capacitance between bulk and source with zero bulk-source voltage and C_{GS0} to mean the overlap capacitance between the gate and source.

1.3   *Analog Signal Processing*[1]

Before beginning an in-depth study of analog-circuit design, it is worthwhile considering the application of such circuits. The general subject of analog signal processing includes most of the circuits and systems that will be presented in this text. Fig. 1.3-1 shows a simple block diagram of a typ-

[1]This section is based in part upon lectures and notes by R.W. Brodersen given at the NATO Advanced Study Institute on Design Methodologies for VLSI in Louvain-le-neuve, Belgium, July, 1980.

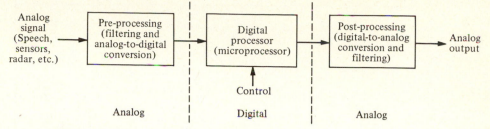

Figure 1.3-1   A block diagram of a typical signal processing system.

ical signal-processing system. In the past, one tried to implement this processing system entirely by digital techniques even though the input and output signals were analog signals. However, the advent of analog sampled-data techniques and MOS technology has made the design of a general signal processor using both analog and digital techniques a viable approach. An example of this approach is the analog signal processor [4].

The first step in the design of an analog signal-processing system is to examine the specifications and decide what part of the system should be analog and what part should be digital. In most cases, the input signal is analog. It could be a speech signal, a sensor output, a radar return, and so on. The first block of Fig. 1.3-1 is a preprocessing block. Typically, this block will consist of filters, an automatic-gain-control circuit, and an analog-to-digital converter. Often, very strict speed and accuracy requirements are placed on the components in this block. The next block of the analog signal processor is essentially a microprocessor. The obvious advantage of this choice is that the function of the processor can easily be controlled and changed. Finally, it may be necessary to have an analog output. In this case, a postprocessing block is necessary. It will typically contain a digital-to-analog converter, amplification, and filtering. The interesting decision on the part of the system designer is where to place the interfaces indicated by the dotted lines.

In a signal processing system, the most important system consideration is probably the bandwidth of the signal to be processed. A graph of the bandwidths of a variety of signals is given in Fig. 1.3-2. The bandwidths in this figure cover an enormous frequency range—10 orders of magnitude. At the low end are seismic signals, which do not extend much below 1 Hz because of the absorption characteristics of the earth. At the other extreme are microwave signals. These are not used much above 30 GHz because of the difficulties in performing even the simplest forms of signal processing at higher frequencies.

To perform signal processing over this range of frequencies, a variety of techniques have been developed that are almost exclusively analog above 10 MHz and digital below 100 Hz, as shown in Fig. 1.3-3. In the overlap region a tradeoff must be made between the accuracy and flexibility of a digital approach and the low cost, power, and size of analog techniques.

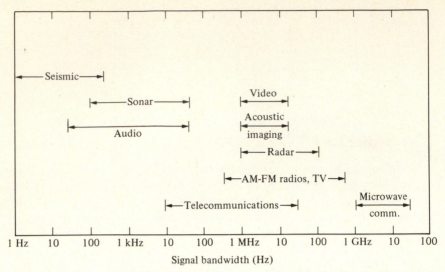

Figure 1.3-2   Bandwidths of signals used in signal processing applications.

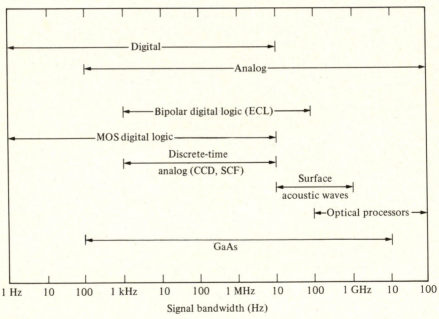

Figure 1.3-3   Signal bandwidths that can be processed by present day technologies.

16

These considerations illustrate some of the advantages of a combined MOS-BJT technology.

By using the fact that VLSI is simply the continuing development of MOS technology, it is possible to extrapolate the trends of the past 10 years to obtain predictions about the capability of MOS technology in the next decade. Some implications of these predictions have been considered, and show that even at the limits of scaling, the processing bandwidth of MOS technology will have improved only to the point where it is equivalent to today's advanced bipolar technologies.

The fastest moving boundary in Fig. 1.3-3 is the upper limit of the MOS-LSI digital logic, which, as the technology progresses to VLSI, should be able to process signals by the year 1990 at bandwidths of 10–100 MHz. This will be accomplished by using greatly increased density along with moderately increased device speeds. However, as can be seen in Fig. 1.3-3, bipolar digital techniques are already able to process at these rates. There-fore, VLSI will not make possible processing rates faster than what can now be achieved with processors built with bipolar technology. Instead, the MOS-VLSI signal processors will offer primarily advantages of reduced cost, size, and power requirements. This will enable signal-processing techniques to make an impact in such cost-sensitive areas as consumer products, areas that until now could not afford the cost of using sophisti-cated signal-processing techniques.

The application–technology–signal bandwidth considerations illus-trated in Figs. 1.3-2 and 1.3-3 are crucial to the future of analog integrated circuit design. It would appear from Fig. 1.3-3 that a BJT technology would enhance the capability of analog circuit performance. However, the tech-nology presently being employed for digital circuits is primarily MOS. Some reasons for this are: lower power dissipation, small device size, and a pro-cess-compatible capacitor. Consequently, analog circuits that are to be incorporated on the same chip as digital circuits will probably use a MOS technology. Since CMOS technology is more compatible with analog cir-cuits, CMOS technology has become the dominant mixed analog-digital technology. Clearly, a combination of bipolar and MOS technology would be able to realize the highest bandwidths for integrated circuits containing both digital and analog circuits, but when and if such a technology becomes available will be strictly determined by the economics of the inte-grated-circuit market.

1.4 *Example of Analog Circuit Design* [5]

Analog-circuit design, like many other complex subjects, is easy to illus-trate but difficult to describe. An example of how the subjects presented in this book might be used to accomplish the design of a 1200 bit/second, quadrature phase-shift keying (QPSK), full-duplex modem is presented. This example will show how analog integrated-circuit design can be hierarchically organized. Fig. 1.4-1 shows a block diagram of the integrated-circuit modem. A modem is a system that converts data in the form of a serial bit string to a form suitable for transmission over a telephone line and then reconverts the transmitted data back to a serial bit string. Designing such a device is a challenging problem because of the limited bandwidth of the telephone system and the desire to transmit at as high a bit rate as possible.

The details of the modem's design will not be considered. Instead, the emphasis will be on how the system is broken into smaller blocks, eventually recognizable as circuits considered in this text. The modem's transmitter is outlined by the upper dotted box in Fig. 1.4-1 and shown in more detail in Fig. 1.4-2. In this diagram blocks begin to emerge, labeled *converter, scrambler, phase encoder, filter, multiplier* ($\times$), and *summer* (Σ). Some of these blocks, such as the converter and phase encoder, are implemented by digital techniques and will not be pursued further. The first part of the receiver channel is indicated by the lower-right dotted box on Fig. 1.4-1 and illustrated in more detail in Fig. 1.4-3. The analog circuit blocks in this figure include the filters, the full-wave rectifier, and the comparator. The comparator is presented in Chapter 7 and the full-wave rectifier in Chapter 11. The filters are described elsewhere [6] but can be further broken into capacitors, switches, and op amps (discussed in Chapters 2, 5, and 8 and 9 respectively).

The next part of the receiver channel is called the data-recovery loop and is implemented by a Costas phase-locked loop that preprocesses the received signal through Hilbert filters. This circuit is identified in Fig. 1.4-1 and illustrated in Fig. 1.4-4. The Hilbert filters consist of two parallel filters, one shifted at 90° to the other. The data-recovery loop consists of many analog circuits, which will be considered in this text. Examples include the multiplier ($\times$), VCO, summer/limiter, and the 6-bit A/D converter.

The adaptive equalizer of the data-recovery loop is shown in more detail in Fig. 1.4-5, where a combination of analog and digital circuits is employed in the implementation. Another part of the receiver channel containing analog circuits is the clock recovery loop shown in Fig. 1.4-6. A portion of this clock recovery loop, the clock, phase-lock loop input circuit, is show in Fig. 1.4-7. Clearly, we have worked our way down the hierarchy of the design to a level where the subjects treated in the following chapters can be used to implement a complex analog/digital system such as a modem.

Fig. 1.4-8 shows a photograph of the integrated-circuit modem. This modem was implemented in an NMOS technology and used about 11,000 transistors. The modem used approximately 60 op amps and had a power

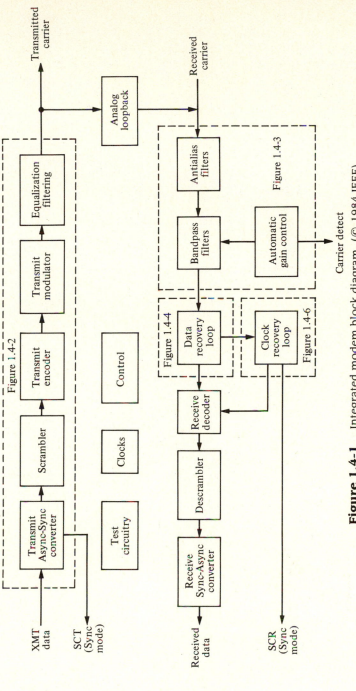

Figure 1.4-1 Integrated modem block diagram. (© 1984 IEEE)

19

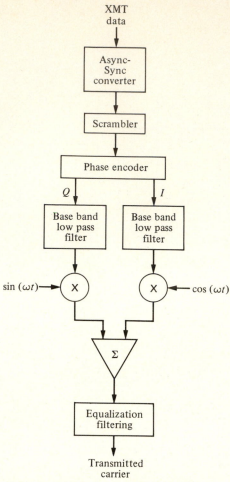

Figure 1.4-2 Differential QPSK transmitter of Fig. 1.4-1. (© 1984 IEEE)

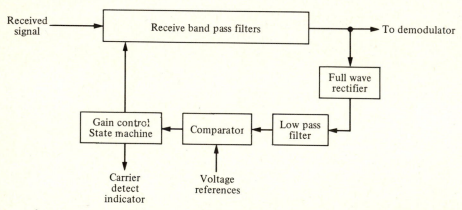

Figure 1.4-3 Automatic gain control circuit of Fig. 1.4-1. (© 1984 IEEE)

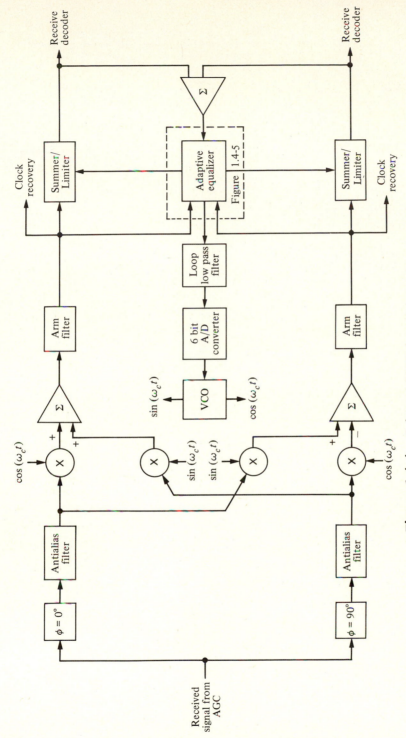

Figure 1.4-4 Coherent QPSK demodulator. (© 1984 IEEE)

21

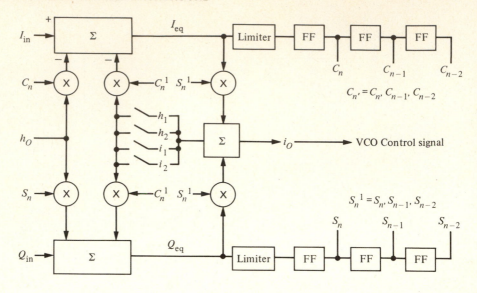

$$I_{eq} = I_{in} - C_{n-2}h_2 - C_{n-1}h_1 - S_{n-2}i_2 - S_{n-1}i_1$$
$$Q_{eq} = Q_{in} - S_{n-2}h_2 - S_{n-1}h_1 + C_{n-2}i_2 + C_{n-1}i_1$$
$$\Delta i_O = K((I_{eq} - C_n h_O)S_n - (Q_{eq} - S_n h_O)C_n)$$
Where $h(t)$ = in phase impulse response
$i(t)$ = cross channel impulse response

Figure 1.4-5   Implementation of the adaptive equalizer of Fig. 1.4-4. (© 1984 IEEE)

dissipation of 750 milliwatts. A CMOS version presently under design is expected to reduce the power consumption significantly.

This example illustrates the hierarchy of a system and its decomposition into circuits and small components. Many other examples exist, such as speech-processing circuits, other telecommunications circuits, and automotive-control systems. As technology advances, such examples will become more numerous and complex. The principles and concepts presented in this text will be embodied in computer-aided design tools to facilitate the design tasks.

1.5   *Summary*

This chapter has presented an introduction to the design of CMOS analog circuits. The first section (1.1) gave a definition of signals in analog circuits and defined analog, digital, and analog sampled-data signals. The difference between analysis and design was discussed. The design differences

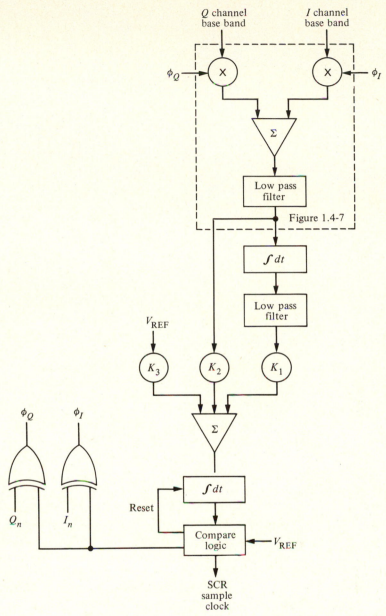

Figure 1.4-6 Implementation of the clock recovery loop of Fig. 1.4-1. (© 1984 IEEE)

23

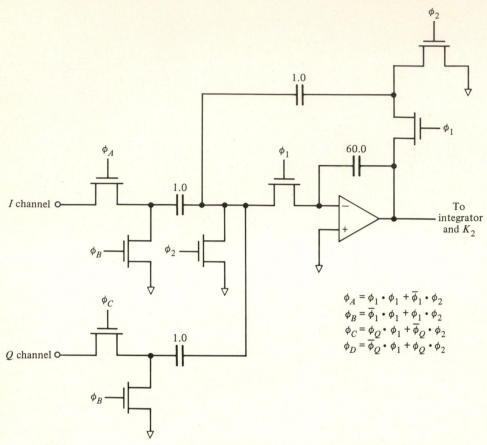

$$\phi_A = \phi_1 \cdot \phi_1 + \overline{\phi}_1 \cdot \phi_2$$
$$\phi_B = \overline{\phi}_1 \cdot \phi_1 + \phi_1 \cdot \phi_2$$
$$\phi_C = \phi_Q \cdot \phi_1 + \overline{\phi}_Q \cdot \phi_2$$
$$\phi_D = \overline{\phi}_Q \cdot \phi_1 + \phi_Q \cdot \phi_2$$

Figure 1.4-7 Implementation of the clock PLL input of Fig. 1.4-6. (The symbology of Fig. 1.2-2 is not used because all transistors are NMOS.) (© 1984 IEEE)

between discrete and integrated analog circuits were primarily due to the designer's control over circuit geometry and the need to computer-simulate rather than build a breadboard. The first section also presented an overview of the text and showed in Table 1.1-2 how the various chapters tied together. It is strongly recommended that the reader refer to Table 1.1-2 at the beginning of each chapter.

The next section (1.2) discussed notation, symbology, and terminology. Understanding these topics is important to avoid confusion in the presentation of the various subjects. The choice of symbols and terminology has been made to correspond with standard practices and definitions. Additional topics concerning the subject in this section will be given in the text at the appropriate place.

An overview of analog signal processing was presented in Section 1.3.

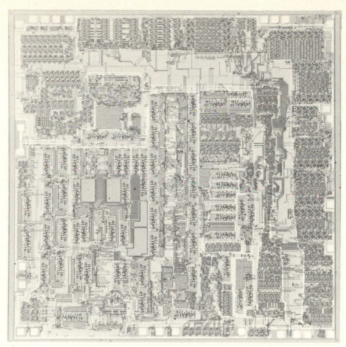

Figure 1.4-8   Photograph of the integrated circuit modem in Fig. 1.4-1. (© 1984 IEEE)

The objective of most analog circuits was seen to be the implementation of some sort of analog signal processing. The important concepts of circuit application, circuit technology, and system bandwidth were introduced and interrelated, and it was pointed out that analog circuits rarely stand alone but are usually combined with digital circuits to accomplish some form of signal processing. The boundaries between the analog and digital parts of the circuit depend upon the application, the performance, and the area.

The last section (1.4) gave an example of the design of an integrated-circuit modem. The example emphasized the hierarchical structure of the design and showed how the subjects to be presented in the following chapters could be used to implement a complex design.

Before beginning the study of the following chapters, the reader may wish to read Appendix A, which presents material that could have been placed at this point in the text. It covers the subject of circuit analysis for analog-circuit design, and some of the problems at the end of this chapter refer to this material. The reader may also wish to review other subjects, such as electronic modeling, computer-simulation techniques, Laplace and z-transform theory, and semiconductor-device theory.

PROBLEMS — *Chapter 1*

1. If the time needed to define and design an integrated circuit is proportional to the number of components per chip and inversely proportional to the feature size, extrapolate Fig. 1.1-4 using the data of Fig. 1.1-3 for the period 1980 to 1990. What will be the time required to define and design a VLSI circuit in 1990 in terms of person-months?

2. If the feature size curve of Fig. 1.1-3 has a steeper slope so that the minimum feature size in 1990 is 0.5 micrometers, what will be the time required to define and design a VLSI circuit in 1990 in terms of person-months?

3. What will be the required time in person-months needed to define and design integrated circuits in the year 2010, as interpolated in Fig. 1.1-4? What will the feature size in microns be in the year 2010 using Fig. 1.1-3?

4. Show how to measure BV_{DSS}, I_{GSS}, and C_{DGO} ($I_D = 1\text{mA}$) for an NMOS enhancement-mode transistor.

5. Show how to measure I_{DSS} and C_{DGO} for a NMOS enhancement-mode transistor.

6. What applications could analog circuits begin to implement in Fig. 1.3-2 if a technology combining MOS digital logic and bipolar digital logic were available?

The following problems refer to material in Appendix A.

7. Use the nodal equation method to find v_{out}/v_{in} of Fig. P1.7.

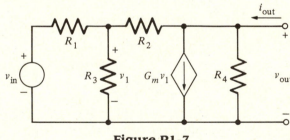

Figure P1-7

8. Use the mesh equation method to find v_{out}/v_{in} of Fig. P1.7.

9. Use the source rearrangement and substitution concepts to simplify the circuit shown in Fig. P1.9 and solve for i_{out}/i_{in} by making chain-type calculations only.

10. Find v_2/v_1 and v_1/i_1 of Fig. P1.10.

11. Use the circuit-reduction technique to solve for v_{out}/v_{in} of Fig. P1.11.

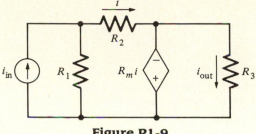

Figure P1-9

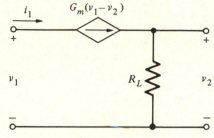

Figure P1-10

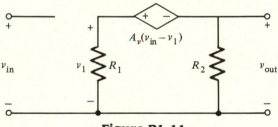

Figure P1-11

12. Use the Miller simplification concept to solve for v_{out}/v_{in} of Fig. A-3 (see Appendix A).

13. Find v_{out}/i_{in} of Fig. A-12 and compare with the results of Example A-1.

14. Use the Miller simplification technique described in Appendix A to solve for the output resistance, v_o/i_o, of Fig. A-4. Calculate the output resistance not using the Miller simplification and compare your results.

15. Consider an ideal voltage amplifier with a voltage gain of $A_v = 0.99$. A resistance $R = 50$ Kilohms is connected from the output back to the input. Find the input resistance of this circuit by applying the Miller simplification concept.

REFERENCES

1. D.A. Hodges, P.R. Gray, and R.W. Brodersen, "Potential of MOS Technologies for Analog Integrated Circuits," *IEEE Journal of Solid-State Circuits,* Vol. SC-13, No. 3, (June 1978) pp. 285–294.
2. J. Cornell, "VLSI: A Semiconductor Perspective," *VLSI Applications—NSF Workshop on the Impact of VLSI on Signal Processing and Communications,"* (Pittsburgh, PA: August 10–11, 1982) Appendix E.
3. G. Moore, "VLSI: Some Fundamental Challenges," *IEEE Spectrum,* Vol. 16, No. 4, (April 1979) pp. 31–37.
4. M. Townsend, M. Hoff, Jr., and R. Holm, "An NMOS Microprocessor for Analog Signal Processing," *IEEE Journal of Solid-State Circuits,* Vol. SC-15, No. 1, (February 1980) pp. 33–38.
5. K. Hanson, W.A. Severin, E.R. Klinkovsky, D.C. Richardson, and J.R. Hochschild, "A 1200 Bit/s QPSK Full Duplex Modem," *IEEE Journal of Solid-State Circuits,* Vol. SC-19, No. 6, (December 1984) pp. 878–887.
6. P.E. Allen and E. Sanchez-Sinencio, *Switched Capacitor Circuits,* (New York: Van Nostrand Reinhold, 1984).

chapter 2
CMOS Technology

The two most popular technologies used for designing integrated circuits are bipolar and MOS. Within each of these families are various subgroups as illustrated in Fig. 2.0-1, which shows a family tree of some of the more widely used silicon integrated-circuit technologies. The majority of the digital and analog circuits designed have been implemented in bipolar, but recently the use of MOS technology has grown widely. This is due in part to the greater density available for circuits implemented in MOS. The need for dense digital circuits such as memories and microprocessors has been the strongest force behind the advancement of MOS for digital applications.

Recognizing the unique capabilities available in MOS circuits, analog designers began designing analog circuits to be implemented with digital circuits on the same integrated circuit. This has been done with great success [1,2,3] and has therefore become a catalyst for the continued improvement of analog MOS capabilities.

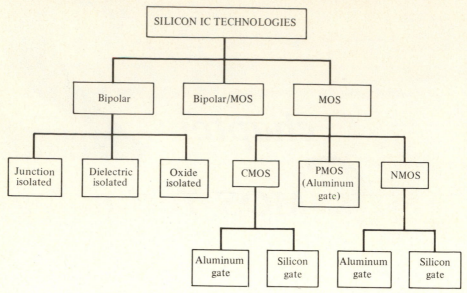

Figure 2.0-1   Categories of silicon technology.

For a time it was not clear which of the MOS technologies would emerge as the dominant one. Due to its efficient use of area and relative processing simplicity, NMOS became the process most preferred. However, the performance of analog circuits designed in NMOS was often lacking (although there are a number of examples of successful NMOS implementations) [1,2,3,4]. As a result, many of the new designs for integration in VLSI have used CMOS [5,6]. It appears that this will be the dominant technology for some time, hence the focus of this book on the use of CMOS for analog circuit design.

There are numerous references that develop the details of the physics of MOS device operation [7,8]. Therefore, this book only covers the aspects of this theory which are pertinent to the viewpoint of the circuit designer. Where possible a qualitative understanding is selected over a quantitative treatment. The objective is to be able to appreciate the limits of the MOS circuit models developed in the next chapter.

This chapter illuminates various aspects of the CMOS process from a physical point of view. In order to understand CMOS technology, a brief review of the basic semiconductor fabrication processes is presented. Next, the pn junction is presented and characterized; and the fabrication and operation of the MOS transistor is given. This discussion is followed by a description of how active and passive components compatible with the MOS transistor are built. Also important are the limitations on the performance of CMOS technology, which include latch-up, temperature, and noise. Finally, this chapter deals with the topological rules employed when drawing the integrated circuit for subsequent fabrication.

2.1 *Basic MOS Semiconductor Fabrication Processes*

Semiconductor technology is based on a number of well-established process steps, which are the means of fabricating semiconductor components. So, in order to understand the fabrication process, it is necessary to understand these steps. The process steps described here include: oxide growth, thermal diffusion, ion implantation, deposition, and etching. The means of defining the area of the semiconductor that will be subject to the various processing steps is called photolithography. Photolithography techniques will be described after the reader has an understanding of the basic process steps.

All processing starts with single-crystal silicon material. There are two methods of growing such crystals [9]. Most of the material is grown by a method based on that developed by Czochralski in 1917. A second method, called the float zone technique, produces crystals of high purity and is often used for power devices. The crystals are normally grown in either a $\langle 100 \rangle$ or $\langle 111 \rangle$ crystal orientation. The resulting crystals are cylindrical and have a diameter of 75–125 mm and a length of 1 m. The cylindrical crystals are sliced into wafers which are approximately 0.8 mm thick. This thickness is determined primarily by the physical strength requirements. When the crystals are grown, they are doped with either an n-type or p-type impurity to form a n- or p-substrate. The substrate is the starting material in wafer form for the fabrication process. The doping level of most substrates is approximately 10^{15} impurity atoms/cm^3, which corresponds to approximately a resistivity of 3–5 ohm-cm for an n-substrate and 14–16 ohm-cm for a p-substrate [10].

There are five basic processing steps which are applied to the doped silicon wafer to fabricate semiconductor components. These steps will be described in their logical order (which is not necessarily their chronological order). The first basic processing step is oxide growth [11]. *Oxide growth* is the process by which a layer of silicon dioxide (SiO$_2$) is formed on the surface of the silicon wafer. The oxide grows both *into* as well as *on* the silicon surface as indicated in Fig. 2.1-1. Typically about 54% of the oxide thickness is above the original surface while about 46% is below the original surface. The oxide thickness, designated t_{ox}, can be grown with either dry or wet techniques, with the former achieving lower defect densities. Typical oxide thicknesses vary from less than 500 angstroms for gate oxides to more than 10,000 angstroms for field oxides. Oxidation takes place at temperatures ranging from 700 °C to 1100 °C with the resulting oxide thickness being proportional to the temperature at which it is grown.

The second basic processing step is diffusion [12]. *Diffusion in semiconductor material is the movement of impurity atoms at the surface of the material into the bulk of the material. Diffusion takes place at temperatures in the range of 800 °C to 1400 °C in the same way as a gas diffuses in air. The two basic types of diffusion are distinguished by the concentration of

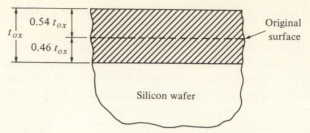

Figure 2.1-1   Oxide growth at the surface of a silicon wafer.

the impurity at the surface of the semiconductor. One type of diffusion assumes that there is an infinite source of impurities at the surface at $t = 0$, (N_o cm^{-3}). As time progresses, the profile of impurity concentration as a function of depth into the material increases above the pre-diffusion concentration N_B, as shown by the impurity concentration profile given in Fig. 2.1-2(a). The second type of diffusion assumes that there is a finite source of impurities at the surface of the material. At $t = 0$ this value is given by N_o. However, as time increases, the impurity concentration at the surface decreases as shown in Fig. 2.1-2(b) where N_B is the pre-diffusion impurity concentration of the semiconductor.

The infinite-source and finite-source diffusions are typical of predeposition and drive-in diffusions, respectively. The object of a predeposition diffusion is to place a large concentration of impurities near the surface of the material. There is a maximum impurity concentration that can be diffused into silicon depending upon the type of impurity. This maximum concentration is due to the solid solubility limit which is in the range of 5×10^{20} to 2×10^{21} atoms/cm^3. The drive-in diffusion follows the deposition diffusion and is used to drive the impurities deeper into the semiconductor. The crossover between the pre-diffusion impurity level and the diffused

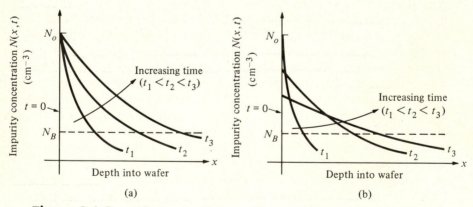

Figure 2.1-2   Diffusion profiles as a function of time for (a) infinite source of impurities at the surface, and (b) a finite source of impurities at the surface.

impurities of the opposite type defines the junction of the semiconductor. This junction is between a p-type and n-type material and is simply called a pn-junction. The distance between the surface of the semiconductor and the junction is called the junction depth. Typical junction depths for diffusion can range from 0.1 micrometers for predeposition type diffusions to 20 micrometers for drive-in type diffusions.

The next basic processing step is called ion implantation and is widely used in the fabrication of MOS components [13,14]. *Ion implantation* is the process by which ions of a particular dopant (impurity) are accelerated by an electric field to a high velocity and physically lodge within the semiconductor material. The average depth of penetration varies from 0.1 to 0.6 micrometers depending on the velocity at which the ions strike the silicon wafer. The path of each ion depends upon the collisions it experiences. This ion-implantation process causes damage to the semiconductor crystal lattice, that can be repaired by an annealing process in which the temperature of the semiconductor after implantation is raised to around 800 °C to allow the ions to become mobile and fit into the semiconductor crystal lattice.

Ion implantation can be used in place of diffusion since in both cases the objective is to insert impurities into the semiconductor material. Ion implantation has several advantages over thermal diffusion. One advantage is the accurate control of doping—to within ±5%. Reproducibility is very good, making it possible to adjust the thresholds of MOS devices or to create precise resistors. A second advantage is that ion implantation is a room-temperature process, although annealing at higher temperatures is required to remove the crystal damage. A third advantage is that it is possible to implant through a thin layer. Consequently, the material to be implanted does not have to be exposed to contaminants. Unlike ion implantation, diffusion requires that the surface be free of oxide or nitride layers. Finally, ion implantation allows control over the profile of the implanted impurities. For example, a concentration peak can be placed below the surface of the wafer if desired.

The fourth basic semiconductor process is called deposition. *Deposition* is the means by which films of various materials may be deposited on the silicon wafer. These films may be deposited by several techniques which include vacuum deposition [9], sputtering [15], and chemical-vapor deposition (CVD) [16,17]. In vacuum deposition, a solid material (aluminum) is placed in a vacuum of at least 10^{-5} torr and heated until it evaporates. The evaporant molecules strike the cooler wafer and condense into a solid film on the wafer surface, approximately 1 μm thick. The sputtering technique uses positive ions to bombard the cathode, which is coated with the material to be deposited. The bombarded or target material is dislodged by direct momentum transfer and deposited on wafers which are placed on the anode. The types of sputtering systems used for depositions in integrated circuits include dc, radio frequency (rf), or magnetron (magnetic field). Sputtering is usually done in a vacuum but with pressures in the range of 25 to 75×10^{-3} torr. Chemical vapor deposition uses a process in which a film is deposited by a chemical reaction or pyrolytic decomposition in the gas phase which occurs in the vicinity of the silicon wafer. This deposition process is typically used to deposit polysilicon, silicon dioxide (SiO_2), or silicon nitride (Si_3N_4). While the chemical vapor deposition is usually performed at atmospheric pressure (760 torr), it can also be done at low pressures of 0.3 to 1 torr, where the diffusivity increases by three orders of magnitude. This technique is called low-pressure chemical-vapor deposition (LPCVD).

The final basic semiconductor fabrication process considered here is etching. _Etching_ is the process of removing unprotected material. How some material is exposed and some is not will be considered next in discussing the subject of photolithography. For the moment, we will assume that the situation illustrated in Fig. 2.1-3(a) exists. Here we see a top layer and an underlying layer. A protective layer covers the top layer except in the area which is to be etched. The objective of etching is to remove just the dotted section of the exposed top layer. However, Fig. 2.1-3(b) shows the typical result. Since etching occurs in all directions, horizontal etching causes an undercut effect. There are preferential etching techniques which can minimize, but not eliminate, this undercutting effect. Also, if the etch rate of the underlying layer is not zero, there will be some etching into the underlying layer simply to insure complete removal of the top layer. To reduce this effect, the etch rate of the top layer should be at least 10 times that of the underlying layer. Materials which are normally etched include polysilicon, silicon dioxide, silicon nitride, and metal.

There are two basic types of etching techniques. _Wet etching_ uses chemicals to remove the material to be etched. Hydrofluoric acid (HF) is

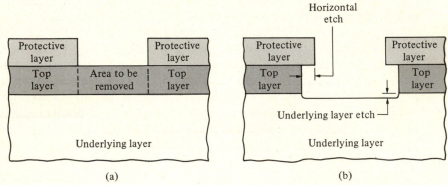

(a)   (b)

Figure 2.1-3   (a) Portion of the top layer ready for etching. (b) Result of etching indicating horizontal etching and etching of the underlying layer.

used to etch silicon dioxide; phosphoric acid (H_3PO_4) is used to remove silicon nitride; nitric acid, acetic acid, or hydrofluoric acid is used to remove polysilicon; and a phosphoric acid mixture is used to remove metal. The wet-etching technique is strongly dependent upon time and temperature, and care must be taken with the acids used in wet etching as they represent a potential hazard. *Dry etching* or *plasma etching* uses ionized gases that are rendered chemically active by an rf-generated plasma. This process requires significant characterization. The variables include pressure, gas flowrate, gas mixture, and rf power. Dry etching is very similar to sputtering and in fact the same equipment can be used. The mixture of dry and wet etching is referred to as reactive-ion etching (RIE) and results in aniso-tropic etching.

Each of the basic semiconductor fabrication processes discussed thus far is only applied to selected parts of the silicon wafer with the exception of deposition. The selection of these parts is accomplished by a process called *photolithography* [11,18,19]. Photolithography refers to the com-plete process of transferring an image from a photographic mask or com-puter database to a wafer. The basic components of photolithography are the photoresist material and the pattern used to expose some areas of the photoresist to ultraviolet (UV) light and to shield the remainder. All inte-grated circuits consist of various layers which form the device or compo-nent. Each distinct layer must be physically defined as an area. This can be done by drawing the layer on a large scale and optically reducing it to the desired size, but the usual technique is to create a computer database that defines the areas of the various layers. Depending upon the design tools, the areas may be restricted to rectangles or may include 45° lines and even circles.

The photoresist is an organic polymer whose characteristics can be altered when exposed to ultraviolet light. Photoresist is classified into pos-itive and negative photoresist. *Positive photoresist* is used to create protec-tive layers where patterns exist (where the mask is opaque to UV light). *Negative photoresist* creates protective layers where patterns do not exist (where the mask is transparent to UV light). The first step in the photolith-ographic process is to apply the photoresist to the surface to be patterned. In order for the photoresist to adhere properly, the surface must be clean. The photoresist is applied to the wafer and the wafer spun at several thou-sand revolutions per minute in order to disperse the photoresist evenly over the surface of the wafer. The thickness of the photoresist depends only upon the angular velocity. The second step is to "soft bake" the wafer at 100 °C to drive off solvents in the photoresist. The next step selectively exposes the wafer to UV light. Using positive photoresist, those areas exposed to UV light can be removed with solvents leaving only those areas that were not exposed. Conversely, if negative photoresist is used, those areas exposed to UV light will be made impervious to solvents while the unexposed areas will be removed. This process of exposing and then selectively removing the photoresist is called *developing.* The developed wafer is then "hard baked" at a temperature slightly greater than 100 °C to achieve maximum adhesion of the remaining photoresist. The hardened photoresist protects selected areas from the acids used to etch polysilicon, oxide, or metal. When its protective function is complete, the photoresist is removed with solvents that will not harm underlying layers. This process must be repeated for each layer of the integrated circuit. Fig. 2.1-4 shows the basic photolithography steps in removing an oxide layer using negative photoresist.

The alteration of the properties of the photoresist on the wafer to make it susceptible or impervious to solvents is called exposure. Several types of exposure systems are used. The simplest is *direct contact printing.* In this method, a glass plate the size of the actual wafer has the desired pattern as an opaque material on the side of the glass that comes in physical contact with the wafer covered with photoresist. This glass plate is commonly called a mask. The system results in high resolution, high throughput, and low cost. Unfortunately, because of the direct contact, the mask wears out and has to be replaced after 10–25 exposures. This method also introduces impurities and defects, because of the physical contact.

A second exposure system is called *projection printing.* In this system, the mask does not come in contact with the wafer. The pattern of the mask is projected onto the wafer through a lens. Because the mask does not contact the wafer, it can be used for many exposures, and defects are reduced, which is extremely important in the yield of integrated circuits. A third exposure system designed to increase the resolution and reduce the defects is called *direct step on wafer* (DSW). It uses a mask which is 1X to 10X larger than the image to be patterned on the wafer. Although the mask must be of higher quality, any defects or dust on the mask will be reduced in size by a factor of 10 (for a 10X mask) on the wafer. The mask in the DSW process contains the pattern of only a single die whereas in other techniques the mask represents an entire array of die (covering the wafer). Therefore, the DSW mask must be exposed and moved repeatedly until the entire wafer has been patterned. As a result, the DSW process achieves much lower throughput than the other techniques discussed. *Electron-beam techniques* are becoming more widely used as an exposure system for high resolution or small geometries. The electron beam is often used to generate the masks for the above exposure systems because of its high resolution (less than 1 micron). However, the electron beam can be used to directly remove photoresist without using a mask. The advantages of using the electron beam as an exposure system are accuracy and the ability to make software changes. The disadvantages are high cost and slow throughput.

1. Grow oxide

2. Apply photoresist.
   Negative resist:
   soluble in certain
   liquid developers
   unless exposed
   to UV light.

3. Selective exposure
   to UV light

4. Remove unexposed
   photoresist by use
   of a developer

5. Remove unexposed
   oxide by use of a
   chemical etch (HF)

6. Remove the exposed
   photoresist (acetone)

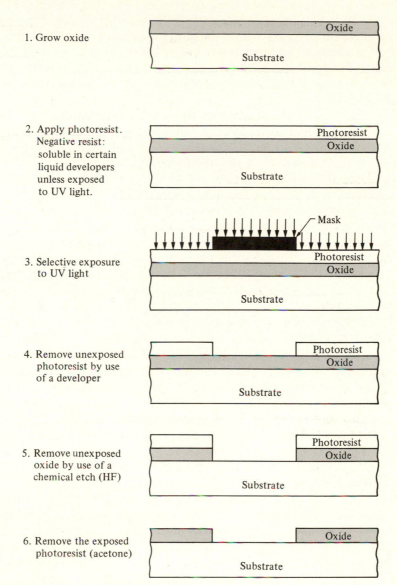

Figure 2.1-4 Basic photolithography steps to remove a portion of an oxide layer.

37

2.2 *The pn Junction*

The pn junction plays an important role in all semiconductor devices. The objective of this section is to develop the concepts of the pn junction that will be useful to us later in our study. These include the depletion-region width, the depletion capacitance, reverse-bias or breakdown voltage, and the diode equation. Further information can be found in the references [20,21].

Fig. 2.2-1(a) shows the physical model of a pn junction. In this model it is assumed that the impurity concentration changes abruptly from N_D donors in the n-type semiconductor to N_A acceptors in the p-type semiconductor. This situation is called a step junction and is illustrated in Fig. 2.2-1(b). The distance x is measured to the right from the metallurgical junction at $x = 0$. When two different types of semiconductor materials are formed in this manner, the free carriers in each type move across the junction by the principle of diffusion. As these free carriers cross the junction, they leave behind fixed atoms which have a charge opposite to the carrier. For example, as the electrons near the junction of the n-type material diffuse across the junction they leave oppositely charged (+), fixed donor atoms near the junction of the n-type material. This is represented in Fig. 2.2-1(c) by the rectangle with a height of qN_D. Similarly, the holes which diffuse across the junction from the p-type material to the n-type material leave behind fixed acceptor atoms that are negatively charged. The electrons and holes that diffuse across the junction quickly recombine with the free majority carriers across the junction. As positive and negative fixed charges are uncovered near the junction by the diffusion of the free carriers, an electric field develops which creates an opposing carrier movement. When the current due to the free carrier diffusion equals the current caused by the electric field, the pn junction reaches equilibrium. In equilibrium, both v_D and i_D of Fig. 2.2-1(a) are zero.

The distance over which the donor atoms have a positive charge (because they have lost their free electron) is designated as x_n in Fig. 2.2-1(c). Similarly, the distance over which the acceptor atoms have a negative charge (because they have lost their free hole) is x_p. In this diagram, x_p is a negative number. The *depletion region* is defined as the region about the metallurgical junction which is depleted of free carriers. The depletion region is defined as

$$x_d = x_n - x_p \qquad (1)$$

Figure 2.2-1 PN junction characterization. (a) Physical structure. (b) Impurity concentration. (c) Depletion charge concentration. (d) Electric field. (e) Electrostatic potential.

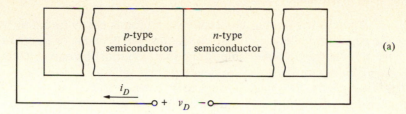

(a)

Impurity concentration (cm^{-3})

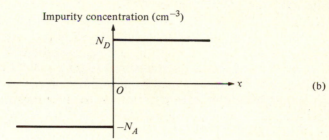

(b)

Depletion charge concentration (cm^{-3})

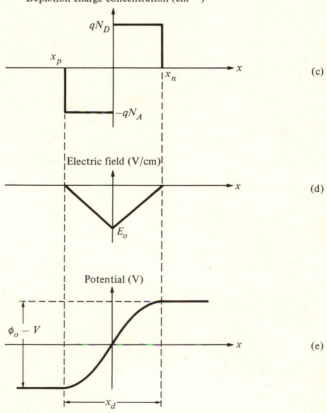

(c)

Electric field (V/cm)

(d)

Potential (V)

(e)

39

The electric field in the depletion region is easily found by integrating the negative depletion-charge concentration. Because of electrical neutrality, the charge on either side of the junction must be equal. Thus,

$$qN_Dx_n = -qN_Ax_p \tag{2}$$

or

$$N_Dx_n = -N_Ax_p \tag{3}$$

where q is the charge of an electron (1.60×10^{-19} C). The result is shown in Fig. 2.2-1(d) where E_o is the maximum electric field that occurs at the junction. From the integral relationship of Gauss's law, we can express E_o as

$$E_o = \frac{qN_Ax_p}{\epsilon_{si}} = \frac{-qN_Dx_n}{\epsilon_{si}} \tag{4}$$

where ϵ_{si} is the dielectric constant of silicon and is $11.7\epsilon_o$ (ϵ_o is 8.85×10^{-14} F/cm).

The voltage drop across the depletion region is shown in Fig. 2.2-1(e). The voltage is found by integrating the negative electric field resulting in

$$\phi_o = v_D = \frac{-E_o(x_n - x_p)}{2} \tag{5}$$

where v_D is an applied external voltage and ϕ_o is called the *barrier potential* and is given as

$$\phi_o = \frac{kT}{q} \ln\left(\frac{N_AN_D}{n_i^2}\right) = V_t \ln\left(\frac{N_AN_D}{n_i^2}\right) \tag{6}$$

Here, k is Boltzmann's constant (1.38×10^{-23} J/°K) and n_i is the intrinsic concentration of silicon which is 1.45×10^{10}/cm³ at 300°K. At room temperature, the value of V_t is 25.9 mV. It is important to note that the notation for kT/q is V_t rather than the conventional V_T. The reason for this is to avoid confusion with V_T which will be used to designate the threshold of the MOS transistor (see Sec. 2.3). Although the barrier voltage exists with $v_D = 0$, it is not available externally at the terminals of the diode. When metal leads are attached to the ends of the diode a metal-semiconductor junction is formed. The barrier potentials of the metal-semiconductor contacts are exactly equal to ϕ_o so that the open circuit voltage of the diode is zero.

Eqs. (3), (4), and (5) can be solved simultaneously to find the width of the depletion region in the n-type and p-type semiconductor. These widths are found as

$$x_n = \left[\frac{2\epsilon_{si}(\phi_o - v_D)N_A}{qN_D(N_A + N_D)} \right]^{1/2} \tag{7}$$

and

$$x_p = - \left[\frac{2\epsilon_{si}(\phi_o - v_D)N_D}{qN_A(N_A + N_D)} \right]^{1/2} \tag{8}$$

The width of the depletion region, x_d, is found from Eqs. (1), (7) and (8) and is

$$x_d = \left[\frac{2\epsilon_{si}(\phi_o - v_D)(N_A + N_D)}{qN_AN_D} \right]^{1/2}$$

$$= \left[\frac{2\epsilon_{si}(N_A + N_D)}{qN_AN_D} \right]^{1/2} (\phi_o - v_D)^{1/2} \tag{9}$$

It can be seen from Eq. (9) that the depletion width for the pn junction of Fig. 2.2-1 is proportional to the square root of the difference between the barrier potential and the externally-applied voltage. It can also be shown that x_d is approximately equal to x_n or x_p if $N_A \gg N_D$ or $N_D \gg N_A$. Consequently, the depletion region will extend further into the lightly-doped semiconductor than it will into the heavily-doped semiconductor.

It is also of interest to characterize the depletion charge Q_j which is equal to the magnitude of the fixed charge on either side of the junction. The depletion charge can be expressed from the above relationships as

$$Q_j = |AqN_Ax_p| = AqN_Dx_n = A \left[\frac{2\epsilon_{si}qN_AN_D}{N_A + N_D} \right]^{1/2} (\phi_o - v_D)^{1/2} \tag{10}$$

where A is the cross-sectional area of the pn junction in cm^2.

The magnitude of the electric field at the junction E_o can be found from Eqs. (4) and (7) or (8). This quantity is expressed as

$$E_o = \left[\frac{2qN_AN_D}{\epsilon_{si}(N_A + N_D)} \right]^{1/2} (\phi_o - v_D)^{1/2} \tag{11}$$

Equations (9), (10), and (11) are key relationships in understanding the pn junction.

The depletion region of a pn junction forms a capacitance called the *depletion-layer capacitance*. It results from the dipole formed by uncovered fixed charges near the junction and will vary with the applied voltage. The depletion-layer capacitance C_j can be found from Eq. (10) using the following definition of capacitance.

$$C_j = \frac{dQ_j}{dv_D} = A \left[\frac{\epsilon_{si}qN_AN_D}{2(N_A + N_D)} \right]^{1/2} \frac{1}{(\phi_o - v_D)^{1/2}} = \frac{C_{jo}}{[1 - (v_D/\phi_o)]^m} \qquad (12)$$

C_{jo} is the depletion-layer capacitance when $v_D = 0$ and m is called a grading coefficient. m is ½ for the case of Fig. 2.2-1 which is called a step junction. If the junction is fabricated using diffusion techniques described in Sec. 2.1, Fig. 2.2-1(b) will become more like the profile of Fig. 2.2-2. It can be shown for this case that m is ⅓. The range of values of the grading coefficient will fall between ⅓ and ½. Fig. 2.2-3 shows a plot of the depletion layer capacitance for a pn junction. It is seen that when v_D is positive and approaches ϕ_o, the depletion-layer capacitance approaches infinity. At this value of

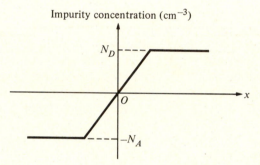

Figure 2.2-2   Impurity concentration profile for diffused pn junction.

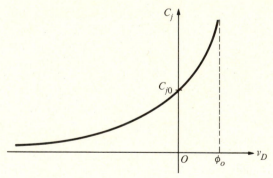

Figure 2.2-3   Depletion capacitance as a function of externally applied junction voltage.

voltage, the assumptions made in deriving the above equations are no longer valid. In particular, the assumption that the depletion region is free of charged carriers is not true. Consequently, the actual curve bends over and C_j decreases as v_D approaches ϕ_o [22].

Example 2.2-1

Characteristics of a pn Junction

Find x_p, x_n, x_d, ϕ_o, C_{j0}, and C_j for an applied voltage of -4 V for a pn diode with a step junction, $N_A = 5 \times 10^{15}/cm^3$, $N_D = 10^{20}/cm^3$, and an area of 10 μm by 10 μm.

At room temperature, Eq. (6) gives the barrier potential as 0.917 V. Eqs. (7) and (8) give $x_n \cong 0$ and $x_p = 1.128$ μm. Thus, the depletion width is approximately x_p or 1.128 μm. Using these values in Eq. (12) we find that C_{j0} is 20.3 fF and at a voltage of -4 V, C_j is 9.18 fF.

The voltage breakdown of a reverse biased ($v_D < 0$) pn junction is determined by the maximum electric field E_{max} that can exist across the depletion region. For silicon, this maximum electric field is approximately 3×10^5 V/cm. If we assume that $|v_D| > \phi_o$, then substituting E_{max} into Eq. (11) allows us to express the maximum reverse-bias voltage or breakdown voltage (*BV*) as

$$BV \cong \frac{\epsilon_{si}(N_A + N_D)}{2qN_AN_D} E_{max}^2 \tag{13}$$

Substituting the values of Example 2.2-1 in Eq. (13) and using a value of 3 $\times$ 10^5 V/cm for E_{max} gives a breakdown voltage of 58.2 volts. However, as the reverse bias voltage starts to approach this value, the reverse current in the pn junction starts to increase. This increase is due to two conduction mechanisms that can take place in a reverse-biased junction between two heavily-doped semiconductors. The first current mechanism is called avalanche multiplication and is caused by the high electric fields present in the pn junction; the second is called Zener breakdown. Zener breakdown is a direct disruption of valence bonds in high electric fields. However, the Zener mechanism does not require the presence of an energetic ionizing carrier. The current in most breakdown diodes will be a combination of these two current mechanisms.

If i_R is the reverse current in the pn junction and v_R is the reverse-bias voltage across the pn junction, then the actual reverse current i_{RA} can be expressed as

$$i_{RA} = Mi_R = \left[\frac{1}{1 - (V_R/BV)^n} \right] i_R \qquad (14)$$

M is the avalanche multiplication factor and n is an exponent which adjusts the sharpness of the "knee" of the curve shown in Fig. 2.2-4. Typically, n varies between 3 and 6. If both sides of the pn junction are heavily doped, the breakdown will take place by tunneling, leading to the Zener break-down, which typically occurs at voltages less than 6 volts. Zener diodes can be fabricated as shown in Fig. 2.2-5 where an n^+ diffusion overlaps with a p^+ diffusion. Note that the Zener diode is compatible with the basic CMOS process although one terminal of the Zener must be either on the lowest power supply, V_{SS}, or the highest power supply, V_{DD}.

The diode voltage-current relationship can be derived by examining the minority-carrier concentrations in the pn junction. Fig. 2.2-6 shows the minority-carrier concentration for a forward-biased pn junction. The majority-carrier concentrations are much larger and are not shown on this figure. The forward bias causes minority carriers to move across the junction where they recombine with majority carriers on the opposite side. The excess of minority-carrier concentration on each side of the junction is shown by the cross-hatched regions. We note that this excess concentration starts at a maximum value at $x = 0$ ($x' = 0$) and decreases to the equilibrium value as x (x') becomes large. The value of the excess concentration at $x = 0$, designated as $p_n(0)$, or $x' = 0$, designated as $n_p(0)$, is

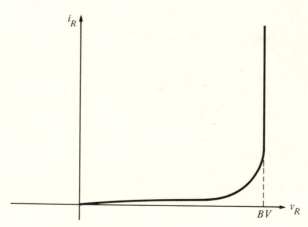

Figure 2.2-4 Reverse-bias voltage-current characteristics of the pn junction illustrating voltage breakdown.

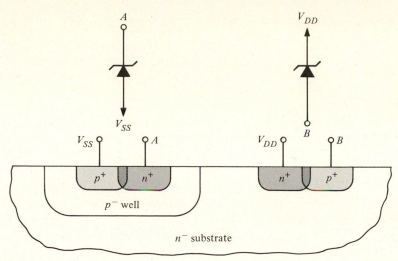

Figure 2.2-5 Implementation of breakdown diodes compatible with a p-well, CMOS technology.

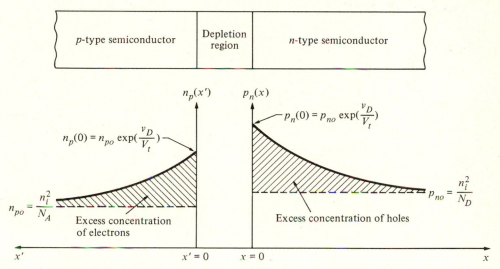

Figure 2.2-6 Minority carrier concentrations in a forward-biased pn junction.

expressed in terms of the forward-bias voltage v_D as

$$p_n(0) = p_{no} \exp\left(\frac{v_D}{V_t}\right) \tag{15}$$

and

$$n_p(0) = n_{po} \exp\left(\frac{v_D}{V_t}\right) \tag{16}$$

where p_{no} and n_{po} are the equilibrium concentrations of the minority carriers in the n-type and p-type semiconductors, respectively. We note that these values are essentially equal to the intrinsic concentration squared divided by the donor or acceptor impurity atom concentration, as shown on Fig. 2.2-6. As v_D is increased, the excess minority concentrations are increased. If v_D is zero, there is no excess minority concentration. If v_D is negative (reverse-biased) the minority-carrier concentration is depleted below its equilibrium value.

The current that flows in the pn junction is proportional to the slope of the excess minority-carrier concentration at $x = 0$ ($x' = 0$). This relationship is given by the diffusion equation expressed below for holes in the n-type material.

$$J_p(x) = -qD_p \left. \frac{dp_n(x)}{dx} \right|_{x=0} \tag{17}$$

where the D_p is the diffusion constant of holes in n-type semiconductor. The excess holes in the n-type material can be defined as

$$p'_n(x) = p_n(x) - p_{no} \tag{18}$$

The decrease of excess minority carriers away from the junction is exponential and can be expressed as

$$p'_n(x) = p'_n(0) \exp\left(\frac{-x}{L_p}\right) = [p_n(0) - p_{no}] \exp\left(\frac{-x}{L_p}\right) \tag{19}$$

where L_p is the diffusion length for holes in an n-type semiconductor. Substituting Eq. (15) into Eq. (19) gives

$$p'_n(x) = p_{no} \left[\exp\left(\frac{v_D}{V_t}\right) - 1 \right] \exp\left(\frac{-x}{L_p}\right) \tag{20}$$

The current density due to the excess-hole concentration in the n-type semiconductor is found by substituting Eq. (20) in Eq. (17) resulting in

$$J_p(0) = \frac{qD_pP_{no}}{L_p}\left[\exp\left(\frac{v_D}{V_t}\right) - 1\right] \tag{21}$$

Similarly, for the excess electrons in the p-type semiconductor we have

$$J_n(0) = \frac{qD_nn_{po}}{L_n}\left[\exp\left(\frac{v_D}{V_t}\right) - 1\right] \tag{22}$$

Assuming negligible recombination in the depletion region leads to an expression for the total current density of the pn junction given as

$$J(0) = J_p(0) + J_n(0) = q\left[\frac{D_pP_{no}}{L_p} + \frac{D_nn_{po}}{L_n}\right]\left[\exp\left(\frac{v_D}{V_t}\right) - 1\right] \tag{23}$$

Multiplying Eq. (23) by the pn junction area A gives the total current as

$$i_D = qA\left[\frac{D_pP_{no}}{L_p} + \frac{D_nn_{po}}{L_n}\right]\left[\exp\left(\frac{v_D}{V_t}\right) - 1\right] = I_s\left[\exp\left(\frac{v_D}{V_t}\right) - 1\right] \tag{24}$$

I_s is a constant called the *saturation current*. Equation (24) is the familiar voltage-current relationship that characterizes the pn junction diode.

Example 2.2-2

Calculation of the Saturation Current

Calculate the saturation current of a pn junction diode with $N_A = 5 \times 10^{15}/cm^3$, $N_D = 10^{20}/cm^3$, $D_n = 20$ cm²/s, $D_p = 10$ cm²/s, $L_n = 10$ μm, $L_p = 5$ μm, and $A = 1000$ μm².
From Eq. (24), the saturation current is defined as

$$I_s = qA\left[\frac{D_pP_{no}}{L_p} + \frac{D_nn_{po}}{L_n}\right]$$

P_{no} is calculated from n_i^2/N_D to get 2.103/cm³; n_{po} is calculated from n_i^2/N_A to get $4.205 \times 10^4/cm^3$. Changing the units of area from μm² to cm² results in a saturation current magnitude of 1.346×10^{-15} A or 1.346 fA.

This section has developed the depletion-region width, depletion capacitance, breakdown voltage, and the voltage-current characteristics of the pn junction. These concepts will be very important in determining the characteristics and performance of MOS active and passive components.

2.3 *The MOS Transistor*

The structure of an n-channel and p-channel MOS transistor using a p-well technology is shown in Fig. 2.3-1. The p-channel device is formed with two heavily-doped p^+ regions diffused into a lighter doped n^- material called the substrate (or bulk). The two p^+ regions are called drain and source, and are separated by a distance, L (referred to as the device length). At the surface between the drain and source lies a gate electrode that is separated from the silicon by a thin dielectric material (silicon dioxide). Similarly, the n-channel transistor is formed by two heavily doped n^+ regions within a lightly doped p^- "well" or "tub" (for p-well processes). It, too, has a gate on the surface between the drain and source separated from the silicon by a thin dielectric material (silicon dioxide). Essentially, both types of transistors are four-terminal devices as shown in Fig. 1.2-2. The B terminal is the bulk, or substrate, which contains the drain and source diffusions. For a p-well process, the n-bulk connection is common throughout the integrated circuit and is connected to V_{DD}. Multiple p-wells can be fabricated on a single circuit, and they can be connected to different potentials in various ways depending upon the application.

The operation of the MOS transistor can be determined from Fig. 2.3-2 for the NMOS transistor. Figure 2.3-2 shows an n-channel transistor with all four terminals connected to ground. At equilibrium, the p^- substrate

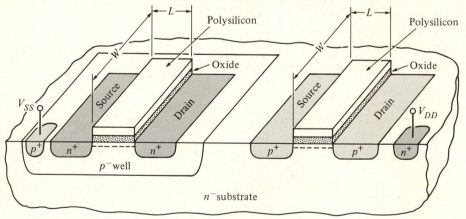

Figure 2.3-1 Physical structure of an NMOS and PMOS transistor in a p-well, CMOS technology.

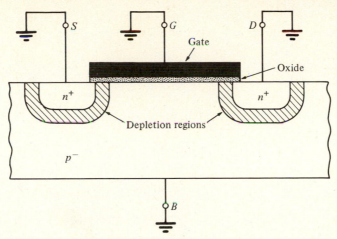

Figure 2.3-2   Cross section of an NMOS transistor with all terminals grounded.

and the n^+ source and drain form a pn junction. Therefore a depletion region exists between the n^+ source and drain and the p^- substrate. Since the source and drain are separated by back-to-back pn junctions, the resistance between the source and drain is very high (approximately 10^{12} ohms). The gate and the substrate of the MOS transistor form the parallel plates of a capacitor with the SiO_2 as the dielectric. This capacitance divided by the area of the gate is designated as C_{ox}. When a positive potential is applied to the gate with respect to the source, positive charge accumulates on the gate and negative charge in the substrate under the gate. The negative charge is created by mobile holes which are pushed down into the substrate leaving behind fixed-donor atoms. The mobile charge dQ of the holes originally contained in an infinitesimal horizontal layer of p-type material below the gate is given by

$$dQ = q(-N_A)dx_d \tag{1}$$

A change in surface potential $d\phi_s$ is required to displace the mobile charge dQ. Thus,

$$d\phi_s = -x_d\, dE = -x_d\left(\frac{dQ}{\epsilon_{si}}\right) = \frac{qN_Ax_d\, dx_d}{\epsilon_{si}} \tag{2}$$

Integrating both sides of Eq. (2) gives

$$\phi_s = \frac{qN_Ax_d^2}{2\epsilon_{si}} + \phi_F \tag{3}$$

where ϕ_F is a constant of integration and is actually the equilibrium electrostatic potential in a semiconductor defined for a p-type semiconductor as

$$\phi_F = V_t \ln(n_i/N_A) \tag{4}$$

and for an n-type semiconductor as

$$\phi_F = V_t \ln(N_D/n_i) \tag{5}$$

Eq. (3) can be solved for x_d assuming that $|\phi_s - \phi_F| \geq 0$ to get

$$x_d = \left[\frac{2\epsilon_{si}|\phi_s - \phi_F|}{qN_A} \right]^{1/2} \tag{6}$$

The immobile charge due to acceptor ions that have been stripped of their mobile holes is given by

$$Q = -qN_A x_d \tag{7}$$

where x_d is the thickness of the region depleted of holes. Substituting Eq. (6) into Eq. (7) gives

$$Q \cong -qN_A \left[\frac{2\epsilon_{si}|\phi_s - \phi_F|}{qN_A} \right]^{1/2} = -\sqrt{2qN_A\epsilon_{si}|\phi_s - \phi_F|} \tag{8}$$

When the gate voltage reaches a value called the *threshold voltage*, designated as V_T, the substrate underneath the gate becomes inverted, i.e. it changes from a p-type to an n-type semiconductor. Consequently, an n-channel exists between the source and drain that allows carriers to flow. In order to achieve this inversion, the surface potential must increase from its original negative value ($\phi_s = -\phi_F$), to zero ($\phi_s = 0$), and then to a positive value ($\phi_s = \phi_F$). The value of gate-source voltage necessary to cause this change in surface potential is defined as the *threshold voltage V_T*. This phenomenon is known as *strong inversion*. The NMOS transistor in this condition is illustrated in Fig. 2.3-3. With the substrate at ground potential, the charge stored in the depletion region between the channel under the gate and the substrate is given by Eq. (8) where ϕ_s has been replaced by $-\phi_F$ to account for the fact that $v_{GS} = V_T$. This charge Q_{b0} is written as

$$Q_{b0} \cong -\sqrt{2qN_A\epsilon_{si}|-2\phi_F|} \tag{9}$$

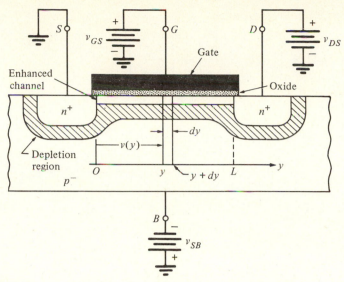

Figure 2.3-3   Cross section of an NMOS transistor with small v_{DS} and $v_{GS} > V_T$.

If a reverse bias voltage v_{BS} is applied across the pn junction, Eq. (9) becomes

$$Q_b \cong \sqrt{2qN_A\epsilon_{si}|-2\phi_F + v_{SB}|} \qquad (10)$$

An expression for the threshold voltage can be developed by breaking it down into several components. First, a voltage term ϕ_{GB} must be included to represent the difference in the work functions between the gate material and bulk silicon in the channel region. For silicon-gate devices, ϕ_{GB} is given by

$$\phi_{GB} = \phi_F(\text{substrate}) - \phi_F(\text{gate}) \qquad (11)$$

and for metal-gate devices, ϕ_{GB} is given by

$$\phi_{GB} = \phi_F(\text{substrate}) - \phi_F(\text{metal}) \qquad (12)$$

where $\phi_F(\text{metal}) = 0.6$ V. Second, a gate voltage of $[-2\phi_F - (Q_b/C_{ox})]$ is required to change the surface potential and offset the depletion layer charge Q_b. Lastly, there is always an undesired positive charge Q_{ss} present in the interface between the oxide and the bulk silicon. This charge is due to impurities and imperfections at the interface and must be compensated by a gate voltage of $-Q_{ss}/C_{ox}$. Thus, the threshold voltage for the MOS transistor can be expressed as

$$V_T = [\phi_{GB}] + \left[-2\phi_F - \frac{Q_b}{C_{ox}} \right] + \left[\frac{-Q_{ss}}{C_{ox}} \right]$$

$$= \phi_{GB} - 2\phi_F - \frac{Q_{b0}}{C_{ox}} - \frac{Q_{ss}}{C_{ox}} - \frac{Q_b - Q_{b0}}{C_{ox}} \qquad (13)$$

The threshold voltage can be redefined as

$$V_T = V_{T0} + \gamma(\sqrt{|-2\phi_F + v_{SB}|} - \sqrt{|-2\phi_F|}) \qquad (14)$$

where

$$V_{T0} = \phi_{GB} - 2\phi_F - \frac{Q_{b0}}{C_{ox}} - \frac{Q_{ss}}{C_{ox}} \qquad (15)$$

and the body-factor, body-effect coefficient or bulk-threshold parameter γ is defined as

$$\gamma = \frac{\sqrt{2q\epsilon_{si}N_A}}{C_{ox}} \qquad (16)$$

The signs of the above analysis can become very confusing. Table 2.3-1 attempts to clarify some of the possible confusion [21].

Table 2.3-1

Sign Convention for the Quantities in the Threshold Voltage Equation.

Parameter	NMOS	PMOS
Substrate	p-type	n-type
ϕ_{GB}		
Metal	−	−
n^+ Si Gate	−	−
p^+ Si Gate	+	+
ϕ_F	−	+
Q_{b0}, Q_b	−	+
Q_{ss}	+	−
V_{SB}	+	−
γ	+	−

Example 2.3-1

Calculation of the Threshold Voltage

Find the threshold voltage and body factor γ for an NMOS transistor with an n^+ silicon gate if $t_{ox} = 0.1 \ \mu m$, $N_A = 3 \times 10^{15}$ cm^{-3}, gate doping $= N_D = 10^{20}$ cm^{-3}, and if the positive charged ions at the oxide-silicon interface per area is 10^{10} cm^{-2}.

From Eq. (4), ϕ_F(substrate) is given as

$$\phi_F(\text{substrate}) = V_t \ln(n_i/N_A)$$
$$= 0.0259 \ln(4.833 \times 10^{-6}) = -0.317 \text{ V}$$

The equilibrium electrostatic potential for the n^+ polysilicon gate is found from Eq. (5) as

$$\phi_F(\text{Gate}) = V_t \ln(N_D/n_i) = 0.587 \text{ V}$$

Eq. (11) gives ϕ_{GB} as ϕ_F(substrate) $- \phi_F$(gate) $= -0.904$ V. The oxide capacitance is given as

$$C_{ox} = \epsilon_{ox}/t_{ox} = (3.9\epsilon_o)/t_{ox} = 3.45 \times 10^{-8} \text{ F/cm}^2$$

The fixed charge in the depletion region, Q_{b0}, is given by Eq. (9) as Q_{b0} $= -[2 \cdot 1.6 \times 10^{-19} \cdot 11.7 \cdot 8.85 \times 10^{-14} \cdot 2 \cdot 0.317 \cdot 3 \times 10^{15}]^{1/2} = -2.511 \times 10^{-8}$ C/cm^2. Dividing Q_{b0} by C_{ox} gives -0.728 V. Finally, Q_{ss}/C_{ox} is given as

$$\frac{Q_{ss}}{C_{ox}} = \frac{10^{10} \cdot 1.60 \times 10^{-19}}{3.45 \times 10^{-8}} = 0.046 \text{ V}$$

Substituting these values in Eq. (13) gives

$$V_{T0} = -0.904 + 0.634 + 0.728 - 0.046 = 0.412 \text{ V}$$

The body factor is found from Eq. (16) as

$$\gamma = \frac{[2 \cdot 1.6 \times 10^{-19} \cdot 11.7 \cdot 8.854 \times 10^{-14} \cdot 3 \times 10^{15}]^{1/2}}{3.45 \times 10^{-8}}$$

$$= 0.914 \text{ V}^{1/2}$$

The above example shows how the value of impurity concentrations can influence the threshold voltage. In fact, the threshold voltage can be set to any value by proper choice of the variables in Eq. (13). Standard practice is to implant the proper type of ions into the substrate in the channel region to adjust the threshold voltage to the desired value. If the opposite impurities are implanted in the channel region of the substrate, the threshold for an NMOS transistor can be made negative. This type of transistor is called a *depletion transistor* and can have current flow between the drain and source for zero values of the gate-source voltage.

When the channel is formed between the drain and source as illustrated in Fig. 2.3-3, a drain current i_D can flow if a voltage v_{DS} exists across the channel. The dependence of this drain current on the terminal voltages of the MOS transistor can be developed by considering the characteristics of an incremental length of the channel designated as dy in Fig. 2.3-3. It is assumed that the width of the MOS (into the page) is W and that v_{DS} is small. The charge per unit area in the channel, $Q_I(y)$, can be expressed as

$$Q_I(y) = C_{ox}[v_{GS} - v(y) - V_T] \tag{17}$$

The resistance in the channel per unit of length dy can be written as

$$dR = \frac{dy}{\mu_n Q_I(y) W} \tag{18}$$

where μ_n is the average mobility of the electrons in the channel. The voltage drop, referenced to the source, along the channel in the y direction is

$$dv(y) = i_D \, dR = \frac{i_D \, dy}{\mu_n Q_I(y) W} \tag{19}$$

or

$$i_D \, dy = W\mu_n Q_I(y) \, dv(y) \tag{20}$$

Integrating along the channel from $y = 0$ to $y = L$ gives

$$\int_0^L i_D \, dy = \int_0^{v_{DS}} W\mu_n Q_I(y) \, dv(y)$$

$$= \int_0^{v_{DS}} W\mu_n C_{ox}[v_{GS} - v(y) - V_T] \, dv(y) \tag{21}$$

Performing the integration results in the desired expression for i_D as

$$i_D = \frac{\mu_n C_{ox} W}{L} \left[(v_{GS} - V_T)v(y) - \frac{v(y)^2}{2} \right] \Bigg|_0^{v_{DS}}$$

$$= \frac{\mu_n C_{ox} W}{2L} [2(v_{GS} - V_T)v_{DS} - v_{DS}^2] \qquad (22)$$

This equation is sometimes called the Sah equation [23] and has been used by Shichman and Hodges [24] as a model for computer simulation. Eq. (22) is only valid when

$$v_{GS} \geqq V_T \quad \text{and} \quad v_{DS} \leqq (v_{GS} - V_T) \qquad (23)$$

The factor $\mu_n C_{ox}$ is often defined as the device-transconductance parameter, given as

$$K' = \mu_n C_{ox} = \frac{\mu_n \epsilon_{ox}}{t_{ox}} \qquad (24)$$

Eq. (22) will be examined in more detail in the next chapter, concerning the modeling of MOS transistors. The operation of the p-channel transistor is essentially the same as that of the n-channel transistor, except that all voltage and current polarities are reversed.

It is important for a circuit designer to understand some of the basic steps involved in fabricating a CMOS circuit. The fabrication steps of one of the more popular CMOS processes, the silicon-gate process will be described in detail. The first step in the p-well silicon-gate CMOS process is to grow a silicon-dioxide region on an n^- substrate (wafer). Subsequent to this, the regions where p-wells are to exist are defined in a masking step by depositing a photoresist material on top of the oxide. After exposing and developing the photoresist, p^- type impurities are implanted or deposited and diffused into the n^- substrate. Fig. 2.3-4(a) shows a cross section of the wafer after the p-well has been implanted. Next, the photoresist and the thin oxide layers are removed and a new thin oxide region is grown. A layer of silicon nitride is deposited over the entire wafer. Photoresist is deposited, patterned, and developed as before, and the silicon nitride is removed from the areas where it has been patterned. Fig. 2.3-4(b) illustrates the result after this step and after the remaining photoresist has been removed. The silicon nitride remains in the areas where active devices will reside. Areas with silicon nitride are called *moat* or *active area*.

In order to insure that parasitic transistors are not formed under various interconnect lines, an implant step is required to heavily dope those regions that should not act like transistors. The area not covered with silicon nitride is called *field*. To accomplish this, the deposition, patterning,

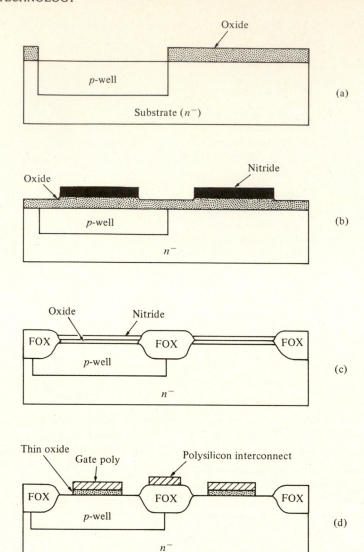

Figure 2.3-4   The major CMOS process steps. (a) Definition of the p-well. (b) Definition of the active device areas. (c) Thick and thin oxide growth. (d) Polysilicon gate and interconnect definition. (e) Diffusion of the n^+ and p^+ regions. (f) Metallization and passivation layer.

and developing is done so that those regions requiring an implant are devoid of photoresist. These regions are then implanted. This must be done in both n^- and p^- areas. Next, a thick silicon-dioxide layer is grown over the entire wafer except where silicon nitride exists (this material impedes oxide growth). Fig. 2.3-4(c) shows the results of this step. Once the thick

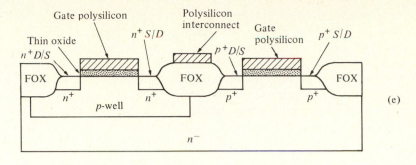

(e)

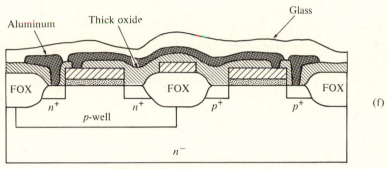

(f)

Figure 2.3-4   (Continued)

field oxide (FOX) is grown, the remaining silicon nitride is removed. The thin oxide is also removed, and a new thin oxide, which will be the gate oxide, is deposited. Next, polysilicon is deposited over the entire wafer and appropriately patterned. After developing the photoresist, the polysilicon is etched, leaving only what is required to make transistor gates and interconnect lines, as shown in Fig. 2.3-4(d).

At this point, the drain and source areas have not been diffused into the substrate. They have been defined in the sense that they will exist in all areas that are not covered by polysilicon or field oxide. To make p^+ sources and drains, photoresist is applied and patterned everywhere p-channel transistors are required; p^+ is also required where metal connections are to be made to p^- material such as the p-well. After developing, the p^+ areas are implanted. The photoresist acts as a barrier to the implant as does the polysilicon. As a result, the p^+ regions that result are properly aligned with the polysilicon gates (this reduces overlaps which can cause a problem in circuit operation). After implantation, the photoresist is removed and reapplied so that the n^+ regions can be patterned and implanted in a manner similar to the p^+ areas. The remaining photoresist, along with the gate oxide covering the source/drain areas, is removed. At this point, as shown in Fig. 2.3-4(e), n- and p-channel transistors are complete except for the necessary terminal connections.

Next, in preparation for the contact step, a new, thick oxide layer is deposited over the entire wafer. Contacts are formed by first defining their location using the photolithographic process applied in previous steps. Next, the oxide areas where contacts are to be made are etched down to the surface of the silicon. The remaining photoresist is removed and metal (aluminum) is deposited on the wafer. Metal interconnect is then defined photolithographically and subsequently etched so that all unnecessary metal is removed. In order to protect the wafer from chemical intrusion or scratching, a passivation layer is applied covering the entire wafer. Pad regions are then defined (areas where wires will be bonded between the integrated circuit and the package containing the circuit) and the passivation layer removed only in these areas. Figure 2.3-4(f) shows a cross section of the final circuit.

There are many other details associated with CMOS processes that have not been described here. Furthermore, there are different variations on the basic CMOS process just described. Some of these provide multiple levels of polysilicon and/or metal interconnect. Others provide good capacitors using either two layers of polysilicon, or polysilicon on top of a heavily implanted (on the same order as a source or drain) diffusion. Density is improved on those processes that provide for a direct connection between diffusion and polysilicon. Still other processes start with a p^- substrate and implant n-wells (rather than p-wells in an n^- substrate). Finally, other processes have both p- and n-wells.

2.4 *Passive Components*

This section examines the passive components that are compatible with fabrication steps used to build the MOS device. These passive components include the capacitor and the resistor. It will be seen in later chapters that the performance of an analog circuit is often limited by the performance of the passive components.

A good capacitor is often required when designing analog integrated circuits. There are basically two types of capacitors suitable for analog-circuit design available on CMOS processes. One type of capacitor is formed using a conducting layer (metal or polysilicon) on top of crystalline silicon separated by a dielectric (silicon dioxide layer). Figure 2.4-1(a) shows an example of this capacitor using polysilicon as the top conducting plate. In order to achieve a low voltage-coefficient capacitor, the bottom plate must be heavily-doped diffusion (similar to that of the source and drain). As the process was described in Sec. 2.3, such heavily-doped diffusion is normally not available underneath polysilicon because the source/drain implant step occurs after polysilicon is deposited and defined. To solve this problem, an extra implant step must be included prior to deposition of the polysilicon layer. The mask-defined implanted region becomes the bottom plate of the

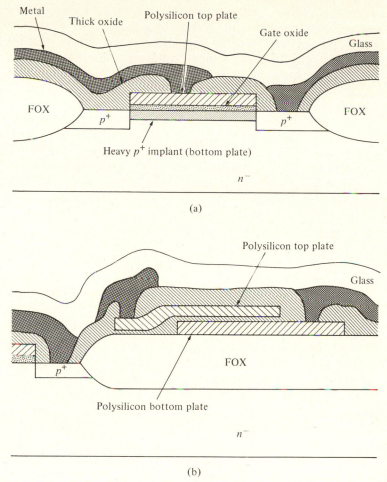

Figure 2.4-1 MOS capacitors. (a) Polysilicon-oxide-channel. (b) Polysilicon-oxide-polysilicon.

capacitor. The capacitance achieved using this technique is normally around 0.35 to 0.5 fF/μm^2.

The second type of capacitor is that formed by two conducting layers (such as metal or polysilicon) separated by a dielectric. Generally those processes that provide this type of capacitor use two polysilicon layers separated by silicon dioxide, as illustrated in Fig. 2.4-1(b). It can be seen that a double polysilicon process is needed to implement this capacitor since both the upper and lower plates are formed with polysilicon. The dielectric is formed by a thin silicon-dioxide layer which can only be produced by using several steps beyond the usual single polysilicon process. One of the advantages of this capacitor is nearly voltage-independent parasitic capacitance. However, even though there is a thick layer of silicon dioxide isolating the polysilicon-oxide-polysilicon capacitance, the parasitics are still not negligible. The typical value of capacitance per area of this type of capacitor is about 0.3 to 0.4 fF/μm^2. This smaller value is due to a slightly larger silicon-dioxide thickness than that underneath the gate polysilicon.

The performance of analog sampled-data circuits can be directly related to the capacitors used in the implementation. Therefore it is important to investigate their characteristics. From the standpoint of analog sampled-data applications the most important characteristics of the capacitor are (1) ratio accuracy, (2) voltage and temperature coefficients, and (3) parasitic capacitances [25]. We shall briefly discuss each of these characteristics.

The value of integrated circuit capacitors is given by

$$C = \frac{\epsilon_{ox}A}{t_{ox}} = C_{ox}A \tag{1}$$

where ϵ_{ox} is the dielectric constant of the silicon dioxide (approximately 3.45×10^{-5} pF/μm), t_{ox} is the thickness of the oxide, and A is the area of the capacitor. The value of the capacitor is seen to depend upon the area A and the oxide thickness t_{ox}. Therefore, errors in the ratio accuracy of two capacitors result from an error in either the ratio of the areas, or the oxide thicknesses. If the error is caused by a uniform linear variation in the oxide thickness, then a common centroid geometry can be used to eliminate its effects [26]. Area related errors result from the inability to precisely define the dimensions of the capacitor on the integrated circuit. This is due to the error tolerance associated with making the mask, the nonuniform etching of the material defining the capacitor plates, and other limitations [27].

In order to match two capacitors as precisely as possible, it is desirable that the errors associated with each are also matched. For example, let C_1' be defined as

$$C_1' = C_1 \pm \Delta C_1 \tag{2}$$

and C_2' be defined as

$$C_2' = C_2 \pm \Delta C_2 \tag{3}$$

The ratio of C_2' to C_1' can be expressed as

$$\frac{C_2'}{C_1'} = \frac{C_2 \pm \Delta C_2}{C_1 \pm \Delta C_1} \tag{4}$$

$$\approx \frac{C_2}{C_1}\left[1 \pm \frac{\Delta C_2}{C_2} + \frac{-\Delta C_1}{C_1} - \frac{(\Delta C_1)(\Delta C_2)}{C_1 C_2}\right] \tag{5}$$

$$\approx \frac{C_2}{C_1}\left[1 \pm \frac{\Delta C_2}{C_2} + \frac{-\Delta C_1}{C_1}\right] \tag{6}$$

where $(\Delta C_1)/C_1$ and $(\Delta C_2)/C_2$ are the error of C_1 and C_2, respectively, and are assumed to be small. Because ΔC is proportional to the periphery of the capacitor and C is proportional to the area of a capacitor, the error of C is proportional to the periphery-area ratio. If the error of C_1 is the same as C_2 then

$$\frac{C_2'}{C_1'} \approx \frac{C_2}{C_1} \tag{7}$$

If two capacitors of the same value are to be matched, then to minimize the ratio error, the capacitors should be identical to one another. This works well when the ratio is one. In order to make ratios other than one, the two capacitors should consist of integer multiples of a unit. Assuming that the errors associated with each unit capacitor are identical to the errors of all other unit capacitors, the ratio will be error free. This principle is illustrated with the following example.

Example 2.4-1

Designing Integer Capacitor Ratios

It is desired to design a capacitor ratio of 3.50. Maximum accuracy is required. To accomplish this, first express the value 3.50 as a ratio.

$$3.5 = \frac{3.5 \text{ pF}}{1 \text{ pF}}$$

Now determine the greatest common multiple of both the numerator and the denominator. In this case it is 0.5 pF. Therefore, the proper capacitor ratio can be achieved by using one capacitor made from seven units of 0.5 pF each, and 2 units of 0.5 pF each. The resulting ratio is seen mathematically as

$$\text{ratio} = \frac{7 \times 0.5 \text{ pF}}{2 \times 0.5 \text{ pF}} = \frac{3.5 \text{ pF}}{1 \text{ pF}} = \text{desired ratio}$$

Maximum accuracy is achieved if the errors of each unit capacitor are identical.

Many times, the unit capacitor value calculated to achieve integer multiple ratios is too small to implement practically. In these cases, the capacitors are made up of many units, with one being slightly larger (or smaller) than the unit size. The example that follows illustrates this technique.

Example 2.4-2

Design of Noninteger Capacitor Ratios

It is desired to design a capacitor ratio of 3.7. The greatest common multiple of 3.7 and 1 is 0.1. This value (0.1 pF) is too small for practical implementation on an integrated circuit. This value must be made from some unit and nonunit capacitors. There are two obvious ways in which this can be done. They are expressed mathematically below:

$$\text{ratio} = \frac{7 \times 0.5 + 0.2 \text{ pF}}{2 \times 0.5 \text{ pF}} = \frac{6 \times 0.5 + 0.7 \text{ pF}}{2 \times 0.5 \text{ pF}}$$

The more appropriate of these two expressions is the second, since a 0.7 pF capacitor is more practical than a 0.2 pF capacitor.

When designing circuitry that depends upon ratios of capacitors, it is desirable to keep the ratios as close to unity as possible. This restriction stems from the fact that ratios close to unity tend to match better, as shown graphically in Fig. 2.4-2 [27]. This figure also shows that the ratios of larger capacitors are more accurate than smaller ones.

The voltage coefficient of integrated capacitors is negative and generally falls within the range of -10 to -200 ppm/V depending upon the doping concentration in the capacitor plates [28]. The temperature coefficient of these capacitors is found to be in the range of 20 to 50 ppm/°C. When considering the ratio of two capacitors on the same substrate, note that the variations on the absolute value of the capacitor due to temperature tend to cancel. Therefore temperature variations have little effect on the matching accuracy of capacitors. When capacitors are switched to different voltages, as in the case of sampled-data circuits, then the voltage coefficient can have a deleterious effect if it is not small.

The parasitic capacitors associated with the capacitors of Fig. 2.4-1 represent a major source of error in analog sampled-data circuits. The upper plate of the capacitor will be designated as the top plate and the lower capacitor plate as the bottom plate. The parasitic capacitor associated with the top plate of the capacitor itself is due primarily to interconnect lines leading to the capacitor. The bottom-plate parasitic capacitance

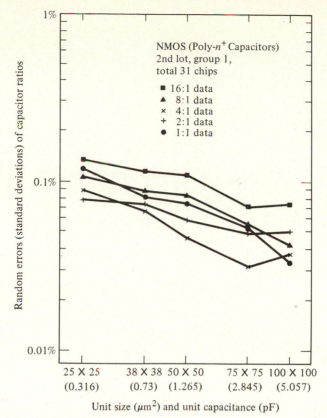

Figure 2.4-2   Relative accuracy of polysilicon MOS capacitors. (© 1984 IEEE)

is primarily due to the capacitance between the bottom plate and the substrate. For polysilicon-oxide-polysilicon capacitors, this parasitic is a polysilicon-oxide-silicon capacitor. It is generally less than one tenth of the value of the desired capacitor. For polysilicon-oxide-silicon capacitors, the bottom plate parasitic is a depletion capacitance between the implanted channel and the substrate with a value on the order of the desired capacitance. Figure 2.4-3 shows a general capacitor with its top and bottom plate parasitics. These parasitic capacitances depend on the capacitor size, layout, and technology, and are generally unavoidable.

The other passive component compatible with MOS technology is the resistor. Even though we shall use primarily circuits containing only MOS active devices and capacitors, some applications, such as digital-to-analog conversion, use the resistor. Resistors compatible with the MOS technology of this section include diffused, polysilicon, ion-implanted, p-well (or n-well), and pinched resistors.

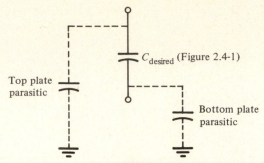

Figure 2.4-3   A model for the capacitors of Fig. 2.4-1.

A diffused resistor is formed using source/drain diffusion and is shown in Fig. 2.4-4(a). The sheet resistance of such resistors is usually in the range of 10 to 100 ohms/square. The fact that the source/drain diffusion is needed as a conductor in integrated circuits conflicts with its use as a resistor. The diffused resistor is found to have a voltage coefficient of resistance in the 100 to 500 ppm/V range. The parasitic capacitance to ground is also voltage dependent in this type of resistor.

A polysilicon resistor is shown in Fig. 2.4-4(b). This resistor is surrounded by thick oxide and has a sheet resistance in the range of 30 to 200 ohms/square. The parasitics of this resistor are very small and voltage independent. The sheet resistance of polysilicon can be increased by a factor of 2 to 3 times if it is shielded from the implantation for the source and the drain. Another advantage of the polysilicon resistor is that it can be trimmed by blowing links by applying either an electric current or a laser.

An ion-implanted resistor can be obtained at the expense of an additional mask step. This resistor is similar to the diffused resistor of Fig. 2.4-4(a). The practical sheet resistances of the ion-implanted resistor fall in the range of 500 to 2000 ohms per square. Unfortunately, the ion-implanted resistor has a large voltage coefficient, voltage-dependent parasitics, and like the other resistors is subject to errors caused by piezoresistance effects induced by nonuniform residual strain in the chip after packaging.

The p-well resistor shown in Fig. 2.4-4(c) is made up of a strip of p-well contacted at both ends with p^+ source/drain diffusion. This type of resistor has a resistance of 1 to 10 KΩ/square and a high value for its voltage coefficient. In cases where accuracy is not required, such as pull-up resistors, or protection resistors, this structure is very useful.

The pinched resistor is a special case of the p-well resistor. It is shown in Fig. 2.4-4(d). An n^+ source/drain region covers the p-well strip and is tied to the substrate (normally the most positive supply). The resulting structure is much like a JFET with the gate tied to positive supply. The voltage coefficient is much higher for this structure than for the p-well resistor.

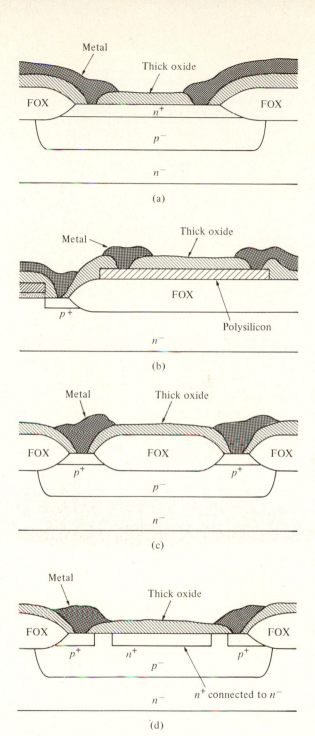

Figure 2.4-4   Resistors. (a) Diffused. (b) Polysilicon. (c) P-well. (d) Pinched.

Table 2.4-1

Approximate Performance Summary of CMOS Passive Components.

Component Type	Range of Values	Relative Accuracy	Temperature Coefficient	Voltage Coefficient	Absolute Accuracy
Poly/poly capacitor	0.3–0.4 fF/μ^2	0.06%	25 ppm/°C	−50 ppm/V	20%
MOS capacitor	0.35–0.5 fF/μ^2	0.06%	25 ppm/°C	−20 ppm/V	10%
Diffused resistor	10–100 ohms/sq.	2% (5 μm width)	1500 ppm/°C	200 ppm/V	35%
Poly resistor	30–200 ohms/sq.	2% (5 μm width)	1500 ppm/°C	100 ppm/V	30%
Ion impl. resistor	0.5–2k ohms/sq.	1% (5 μm width)	400 ppm/°C	800 ppm/V	5%
p-well resistor	1–10k ohms/sq.	2%	8000 ppm/°C	10k ppm/V	40%
pinch resistor	5–20k ohms/sq.	10%	10k ppm/°C	20k ppm/V	50%

Care must be taken in using the pinched resistor in high voltage circuits because it may become pinched off. These problems can be avoided with proper modeling and characterization.

Other types of resistors are possible if the process is altered. The five categories above represent those most compatible with standard MOS technology. Table 2.4-1 summarizes the characteristics of the passive components hitherto discussed.

2.5   *Other Considerations of CMOS Technology*

In the previous two sections, the active and passive components of the basic CMOS process have been presented. In this section we wish to consider some other components that are also available from the basic CMOS process but that are not used as extensively as the previous components We will further consider some of the limitations of CMOS technology, including latch-up, temperature, and noise. This information will become useful later, when the performance of CMOS circuit is characterized.

So far we have seen that it is possible to make resistors, capacitors, and pn diodes that are compatible with the basic single-well CMOS fabrication process illustrated in Fig. 2.3-1. However, it is possible to implement a bipolar junction transistor (BJT) that is also compatible with this process, even though the collector terminal is constrained to V_{DD} (or V_{SS}). Fig. 2.5-1 shows how the BJT is implemented for a p-well process. The emitter is the source or drain diffusion of an NMOS device, the base is the p-well (with a base width of W_B) and the n$^-$ substrate is the collector. Because the pn

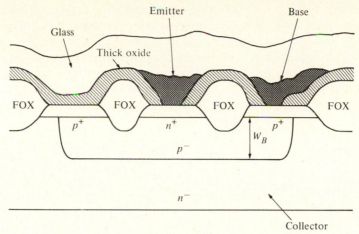

Figure 2.5-1   Substrate BJT available from a bulk CMOS process.

junction between the p-well and the n^- substrate must be reverse biased, the collector must always be connected to the most positive power-supply voltage, V_{DD}. The BJT will still find many useful applications even though the collector is constrained. The BJT illustrated in Fig. 2.5-1 is often called a *substrate BJT*. The substrate BJT functions like the BJT fabricated in a process designed for BJTs. The only difference is that the collector is constrained and the base width is not well controlled, resulting in a wide variation of current gains.

Fig. 2.5-2 shows the minority-carrier concentrations in the BJT. Normally, the base-emitter (*BE*) pn junction is forward biased and the collec-

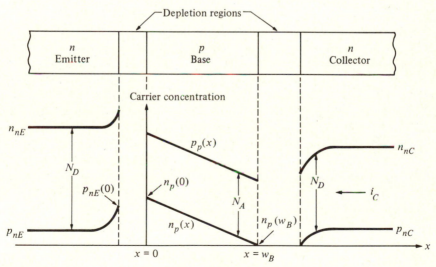

Figure 2.5-2   Minority carrier concentrations for a bipolar junction transistor.

tor-base (CB) pn junction is reverse biased. The forward-biased EB junction causes free electrons to be injected into the base region. If the base width W_B is small, most of these electrons reach the CB junction and are swept into the collector by the reverse-bias voltage. If the minority-carrier concentrations are much less than the majority-carrier concentrations, then the collector current can be found by solving for the current in the base region. In terms of current densities, the collector current density is

$$J_C = -J_n \bigg|_{base} = -qD_n \frac{dn_p(x)}{dx} = qD_n \frac{n_p(0)}{W_B} \tag{1}$$

From Eq. (16) of Sec. 2.2 we can write

$$n_p(0) = n_{po} \exp\left(\frac{v_{BE}}{V_t}\right) \tag{2}$$

Combining Eqs. (1) and (2) and multiplying by the area of the BE junction A gives the collector current as

$$i_C = AJ_C = \frac{qAD_n n_{po}}{W_B} \exp\left(\frac{v_{BE}}{V_t}\right) = I_s \exp\left(\frac{v_{BE}}{V_t}\right) \tag{3}$$

where I_s is defined as

$$I_s = \frac{qAD_n n_{po}}{W_B} \tag{4}$$

As the electrons travel through the base, a small fraction will recombine with holes which are the majority carriers in the base. As this occurs, an equal number of holes must enter the base from the external base circuit in order to maintain electrical neutrality in the base region. Also, there will be injection of the holes from the base to the emitter due to the forward-biased BE junction. This injection is much smaller than the electron injection from the emitter because the emitter is more heavily doped than the base. The injection of holes into the emitter and the recombination of electrons with holes in the base both constitute the external base current i_B that flows into the base. The ratio of collector current to base current, i_C/i_B is defined as β_F or the common-emitter current gain. Thus, the base current is expressed as

$$i_B = \frac{i_C}{\beta_F} = \frac{I_s}{\beta_F} \exp\left(\frac{v_{BE}}{V_t}\right) \tag{5}$$

The emitter current can be found from the base current and the collector current because the sum of all three currents must equal zero. Although β_F has been assumed constant it varies with i_C, having a maximum for moderate currents and falling off from this value for large or small currents.

In addition to the substrate BJT, it is also possible to have a lateral BJT. Fig. 2.3-1 can be used to show how the lateral BJT can be implemented. The emitter could be the p^+ source of the PMOS device, the base the n^- substrate, and the collector the p^- well. Although the base is constrained to the substrate potential of the chip, the emitter and collector can have arbitrary voltages. Unfortunately the lateral BJT is not very useful because of the large base width. In fact the lateral BJT is considered more as a parasitic transistor. However, this lateral BJT becomes important in the problem of latch-up of CMOS circuits which is discussed next [29].

Latch-up in integrated circuits may be defined as a high current state accompanied by a collapsing or low-voltage condition. Upon application of a radiation transient or certain electrical excitations, the latched or high current state can be triggered. Latch-up can be initiated by at least three regenerative mechanisms. They are: (1) the four-layer, silicon-controlled-rectifier (SCR), regenerative switching action; (2) secondary breakdown; and (3) sustaining voltage breakdown. Because of the multiple p and n diffusions present in CMOS, they are susceptible to SCR latch-up.

Fig. 2.5-3(a) shows a cross-section of Fig. 2.3-1 and how the PNPN SCR is formed. The schematic equivalent of Fig. 2.5-3(a) is given in Fig. 2.5-3(b). Here the SCR action is clearly illustrated. The resistor R_{N-} is the resistance from the base of the lateral *PNP* (Q1) and the collector of the vertical *NPN* (Q2) to V_{DD}. The resistor R_{P-} is the resistance from the collector of the lateral *PNP* (Q1) and the base of the vertical *NPN* (Q2) to V_{SS}.

Regeneration occurs when three conditions are satisfied. The first condition is that the loop gain must exceed unity. This condition is stated as

$$\beta_{NPN}\beta_{PNP} \geq 1 \qquad (6)$$

where β_{NPN} and β_{PNP} are the common-emitter, current-gain ratios of Q2 and Q1, respectively. The second condition is that both of the base-emitter junctions must become forward biased. The third condition is that the circuits connected to the emitter must be capable of sinking and sourcing a current greater than the holding current of the *PNPN* device.

To prevent latch-up, several standard precautions are taken. One approach is to keep the source/drain of the p-channel device as far away from the p-well as possible. This reduces the value of β_{PNP} and helps to prevent latch-up. Unfortunately, this is very costly in terms of area. A second approach is to reduce the values of R_{N-} and R_{P-}. Smaller resistor values are helpful because more current has to flow through them in order to forward bias the base-emitter regions of Q1 and Q2. These resistances can be

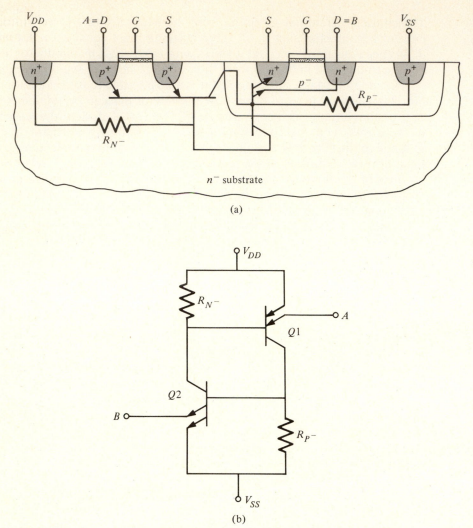

Figure 2.5-3 (a) Parasitic lateral PNP and vertical NPN bipolar transistor in CMOS integrated circuits. (b) Equivalent circuit of the SCR formed from the parasitic bipolar transistors.

reduced by surrounding the p-channel devices with a n^+ guard ring connected to V_{DD} and by diffusing a p^+ guard ring into the perimeter of the p-well as shown in Fig. 2.5-4. (For an n-well technology, a p^+ guard ring would surround the n-channel devices and be connected to V_{SS} and an n^+ guard ring would be diffused into the perimeter of the n-well.) Not only does this reduce the value of R_{N-} and R_{P-}, it also helps to prevent the possibility of an n^+ diffusion in the p-well from being too close to the edge and

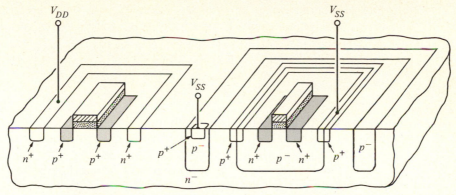

Figure 2.5-4   Preventing latch-up in a p-well technology.

creating a V_{SS} to V_{DD} short. We are assuming that the source of the n-channel devices will typically be at V_{SS}. A third approach is to make a p^- diffusion outside of the p-well. This will short the collector of Q1 of Fig. 2.5-3 (b) to ground. Unfortunately, this must be a deep diffusion and therefore requires a significant amount of area.

Latch-up can also be prevented by keeping the potential of the source/drain of the p-channel device [A in Fig. 2.5-3 (b)] from being higher than V_{DD} or the potential of the source/drain of the n-channel device [B in Fig. 2.5-3 (b)] from going below V_{SS}. By careful design and layout, latch-up can be avoided in most cases. In the design of various circuits, particularly those that have high currents, one must use care to avoid circuit conditions that will initiate latch-up.

Another important consideration of CMOS technology is the electrostatic discharge protection of the gates of transistors which are externally accessible. To prevent accidental destruction of the gate oxide, a resistance and two reverse-biased pn junction diodes are employed to form an input protection circuit. One of the diodes is connected with the n side to the highest circuit potential (V_{DD}) and the p side to the gate to be protected. The other diode is connected with the n side to the gate to be protected and the p side to the lowest circuit potential (V_{SS}).

For a p-well process, the first diode is usually made by a p^+ diffusion into the n^- substrate. The second diode is made by a n^+ diffusion into a p^-well. The resistor is connected between the external contact and the junction between the diodes and the gate to be protected. If a large voltage is applied to the input, one of the diodes will breakdown depending upon the polarity of the voltage. If the resistor is large enough, it will limit the breakdown current so that the diode is not destroyed. This circuit should be used whenever the gates of a transistor (or transistors) are taken to external circuits.

The temperature dependence of CMOS components is an important performance characteristic in analog circuit design. The temperature behavior of passive components is usually expressed in terms of a *fractional temperature coefficient* TC_F defined as

$$TC_F = \frac{1}{X} \cdot \frac{dX}{dT} \tag{7}$$

where X can be the resistance or capacitance of the passive component. Generally, the fractional temperature coefficient is multiplied by 10^6 and expressed in units of parts per million per °C or ppm/°C. The fractional temperature coefficient of various CMOS passive components has been given in Table 2.4-1.

The temperature dependence of the CMOS device can be found from the expression for drain current given in Eq. (22) of Sec. 2.3. The primary temperature-dependent parameters are the mobility μ and the threshold voltage V_T. The temperature dependence of the carrier mobility μ is given as [30],

$$\mu = K_\mu T^{-1.5} \tag{8}$$

The temperature dependence of the threshold voltage can be approximated by the following expression [31]

$$V_T(T) = V_T(T_0) - \alpha(T - T_0) \tag{9}$$

where α is approximately 2.3 mV/°C. This expression is valid over the range of 200 to 400 °K, with α depending on the substrate doping level and the dosages of the implants used during fabrication. These expressions for the temperature dependence of mobility and threshold voltage will be used later to determine the temperature performance of CMOS circuits and are valid only for limited ranges of temperature variation about room temperature. Other modifications are necessary for extreme temperature ranges.

The temperature dependence of the pn junction is also important in this study. For example, the pn-junction diode can be used to create a reference voltage whose temperature stability will depend upon the temperature characteristics of the pn-junction diode. We shall consider the reverse-biased pn-junction diode first. Eq. (24) of Sec. 2.2 shows that when $v_D < 0$, that the diode current is given as

$$-i_D \cong I_s = qA\left[\frac{D_p p_{no}}{L_p} + \frac{D_n n_{po}}{L_n}\right] \cong \frac{qAD}{L}\frac{n_i^2}{N} = KT^3 \exp\left(\frac{-V_{Go}}{V_t}\right) \tag{10}$$

where it has been assumed that one of the terms in the brackets is dominant and that L and N correspond to the diffusion length and impurity concentration of the dominant term. Also T is the absolute temperature in °K and V_{Go} is the bandgap voltage of silicon at 300 °K (1.205 V). Differentiating Eq. (10) with respect to T results in

$$\frac{dI_s}{dT} = \frac{3KT^3}{T} \exp\left(\frac{-V_{Go}}{V_t}\right) + \frac{qKT^3 V_{Go}}{kT^2} \exp\left(\frac{-V_{Go}}{V_t}\right) = \frac{3I_s}{T} + \frac{I_s}{T}\frac{V_{Go}}{V_t} \tag{11}$$

The TC_F for the reverse diode current can be expressed as

$$\frac{1}{I_s}\frac{dI_s}{dT} = \frac{3}{T} + \frac{1}{T}\frac{V_{Go}}{V_t} \tag{12}$$

The reverse diode current is seen to double approximately every 5 °C increase as illustrated in the following example.

Example 2.5-1

Calculation of the Reverse Diode Current Temperature Dependence and TC_F

Assume that the temperature is 300 °K (room temperature) and calculate the reverse diode current change and the TC_F for a 5 °K increase.

The TC_F can be calculated from Eq. (12) as

$$TC_F = 0.01 + 0.155 = 0.165$$

Since the TC_F is change per degree, the reverse current will increase by 1.165 for every °K (or °C) change in temperature. Multiplying by 1.165 five times gives an increase of approximately 2. This implies that the reverse saturation current will approximately double for every 5 °C temperature increase. Experimentally, the reverse current doubles for every 8 °C increase in temperature because the reverse current is in part leakage current.

The forward biased pn-junction diode current is given by

$$i_D \cong I_s \exp\left(\frac{v_D}{V_t}\right) \tag{13}$$

Differentiating this expression with respect to temperature and assuming that the diode voltage is a constant ($v_D = V_D$) gives

$$\frac{di_D}{dT} = \frac{i_D}{I_s} \cdot \frac{dI_s}{dT} - \frac{1}{T} \cdot \frac{V_D}{V_t} i_D \tag{14}$$

The fractional temperature coefficient for i_D results from Eq. (14) as

$$\frac{1}{i_D} \cdot \frac{di_D}{dT} = \frac{1}{I_s} \cdot \frac{dI_s}{dT} - \frac{V_D}{TV_t} = \frac{3}{T} + \left[\frac{V_{Go} - V_D}{TV_t} \right] \tag{15}$$

If V_D is assumed to be 0.6 volts, then the fractional temperature coefficient is equal to $0.01 + (0.155 - 0.077) = 0.0879$. It can be seen that the forward diode current will double for approximately a 10°C increase in temperature.

The above analysis for the forward-bias pn-junction diode assumed that the diode voltage v_D was held constant. If the forward current is held constant ($i_D = I_D$), then the fractional temperature coefficient of the forward diode voltage can be found. From Eq. (13) we can solve for v_D to get

$$v_D = V_t \ln \left(\frac{I_D}{I_s} \right) \tag{16}$$

Differentiating Eq. (16) with respect to temperature gives

$$\frac{dv_D}{dT} = \frac{V_D}{T} - V_t \left(\frac{1}{I_s} \cdot \frac{dI_s}{dT} \right) = \frac{V_D}{T} - \frac{3V_t}{T} - \frac{V_{Go}}{T}$$

$$= - \left[\frac{V_{Go} - V_D}{T} \right] - \frac{3V_t}{T} \tag{17}$$

Assuming that $v_D = V_D = 0.6$ V the temperature dependence of the forward diode voltage at room temperature is approximately -2.3 mV/°C.

Another limitation of CMOS components is noise. Noise is a phenomenon caused by small fluctuations of the analog signal within the components themselves. Noise results from the fact that electrical charge is not continuous but the result of quantized behavior and is associated with the fundamental processes in a semiconductor component. In essence, noise acts like a random variable and is often treated as one. Our objective is to introduce the basic concepts concerning noise in CMOS components. More detail can be found in several excellent references [20,32].

Several sources of noise are important in CMOS components. *Shot noise* is associated with the dc current flow across a pn-junction. It typically has the form of

$$\bar{i}^2 = 2qI_D\Delta f \text{ (Amperes}^2) \tag{18}$$

where $\bar{i}^2$ is the mean-square value of the noise current, q is the charge of an electron, I_D is the average dc current of the pn junction, and Δf is the bandwidth in hertz. Noise-current spectral density can be found by dividing $\bar{i}^2$ by Δf. The noise-current spectral density is denoted as $\bar{i}^2/\Delta f$.

Another source of noise, called *thermal noise,* is due to random thermal motion of the electron and is independent of the dc current flowing in the component. It typically has the form of

$$\bar{v}^2 = 4kTR\Delta f \tag{19}$$

where k is Boltzmann's constant and R is the resistor or equivalent resistor in which the thermal noise is occurring.

An important source of noise for CMOS components is the *flicker noise* or the *1/f noise.* This noise is associated with carrier traps in semiconductors which capture and release carriers in a random manner. The time constants associated with this process give rise to a noise signal with energy concentrated at low frequency. The typical form of the $1/f$ noise is given as

$$\bar{i}^2 = K_f \left[\frac{I^a}{f^b} \right] \Delta f \tag{20}$$

where K_f is a constant, a is a constant (0.5 to 2), and b is a constant ($\cong 1$). The current-noise spectral density for typical $1/f$ noise is shown in Fig. 2.5-5. Other sources of noise exist, such as burst noise and avalanche noise, but are not important in CMOS components and are not discussed here.

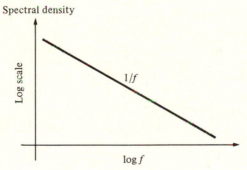

Figure 2.5-5   Illustration of $1/f$ noise spectrum.

2.6 *Geometrical Considerations of Integrated Circuits*

The last subject in this chapter concerns the geometrical characteristics of integrated circuits. A unique aspect of integrated-circuit design is that it gives the designer control over the circuit's geometrical design as well as its electrical design. This feature not only allows the designer an additional degree of design freedom, it requires familiarity with design or layout rules. These geometrical constraints imposed on the features defining various elements of the integrated circuit help guarantee its proper operation. Layout rules are governed primarily by two basic factors—photolithographic resolution and electrical parameters.

Each layer implemented in the integrated-circuit process requires a mask that defines where the wafer is to be implanted, etched, oxidized, etc. A mask is generated from a computerized drawing of the circuit using a photolithographic process. This process is similar in concept to that used to make black and white prints from exposed film. A feature on the mask (a clear or dark area) will ultimately be a feature on the integrated circuit. Because of this, limitations in generating a geometry on the mask are also imposed on the integrated circuit.

Another limitation on geometry definition arises from the fact that at each mask step in the process, features of the next mask must be aligned to features previously defined on the integrated circuit. Even when using precision automatic alignment tools, there is still some error in alignment. In some cases, alignment of two layers is critical to circuit operation. As a result, alignment tolerances impose a limitation of feature size and orientation with respect to other layers on the circuit.

The above concepts can be illustrated by reviewing Figs. 2.3-4, 2.4-1, and 2.4-4 which have been drawn approximately to scale. It can be seen that as the number of steps increase, the surface of the integrated circuit is far from flat. A scanning electron microscope photograph of an NMOS transistor is shown in Fig. 2.6-1. The n^+ diffusions are seen as the light grey areas in the substrate which is dark. One should note the lateral diffusion under the polysilicon gate. To the left of the picture is a metal contact to the n^+ diffusion. While the details of this photograph are more of interest to the device designer, it is obvious that the distinct lines used in the previously cited figures do not exist. Consequently, one of the reasons for design rules is to maximize the probability of a circuit not having a failure due to processing.

Electrical performance requirements also dictate feature size and orientation with respect to other layers. A good example of this is the allowable distance between diffusions supporting a particular voltage differential. Understanding the rules associated with electrical performance is most important to the designer if circuits are to be designed that challenge the limits of the technology. The limits for these rules are constrained by the

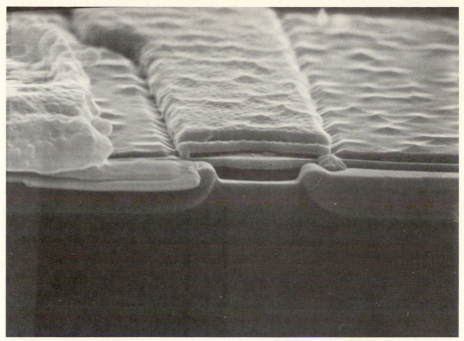

Figure 2.6-1   Scanning electron microscope photograph of a cut-away cross-section of an NMOS transistor. (By permission from Van Nostrand Reinhold Co.)

process (doping concentration, junction depth, etc.) characterized under a specific set of conditions. A good circuit designer will understand these design rules and be able to modify them to meet a specific circuit requirement not addressed by the design rules being used.

In recent years, there have been some important philosophical changes in approaching the design of an integrated circuit, changes that have resulted in new automated approaches to implementing a system function on an integrated circuit. In earlier years, as integrated-circuit technology was developing, a great amount of effort was applied to understanding the smallest detail of the physics and electrical aspects of the circuit. Indeed, the early IC designer required a working knowledge of the physics of semiconductors and details of the process being used. The circuits designed in this era were small in size and had few active devices. As IC technology matured, the circuits have grown and many active devices are now implemented on a single chip.

The above result has brought about some major changes in the methodology used to design integrated circuits. If reasonable design schedules are to be maintained, naturally less time can be spent thinking about the details of implementing a single active device on a chip. Therefore, the design methodology must account for this by structuring the design process under rigid rules (layout and electrical) that insure proper device operation. Under this structured approach, the designer no longer thinks about the individual pieces of an active device (drain/source, contact size, etc.), but rather thinks more about how this device is to be connected together with others to form, say, an integrator or summer.

A further step in the development of a design methodology is achieved when one no longer considers how to connect transistors to form a larger subcircuit, but considers how to connect various subcircuits (integrators, summers, etc.) to form an entire system function, e.g. codec [33]. If the methodology is structured and well regulated, then IC design can be performed by a system designer who has no previous IC experience. There is a great thrust in the high-technology community to achieve this step in the growth of integrated circuit design technology. There are many computer aids to design (CAD) being developed that will help move the capability of IC design into the hands of systems designers.

This design philosophy has an impact on the layout rules. Since many designers from different disciplines will be designing circuits to be fabricated at various silicon foundries, a consistency must be maintained in the rules for layout. This permits process-independent circuit design using generalized CAD tools. The layout rules described here are done with this generality in mind.

The following set of design rules are based upon the minimum dimension resolution λ (lambda, not to be confused with the channel length modulation parameter λ which will be introduced in Chapter 3). The minimum grid resolution λ is typically one micrometer or less.

The basic layout levels needed to define a p-well, silicon gate CMOS circuit include x-well (p^- or n^-), moat (active area), x^+ guard ring (p^+ or n^+ implant), x-gate (polysilicon), capacitor plates, contact, metal, and pad opening. The symbols for these levels are shown in Fig. 2.6-2. Table 2.6-1 gives the simplified design rules for a polysilicon-gate, bulk CMOS process. Fig. 2.6-2 illustrates these rules. Fig. 2.6-3 is a circuit schematic of the actual circuit geometrically illustrated in Fig. 2.6-2.

In most cases (though not consistent with the generalized design methodology previously described), design rules are unique to each wafer manufacturer. The design rules for the particular wafer manufacturer should be obtained before the design is begun and consulted during the design. This is especially important in the design of state-of-the-art analog CMOS. However, the principles developed here should remain unaltered while translated to specific processes.

The geometries used to describe the physical layout of MOS circuits usually consist of rectangles and/or polygons. Fig. 2.6-4 shows a simple rectangle. Often the designer wants to know what the resistance of such a structure might be. This characteristic is generally given in terms of sheet

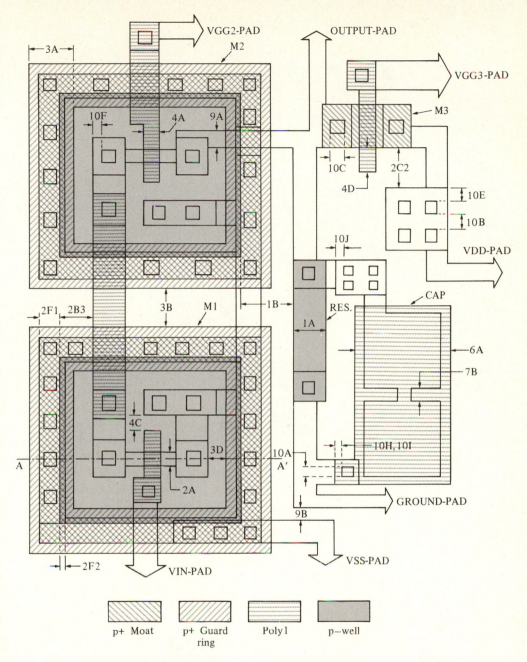

Figure 2.6-2(a) Illustration of the design rules of Table 2.6-1 for the circuit of Fig. 2.6-3. (continued on next page)

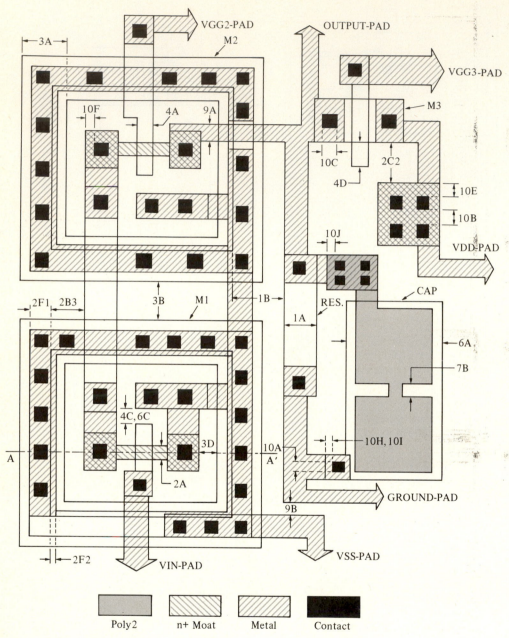

VGG2-PAD
OUTPUT-PAD
VGG3-PAD
M2
3A
M3
10F 4A 9A
10C 2C2
4D
10E
10B
VDD-PAD
10J
2F1 2B3 3B M1 1B RES. CAP
1A
4C,6C 6A
3D 7B
2A 10A
A A' 10H,10I
9B GROUND-PAD
2F2 VIN-PAD VSS-PAD

| Poly2 | n+ Moat | Metal | Contact |

Figure 2.6-2(b)   (Continued)

80

Table 2.6-1

Design Rules for a Polysilicon-Gate, Bulk CMOS Process.

Layer	Minimum Dimension (λ)
1. X-well	
1A. width	4
1B. spacing	7
2. Moat	
2A. width	2
2B. spacing to x-well	
2B1. x-moat contained in x-well	2
2B2. x-moat not touching x-well	5
2B3. xb-moat contained in x-well	3
2B4. xb-moat not touching x-well	5
2C. spacing to other moat (inside or outside well)	
2C1. x-moat to x-moat	3
2C2. xb-moat to x-moat	3
2C3. xb-moat to xb-moat	3
2D. spacing to resistor	2
2E. source-drain overhang of x-gate (or xb)	3
2F. x^+ guard ring structure	
2F1. extends outside x-well	3
2F2. extends inside x-well	1
3. X^+ Guard Ring	
3A. width	6
3B. spacing	2
3C. spacing to x-moat (or xb) outside x-well	3
3D. spacing to x-moat (or xb) inside x-well	3
4. X-Gate	
4A. width	2
4B. spacing	2
4C. spacing to Moat	
4C1. x-moat (x-gate in x-moat)	2
4C2. x-moat (x-gate in field)	2
4C3. xb-moat (x-gate in xb-moat)	2
4C4. xb-moat (x-gate in field)	2
4D. overhang of x-gate	2
5. Xb-Gate	
5A. width	2
5B. spacing	2
5C. spacing to Moat	
5C1. x-moat (xb-moat in x-moat)	2
5C2. x-moat (xb-gate in field)	2
5C3. xb-moat (xb-gate in xb-moat)	2
5C4. xb-moat (xb-gate in field)	2
6. Capacitor bottom plate	
6A. width	2
6B. spacing	2
6C. spacing to x-moat (or xb)	2
6D. overhang of capacitor top plate	2

Table 2.6-1 (Continued)
Design Rules for a Polysilicon-Gate, Bulk CMOS Process.

Layer	Minimum Dimension (λ)
7. Capacitor top plate	
7A. width	2
7B. spacing	2
7C. spacing to x-moat (or xb)	2
8. Resistor	
8A. width	2
8B. spacing	2
8C. spacing to x-moat (or xb)	2
8D. spacing to x-gate (or xb)	2
9. Metal	
9A. width	2
9B. spacing	2
9C. Bonding Pad	
9C1. size	46x46
9C2. spacing (pad metal to pad metal)	20
9C3. spacing to x-moat	12
9C4. spacing to xb-moat	12
9C5. spacing to metal circuitry	12
9C6. spacing to x-gate (or xb)	12
9C7. spacing to resistor	12
9C8. spacing to capacitor bottom plate	12
10. Contacts	
10A. size	2x2
10B. spacing	3
10C. spacing to x-gate (or xb)	2
10D. spacing to x-moat (or xb)	2
10E. metal overlap of contact	1
10F. x-moat (or xb) overlap of contact	1
10G. resistor overlap of contact	1
10H. x-gate (or xb) overlap of contact	1
10I. capacitor bottom plate overlap of contact	1
10J. capacitor top plate overlap of contact	1
10K. metal overlap of pad opening	3
11. Pad opening	
11A. Bonding pad opening	40x40
11B. Probe pad opening	26x26
12. Mask	
12A. spacing to x-gate (or xb)	2
12B. overhang of resistor	3

Note: If x is p (or n) type, xb is n (or p) type.

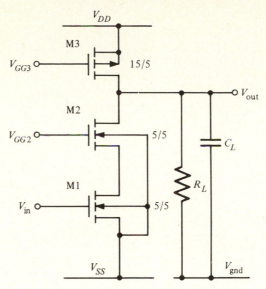

Figure 2.6-3 Simple cascode amplifier (W/L ratios are $\mu m/\mu m$)

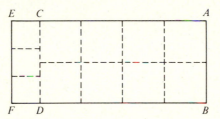

Figure 2.6-4 Rectangular geometry used to demonstrate the calculation of the equivalent number of squares.

resistance (ohms/square). One can determine the resistance of a rectangular structure simply by constructing squares. An example demonstrates the method.

Example 2.6-1

Calculation of the Resistance of a Rectangle

Find the equivalent number of squares for the rectangular geometry shown in Fig. 2.6-4. Fig. 2.6-4 has been subdivided into squares starting from right to left. When the larger squares do not fit into the remaining area, one uses smaller squares. The resistance of this

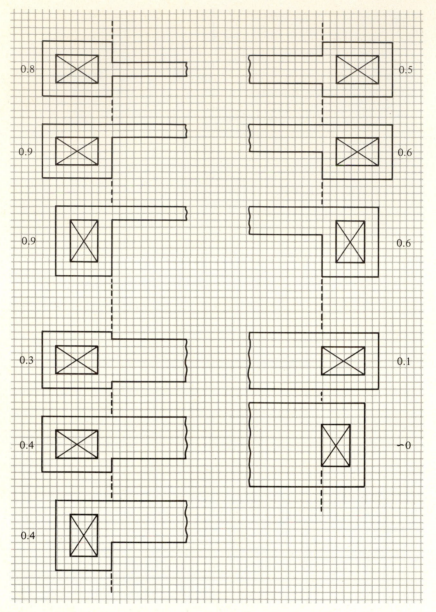

Figure 2.6-5 Equivalent number of squares for various types and sizes of contacts. (By permission from Van Nostrand Reinhold Co.)

geometry from side *AB* to side *EF* can be determined as follows. Between side *AB* and line *CD* there are four series squares in parallel. Thus there is an effective number of squares between *AB* and *CD* of two. Between line *CD* and side *EF*, there are three smaller squares in parallel resulting in an effective number of squares of one-third. Therefore, the total number of squares of the geometry in Fig. 2.6-4 is two and one-third. When this number is multiplied by the sheet resistance (ohms/square), the resistance between sides *AB* and *EF* will be known.

The above technique works well for rectangular geometries where ideal contact can be assumed at the edges *AB* and *EF*. Unfortunately, such ideal edge contacts do not exist in integrated circuits. The sheet resistance of metal (aluminum) is several orders of magnitude smaller than polysilicon or diffusion. The situation of interest is when metal is used to contact either polysilicon or diffusion. Fig. 2.6-5 shows some examples of the resistance contribution of various methods of making metal contacts to polysilicon or diffusion [34]. One calculates the resistance of the entire structure in the following manner. Up to the dotted line, the calculation is similar to that used in Ex. 2.6-1. The number shown by the geometry is the approximate number of squares from the dotted line to the metal contact. To find the total number of squares, one adds the number in Fig. 2.6-5 to the number of squares for the interior portion of the geometry.

Example 2.6-2

Effective Number of Squares for a Complete Resistor

Find the resistance of the polysilicon link shown in Fig. 2.6-6 if the sheet resistance of polysilicon is 100 ohms/square. It can be seen that the interior rectangle has 7.5 squares. Adding to this figure 0.5 squares for each contact gives a total of 8.5 squares. Thus, the resistance of the polysilicon link is 850 ohms.

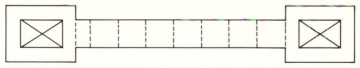

Figure 2.6-6   A polysilicon resistor.

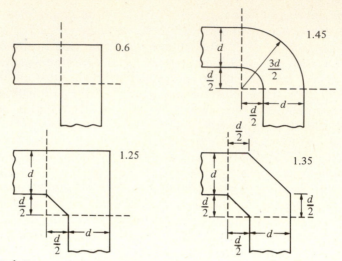

Figure 2.6-7 Examples of the equivalent number of squares for corners or bends.

Another problem concerning the identification of the correct number of squares involves bends or corners. Fig. 2.6-7 shows several examples of corners or bends. The equivalent number of squares between the dotted lines is shown by the various examples.

This section has presented some of the geometrical aspects of integrated circuits. The reasons behind the design rules are found in the desire for an integrated circuit with a high yield. If the design rules are not respected, then problems such as misalignment, incomplete metal coverage, etc., can occur. Also, in laying out circuits the designer must be able to predict resistances and W/L ratios of odd-shaped geometries.

2.7 *Summary*

This chapter has introduced CMOS technology from the viewpoint of its use to implement analog circuits. The basic semiconductor fabrication processes were described in order to understand the fundamental elements of this technology. The basic fabrication steps include diffusion, implantation, deposition, etching, and oxide growth. These steps are implemented by the use of photolithographic methods which limit the processing steps to certain physical areas of the silicon wafer.

The pn junction was reviewed following the introduction to CMOS technology because it plays an important role in all semiconductor devices. This review examined a step pn junction and developed the physical dimensions, the depletion capacitance, and the voltage-current characteristics of the pn junction.

Next, the MOS transistor was introduced and characterized with respect to its behavior. It was shown how the channel between the source and drain is formed and the influence of the gate voltage upon this channel was discussed. The MOS transistor is physically a very simple component. Finally, the steps necessary to fabricate the transistor were presented.

The processing steps to fabricate an MOS transistor defined the possible passive components, which were discussed next. These components include only resistors and capacitors. The absolute accuracy of these components depends on their edge uncertainties and improves as the components are made physically larger. The relative accuracies of the passive components were seen to be better than 1% in many cases.

The next section discussed further considerations of CMOS technology. These considerations included: the substrate and lateral BJTs compatible with the CMOS process; latch-up, which occurs under certain high-current conditions; the temperature dependence of CMOS components; and the noise sources in these components.

The last section covered the geometrical definition of CMOS devices. This focused on the physical constraints that insure that the devices will work correctly after fabrication. This material will lead naturally to the next chapter where circuit models are developed to be used in analyzing and designing circuits.

PROBLEMS — *Chapter 2*

1. List the five basic MOS fabrication processing steps and give the purpose or function of each step.
2. What is the difference between positive and negative photoresist and how is photoresist used?
3. Repeat Example 2.2-1 if the applied voltage is -1 V.
4. Develop Eq. (9) of Sec. 2.2 using Eqs. (1), (7), and (8) of the same section.
5. Redevelop Eqs. (7) and (8) if the impurity concentration of a pn junction is given by the dotted line of Fig. 2.2-2 rather than the step junction of Fig. 2.2-1(b).
6. Plot the normalized reverse current, i_{RA}/i_R, versus the reverse voltage v_R of a silicon pn diode which has $BV = 10$ V and $n = 6$.
7. What is the breakdown voltage of a pn junction with $N_A = N_D = 10^{17}/$ cm^3?
8. What change in v_D of a silicon pn diode will cause an increase of 10 (an order of magnitude) in the forward diode current?
9. Explain in your own words why the threshold voltage in Eq. (4) of Sec. 2.3 decreases as the value of the source-bulk voltage increases.

10. If $V_{SB} = 5$ V, find the value of V_T for the NMOS transistor of Ex. 2.3-1.

11. List two advantages and two disadvantages of CMOS technology compared with NMOS technology.

12. If the mobility of an electron is 550 cm²/(volt·sec) and the mobility of a hole is 225 cm²/(volt·sec), compare the performance of an NMOS with a PMOS transistor. In particular, consider the value of the transconductance parameter and speed of the MOS transistor.

13. What is the function of silicon nitride in the CMOS fabrication process described in Fig. 2.3-4?

14. A top view of an NMOS transistor with a $W = 10$ μm fabricated in a p⁻ substrate is shown in Fig. P2.14. Draw the cross section, AA' as close to scale as possible if the n⁺ diffusion depth and lateral diffusion are both 1 μm. The various thicknesses of the process are: field oxide $= 1$ μm, intermediate oxide $= 1$ μm, thin oxide $= 0.1$ μm, polysilicon $= 0.5$ μm, and metal $= 1$ μm. Be sure to label the layers in your cross-sectional view.

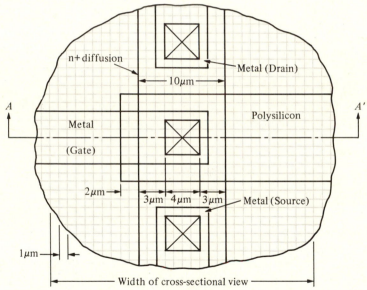

Figure P2.14

15. Give typical thicknesses for the field oxide (FOX), thin oxide (TOX), n⁺ or p⁺, p-well, and metal in terms of micrometers (μm).

16. Fig. P2.16 shows the layout of a PMOS transistor where the channel area has not been defined (due to a mistake). Draw the cross section AA' approximately to scale if the junction depth of the p⁺ diffusion is 0.4 μm and the various thicknesses of the process are field oxide = 1 μm, intermediate oxide = 1 μm, thin oxide = 0.1 μm, polysilicon =

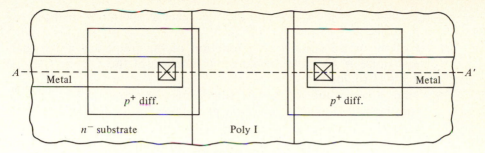

Figure P2.16

0.5 μm, and metal = 1 μm. Discuss how the performance of this transistor will be influenced by this mistake.

17. List two sources of error that can make the actual capacitor, fabricated using a CMOS process, differ from its designed value.

18. What is the purpose of the p$^+$ implantation in the capacitor of Fig. 2.4-1(a)?

19. Plot the random error of capacitor ratios as a function of the capacitor ratio using the data of Fig. 2.4-2 if the unit size of the capacitor is 50 μm $\times$ 50 μm. Repeat if the unit size of the capacitor is 100 μm $\times$ 100 μm.

20. Why is the capacitor/unit area value of the poly-poly capacitor typically less than that of a MOS capacitor?

21. Draw the approximate voltage-current characteristics of the pinched resistor of Fig. 2.4-4(d).

22. A second-order low-pass filter is shown in Fig. P2.22. It is important that the gain of this filter be 1 $\pm$ 0.005 while the cutoff frequency (f_o = 1/RC) can vary by $\pm$20%. Can this circuit be implemented successfully in CMOS technology? Why or why not?

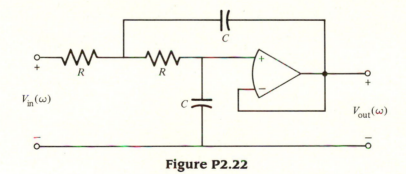

Figure P2.22

23. Derive the fractional temperature coefficient of the forward-biased current in a silicon pn-diode as given in Eq. (15) of Sec. 2.5.

24. Assume $v_D = 0.7$ V and find the fractional temperature coefficient of I_s and v_D.

25. Plot the noise voltage as a function of the frequency if the thermal noise is 100 nV/$\sqrt{\text{Hz}}$ and the junction of the $1/f$ and thermal noise is 10,000 Hz.

26. Find the resistance of the polysilicon resistor shown in Fig. P2.26, if the sheet resistance is 100 ohms per square.

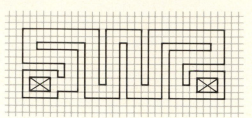

Figure P2.26

27. Find all the design rule violations in Fig. P2.27 for the design rules of Table 2.6-1. Assume that the grid size of Fig. P2.27 corresponds to the minimum process resolution (λ).

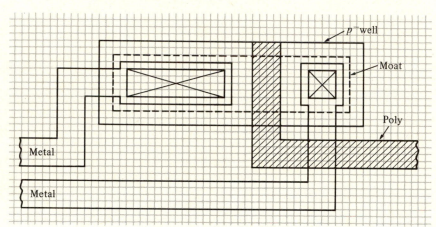

Figure P2.27

28. Find the resistance R_{AB}, R_{BC}, and R_{AC} for the polysilicon resistor shown in Fig. P2.28, if the sheet resistance is 50 ohms/square.
    Fig. P2.29 has been suggested as a means of measuring contact resistance. The procedure is to measure the resistance between terminals 1 and 2 (R_{12}), terminals 1 and 3 (R_{13}), and terminals 1 and 4 (R_{14}). If these measurements result in $R_{12} = 70$ Ω, $R_{13} = 120$ Ω, and $R_{14} = 170$ Ω what is the value of the resistance of a single contact R_C from metal to moat? Assume that the resistance of the metal is negligibly small.

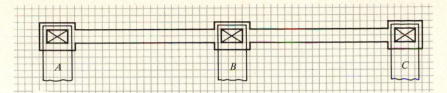

Figure P2.28

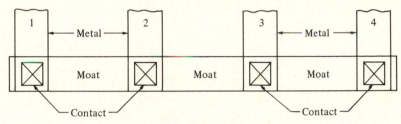

Figure P2.29

30. Identify each of the transistors and R_L and C_L of Fig. 2.6-3 on the layout of Fig. 2.6-2.

31. Assume that the design rules of Table 2.6-1 correspond to a technology with the following process target parameters for an n-type, 100 orientation, 4–6 ohm-cm substrate:

p-well XJ (field) = 8 μm   p-well XJ (active) = 7 μm

LD(n^+) = 0.8 μm   LD(p^+) = 1.2 μm

FOX thickness = 12,000 Angstroms   TOX = 800 Angstroms

Metal thickness = 13,000 Angstroms

Polysilicon thickness = 5000 Angstroms

p^+ diffusion = 67 ohms/square

Polysilicon (n or p) = 25 ohms/square

n^+ diff = 15 ohms/square

p-well (active) = 7 kilohms/square

p-well (field) = 3 Kilohms/square

Intermediate oxide = 10,000 Angstroms

Plot to scale the cross-section indicated by AA' on Fig. 2.6-2.

32. Fig. P2.32 shows a transistor that bends the channel in a serpentine manner in order to fit large W/L transistors into a more rectangular shape. Find the equivalent W/L of this transistor assuming that λ is

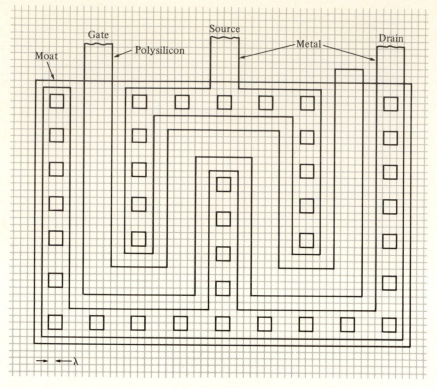

Figure P2.32

2.5 micrometers. Explain the reason behind the large number of metal contacts that are used to connect the source or drain to the source or drain metal.

REFERENCES

1. Y.P. Tsividis and P.R. Gray, "A Segmented $\mu255$ Law PCM Voice Encoder Utilizing NMOS Technology," *IEEE Journal of Solid-State Circuits*, Vol. SC-11, (December 1976) pp. 740–747.
2. B. Fotohouhi and D.A. Hodges, "High-Resolution A/D Conversion in MOS/LSI," *IEEE Journal of Solid-State Circuits*, Vol. SC-14, (December 1979) pp. 920–926.
3. J.T. Caves, C.H. Chan, S.D. Rosenbaum, L.P. Sellers and J.B. Terry, "A PCM Voice Codec with On-Chip Filters," *IEEE Journal of Solid-State Circuits*, Vol. SC-14, (February 1979) pp. 65–73.
4. Y.P. Tsividis and P.R. Gray, "An Integrated NMOS Operational Amplifier with Internal Compensation," *IEEE Journal of Solid-State Circuits*, Vol. SC-11, (December 1976) pp. 748–754.
5. B.K. Ahuja, P.R. Gray, W.M. Baxter, and G.T. Uehara, "A Programmable CMOS

Dual Channel Interface Processor for Telecommunications Applications," *IEEE Journal of Solid-State Circuits,* Vol. SC-19, (December 1984) pp. 892–899.

6. H. Shirasu, M. Shibukawa, E. Amada, Y. Hasegawa, F. Fujii, K. Yasunari, and Y. Toba, "A CMOS SLIC with an Automatic Balancing Hybrid," *IEEE Journal of Solid-State Circuits,* Vol. SC-18, (December 1983) pp. 678–684.

7. A.S. Grove, *Physics and Technology of Semiconductor Devices* (New York: John Wiley & Sons, 1967).

8. R.S. Muller and T.I. Kamins, *Device Electronics for Integrated Circuits* (New York: John Wiley & Sons, 1977).

9. R.C. Colclaser, *Microelectronics Processing and Device Design* (New York: John Wiley & Sons, 1977).

10. J.C. Irvin, "Resistivity of Bulk Silicon and Diffused Layers in Silicon," *Bell System Technical Journal,* Vol. 41, (March 1962) pp. 387–410.

11. D.G. Ong, *Modern MOS Technology—Processes, Devices, & Design,* (New York: McGraw-Hill, 1984).

12. D.J. Hamilton and W.G. Howard, *Basic Integrated Circuit Engineering,* (New York: McGraw-Hill, 1975), Chapter 2.

13. D.H. Lee and J.W. Mayer, "Ion Implanted Semiconductor Devices," *Proceeding of IEEE,* (September 1974) pp. 1241–1255.

14. J.F. Gibbons, "Ion Implantation in Semiconductors," *Proceeding of IEEE,* Part I, Vol. 56, (March 1968) pp. 295–319, Part II, Vol. 60, (September 1972) pp. 1062–1096.

15. A.B. Glaser and G.E. Subak-Sharpe, *Integrated Circuit Engineering—Design, Fabrication, and Applications,* (Reading, MA: Addison-Wesley Publishing Co., 1977).

16. J.L. Vossen and W. Kern (eds.), *Thin Film Processes,* Part III-2, (Academic Press, New York: 1978).

17. P.E. Gise and R. Blanchard, *Semiconductor and Integrated Circuit Fabrication Technique,* (Reston, VA: Reston Publishers, 1979) Chapters 5, 6, 10, and 12.

18. R.W. Hon and C.H. Sequin, "A Guide to LSI Implementation," 2nd Ed., (Xerox Palo Alto Research Center, 3333 Coyote Hill Rd., Palo Alto, CA 94304: January 1980) Chapter 3.

19. D.J. Elliot, *Integrated Circuit Fabrication Technology* (New York: McGraw-Hill, 1982).

20. P.R. Gray and R.G. Meyer, *Analysis and Design of Analog Integrated Circuits,* Second Edition (New York: John Wiley & Sons, 1984).

21. D.A. Hodges and H.G. Jackson, *Analysis and Design of Digital Integrated Circuits* (New York: McGraw-Hill, 1983).

22. B.R. Chawla and H.K. Gummel, "Transition Region Capacitance of Diffused pn Junctions," *IEEE Transactions on Electron Devices,* Vol. ED-18, (March 1971) pp. 178–195.

23. C.T. Sah, "Characteristics of the Metal-Oxide-Semiconductor Transistor," *IEEE Transactions Electron Devices,* Vol. ED-11, (July 1964) pp. 324–345.

24. H. Shichman and D. Hodges, "Modelling and Simulation of Insulated-Gate Field-Effect Transistor Switching Circuits," *IEEE Journal of Solid-State Circuits,* Vol. SC-13, No. 3, (September 1968) pp. 285–289.

25. R.W. Brodersen, P.R. Gray, and D.A. Hodges, "MOS Switched-Capacitor Filters," *Proceedings of the IEEE,* Vol. 67, No. 1, (January 1979) pp. 61–75.

26. J.L. McCreary and P.R. Gray, "All-MOS Charge Redistribution Analog-to-Digital Conversion Techniques—Part I", *IEEE Journal of Solid-State Circuits*, Vol. SC-10, No. 6, (December 1975) pp. 371–379.

27. J.B. Shyu, G.C. Temes, and F. Krummenacher, "Random Error Effects in Matched MOS Capacitors and Current Sources," *IEEE Journal of Solid-State Circuits*, Vol. SC-19, (December 1984) pp. 948–955.

28. J.L. McCreary, "Matching Properties, and Voltage and Temperature Dependence of MOS Capacitors," *IEEE Journal of Solid-State Circuits*, Vol. SC-16, (December 1981) pp. 608–616.

29. D.B. Estreich and R.W. Dutton, "Modeling Latch-Up in CMOS Integrated Circuits and Systems," *IEEE Transactions on CAD*, Vol. CAD-1, No. 4, October 1982, pp. 157–162.

30. S.M. Sze, *Physics of Semiconductor Devices*, 2nd Ed., (New York: John Wiley & Sons, 1981), p. 28.

31. R.A. Blauschild, P.A. Tucci, R.S. Muller, and R.G. Meyer, "A New Temperature-Stable Voltage Reference," *IEEE Journal of Solid-State Circuits*, Vol. SC-19, No. 6, Dec. 1978, pp. 767–774.

32. C.D. Motchenbacher and F.G. Fitchen, *Low-Noise Electronic Design*, (New York: John Wiley and Sons, 1973).

33. D.A. Hodges, P.R. Gray, and R.W. Brodersen, "Potential of MOS Technologies for Analog Integrated Circuits," *IEEE Journal of Solid-State Circuits*, Vol. SC-8, No. 3, (June 1978) pp. 285–294.

34. H.R. Camenzind, *Electronic Integrated Systems Design* (New York: Van Nostrand Reinhold, 1972) Chapter 4.

OTHER REFERENCES

35. T. Redfern, "C-MOS Offers Alternatives to Linear-IC Design," *Electronics*, (April 5, 1984) pp. 142–144.

36. W.M. Penny and L. Lau, Eds., *MOS Integrated Circuits* (New York: Van Nostrand Reinhold, 1972).

37. Engineering Staff of American Microsystems, *MOS Integrated Circuits—Theory, Fabrication, Design and Systems Application MOS/LSI*, (New York: Van Nostrand Reinhold, 1972).

3

chapter

CMOS Analog Circuit Modeling

One of the most important functions of analog integrated-circuit design is that of predicting or verifying the performance of the circuit or system. This function is accomplished through the use of modeling. Modeling is defined as the process by which the electrical properties of a semiconductor device or a group of interconnected devices are represented by means of mathematical equations, circuit representations, or tables.[1] Most of the modeling used in this text will focus on the active and passive devices discussed in the previous chapter as opposed to a higher-level modeling sometimes called macromodeling.

[1] A more light-hearted definition of modeling proposed by Alva Archer of Datel Linear states that "a model is an artifice that gives one the illusion of knowing more about a process than one actually does."

The ability to model MOS devices by characterizing them in terms of terminal voltages and currents is used to characterize both MOS devices and components for the purposes of simulation. Accurate modeling is much more important in analog circuits than in digital circuits. One of the difficulties in using the state of the art CMOS technologies for analog circuit design is that good models and modeling information are not available [1]. As the geometries of CMOS technology shrink, the difficulty of obtaining good models increases. If good models are not available to the designer, then it is not possible to fully utilize the capabilities of the technology.

In this chapter the large signal model of the MOS device described in Sec. 2.3 will be developed. Referring to Table 1.1-2, we see that this material is of crucial importance at the device level of analog circuit design. A simple model, good for large devices, is presented first. This model is suitable for hand, calculator, or computer analysis. Next, the model for the MOS device will be extended to include capacitance, noise, and ohmic resistance. The resulting model will represent a reasonably complete model for the large-signal behavior of the MOS device.

The primary application of the large-signal model is to simulate or solve the large-signal behavior. This behavior includes the biasing of the active devices. Once the bias points have been established, a small-signal model can be used to determine the small-signal performance. This small-signal model is a linear model. This chapter will show how to develop the small-signal model from the large-signal model. Since most of the parameters of the small signal model depend on the large-signal voltages and currents, the small-signal model depends heavily upon the large-signal variables.

A large-signal, strong-inversion model that includes second-order effects is presented next. In many cases the increased accuracy of the second-order model is necessary. Typical examples would be for short-channel devices or for high currents. In many cases of CMOS analog-circuit design, very low-power dissipation circuits are desired. This results in operating the devices at the subthreshold region where the strong-inversion model is no longer appropriate. A large-signal model suitable for subthreshold operation of MOS devices is presented.

The last section shows how to combine the computer with the device models to simulate CMOS analog circuits. A hand calculator program for solving a large signal model of the MOS device and for finding the small signal transconductances is given in Appendix B. This program is written for an HP-41 programmable calculator and is useful for design purposes.

3.1 *Simple MOS Large-Signal Model*

All large-signal models will be developed for the n-channel MOS device with the positive polarities of voltages and currents shown in Fig. 3.1-1(a). The same models can be used for the p-channel MOS device if all voltages

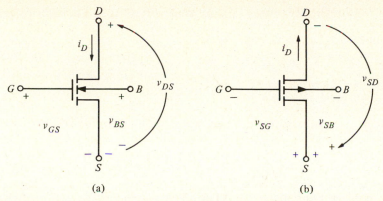

Figure 3.1-1   Positive sign convention for (a) n-channel, and (b) p-MOS transistor.

and currents are multiplied by -1 and the absolute value of the p-channel threshold is used. This is equivalent to using the voltages and currents defined by Fig. 3.1-1(b) which are all positive quantities. As mentioned in Chapter 1, lower-case variables with capital subscripts will be used for the variables of large-signal models and lower-case variables with lower-case subscripts will be used for the variables of small-signal models. When the voltage or current is a model parameter, such as threshold voltage, it will be designated by an upper-case variable and an upper-case subscript.

When the length or width of the MOS device is greater than about 10 μm, the substrate doping is low, and when a simple model is desired, the model suggested by Sah [2] and used in SPICE by Shichman and Hodges [3] is very appropriate. This model was developed in Eq. (22) of Sec. 2.3. It has been modified to include the effects of channel modulation. The result is

$$i_D = \frac{\mu_o C_{ox} W}{L} [(v_{GS} - V_T) - (v_{DS}/2)]v_{DS}(1 + \lambda v_{DS}) \tag{1}$$

The terminal voltages and currents have been defined in the previous chapter. The various parameters of (1) are defined as

μ_o = surface mobility of the channel for the NMOS or PMOS device (cm²/volt·seconds)

$C_{ox} = \dfrac{\epsilon_{ox}}{t_{ox}}$ = capacitance per unit area of the gate oxide (F/cm₂)

W = effective channel width

L = effective channel length

λ = channel length modulation parameter (volts⁻¹)

The threshold voltage V_T is given by Eq. (14) of Sec. 2.3 for an n-channel transistor

$$V_T = V_{T0} + \gamma[\sqrt{2|\phi_F| + v_{SB}} - \sqrt{2|\phi_F|}] \tag{2}$$

$$V_{T0} = V_T (v_{SB} = 0) = V_{FB} + 2|\phi_F| + \frac{\sqrt{2q\epsilon_{si}N_{SUB}2|\phi_F|}}{C_{ox}} \tag{3}$$

$$\gamma = \text{bulk threshold parameter (volts}^{1/2}) = \frac{\sqrt{2\epsilon_{si}qN_{SUB}}}{C_{ox}} \tag{4}$$

$$\phi_F = \text{strong inversion surface potential (volts)} = \frac{kT}{q} \ln\left(\frac{N_{SUB}}{n_i}\right) \tag{5}$$

$$V_{FB} = \text{flatband voltage (volts)} = \phi_{GB} - \frac{Q_{ss}}{C_{ox}} \tag{6}$$

These new parameters and constants are defined as

$$\phi_{GB} = \phi_F \text{ (substrate)} - \phi_F \text{ (gate) [Eq. (11) of Sec. 2.3]}$$

$$\phi_F\text{(substrate)} = \frac{kT}{q} \ln\left(\frac{n_i}{N_{SUB}}\right) \text{ [NMOS with p-substrate]}$$

$$\phi_F\text{(gate)} = \frac{kT}{q} \ln\left(\frac{N_{GATE}}{n_i}\right) \text{ [NMOS with n}^+ \text{ polysilicon gate]}$$

$$Q_{ss} = \text{oxide charge} = N_{ss}q$$

$$k = \text{Boltzmann's constant } (1.381 \times 10^{-23} \text{ J/°K})$$

$$T = \text{temperature (°K)}$$

$$n_i = \text{intrinsic carrier concentration } (1.45 \times 10^{10} \text{ cm}^{-3})$$

Table 3.1-1 gives some of the pertinent constants for silicon. A unique aspect of the MOS device is its dependence upon the voltage from the source to bulk as shown by Eq. (2). This dependence means that the MOS device must be treated as a four-terminal element. It will be shown later

Table 3.1-1

Constants for Silicon.

Constant Symbol	Constant Description	Value	Units
V_{Go}	Silicon bandgap (27°C)	1.205	volts
k	Boltzmann's constant	1.381×10^{-23}	Joules/°K
n_i	Intrinsic carrier concentration (27°C)	1.45×10^{10}	cm^{-3}
ϵ_{si}	Permittivity of silicon	1.0359×10^{-12}	Farads/cm
ϵ_{ox}	Permittivity of SiO$_2$	3.45×10^{-13}	Farads/cm

how this behavior can influence both the large- and small-signal performance of MOS circuits.

In the realm of circuit design, it is more desirable to express the model
equations in terms of electrical rather than physical parameters. For this
reason, the drain current is often expressed as

$$i_D = \beta \left[(v_{GS} - V_T) - \frac{v_{DS}}{2} \right] v_{DS}(1 + \lambda v_{DS})$$

$$= K' \frac{W}{L} \left[(v_{GS} - V_T) - \frac{v_{DS}}{2} \right] v_{DS}(1 + \lambda v_{DS}) \qquad (7)$$

where the transconductance parameter β is given in terms of physical
parameters as

$$\beta = (K') \frac{W}{L} \cong (\mu_o C_{ox}) \frac{W}{L} \text{ (amps/volt}^2) \qquad (8)$$

When devices are characterized in the nonsaturation region the value for
K' is approximately equal to $\mu_o C_{ox}$ in the simple model. This is not the case
when devices are characterized in the saturation region, in that K' is usually
smaller. Typical values for the model parameters of Eq. (7) are given in
Table 3.1-2.

Table 3.1-2
**Model Parameters for a Typical CMOS Bulk Process Suitable for Hand
Calculations Using the Simple Model. These Values Are Based upon a 5 μm
Silicon-Gate Bulk CMOS p-Well Process.**

Parameter Symbol	Parameter Description	Typical Parameter Value		Units
		NMOS	PMOS	
V_{TO}	Threshold Voltage ($V_{BS} = 0$)	1 ± 0.2	-1 ± 0.2	volts
K'_{sat}	Transconductance Parameter (in saturation)	$17.0 \pm 10\%$	$8.0 \pm 10\%$	μA/volt^2
K'_{nonsat}	Transconductance Parameter (in nonsaturation)	$25.0 \pm 10\%$	$10.0 \pm 10\%$	μA/volt^2
γ	Bulk threshold parameter	1.3	0.6	(volts)$^{1/2}$
λ	Channel length modulation parameter	0.01 (L = 10 μm) 0.004 (L = 20 μm)	0.02 (L = 10 μm) 0.008 (L = 20 μm)	(volts)$^{-1}$
$2\|\phi_F\|$	Surface potential at strong inversion	0.7	0.6	volts

There are various regions of operation of the MOS transistor based on the model of Eq. (1). These regions of operation depend upon the value of $v_{GS} - V_T$. If $v_{GS} - V_T$ is zero or negative, then the MOS device is in the cutoff region and Eq. (1) becomes

$$i_D = 0, \qquad v_{GS} - V_T \leqq 0 \tag{9}$$

In this region, the channel acts like an open circuit.

A plot of Eq. (1) with $\lambda = 0$ as a function of v_{DS} is shown in Fig. 3.1-2 for various values of $v_{GS} - V_T$. At the maximum of these curves the MOS transistor is said to saturate. The value of v_{DS} at which this occurs is called the saturation voltage and is given as

$$v_{DS}(\text{sat.}) = v_{GS} - V_T \tag{10}$$

Thus, $v_{DS}(\text{sat.})$ defines the boundary between the remaining two regions of operation. If v_{DS} is less than $v_{DS}(\text{sat.})$, then the MOS transistor is in the non-saturated region and Eq. (1) becomes

$$i_D = \frac{\mu_o C_{ox} W}{L} \left[(v_{GS} - V_T) - \frac{v_{DS}}{2} \right] v_{DS} (1 + \lambda v_{DS}),$$

$$0 < v_{DS} \leqq (v_{GS} - V_T) \tag{11}$$

In Fig. 3.1-2, the nonsaturated region lies between the vertical axis ($v_{DS} = 0$) and $v_{DS} = v_{GS} - V_T$ curve.

The third region occurs when v_{DS} is greater than $v_{DS}(\text{sat.})$ or $v_{GS} - V_T$.

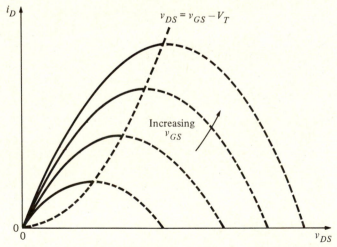

Figure 3.1-2 Graphical illustration of the modified Sah equation.

At this point the current i_D becomes independent of v_{DS}, ignoring for the moment the channel-length modulation effect modeled by λ. Therefore, v_{DS} in Eq. (1) is replaced by v_{DS}(sat.) of Eq. (10) to get

$$i_D = \frac{\mu_o C_{ox} W}{2L} (v_{GS} - V_T)^2 (1 + \lambda v_{DS}), \ 0 < (v_{GS} - V_T) \leqq v_{DS} \qquad (12)$$

where v_{DS} of the $(1 + \lambda v_{DS})$ factor used to model the channel-length modulation behavior in the saturated region has not been replaced by Eq. (10).

The output characteristics of the MOS transistor can be developed from Eqs. (9), (11), and (12). Fig. 3.1-3 shows these characteristics plotted on a normalized basis. These curves have been normalized to the upper curve where V_{GS0} is defined as the value of v_{GS} which causes a drain current of I_{D0} in the saturation region. The entire characteristic is developed by extending the solid curves of Fig. 3.1-2 horizontally to the right from the maximum points. The solid curves of Fig. 3.1-3 correspond to $\lambda = 0$. If $\lambda \neq 0$, then the curves are the dashed lines.

Another important characteristic of the MOS transistor can be obtained by plotting i_D versus v_{GS} using Eq. (12). Fig. 3.1-4 shows this result. This

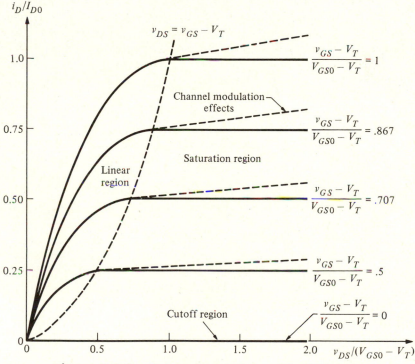

Figure 3.1-3   Output characteristics of the MOS device.

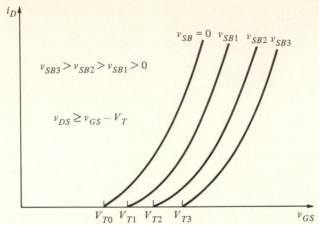

Figure 3.1-4   Transconductance characteristic of the MOS transistor as a function of the bulk-source voltage v_{SB}.

characteristic of the MOS transistor is called the transconductance characteristic. We note that the transconductance characteristic in the saturation region can be obtained from Fig. 3.1-3 by drawing a vertical line to the right of the parabolic dashed line and plotting values of i_D versus v_{GS}. Fig. 3.1-4 is also useful for illustrating the effect of the source-bulk voltage, v_{SB}. As the value of v_{SB} increases, the value of V_T increases for the enhancement, n-channel devices (for a p-channel device, $|V_T|$ increases as v_{BS} increases). V_T also increases positively for the n-channel depletion device, but since V_T is negative, the value of V_T approaches zero from the negative side. If v_{SB} is large enough, V_T will actually become positive and the depletion device becomes an enhancement device.

Since the MOS transistor is a bidirectional device, determining which physical node is the drain and which the source may seem arbitrary. This is not really the case. For an n-channel transistor, the source is always at the lower potential of the two nodes. For the p-channel transistor, the source is always at the higher potential. It is obvious that the drain and source designations are not constrained to a given node of a transistor but can switch back and forth depending upon the external voltages applied to the transistor.

A circuit version of the large-signal model of the MOS transistor consists of a current source, connected between the drain and source terminals, that depends on the drain, source, gate, and bulk terminal voltages defined by the simple model described in this section. This simple model has five electrical and process parameters that completely define it. These parameters are K', V_T, γ, λ, and $2\phi_F$. The subscript n or p will be used when the parameter refers to an NMOS or PMOS device, respectively. They constitute the Level I model parameters of SPICE2 [4]. Typical values for these model parameters are given in Table 3.1-2.

The function of the large-signal model is to solve for the drain current given the voltages of the MOS device. An example will help to illustrate this as well as show how the model is applied to the p-channel device.

Example 3.1-1

Application of the Simple MOS Large Signal Model

Assume that the transistors in Fig. 3.1-1 have a *W/L* ratio of 100 μm/10 μm and that the large signal model parameters are those given in Table 3.1-2. If the drain, gate, source, and bulk voltages of the NMOS transistor is 5 V, 3 V, 0 V, and 0 V, respectively, find the drain current. Repeat for the PMOS transistor if the drain, gate, source, and bulk voltages are -5 V, -3 V, 0 V, and 0 V, respectively.

We must first determine in which region the transistor is operating. Eq. (10) gives v_{DS}(sat) as 3 V $-$ 1 V $=$ 2 V. Since v_{DS} is 5 V, the NMOS transistor is in the saturation region. Using Eq. (12) and the values from Table 3.1-2 we have

$$i_D = \frac{K'_N W}{2L} (v_{GS} - V_{TN})^2 (1 + \lambda_N v_{DS})$$

$$= \frac{17 \times 10^{-6}(100 \ \mu m)}{2(10 \ \mu m)} (3 - 1)^2 (1 + 0.01 \times 5) = 357 \ \mu A$$

Evaluation of Eq. (10) for the PMOS transistor is given as

$$v_{SD}(\text{sat}) = v_{SG} - |V_{TP}| = 3 \text{ V} - 1 \text{ V} = 2 \text{ V}.$$

Since v_{SD} is 5 V, the PMOS transistor is also in the saturation region, Eq. (12) is applicable. The drain current of Fig. 3.1-1(b) can be found using the values from Table 3.1-2 as

$$i_D = \frac{K'_P W}{2L} (v_{SG} - |V_{TP}|)^2 (1 + \lambda_P v_{SD})$$

$$= \frac{8 \times 10^{-6}(100 \ \mu m)}{2(10 \ \mu m)} (3 - 1)^2 (1 + 0.02 \times 5) = 176 \ \mu A$$

3.2 *Other MOS Large-Signal Model Parameters*

The large-signal model also includes several other characteristics such as the source/drain bulk junctions, source/drain ohmic resistances, various capacitors, and noise. The complete version of the large-signal model is given in Fig. 3.2-1.

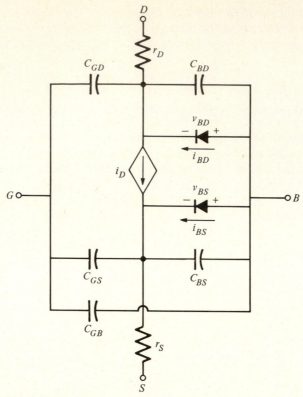

Figure 3.2-1   Complete large-signal model for the MOS transistor.

The diodes of Fig. 3.2-1 represent the pn junctions between the source and substrate and the drain and substrate. For proper transistor operation, these diodes must always be reverse biased. Their purpose in the dc model is primarily to model leakage currents. These currents are expressed as

$$i_{BD} = I_s[\exp(qv_{BD}/kT) - 1] \qquad (1)$$

and

$$i_{BS} = I_s[\exp(qv_{BS}/kT) - 1] \qquad (2)$$

where I_s is the reverse saturation current of a pn junction, q is the charge of an electron, k is Boltzmann's constant, and T is temperature in degrees Kelvin.

The resistors r_D and r_s represent the ohmic resistance of the drain and source, respectively. Typically, these resistors may be 50 to 100 ohms and do not have much influence upon the MOS transistor.

The capacitance of Fig. 3.2-1 can be separated into three types. The first type includes capacitors C_{BD} and C_{BS} which are associated with the back-biased depletion region between the drain and substrate and the source and substrate. The second type includes capacitors C_{GD}, C_{GS}, and C_{GB} which are all common to the gate and are dependent upon the operating condition of the transistor. The third type includes parasitic capacitors which are independent of the operating conditions.

The depletion capacitors are a function of the voltage across the pn junction. The expression of this junction-depletion capacitance is divided into two regions to account for the high injection effects. The first is given as

$$C_{BX} = C_{BX0}A_{BX}[1 - (v_{BX}/PB)]^{-MJ}, \qquad v_{BX} \leqq (FC)(PB) \qquad \textbf{(3)}$$

where

$$X = D \text{ for } C_{BD} \text{ or } X = S \text{ for } C_{BS}$$

A_{BX} = junction areas

$C_{BX0} = C_{BX}$ (when $v_{BX} = 0$) $\cong \sqrt{(q\epsilon_{si}N_{SUB})/(PB)}$

PB = bulk junction potential (same as ϕ_0 given in Eq. (6), sec. 2.2)

FC = forward-bias nonideal junction-capacitance coefficient ($\cong 0.5$)

MJ = bulk-junction grading coefficient (½ for step junctions and ⅓ for graded junctions)

The second region is given as

$$C_{BX} = \frac{C_{BX0}A_{BX}}{(1 - FC)^{1+MJ}} \left[1 - (1 + MJ)FC + MJ\frac{v_{BX}}{PB} \right], \qquad v_{BX} > (FC)(PB) \quad \textbf{(4)}$$

Fig. 3.2-2 illustrates how the junction-depletion capacitances of Eqs. (3) and (4) are combined to model the large signal capacitances C_{BD} and C_{BS}. It is seen that Eq. (4) prevents C_{BX} from approaching infinity as v_{BX} approaches PB.

Since the drain-bulk and source-bulk junctions are identical in most MOS devices, $C_{BS0} = C_{BD0}$. Often the simpler notation of CJ is used for either C_{BS0} or C_{BD0}. A closer examination of the depletion capacitors in Fig. 3.2-3 shows that this capacitor is like a tub. It has a bottom with an area equal to the area of the drain or source. However, there are the sides that are also part of the depletion region. This area is called the sidewall. A_{BX} in Eqs. (3) and (4) should include both the bottom and sidewall assuming the zero-

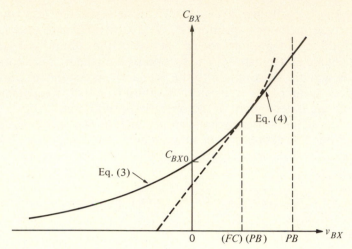

Figure 3.2-2   Example of the method of modeling the voltage dependence of the bulk junction capacitances.

bias capacitances of the two regions are similar. To more closely model the depletion capacitance, break it into the bottom and sidewall components, given as follows.

$$C_{BX} = \frac{(CJ)(AX)}{\left[1 - \left(\dfrac{V_{BX}}{PB}\right)\right]^{MJ}} + \frac{(CJSW)(PX)}{\left[1 - \left(\dfrac{V_{BX}}{PB}\right)\right]^{MJSW}}, \qquad V_{BX} \leqq (FC)(PB) \qquad \textbf{(5)}$$

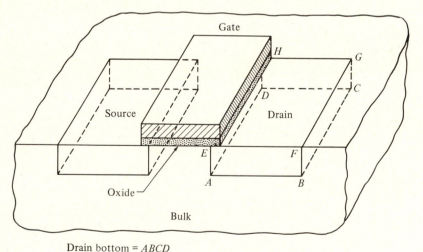

Drain bottom = $ABCD$
Drain sidewall = $ABEF + BCFG + DCGH + ADEH$

Figure 3.2-3   Illustration showing the bottom and sidewall components of the bulk junction capacitors.

and

$$C_{BX} = \frac{(CJ)(AX)}{(1 - FC)^{1+MJ}} \left[1 - (1 + MJ)FC + MJ \frac{V_{BX}}{PB} \right]$$
$$+ \frac{(CJSW)(PX)}{(1 - FC)^{1+MJSW}} \left[1 - (1 + MJSW)FC + \frac{V_{BX}}{PB}(MJSW) \right],$$
$$V_{BX} \geq (FC)(PB) \tag{6}$$

where

$$AX = \text{area of the source } (X = S) \text{ or drain } (X = D)$$

$$PX = \text{perimeter of the source } (X = S) \text{ or drain } (X = D)$$

$$CJSW = \text{zero-bias, bulk–source/drain sidewall capacitance}$$

$$MJSW = \text{bulk–source/drain sidewall grading coefficient}$$

Table 3.2-1 gives the values for CJ, $CJSW$, MJ, and $MJSW$ for an MOS device which has an oxide thickness of 800 Angstroms or a $C_{ox} = 4.3 \times 10^{-4}$ F/m². It can be seen that the depletion capacitors cannot be accurately modeled until the geometry of the device is known, i.e. the area and perimeter of the source and drain. However, values can be assumed for the purpose of design. For example, one could consider a typical source or drain to be 10 micrometers by 10 micrometers. Thus a value for C_{BX} of 50 fF and 31 fF results, for NMOS and PMOS devices respectively, for $V_{BX} = 0$.

The large-signal, charge-storage capacitors of the MOS device consist of the gate-to-source (C_{GS}), gate-to-drain (C_{GD}), and gate-to-bulk (C_{GB}) capacitances. In addition, C_{GS} and C_{GD} can be divided into intrinsic voltage-dependent capacitances (C'_{GS} and C'_{GD}) and extrinsic overlap capacitances (C''_{GS} and C''_{GD}). Fig. 3.2-4 shows a cross section of the various capacitances that constitute the charge-storage capacitors of the MOS device. C_{BS} and C_{BD}

Table 3.2-1

Capacitance Values and Coefficients for the MOS Model.

Type	P-Channel	N-Channel	Units
CGSO	350×10^{-12}	350×10^{-12}	F/m
CGDO	350×10^{-12}	350×10^{-12}	F/m
CGBO	200×10^{-12}	200×10^{-12}	F/m
CJ	150×10^{-6}	300×10^{-6}	F/m²
CJSW	400×10^{-12}	500×10^{-12}	F/m
MJ	0.5	0.5	
MJSW	0.25	0.3	

Based on an oxide thickness of 800 Angstroms or $C_{ox} = 4.3 \times 10^{-4}$ F/m².

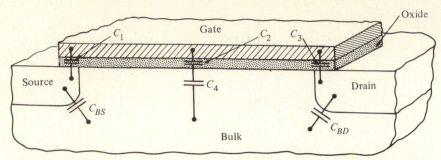

Figure 3.2-4   Large-signal, charge-storage capacitors of the MOS device.

are the source-to-bulk and drain-to-bulk capacitors discussed above. The following discussion represents a heuristic development of a model for the large-signal charge-storage capacitors.

C_1 and C_3 are overlap capacitances and are due to an overlap of two conducting surfaces separated by a dielectric. The overlapping capacitors are shown in more detail in Fig. 3.2-5. The amount of overlap is designated as LD. This overlap is due to the lateral diffusion of the source and drain underneath the polysilicon gate. For example, a 5 micrometer CMOS process might have a diffusion depth of 1 micrometer and thus LD is approximately 0.8 micrometers. The overlap capacitances can be approximated as

$$C_1 = C_3 \simeq (LD)(W_{\text{eff}})C_{ox} = (CGXO)W_{\text{eff}} \tag{7}$$

where W_{eff} is the effective channel width and $CGXO$ ($X = S$ or D) is the overlap capacitance in F/m for the gate-source or gate-drain overlap. The difference between the mask W and actual W is due to the encroachment of the field oxide under the silicon nitride. Table 3.2-1 gives a value for $CGSO$ and $CGDO$ based on a device with an oxide thickness of 800 Angstroms or $C_{ox} = 4.3 \times 10^{-4}$ F/m^2. A third overlap capacitance that can be significant is the overlap between the gate and the bulk. Fig. 3.2-6 shows this overlap capacitor (C_5) in more detail. This is the capacitance that occurs between the gate and bulk at the edges of the channel and is a function of the effective length of the channel, L_{eff}. Table 3.2-1 gives a typical value for $CGBO$ for a device based on an oxide thickness of 800 Angstroms or $C_{ox} = 4.3 \times 10^{-4}$ F/m^2.

The channel of Fig. 3.2-4 is shown for the saturated state and would extend completely to the drain if the MOS device were in the nonsaturated state. C_2 is the gate-to-channel capacitance and is given as

$$C_2 = W_{\text{eff}}(L - 2LD)C_{ox} = W_{\text{eff}}(L_{\text{eff}})C_{ox} \tag{8}$$

The term L_{eff} is the effective channel length resulting from the mask-defined length being reduced by the amount of lateral diffusion (note that up until

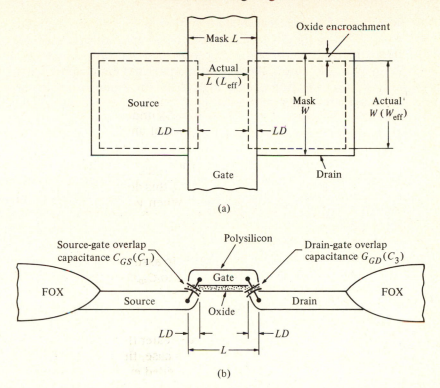

(a)

(b)

Figure 3.2-5 Overlap capacitances of an MOS transistor. (a) Top view showing the overlap between the source or drain and the gate. (b) Side view.

now, the symbols L and W were used to refer to "effective" dimensions whereas now these have been changed for added clarification). C_4 is the channel-to-bulk capacitance which is a depletion capacitance that will vary with voltage like C_{BS} or C_{BD}.

It is of interest to examine C_{GB}, C_{GS}, and C_{GD} as v_{DS} is held constant and v_{GS} is increased from zero. To understand the results, one can imagine following a vertical line on Fig. 3.1-3 at say, $v_{DS} = 0.5(V_{GS0} - V_T)$, as v_{GS} increases from zero. The MOS device will first be off until v_{GS} reaches V_T.

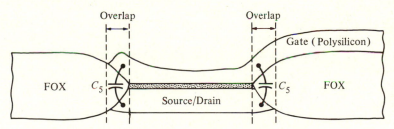

Figure 3.2-6 Gate-bulk overlap capacitances.

Next, it will be in the saturated region until v_{GS} becomes equal to $v_{DS}(\text{sat.})$ + V_T. Finally, the MOS device will be in the nonsaturated region. The approximate variation of C_{GB}, C_{GS}, and C_{GD} under these conditions is shown in Fig. 3.2-7. In cutoff, there is no channel and C_{GB} is approximately equal to $C_2 + 2C_5$. As v_{GS} approaches V_T from the off region, a thin depletion layer is formed, creating a large value of C_4. Since C_4 is in series with $C_2 + 2C_5$, little effect is observed. As v_{GS} increases, this depletion region widens, causing C_4 to decrease and reducing C_{GB}. When $v_{GS} = V_T$, an inversion layer is formed which prevents further decreases of C_4 (and thus C_{GB}).

C_1, C_2, C_3, and C_5 constitute C_{GS} and C_{GD}. The problem is how to allocate C_2 to C_{GS} and C_{GD}. The approach used is to assume in saturation that approximately ⅔ of C_2 belongs to C_{GS} and none to C_{GD}. This is, of course, an approximation. However, it has been found to give reasonably good results. Fig. 3.2-7 shows how C_{GS} and C_{GD} change values in going from the off to the saturation region. Finally, when v_{GS} is greater than $v_{DS} + V_T$, the MOS device enters the nonsaturated region. In this case, the channel extends from the drain to the source and C_2 is simply divided evenly between C_{GD} and C_{GS} as shown in Fig. 3.2-7.

As a consequence of the above considerations, we shall use the following formulas for the charge-storage capacitances of the MOS device in the indicated regions.

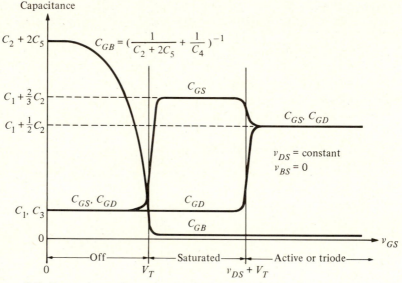

Figure 3.2-7 Voltage dependence of C_{GS}, C_{GD}, and C_{GB} as a function of v_{GS} with v_{DS} constant and $v_{BS} = 0$.

Off

$$C_{GB} = C_2 + 2C_5 = C_{ox}(W_{eff})(L_{eff}) + 2CGBO(L_{eff}) \tag{9a}$$

$$C_{GS} = C_1 \simeq C_{ox}(LD)(W_{eff}) = CGSO(W_{eff}) \tag{9b}$$

$$C_{GD} = C_3 \simeq C_{ox}(LD)(W_{eff}) = CGDO(W_{eff}) \tag{9c}$$

Saturation

$$C_{GB} = \frac{(C_2 + 2C_5)C_4}{C_2 + 2C_5 + C_4} \simeq C_4 \simeq 0 \tag{10a}$$

$$\begin{aligned} C_{GS} = C_1 + (\tfrac{2}{3})C_2 &= C_{ox}(LD + 0.67L_{eff})(W_{eff}) \\ &= CGSO(W_{eff}) + 0.67C_{ox}(W_{eff})(L_{eff}) \end{aligned} \tag{10b}$$

$$C_{GD} = C_3 \simeq C_{ox}(LD)(W_{eff}) = CGDO(W_{eff}) \tag{10c}$$

Nonsaturated

$$C_{GB} = \frac{(C_2 + 2C_5)C_4}{C_2 + 2C_5 + C_4} \simeq C_4 \simeq 0 \tag{11a}$$

$$\begin{aligned} C_{GS} = C_1 + 0.5C_2 &= C_{ox}(LD + 0.5L_{eff})(W_{eff}) \\ &= (CGSO + 0.5C_{ox}L_{eff})W_{eff} \end{aligned} \tag{11b}$$

$$\begin{aligned} C_{GD} = C_3 + 0.5C_2 &= C_{ox}(LD + 0.5L_{eff})(W_{eff}) \\ &= (CGDO + 0.5C_{ox}L_{eff})W_{eff} \end{aligned} \tag{11c}$$

Equations which provide a smooth transition between the three regions can be found in the literature [5].

Other capacitor parasitics include the thick oxide sandwiched between polysilicon (or metal) and the field (substrate) and the thick oxide between metal and the first polysilicon or the second polysilicon layers (if the process has two layers of polysilicon). This type of capacitance typically constitutes the major portion of C_{GB} in the nonsaturated and saturated regions. Gate connections to other places in the integrated circuit will cause significant parasitic capacitance to exist between the gate and the substrate. All of these parasitics are very important and should be considered in the design of CMOS circuits.

Another important aspect of modeling the CMOS device is noise. The existence of noise is due to the fact that electrical charge is not continuous but is carried in discrete amounts equal to the charge of an electron. In electronic circuits, noise manifests itself by representing a lower limit below which electrical signals cannot be amplified without significant deterioration in the quality of the signal. Noise can be modeled by a current source connected in parallel with i_D of Fig. 3.2-1. This current source represents two sources of noise, called thermal noise and flicker noise [6,7]. These sources of noise were discussed in Sec. 2.5. The mean-square current-noise source is defined as

$$\overline{i_N^2} = \left[\frac{8kTg_m(1 + \eta)}{3} + \frac{(KF)I_D}{fC_{ox}L^2} \right] \Delta f \; (\text{amperes}^2) \tag{12}$$

where

Δf = bandwidth at a frequency f

$\eta = g_{mbs}/g_m$ (see Eq. 8 of Section 3.3)

k = Boltzmann's constant

T = temperature (°K)

g_m = small-signal transconductance from gate to channel (see Eq. 6 of Section 3.3)

KF = flicker-noise coefficient (Farad·Amperes)

f = frequency (Hertz)

KF has a typical value of 10^{-28} (Farad·Amperes). Both sources of noise are process dependent and the values are usually different for enhancement and depletion mode FETs.

The mean-square current noise can be reflected to the gate of the MOS device by dividing Eq. (12) by g_m^2, giving

$$\overline{V_{eq}^2} = \frac{\overline{i_N^2}}{g_m^2} = \left[\frac{8kT(1 + \eta)}{3g_m} + \frac{KF}{2fC_{ox}WLK'} \right] \Delta f \tag{13}$$

The equivalent input-mean-square voltage-noise form of Eq. (13) will be useful for analyzing the noise performance of CMOS circuits in later chapters.

The experimental noise characteristics of NMOS and PMOS devices are shown in Figures 3.2-8(a) and 3.2-8(b) [8]. These devices were fabricated using a 6 μm, silicon-gate, p-well, CMOS process. The oxide thickness was 750 Angstroms and the drain current for all measurements was 45 μA. The data in Figs. 3.2-8(a) and 3.2-8(b) are typical for MOS devices and show that the $1/f$ noise is the dominant source of noise for frequencies below 1000 Hz. Consequently, in many practical cases, the equivalent input-mean-square voltage noise of Eq. (13) is simplified to

$$\overline{V_{eq}^2} \cong \left[\frac{KF}{2fC_{ox}WLK'} \right] \Delta f \tag{14}$$

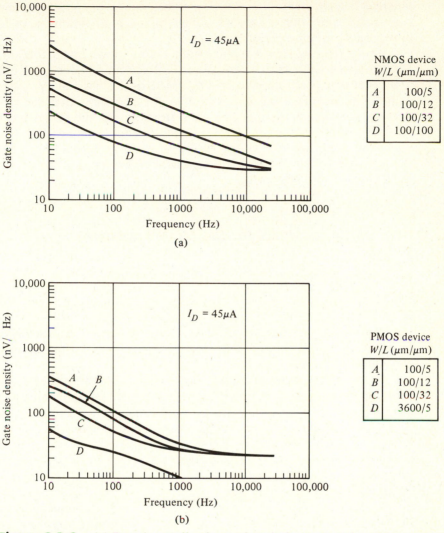

Figure 3.2-8 (a) Experimentally observed equivalent gate noise of four NMOS transistors. (b) Experimentally observed equivalent gate noise of four PMOS transistors. (© 1982 IEEE)

or in terms of the input-voltage-noise spectral density we can rewrite Eq. (14) as

$$\overline{e}^2_{eq} = \frac{\overline{V}^2_{eq}}{\Delta f} = \frac{KF}{2fC_{ox}WLK'} = \frac{B}{fWL} \tag{15}$$

where B is a constant for a NMOS or PMOS device of a given process. The right-hand expression of Eq. (15) will be important in optimizing the design with respect to noise performance.

Comparison of Figs. 3.2-8(a) and 3.2-8(b) shows that a 100 μm/5 μm NMOS device exhibits a mean-square voltage-noise density of 680 nV/$\sqrt{\text{Hz}}$ at 100 Hz while a corresponding PMOS device has 120 nV/$\sqrt{\text{Hz}}$ at the same frequency. Figs. 3.2-8(a) and 3.2-8(b) also show that the $1/f$ noise decreases with increasing device area. In summary, the equivalent input-referred voltage noise of a PMOS device is about 5 times less than the equivalent input-referred voltage noise of a NMOS device ($B_n \cong 5B_p$) and the equivalent input-referred voltage noise of a given device will decrease as the device area increases.

3.3 *Small-Signal Model for the MOS Transistor*

Up to this point, we have been considering the large-signal model of the MOS transistor shown in Fig. 3.2-1. However, after the large-signal model has been used to find the dc conditions, the small-signal model becomes important. The small-signal model is a linear model which helps to simplify calculations. It is only valid over voltage or current regions where the large-signal voltage and currents can be adequately represented by a straight line.

Fig. 3.3-1 shows a linearized small-signal model for the MOS transistor. The parameters of the small-signal model will be designated by lower case subscripts. The various parameters of this small-signal model are all related to the large-signal model parameters and dc variables. The normal relationship between these two models assumes that the small-signal parameters are defined in terms of the ratio of small perturbations of the large-signal variables or as the partial differentiation of one large-signal variable with respect to another.

The conductances g_{bd} and g_{bs} are the equivalent conductances of the bulk-to-drain and bulk-to-source junctions. Since these junctions are normally reverse biased, the conductances are very small. They are defined as

$$g_{bd} = \frac{\partial i_{BD}}{\partial v_{BD}} \text{ (at the quiescent point)} \simeq 0 \tag{1}$$

and

$$g_{bs} = \frac{\partial i_{SB}}{\partial v_{SB}} \text{ (at the quiescent point)} \simeq 0 \tag{2}$$

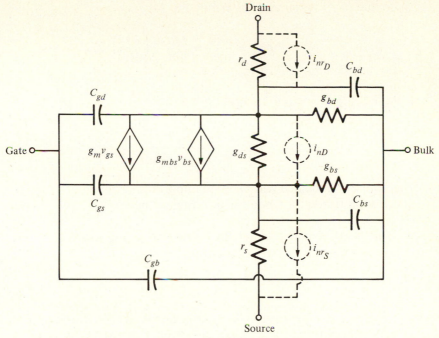

Figure 3.3-1   Small-signal model of the MOS transistor.

The channel conductances, g_m, g_{mbs}, and g_{ds} are defined as

$$g_m = \frac{\partial i_D}{\partial v_{GS}} \text{ (at the quiescent point)} \qquad (3)$$

$$g_{mbs} = \frac{\partial i_D}{\partial v_{BS}} \text{ (at the quiescent point)} \qquad (4)$$

and

$$g_{ds} = \frac{\partial i_D}{\partial v_{DS}} \text{ (at the quiescent point)} \qquad (5)$$

The values of these small signal parameters depend on which region the quiescent point occurs in. For example, in the saturated region g_m can be found from Eq. (12) of Section 3.1 as

$$g_m = \sqrt{(2K'W/L)|I_D|}(1 + \lambda V_{DS}) \cong \sqrt{(2K'W/L)|I_D|} \qquad (6)$$

which emphasizes the dependence of the small-signal parameters upon the large-signal operating conditions. The small-signal channel transconductance due to v_{SB} is found by rewriting Eq. (4) as

$$g_{mbs} = \frac{\partial i_D}{\partial v_{SB}} = \left(\frac{\partial i_D}{\partial V_T}\right)\left(\frac{\partial V_T}{\partial v_{SB}}\right) \tag{7}$$

Using Eq. (2) of Section 3.1 and noting that $\dfrac{\partial i_D}{\partial V_T} = \dfrac{-\partial i_D}{\partial v_{GS}}$, we get

$$g_{mbs} = g_m \frac{\gamma}{2(2|\phi_F| + V_{SB})^{1/2}} = \eta g_m \tag{8}$$

This transconductance will become important in our small-signal analysis of the MOS transistor when the ac value of the source-bulk potential v_{sb} is not zero.

The small-signal channel conductance, g_{ds} (g_o), is given as

$$g_{ds} = g_o = \frac{I_D \lambda}{1 + \lambda V_{DS}} \simeq I_D \lambda \tag{9}$$

The channel conductance will be dependent upon L through λ which is inversely proportional to L. We have assumed the MOS transistor is in saturation for the results given by Eqs. (6), (8), and (9).

The important dependence of the small-signal parameters upon the large-signal model parameters and dc voltages and currents is illustrated in Table 3.3-1. In this Table we see that the three small-signal model parameters of g_m, g_{mbs}, and g_{ds} have several alternate forms. An example of the typical values of the small-signal model parameters follows.

Example 3.3-1

Typical Values of Small Signal Model Parameters

Find the values of g_m, g_{mbs}, and g_{ds} using the large signal model parameters in Table 3.1-2 for both an NMOS and PMOS device if the dc value of the magnitude of the drain current is 50 μA and the magnitude of the dc value of the source-bulk voltage is 5 V. Assume that the W/L ratio is 10 μm/10 μm.

Using the values of Table 3.1-2 and Eqs. (6), (8), and (9) gives g_m = 41.2 μA/V, g_{mbs} = 11.2 μA/V, and $r_{ds} \cong$ 2 MΩ for the NMOS device and g_m = 28.3 μA/V, g_{mbs} = 3.59 μA/V, and $r_{ds} \cong$ 1 MΩ for the PMOS device.

Table 3.3-1

Relationships of the Small Signal Model Parameters upon the DC Values of Voltage and Current in the Saturation Region.

Small Signal Model Parameters	DC Current	DC Current and Voltage	DC Voltage
g_m	$\approx (2K'I_DW/L)^{1/2}$	—	$\approx \dfrac{K'W}{L}(V_{GS} - V_T)$
g_{mbs}	—	$\dfrac{\gamma(2I_D\beta)^{1/2}}{2(2|\phi_F| + V_{SB})^{1/2}}$	$\dfrac{\gamma[\beta(V_{GS} - V_T)]^{1/2}}{2(2|\phi_F| + V_{SB})^{1/2}}$
g_{ds}	$\approx \lambda I_D$	—	—

Although the MOS devices are not often used in the nonsaturation region in analog circuit design, the relationships of the small-signal model parameters in the nonsaturation region are given as

$$g_m = \frac{\partial i_d}{\partial v_{GS}} = \beta V_{DS}(1 + \lambda V_{DS}) \cong \beta V_{DS} \qquad (10)$$

$$g_{mbs} = -\frac{\partial i_D}{\partial v_{BS}} = \frac{\beta\gamma V_{DS}}{2(2|\phi_F| + V_{SB})^{1/2}} \qquad (11)$$

and

$$g_{ds} = \beta(V_{GS} - V_T - V_{DS})(1 + \lambda V_{DS}) + \frac{I_D\lambda}{1 + \lambda V_{DS}} \qquad (12)$$

$$\cong \beta(v_{GS} - V_T - V_{DS})$$

Table 3.3-2 summarizes the dependence of the small-signal model parameters on the large-signal model parameters and dc voltages and currents for the nonsaturated region. The typical values of the small-signal model parameters for the nonsaturated region are illustrated in the following example.

Table 3.3-2

Relationships of the Small-Signal Model Parameters upon the DC Values of Voltage and Current in the Nonsaturation Region.

Small Signal Model Parameters	DC Voltage and/or Current Dependence
g_m	$\approx \beta V_{DS}$
g_{mbs}	$\dfrac{\beta\gamma V_{DS}}{2(2|\phi_F| + V_{SB})^{1/2}}$
g_{ds}	$\approx \beta(V_{GS} - V_T - V_{DS})$

Example 3.3-2

Typical Values of the Small-Signal Model Parameters in the Nonsaturated Region

Find the values of the small-signal model parameters in the non-saturation region for an NMOS and PMOS transistor if $V_{GS} = 5$ V, $V_{DS} = 1$ V, and $V_{BS} = -5$ V. Assume that the W/L ratios of both transistors is 10 μm/10 μm.

First it is necessary to calculate the threshold voltage of each transistor using Eq. (2) of Sec. 3.1. The results are a V_T of 3.016 V for the NMOS and -1.955 V for the PMOS. This gives a dc current of 37.5 μA and 25.96 μA, respectively. Using Eqs. (10), (11), and (12), we get $g_m = 25.3$ μA/V, $g_{mbs} = 6.81$ μA/V, and $r_{ds} = 39.7$ KΩ for the NMOS transistor and $g_m = 10.2$ μA/V, $g_{mbs} = 1.27$ μA/V, and $r_{ds} = 46.3$ KΩ for the PMOS transistor.

The values of r_d and r_s are assumed to be the same as r_D and r_S of Fig. 3.2-1. Likewise, for small-signal conditions C_{gs}, C_{gd}, C_{gb}, C_{bd}, and C_{bs} are assumed to be the same as C_{GS}, C_{GD}, C_{GB}, C_{BD}, and C_{BS}, respectively. A very useful approximation for finding C_{bs} or C_{bd} knowing C_{gs} or C_{gd} is given as [9]

$$C_{bs} \simeq (g_{mbs}/g_m)C_{gs} = \eta C_{gs} \tag{13}$$

and

$$C_{bd} \simeq (g_{mbs}/g_m)C_{gd} = \eta C_{gd} \tag{14}$$

If the noise of the MOS transistor is to be modeled, then three additional current sources are added to Fig. 3.3-1 as indicated by the dashed lines. The values of the mean-square noise-current sources are given as

$$\bar{i}_{nrD}^2 = \left(\frac{4kT}{r_D} \right) \Delta f \text{ (amperes}^2) \tag{15}$$

$$\bar{i}_{nrS}^2 = \left(\frac{4kT}{r_S} \right) \Delta f \text{ (amperes}^2) \tag{16}$$

and

$$\bar{i}_{nD}^2 = \left[\frac{8kTg_m(1 + \eta)}{3} + \frac{(KF)I_D}{fC_{ox}L^2} \right] \Delta f \text{ (amperes}^2) \tag{17}$$

The various parameters for these equations have been previously defined. With the noise modeling capability, the small-signal model of Fig. 3.3-1 is a very general model.

It will be important to be familiar with the small-signal model for the saturation region developed in this section. This model, along with the circuit simplification techniques given in Appendix A, will be the key element in analyzing the circuits in the following chapters.

3.4 Second-Order Model Effects

The large-signal model of the MOS device previously discussed neglects many important second-order effects. Most of these second-order effects are due to narrow or short channel dimensions (less than 20 micrometers). We shall also consider the effects of temperature upon the parameters of the MOS large signal model.

We first consider second-order effects due to small geometries. When v_{GS} is greater than V_T, the drain current for a small device can be given as [4,5]

$$i_D = \frac{\mu_s C_{ox} W}{L_{mod}} \left\{ \left[v_{GS} - V_{BIN} - \frac{\theta v_{DS}}{2} \right] v_{DS} - \frac{2}{3} \gamma_s \left[(2|\phi_F| \right. \right.$$

$$\left. \left. + v_{DS} + v_{SB})^{1.5} - (2|\phi_F| + v_{SB})^{1.5}] \right\} \tag{1}$$

where μ_s accounts for degradation of the surface mobility μ_o due to increasing electric fields and is expressed empirically as

$$\mu_s = \mu_o \left[\frac{(UCRIT)\epsilon_{si}}{C_{ox}[v_{GS} - V_T - (UTRA)v_{DS}]} \right]^{UEXP} \tag{2}$$

when

$$\frac{(UCRIT)\epsilon_{si}}{C_{ox}} < v_{GS} - V_T - (UTRA)v_{DS} \tag{3}$$

otherwise $\mu_s = \mu_o$. The parameters of Eq. (3) are defined as

UCRIT = critical field for mobility degradation and is the limit at which μ_s starts decreasing according to Eq. (2).

UTRA = transverse field coefficient effecting mobility.

UEXP = critical field exponent for mobility degradation.

V_{BIN} of Eq. (1) is defined as

$$V_{BIN} = V_{FB} + 2|\phi_F| + \frac{\pi\epsilon_{si}}{4C_{ox}W}(2|\phi_F| + v_{SB}) \qquad (4)$$

The parameter θ of Eq. (1) is defined as

$$\theta = 1 + \frac{\pi\epsilon_{si}}{4C_{ox}W} \qquad (5)$$

Finally, the parameter γ_s of Eq. (1) is the bulk threshold parameter corrected for small geometries and is given as

$$\gamma_S = \gamma(1 - \alpha_S - \alpha_D) \qquad (6)$$

where

$$\alpha_S = \frac{XJ}{2L}[\sqrt{1 + (2W_S/XJ)} - 1] \qquad (7)$$

and

$$\alpha_D = \frac{XJ}{2L}[\sqrt{1 + (2W_D/XJ)} - 1] \qquad (8)$$

where XJ is the metallurgical junction depth and W_S and W_D are the depletion widths of the source and drain, respectively, defined as

$$W_S = \sqrt{(2\epsilon_{si}/qN_{SUB})}\sqrt{(2|\phi_F| + v_{SB})} \qquad (9)$$

and

$$W_D = \sqrt{(2\epsilon_{si}/qN_{SUB})}\sqrt{(2|\phi_F| + v_{SB} + v_{DS})} \qquad (10)$$

Fig. 3.4-1 shows the geometrical implications of the above relationships. Table 3.4-1 gives typical values of the above parameters for the second-order model.

The second-order model for i_D given in Eq. (1) is for the nonsaturated region. In this case L_{eff} is given as

$$L_{eff} = L - 2(LD) \qquad (11)$$

where LD is the lateral diffusion and L is the mask-defined length of the channel. The saturation voltage, v_{DS}(sat.), corrected for small size effects is

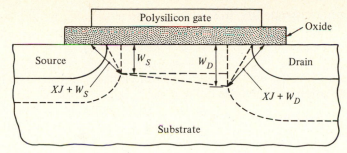

Figure 3.4-1 Illustration of the short-channel effects in the MOS transistor.

given as

$$v_{DS}(\text{sat.}) = \frac{V_{GS} - V_{BIN}}{\theta}$$

$$+ \frac{1}{2} \left(\frac{\gamma_S}{\theta} \right)^2 \left\{ 1 - \left[1 + \left(\frac{2\theta}{\gamma_S} \right)^2 \left(\frac{V_{GS} - V_{BIN}}{\theta} + 2|\phi_F| + v_{SB} \right) \right]^{1/2} \right\} \qquad \textbf{(12)}$$

When v_{DS} exceeds $v_{DS}(\text{sat.})$, the model for i_D is obtained by replacing v_{DS} in Eq. (1) by $v_{DS}(\text{sat.})$. In addition, there is an increase in the output conductance due to the spread of the pinch-off region into the channel (called channel-length modulation) and therefore reducing the effective channel

Table 3.4-1

Typical Model Parameters Suitable for SPICE2 Simulations Using Level-2 Model (Extended Model). These Values Are Based upon a 5μm Si-Gate Bulk CMOS p-Well Process.

Parameter Symbol	Parameter Description	Typical Parameter Value NMOS	PMOS	Units
V_{TO}	Threshold voltage	1 ± 0.2	1 ± 0.2	volts/cm
μ_o	mobility	580	230	cm²/V-s
UCRIT	Critical electric field for mobility degradation	80k	80k	volts/cm
UTRA	Transverse field coefficient	0.1	0.1	—
UEXP	Critical field exponent	0.1	0.1	—
NSUB	Substrate doping	10×10^{15}	2×10^{15}	cm⁻³
T_{ox}	Oxide thickness	800	800	angstroms
XJ	Metallurgical junction depth	1.0	1.0	μm
LD	Lateral diffusion	0.8	0.8	μm
NFS	Parameter for weak inversion modeling	2×10^{11}	1.5×10^{11}	cm⁻²

length. Consequently, in saturation, the modulated length, L_{mod}, is given as

$$L_{mod} = L_{eff}(1 - \lambda v_{DS}) \tag{13}$$

where v_{DS} is the actual value of drain-source voltage. The channel-length modulation parameter is defined as

$$\lambda = \frac{1}{L_{eff}V_{DS}}\left[\frac{2\epsilon_{si}}{qN_{SUB}}\right]^{1/2}\left\{\frac{V_{DS} - V_{DS}(\text{sat.})}{4} + \left[1 + \left(\frac{V_{DS} - V_{DS}(\text{sat.})}{4}\right)^2\right]^{1/2}\right\} \tag{14}$$

Numerical problems can occur with Eq. (13) when λv_{DS} approaches unity. In most model simulations, L_{mod} is redefined so it will always be positive [5].

The channel-length modulation parameter can also be influenced because of current saturation occurring because the charge carriers reach their maximum scattering velocity before pinch off. In this case, a lower current results from a longer channel device of identical geometric ratios, processing, and biasing conditions. This effect has been modeled and can be found in more detail elsewhere [5]. Table 3.4-2 shows a summary of the simple and extended large-signal strong-inversion model equations.

Temperature represents another important second-order effect. The temperature-dependent variables in the models developed so far include the: Fermi potential, ϕ_F (Eq. (5) of Section 3.1), energy gap E_g, bulk junction potential of the source-bulk and drain-bulk junctions (PB of Eqs. (3) and (4) of Section 3.2 or ϕ_o of Eq. (6) of Section 2.2), the reverse currents of the pn junctions (I_s of Eqs. (1) and (2) of Section 3.2), and the dependence of mobility upon temperature. The temperature dependence of most of these variables is found in the equations given previously or from well-known expressions. The dependence of mobility upon temperature is given as

$$\mu(T) \propto T^{-1.5} \tag{15}$$

One must be careful in extreme temperature ranges to incorporate temperature effects that are not modeled in the above considerations [10]. The temperature dependence of some of the dc model parameters for the MOS device can be illustrated by the following relationships [11]

$$K'(T) = K'(T_0)\left(\frac{T}{T_0}\right) T_{K'0} \tag{16}$$

$$V_T(T) = V_T(T_0) + [\phi(T) - \phi(T_0)] + \gamma[\sqrt{\phi(T)} - \sqrt{\phi(T_0)}] \tag{17}$$

Table 3.4-2
Summary of the Large Signal Model Expressions for the Simple and Second-Order MOS Models.

Simple Model (Section 3.1)	*Eq. #*

$$i_D = K'\left(\frac{W}{L}\right)[(v_{GS} - V_T) - (v_{DS}/2)]v_{DS}(1 + \lambda v_{DS}) \tag{1}$$

$$V_T = V_{TO} + \gamma[\sqrt{2|\phi_F| + v_{SB}} - \sqrt{2|\phi_F|} \tag{2}$$

Extended Model (Section 3.4)	*Eq. #*

$$i_D = \frac{\mu_s C_{ox} W}{L_{mod}}\left\{\left[v_{GS} - V_{BIN} - \frac{\theta v_{DS}}{2}\right]v_{DS}\right. \tag{1}$$

$$\left. -\frac{2}{3}\gamma_s[(2|\phi_F| + v_{DS} + v_{SB})^{1.5} - (2|\phi_F| + v_{SB})^{1.5}]\right\}$$

$$\mu_s = \mu_o\left[\frac{(UCRIT)\epsilon_{si}}{C_{ox}[v_{GS} - V_T - (UTRA)v_{DS}]}\right]^{UEXP} \tag{2}$$

$$V_{BIN} = V_{FB} + 2|\phi_F| + \frac{\pi\epsilon_{si}}{4C_{ox}W}(2|\phi_F| + v_{SB}) \tag{4}$$

$$\theta = 1 + \frac{\pi\epsilon_{si}}{4C_{ox}W} \tag{5}$$

$$\gamma_s = \gamma(1 - \alpha_s - \alpha_D) \tag{6}$$

$$\alpha_s = \frac{XJ}{2L}[\sqrt{1 + (2W_s/XJ)} - 1] \tag{7}$$

$$\alpha_D = \frac{XJ}{2L}[\sqrt{1 + (2W_D/XJ)} - 1] \tag{8}$$

$$W_s = \sqrt{(2\epsilon_{si}/qN_{SUB})}\sqrt{(2|\phi_F| + v_{SB})} \tag{9}$$

$$W_D = \sqrt{(2\epsilon_{si}/qN_{SUB})}\sqrt{2|\phi_F| + v_{SB} + v_{DS})} \tag{10}$$

$$v_{DS}(sat.) = \frac{V_{GS} - V_{BIN}}{\theta} \tag{12}$$

$$+\frac{1}{2}\left(\frac{\gamma_s}{\theta}\right)^2\left\{1 - \left[1 + \left(\frac{2\theta}{\gamma_s}\right)^2\left(\frac{V_{GS} - V_{BIN}}{\theta} + 2|\phi_F| + v_{SB}\right)\right]^{1/2}\right\}$$

$$L_{mod} = L_{eff}(1 - \lambda v_{DS}) \tag{13}$$

$$L_{eff} = L - 2(LD) \tag{11}$$

$$\lambda = \frac{1}{L_{eff}V_{DS}}\left[\frac{2\epsilon_{si}}{qN_{SUB}}\right]^{1/2}\left\{\frac{V_{DS} - v_{DS}(sat.)}{4} + \left[1 + \left(\frac{V_{DS} - v_{DS}(sat.)}{4}\right)^2\right]^{1/2}\right\} \tag{14}$$

$$\phi(T) = \phi(T_0)\frac{T}{T_0} - \frac{2kT}{q}\log\left[\frac{n_i(T)}{n_i(T_0)}\right] \tag{18}$$

$$PB(T) = PB(T_0)\frac{T}{T_0} - \frac{2kT}{q}\log\left[\frac{n_i(T)}{n_i(T_0)}\right] \tag{19}$$

and

$$I_S(T) = I_S(T_0)\left[\frac{n_i(T)}{n_i(T_0)}\right]^2 \tag{20}$$

where

$$\frac{n_i(T)}{n_i(T_0)} = \left[\frac{T}{T_0}\right]^{1.5}\exp\left\{-\left(\frac{qE_g}{2k}\right)\left[\frac{1}{T} - \frac{1}{T_0}\right]\right\} \tag{21}$$

$T_{K'0}$ is the temperature coefficient of surface mobility. An alternate form of the temperature dependence of the MOS model can be found elsewhere [12].

Other improvements to the MOS device model include conservation of the charge in the channel and the charge in the source and drain [13]. This becomes important as the devices and capacitors become small.

3.5 *Subthreshold MOS Model*

The models discussed in previous sections predict that no current will flow in a device when the gate-source voltage is at or below the threshold voltage. In reality, this is not the case. As v_{GS} approaches V_T, the $i_D - v_{GS}$ characteristics change from square-law to exponential. Whereas the region where v_{GS} is above the threshold is called the *strong inversion* region, the region below (actually, the transition between the two regions is not well defined as will be explained later) is called the subthreshold, or *weak inversion* region. This is illustrated in Fig. 3.5-1 where a MOSFET transconductance characteristic is shown with the square root of current plotted as a function of the gate-source voltage. When the gate-source voltage reaches the value designated as V_{ON} (this relates to the SPICE2 model formulation), the current changes from square-law to an exponential-law behavior. It is the objective of this section to present two models suitable for the subthreshold region. The first is the SPICE2 model derived from the model given in reference [14], while the second is a formulation given in reference [17].

In the SPICE2 model, the transition point from the region of strong inversion to the weak inversion characteristic of the MOS device is designated as V_{ON} and is greater than V_T. V_{ON} is given by [4]

$$V_{ON} = V_T + \frac{nkT}{q} \tag{1}$$

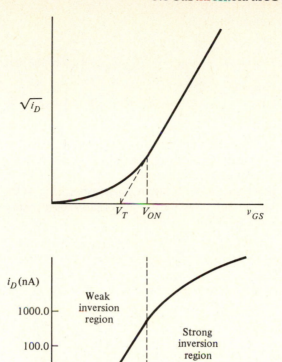

Figure 3.5-1   Weak-inversion characteristics of the MOS transistor as modeled by Eq. (4).

where

$$n = 1 + \frac{q(NFS)}{C_{ox}} + \frac{C_D}{C_{ox}} \qquad (2)$$

and

$$C_D = \frac{\partial Q_B}{\partial v_{SB}} = \left[-\gamma_s \frac{\partial (2|\phi_F| + v_{SB})^{1/2}}{\partial v_{SB}} \right.$$

$$\left. -\frac{\partial \gamma_s}{\partial v_{SB}} (2|\phi_F| + v_{SB})^{1/2} + \frac{\pi \epsilon_{si}}{4 C_{ox} W} \right] C_{ox} \qquad (3)$$

NFS is a parameter used in the evaluation of V_{ON} and can be extracted from measurements. The drain current in the weak inversion region, v_{GS} less than

V_{ON}, is given as

$$i_D = \frac{\mu_s C_{ox} W}{L_{mod}} \left\{ \left(V_{ON} - V_{BIN} - \frac{\theta v_{DS}}{2} \right) v_{DS} - \left(\frac{2}{3} \right) \gamma_s [(2|\phi_F| + v_{SB} + v_{DS})^{1.5} \right.$$

$$\left. - (2|\phi_F| + v_{SB})^{1.5}] \right\} \exp \left[\frac{q}{nkT} (v_{GS} - V_{ON}) \right] \quad (4)$$

A saturation voltage is determined by substituting $v_{GS} = V_{ON}$ into the $v_{DS}(\text{sat})$ equation given in Eq. (12) of Sec. 3.4. The transistor is in saturation when v_{DS} is greater than $v_{DS}(\text{sat})$.

An alternate model for transistors operating in the subthreshold region is given below [17]. Each voltage is taken with respect to the substrate.

$$i_D = \frac{W}{L} I_{DO} \exp \left(\frac{q v_G}{nkT} \right) \left[\exp \left(\frac{-q v_S}{kT} \right) - \exp \left(\frac{-q v_D}{kT} \right) \right] \quad (5)$$

The transition point where Eq. (5) is valid occurs at

$$i_D \lesssim \frac{n-1}{e^2} \left(\frac{W}{L} \right) \mu C_{ox} \left(\frac{kT}{q} \right)^2 \quad (6)$$

when $v_{DS} > 3kT/q$ (saturation region). This model is more appropriate for hand calculations but it does not accommodate a smooth transition into the strong-inversion region. Eq. (5) can be simplified if we assume that qv_D/kT is much greater than unity and $v_S = 0$ (with respect to the substrate), yielding

$$i_D \cong \frac{W}{L} I_{DO} \exp \left(\frac{q v_{GS}}{nKT} \right) \quad (7)$$

The term n is the subthreshold slope factor, and I_{DO} is a process-dependent parameter which is dependent also on v_{SB} and V_T. These two terms are best extracted from experimental data.

Unfortunately, neither of the two model equations given here properly model the transistor as it makes the transition from strong to weak inversion. In reality, there is a transition region of operation between strong and weak inversion called the "moderate inversion" region [15]. This is illustrated in Fig. 3.5-2. A complete treatment of the operation of the transistor through this region is given in the literature [15,16].

It is important to consider the temperature behavior of the MOS device operating in the subthreshold region. As is the case for strong inversion, the temperature coefficient of the threshold voltage is negative in the subthreshold region. The variation of current due to temperature of a device

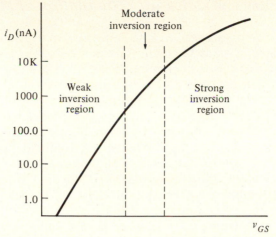

Figure 3.5-2   The three regions of operation of an MOS transistor.

operating in weak inversion is dominated by the negative temperature coefficient of the threshold voltage. Therefore, for a given gate-source voltage, subthreshold current increases as the temperature increases. This is illustrated in Fig. 3.5-3 [21].

Operation of the MOS device in the subthreshold region is very important when low-power circuits are desired. A whole class of CMOS circuits have been developed based on the weak-inversion operation characterized by the above model [17,18,19,20]. We will consider some of these circuits in later chapters.

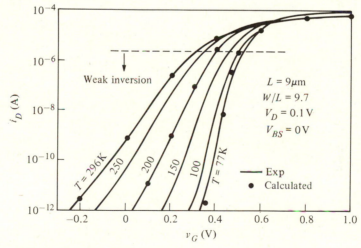

Figure 3.5-3   Transfer characteristics of a long-channel device as a function of temperature. (© 1977 IEEE)

3.6 *Computer and Calculator Simulation of MOS Circuits*

The objective of this section is to show how to use simulators to verify the performance of an MOS circuit. The models of the previous section were developed to be used in simulation. Because of the complexity of analog integrated circuits, it is not always possible to physically breadboard the circuit in order to verify or predict its performance. This breadboarding capability is very important because the designer has no way other than fabricating to know whether or not his design is correct. Fabrication without simulation can be an expensive proposition if the circuit does not work because of a mistake that could have been avoided by simulation, so the economics of this situation have strongly favored the development of simulation capabilities rather than trial-and-error fabrication cycles.

Simulation is basically the application of the models of previous sections together with computer analysis techniques to simulate the performance of a nonlinear electronic circuit. We have seen that there are two categories of models: large signal and small signal. Simulation techniques use the large-signal model to find the dc variables. The dc variables are then used to find the small signal model parameters, which in turn are used to analyze the circuit. Types of analyses that can be done by simulation include dc operating point, nonlinear time or frequency domain response, and linear time or frequency domain response.

Fig. 3.6-1 shows how a simulator of analog circuits might be organized. The typical input to the simulator would be a topological description of the circuit on a branch-by-branch basis. Each branch would be identified as to the type of component, the nodes between which it is connected, and other pertinent information. This is the point at which the parameters of the models for nonlinear devices are described to the simulator. In addition, the designer must identify all dependent sources applied to the circuit. The simulator will use this information to solve for the dc values of voltages and currents. Most simulators today use algorithms that guess the node voltages and use the models to calculate all branch currents. These branch currents are summed at every node to see if they satisfy Kirchhoff's current law. If they do not sum to zero within a specified limit, the simulator re-guesses the voltage using various algorithms until the nodal voltages converge on the values that cause all branch currents to sum to zero at each node within the specified tolerance. The perceptive reader can appreciate that how the simulator is used may influence the ease with which it converges to a solution. One of the objectives of this section is to suggest ways that the designer can enhance the probability of convergence in circuit simulation.

It should be noted that all the models developed in this chapter assume that voltages are the independent variables and currents the dependent variables. Once the circuit converges to a solution of the nodal voltages

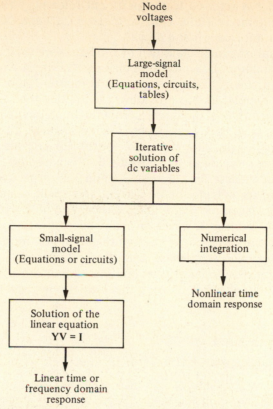

Figure 3.6-1   General organization of a simulator for analog circuits.

and currents, the simulation can take two paths. If the desired analysis is small signal, then the small-signal models are evaluated using the relationships of Secs. 3.2 and 3.3. These models are linear and thus linear-matrix solution methods are used to evaluate the time or frequency domain response. Methods for solving these linear equations are well known [24–27]. The second path that can be taken is the use of numerical integration to find the large-signal time or frequency domain response. The designer must then describe the time characteristics of dependent sources applied to the circuit. This method uses the dc values as the initial starting point. A time step is defined in which the simulator uses various numerical integration algorithms to find or predict the value of the nodal voltages at the end of the time step. These algorithms are iterative and may be subject to convergence problems. Most simulators can vary the time step in order to try to enhance the tradeoff between convergence and computational efficiency. The details of the algorithms for solving the nonlinear equations and for numerical integration are beyond the scope of this text. The interested reader is referred to the pertinent references [5,22–27] and to the technical literature *(IEEE Transactions on CAD and Circuits and Systems)* for further information and details.

Simulators can be classified by the hardware used to implement them. Today, simulators fall into three classes. The most widely used are those based on a small- to moderate-sized computer. More recently, personal-computer simulators are providing capabilities that approach and sometimes exceed that of the larger computer-implemented simulators. Lastly, simulators have been implemented using some of the more sophisticated calculators. A calculator simulator for the HP 41C has been written and is described in more detail in Appendix B. It is based on the level 1 and level 2 MOS models described in Secs. 3.1 and 3.3, respectively. This program is only useful for solving the dc current and small signal parameters of g_m, g_{mbs}, and g_{ds} given the drain, gate, source, and bulk terminal voltages for a single MOS device. However, in many cases the designer can use such a program more efficiently for design than a larger, more complex simulator simply because it is quick to use. This calculator also has provision for model libraries. The library can contain 3 models for level 1 and 2 models for level 2. An example of using this program is given below.

Example 3.6-1

Use of a Calculator Simulator to Analyze a MOSFET

Use the calculator program in Appendix B to solve for the dc value of currents in M1, M2, and M3 of Fig. 3.6-2 and the dc value of voltage which will cause V_{OUT} to be zero. Use the level 1 model and the parameters described in Table 3.1-2.

Because the calculator program will only handle one transistor at a time, the designer must use a little cleverness. First solve for the current in M3 by iteratively solving the equation, $10 = V_{GS3} + 10K(I_{D3})$. Assume a value for V_{GS3} and solve for I_{D3} and evaluate the above equation. After several iterations using the calculator program in Appendix B, the value of V_{GS3} was found as -0.5927 V. Applying this voltage to the gate of M2 with the drain assumed to be at 0 V gave a current of 43.87 μA. Finally, the value of v_{IN} is iterated with the drain of M1 at 0 V until the current in M1 equals 43.87 μA. After several iterations, the value of $V_{IN} = -2.432$ V gave a current of 43.87 μA. Better modeling could be obtained by repeating this approach with the level 2 model.

The SPICE2 simulator [22,23] has become the dominant simulator implemented on personal and large computers. This simulator is the result

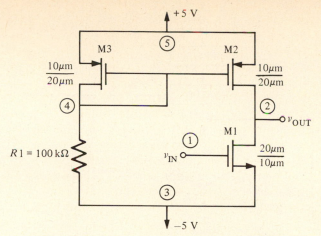

Figure 3.6-2   An MOS inverting amplifier.

of many years of effort by many researchers. An excellent account of the development of circuit simulators can be found in an historical review of circuit simulation written by Prof. D.O. Pederson [28]. This review shows the extensive amount of effort that has resulted in the simulator now known as SPICE2. The modeling topics discussed in this chapter have been oriented to those models used in the SPICE2 program. Consequently, it should be easy for the user who is unfamiliar with SPICE2 to be able to use it as a simulation tool for MOS circuits.

The SPICE2 program uses an iterative approach of guessing voltages and solving for the nodal currents to solve the network nodal voltages. This approach has been embodied in several sophisticated algorithms which try to avoid convergence problems. Once SPICE2 has converged to the nodal voltages that will permit the nodal currents to sum to less than some specified small value, then the small-signal parameters are calculated and a linear analysis can be performed. A simplified functional flowchart of SPICE2 is shown in Fig. 3.6-3. The flow of simulation shown in Fig. 3.6-1 can be roughly identified with the program flow shown in Fig. 3.6-3. Over many years of development, SPICE2 has become a sophisticated program using techniques such as dynamic memory allocation and sparse-matrix techniques. Programming details of SPICE2 can be found in the literature [22].

One of the problems the designer will face is that of failure to converge. Based upon the cursory review above of how the simulator works, the designer should be able to use the simulator in a way that enhances the chances of success. For example, if a circuit is in its high-impedance state, then the currents will be small. Because these currents are summed at each node to determine the validity of the guesses for the nodal voltages, small currents may lead to convergence problems. The designer would be better

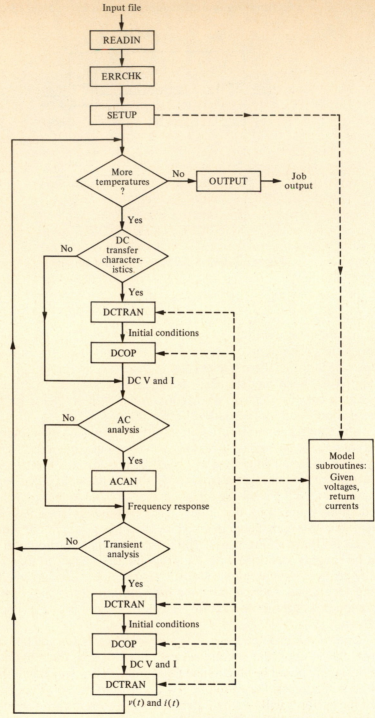

Input file

READIN

ERRCHK

SETUP

More temperatures? — No → OUTPUT → Job output

Yes

DC transfer characteristics — No

Yes

DCTRAN

Initial conditions

DCOP

DC V and I

AC analysis — No

Yes

ACAN

Frequency response

Transient analysis — No

Yes

DCTRAN

Initial conditions

DCOP

DC V and I

DCTRAN

$v(t)$ and $i(t)$

Model subroutines: Given voltages, return currents

Figure 3.6-3   SPICE2 simplified functional flowchart.

advised to start the simulation at a point where most transistors are on or conducting. This allows the simulator to solve quickly for the dc operating point. If the analysis proceeds to the region where the devices become off, the simulator should still converge since it is starting from previous solutions. A good example is the inverting amplifier of Fig. 3.6-2. If one desires to plot v_{OUT} as a function of v_{IN}, it is better to start at v_{IN} at $+5$ V and go to -5 V because at 5 V all transistors are on. In addition to its use as a simulator, SPICE2 has many user-activated options—such as defining node voltages, changing the convergence tolerance, changing the algorithm for iteration, and so on—that can help to overcome problems.

SPICE2 is capable of many different forms of analyses including dc, transient, frequency response, distortion, noise, and so forth. The circuits that can be analyzed by SPICE2 include resistors, capacitors, inductors, mutual inductors, independent voltage and current sources, four types of dependent sources, transmission lines, and the four most common semiconductor devices: diodes, BJTs, JFETs, and MOSFETs. SPICE2 can provide output information such as device currents and voltages, model parameters calculated by the simulator, and tables and plots of the various analyses. Further information is available from the references [5,22,23].

The successful use of SPICE2 requires that the model information given in this chapter be correctly applied. From this viewpoint, we shall focus on the MOS model and show how and when the parameters are entered. The general hierarchy is shown in Fig. 3.6-4. There are device cards (lines) and model cards (lines). All information on all cards is delim-

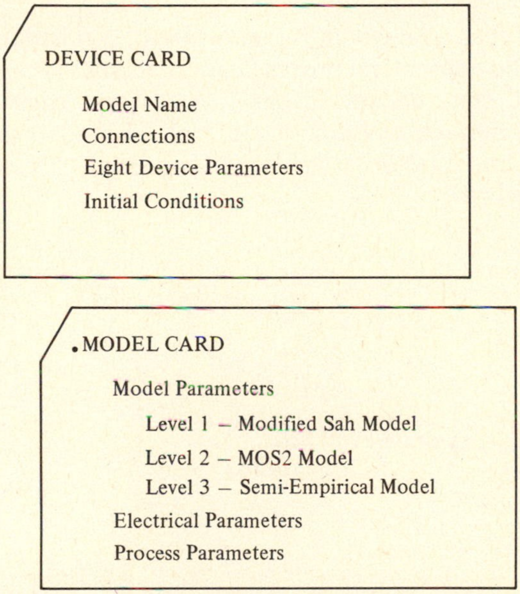

DEVICE CARD

    Model Name

    Connections

    Eight Device Parameters

    Initial Conditions

.MODEL CARD

    Model Parameters

        Level 1 — Modified Sah Model

        Level 2 — MOS2 Model

        Level 3 — Semi-Empirical Model

    Electrical Parameters

    Process Parameters

Figure 3.6-4   General hierarchy for device model information.

ited by spaces. The purpose of the device card is to identify the model name, describe the topological connection of the device, provide 8 geometric parameters, and give the initial conditions. The information on the device card is specific to a given device. The 8 geometric parameters include the width of the channel in meters (W), the length of the channel in meters (L), the area of the drain in square meters (AD), the area of the source in square meters (AS), the perimeter of the drain in meters (PD), the perimeter of the source in meters (PS), the equivalent number of squares for the drain (NRD), and the equivalent number of squares for the source (NRS). Obviously, this information cannot be entered until the device is geometrically defined. In the early phases of simulation, only W and L are entered.

The information on the model card is much more extensive and will be covered in this and the following paragraphs. The model card (line) is preceded by a period to flag the program that this card (line) is not a component. The model card (line) identifies the model level (1, 2, or 3) and provides the electrical and process parameters. If the user does not input the various parameters, default values are used. These default values are indicated in the SPICE2 Users Guide that can be printed from any SPICE2 program. The level 1 model parameters were covered in Sec. 3.1 and are: the zero-bias threshold voltage in volts extrapolated to $i_D = 0$ for large devices (VTO), the intrinsic transconductance parameter in amperes/volt^2 (KP), the bulk threshold parameter in volts$^{1/2}$ (GAMMA), the surface potential at strong inversion in volts (PHI), and the channel-length modulation parameter in volts^{-1} (LAMBDA). Values for these parameters can be found in Table 3.1-2.

Sometimes, one would rather let SPICE2 calculate the above parameters from the appropriate process parameters. This can be done by entering the surface state density in cm^{-2} (NSS), the oxide thickness in meters (TOX), the surface mobility in cm^2/volt·second (UO), and the substrate doping in cm^{-3} ($NSUB$). The equations used to calculate the electrical parameters are

$$VTO = \phi_{GB} - \frac{q(NSS)}{(\epsilon_{ox}/TOX)} + \frac{2(q\epsilon_{si}(NSUB|\phi_F|)^{1/2}}{(\epsilon_{ox}/TOX)} \tag{1}$$

$$KP = UO\frac{\epsilon_{ox}}{TOX} \tag{2}$$

$$GAMMA = \frac{(2q\epsilon_{si}NSUB)^{1/2}}{(\epsilon_{ox}/TOX)} \tag{3}$$

and

$$PHI = 2\phi_F = \frac{2kT}{q}\ln\left[\frac{NSUB}{n_i}\right] \tag{4}$$

LAMBDA is not calculated from the process parameters for the level 1 model. The constants for silicon, given in Table 3.1-1, are contained within the SPICE2 program and do not have to be entered.

The next model parameters considered are those that were considered in Sec. 3.2. The first parameters considered were associated with the bulk-drain and bulk-source pn junctions. These parameters include the reverse current of the drain-bulk or source-bulk junctions in amperes (*IS*) or the reverse-current density of the drain-bulk or source-bulk junctions in amperes/meter^2 (*JS*). *JS* requires the specification of *AS* and *AD* on the model card. If *IS* is specified, it overrides *JS*. The default value of *IS* is usually 10^{-14} amperes. The next parameters considered in Sec. 3.2 were the drain ohmic resistance in ohms (*RD*), the source ohmic resistance in ohms (*RS*), and the sheet resistance of the source and drain in ohms/square (*RSH*). *RSH* is overridden if *RD* or *RS* are entered. To use *RSH*, the values of *NRD* and *NRS* must be entered on the model card.

The drain-bulk and source-bulk depletion capacitors can be specified by the zero-bias bulk junction bottom capacitance in farads per meter^2 of junction area (*CJ*). *CJ* requires *NSUB* and assumes a step junction using a formula similar to Eq. (12) of Sec. 2.2. Alternately, the drain-bulk and source-bulk depletion capacitances can be specified using Eqs. (5) and (6) of Sec. 3.2. The necessary parameters include the zero-bias bulk-drain junction capacitance in farads (*CBD*), the zero-bias bulk-source junction capacitance in farads (*CBS*), the bulk junction potential in volts (*PB*), the coefficient for forward-bias depletion capacitance (*FC*), the zero-bias bulk junction sidewall capacitance in farads per meter of junction perimeter (*CJSW*), and the bulk junction sidewall capacitance grading coefficient (*MJSW*). If *CBD* or *CBS* is specified, then *CJ* is overridden. The values of *AS*, *AD*, *PS*, and *PD* must be given on the device card to use the above parameters. Typical values of these parameters are given in Table 3.2-1.

The next parameters discussed in Sec. 3.2 were the gate overlap capacitances. These capacitors are specified by the gate-source overlap capacitance in farads/meter (*CGSO*), the gate-drain overlap capacitance in farads/meter (*CGDO*), and the gate-bulk overlap capacitance in farads/meter (*CGBO*). Typical values of these overlap capacitances can be found in Table 3.2-1. Finally, the noise parameters include the flicker noise coefficient (*KF*) and the flicker noise exponent (*AF*). Typical values of these parameters are 10^{-28} and 1, respectively.

The above parameters constitute the complete level 1 model as described in Sec. 3.1 and 3.2. In Sec. 3.4 a second-order model was introduced. This model is used if LEVEL = 2 on the model card (if LEVEL is not specified then LEVEL = 1 is assumed). In Sec. 3.4 it was shown that this model required the critical field for mobility degradation in volts/cm (UCRIT), the transverse field coefficient for mobility degradation (UTRA), the critical field exponent for mobility degradation (UEXP), and the lateral diffusion in the direction of the channel length in meters (LD). Additional parameters not discussed in Sec. 3.4 include the type of gate material (*TPG*), the thin oxide capacitance model flag, coefficient of channel charge allocated to the drain (*XQC*), the metallurgical junction depth in meters (*XJ*), and the oxide encroachment in the direction of the channel width in meters (*WD*). The choices for *TPG* are $+1$ if the gate material is opposite to the substrate, -1 if the gate material is the same as the substrate, and 0 if the gate material is aluminum. A charge controlled model is used in the SPICE2 simulator if the value of the parameter *XQC* has a value smaller than or equal to 0.5. This model attempts to keep the sum of charge associated with each node equal to zero. If *XQC* is larger than 0.5, charge conservation is not guaranteed.

If LEVEL = 3, an empirical model described in the references [4], is used. The additional parameters include the static feedback effect parameter (ETA), the empirical mobility modulation parameter in volts^{-1} (THETA), and the field correlation factor used to adjust the electric field across the surface depletion region of a device in saturation (KAPPA).

Although much more could be said about the use of the SPICE2 simulation program, the above information along with the User's Guide [23] covers the necessary background. In order to illustrate its use and to provide examples for the novice user to follow, several examples will be given showing how to use SPICE2 to perform various simulations. The SPICE2 simulator used in these and other examples of the book is called PSPICE[1] and is implemented on an IBM personal computer with 512K of memory and an 8087 math coprocessor. The plots were obtained using an HP 7470A plotter driven by the graphics output (PROBE) of PSPICE.

Example 3.6-2
Use of SPICE2 to Simulate MOS Output Characteristics

Use SPICE2 to obtain the output characteristics of the NMOS transistor shown in Fig. 3.6-5 using the level 1 model and the parameter values of Table 3.1-2. The output curves are to be plotted for drain-source voltages from 0 to 10 V and for gate-source voltages of 2, 3, 4, 5, 6, 7, and 8 volts. Assume that the bulk voltage is zero. Table 3.6-1 shows the input file for SPICE2 to solve this problem. The first line is a title for the simulation file and must be present. The second line defines the width of the input and output. The lines not preceded by "." define the topology of the circuit. The third line describes how the transistor is connected, defines the model to be used, and gives the *W*

[1]PSPICE is a product of MicroSim Corporation, 23175 La Cadena Drive, Laguna Hills, CA 92653, USA.

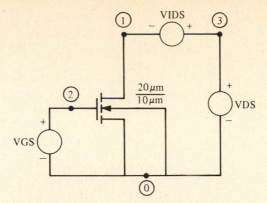

Figure 3.6-5   Circuit for Example 3.6-2.

Table 3.6-1

SPICE2 Input File for Example 3.6-2.

Ex. 3.6-2 Use of SPICE2 to Simulate MOS Output
.WIDTH IN = 72 OUT = 80
M1 2 2 0 0 MOS1 W = 20U L = 10U
VDS 3 0
VGS 2 0
VIDS 3 1 DC 0.0
.MODEL MOS1 NMOS VTO = 1.0 KP = 17U GAMMA = 1.3 LAMBDA = 0.01 PHI = 0.7
.DC VDS 0 10 0.5 VGS 2 8 1
.PRINT DC I(VIDS)
.PROBE
.END

and L values. Note that because the units are meters, the suffix "U" is used to convert to micrometers. The fourth through sixth lines describe the independent voltages. VGS and VDS are used to bias the MOSFET while VIDS is a zero-value voltage source used to measure the current through the drain. Some versions of SPICE2 have the ability to measure the drain current directly and thus VIDS would not be needed.

The seventh line is the model description for $M1$. It must have the name of the model specified in the second line. In addition it must indicate whether the transistor is NMOS or PMOS. Next, the pertinent parameters are entered. If the information is too long for a single line, a continuation is indicated by a "+" at the beginning of the continued line. If the beginning of a line has an "*", SPICE2 ignores the line. This is a good way to insert comments or to remove an element.

What the simulator is to do is defined by the ninth and tenth lines. ".DC" asks for a dc sweep. In this particular case, a nested dc sweep is specified in order to avoid seven consecutive analyses. The ".DC. . . ." line will set VGS = 2 V and then sweep VDS from 0 to 10 V in 0.5 increments. Next, it will increment VGS to 3 V and repeat the VDS sweep. This is continued until seven VDS sweeps have been made with the desired values of VGS. The ".PRINT. . . ." line directs the program to print the values of the dc sweeps. This is necessary to use the plotting program called by .PROBE which is unique to PSPICE. Normally, SPICE2 would plot the output on the line printer with the statement ".PLOT DC I(VIDS)". The quality of line printer plots leaves a little to be desired but gives the designer a gross feeling for the simulation. The printed values can be used to examine the results more closely. Many facilities which support SPICE2 also have alternate plotting capabilities which can provide better output plots. The last line of every SPICE input file must be .END which is line eleven. Fig. 3.6-6 shows the output plot of this analysis using the plotting features of the PSPICE program.

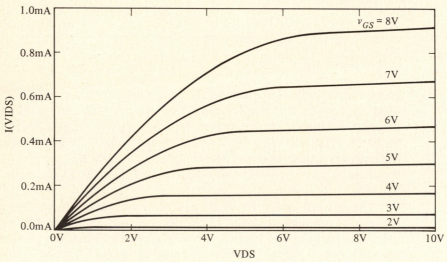

Figure 3.6-6   Output of Example 3.6-2.

One must be careful in the use of SPICE2 not to exceed the internal defaults of the program. For example, the default for the number of points that SPICE2 will calculate is 201. In the above example we see that each sweep takes 20 points and is repeated 7 times to give a total calculated points of 140. If the designer had asked for the sweep to have 0.1 volt steps, the total points calculated would be 700 which would create an error. In this case, the designer can increase the number of calculations through the command ".OPTIONS LIMPTS = 750". This example points out that the user of SPICE2 should be familiar with its limits, limits that are discussed in more detail in reference [23].

Example 3.6-3
dc Analysis of Fig. 3.6-2.

Use the SPICE2 simulator to obtain a plot of the value of v_{OUT} as a function of v_{IN} of Fig. 3.6-2. Identify the dc value of v_{IN} which gives $v_{OUT} = 0$ V.

The input file for SPICE2 is shown in Table 3.6-2. It follows the same format as the previous example except that two types of transistors are used. These models are designated by MOSN and MOSP. A dc sweep is requested starting from $v_{IN} = -5$ V and going to $+5$ V. Although this is contradictory to our previous considerations, we will be able to obtain convergence because the circuit is simple. Figure 3.6-7 shows the resulting output of the dc sweep. At $v_{IN} = -5$ V, the output is at 5 V. As v_{IN} increases to -4 V, the output starts to decrease. At v_{IN} approximately -2.5 V, the output makes a rapid transition from 1.5 V to -4 V. This is the region in which the circuit should be biased to obtain a high-gain small-signal inverting amplifier. For values of v_{IN} above -2.5 V, the output slowly approaches -5 V. This curve gives a great deal of information about the capabilities of Fig. 3.6-2 which will be developed in more detail in Sec. 6.1.

Figure 3.6-7 allows the designer to get a coarse picture of the dc transfer function of Fig. 3.6-2. To examine the transition region in more detail, the .DC VIN -5 5 0.1 line was replaced by .DC VIN -3 -2 0.01. The result is shown in Fig. 3.6-8. In order to operate in the linear region it is necessary to make $v_{OUT} \approx 0$. Although the printed output gives more detail, the designer can see that a value of v_{IN} of a little less than -2.4 V gives a value of $v_{OUT} = 0$ V. The actual value selected was $v_{IN} = -2.42$ V which gives $v_{OUT} = 0.9$ V and is approximately in the middle of the transition region.

Table 3.6-2

SPICE2 Input File for Example 3.6-3.

Ex. 3.6-3 DC Analysis of Fig. 3.6-2.
.WIDTH IN = 72 OUT = 80
M1 2 1 3 3 MOSN W = 20U L = 10U
M2 2 4 5 5 MOSP W = 10U L = 20U
M3 4 4 5 5 MOSP W = 10U L = 20U
R1 4 3 100K
VDD 5 0 DC 5.0
VSS 0 3 DC 5.0
VIN 1 0
.MODEL MOSN NMOS VTO = 1.0 KP = 17U GAMMA = 1.3 LAMBDA = 0.01 PHI = 0.7
.MODEL MOSP PMOS VTO = −1.0 KP = 8U GAMMA = 0.6 LAMBDA = 0.008 PHI = 0.6
.DC VIN −5 5 0.1
.PRINT DC V(2)
.PROBE
.END

Example 3.6-4

ac Analysis of Fig. 3.6-2

Use SPICE2 to obtain a small signal frequency response of $V_{OUT}(\omega)/V_{IN}(\omega)$ when the amplifier is biased in the transition region. Assume that a 5 pF capacitor is attached to the output of Fig. 3.6-2 and find the magnitude and phase response over the frequency range of 100 Hz to 100 MHz.

The SPICE2 input file for this example is shown in Table 3.6-3. It is important to note that v_{IN} has been defined as both an ac and dc voltage source with a dc value of −2.42 V. If the dc voltage were not included, SPICE2 would find the dc solution for $v_{IN} = 0$ V which is not in the transition region. Therefore, the small signal solution would not be evaluated in the transition region. Once the dc solution has been evaluated, the amplitude of the signal applied as the ac input has no influence on the simulation. Thus, it is convenient to use ac inputs of unity in order to treat the output as a gain quantity. Here, we have assumed an ac input of 1.0 volt peak. In the actual circuit, such an input centered at −2.42 V would saturate the output of the amplifier.

The simulation desired is defined by the .AC DEC 20 100 100MEG line. This line directs SPICE2 to make an ac analysis over a log frequency with 20 points per decade from 100 Hz to 100 MHz. The .OP option has been added to print out the dc voltages of all circuit nodes in order to verify that the ac solution is in the desired region. This

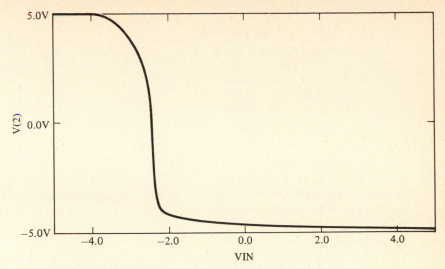

Figure 3.6-7   Output of Example 3.6-3 for -5 V $< v_{IN} < 5$ V.

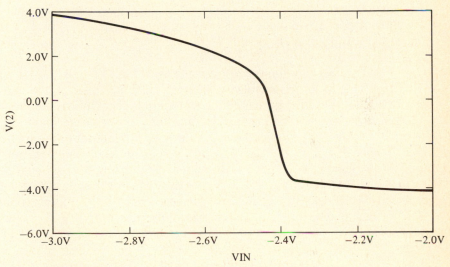

Figure 3.6-8   Output of Example 3.6-3 for -3 V $< v_{IN} < -2$.

option also gives a total circuit power of 0.88 milliwatts. The program will calculate the linear magnitude, dB magnitude, and phase of the output voltage. Figures 3.6-9(a) and (b) show the magnitude (dB) and the phase of this simulation.

Table 3.6-3

SPICE2 Input File for Example 3.6-4.

Ex. 3.6-4 AC Analysis of Fig. 3.6-2.
.WIDTH IN = 72 OUT = 80
M1 2 1 3 3 MOSN W = 20U L = 10U
M2 2 4 5 5 MOSP W = 10U L = 20U
M3 4 4 5 5 MOSP W = 10U L = 20U
CL 2 0 5P
R1 4 3 100K
VDD 5 0 DC 5.0
VSS 0 3 DC 5.0
VIN 1 0 DC −2.42 AC 1.0
.MODEL MOSN NMOS VTO = 1.0 KP = 17U GAMMA = 1.3 LAMBDA = 0.01 PHI = 0.7
.MODEL MOSP PMOS VTO = − 1.0 KP = 8U GAMMA = 0.6 LAMBDA = 0.008 PHI = 0.6
.AC DEC 20 100 100MEG
.OP
.PRINT AC VM(2) VDB(2) VP(2)
.PROBE
.END

Example 3.6-5

Transient Analysis of Fig. 3.6-2

The last simulation to be made with Fig. 3.6-2 is the transient response to an input pulse. The input pulse is shown in Fig. 3.6-10. This simulation will include the 5 pF output capacitor of the previous example and will be made from time zero to 6 microseconds.

Table 3.6-4 shows the SPICE2 input file. Note that the number of points has been increased to 301 using the .OPTIONS feature. The input pulse of Fig. 3.6-10 is described using the piecewise linear capability (PWL) of SPICE2. The output desired is defined by .TRAN 0.03U 6U which asks for a transient analysis from 0 to 6 microseconds at points spaced every 0.03 microseconds. The output will consist of both $v_{IN}(t)$ and $v_{OUT}(t)$ and is shown in Fig. 3.6-11.

The above examples will serve to introduce the reader to the basic ideas and concepts of using the SPICE2 program. In addition to what the reader has distilled from these examples, a useful set of guidelines is offered which has resulted from extensive experience in using SPICE2. These guidelines are listed as:

1. Never use a simulator unless you know the range of answers beforehand.

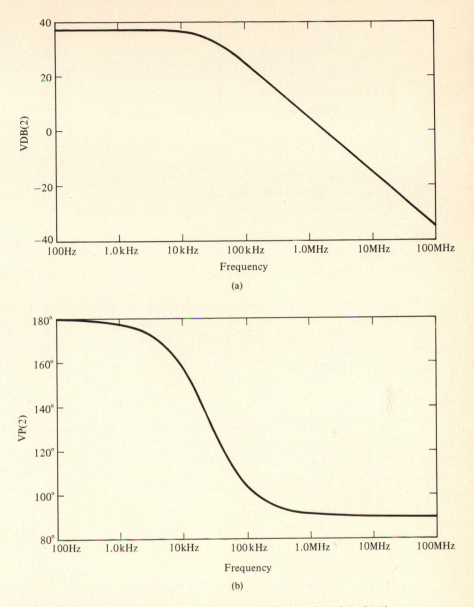

Figure 3.6-9 (a) Magnitude (dB) response of Example 3.6-4. (b) Phase response of Example 3.6-4.

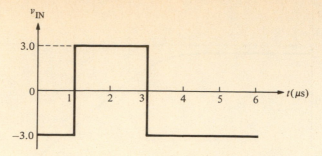

Figure 3.6-10   Input pulse for Example 3.6-5.

2.   Never simulate more of the circuit than is necessary.
3.   Always use the simplest model that will do the job.
4.   Always start a dc solution from the point at which the majority of the devices are on.
5.   Use a simulator in exactly the same manner as you would make the measurement on the bench.
6.   Never change more than one parameter at a time when using the simulator for design.
7.   Learn the basic operating principles of the simulator so that you can enhance its capability. Know how to use its options.
8.   Watch out for syntax problems like O and 0.
9.   Use the correct multipliers for quantities.
10.   Use common sense.

Table 3.6-4

SPICE2 Output for Example 3.6-5.

Ex. 3.6-5 Transient Analysis of Fig. 3.6-2.
.WIDTH IN = 72 OUT = 80
.OPTION LIMPTS = 301
M1 2 1 3 3 MOSN W = 20U L = 10U
M2 2 4 5 5 MOSP W = 10U L = 20U
M3 4 4 5 5 MOSP W = 10U L = 20U
CL 2 0 5P
R1 4 3 100K
VDD 5 0 DC 5.0
VSS 0 3 DC 5.0
VIN 1 0 PWL(0 −3V 1U −3V 1.05U 3V 3U 3V 3.05U −3V 6U −3V)
.MODEL MOSN NMOS VTO = 1.0 KP = 17U GAMMA = 1.3 LAMBDA = 0.01 PHI = 0.7
.MODEL MOSP PMOS VTO = −1.0 KP = 8U GAMMA = 0.6 LAMBDA = 0.008 PHI = 0.6
.TRAN 0.03U 6U
.PRINT TRAN V(2) V(1)
.PROBE
.END

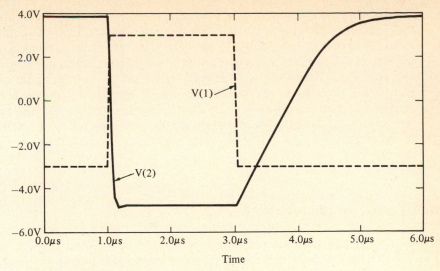

Figure 3.6-11   Transient response of Example 3.6-5.

Most problems with simulators can be traced back to a violation of one or more of these guidelines.

3.7   *Summary*

This chapter has tried to give the reader the background necessary to be able to simulate CMOS circuits. The approach used has been based on the SPICE2 simulation program. This program normally has three levels of MOS models which are available to the user. The function of models is to solve for the dc operating conditions and then use this information to develop a linear small-signal model. Sec. 3.1 described the level 1 model used by SPICE2 to solve for the dc operating point. This model also uses the additional model parameters presented in Sec. 3.2. These parameters include bulk resistance, capacitance, and noise. A small-signal model that was developed from the large-signal model was described in Sec. 3.3. These three sections represent the basic modeling concepts for MOS transistors.

A more complicated large-signal model was presented in order to model effects such as short channel, mobility degradation, temperature and subthreshold operation. The large-signal model presented was based on the level 2 model of the SPICE2 program. Large signal models suitable for weak inversion were also described. Further details of these models and other models are found in the references for this chapter. A brief background of simulation methods was presented in Sec. 3.6. Simulation of MOS circuits using the calculator, personal computer, and larger computer were discussed. After reading this chapter, the reader should be able to use the model information presented along with a SPICE2 simulator to analyze MOS circuits. This ability will be very important in the remainder of this text. It will be used to verify intuitive design approaches and to perform analyses beyond the scope of the techniques presented. One of the important aspects of modeling is to determine the model parameters which best fit the MOS process that is being used. The next chapter will be devoted to this subject.

PROBLEMS — *Chapter 3*

1. Sketch to scale the output characteristics of an enhancement NMOS device if $V_T = 1$ volt and $I_D = 500$ microamperes when $V_{GS} = 6$ volts in saturation. Choose values of $V_{GS} = 1, 2, 3, 4, 5,$ and 6 volts. Assume that the channel modulation parameter is zero.

2. Sketch to scale the output characteristics of an enhancement PMOS device if $V_T = 1$ volt and $I_D = -500$ microamperes when $V_{GS} = -1, -2, -3, -4, -5,$ and -6 volts. Assume that the channel modulation parameter is zero.

3. In Table 3.1-2, why is γ_N greater than γ_P for a p-well, CMOS technology? Why is K'_{nonsat} greater than K'_{sat}?

4. A large-signal model for the MOSFET which features symmetry for the drain and source is given as

$$i_D = K' \frac{W}{L} \{[(v_{GS} - V_{TS})^2 u(v_{GS} - V_{TS})] - [(v_{GD} - V_{TD})^2 u(v_{GD} - V_{TD})]\}$$

where $u(x)$ is 1 if x is greater than or equal to zero and 0 if x is less than zero (step function) and V_{TX} is the threshold voltage evaluated from the gate to X where X is either S (Source) or D (Drain). Sketch this model in the form of i_D versus v_{DS} for a constant value of v_{GS} ($v_{GS} > V_{TS}$) and identify the saturated and nonsaturated regions. Be sure to extend this sketch for both positive and negative values of v_{DS}. Repeat the sketch of i_D versus v_{DS} for a constant value of v_{GD} ($v_{GD} > V_{TD}$). Assume that both V_{TS} and V_{TD} are positive.

5. Let Eqs. (11) and (12) of Sec. 3.1 be given as

$$i_{D1} = \frac{\mu C_{ox} W}{L} [(v_{GS} - V_T)v_{DS} - 0.5v_{DS}^2], \qquad v_{GS} - V_T > v_{DS}$$

and

$$i_{D1} = \frac{\mu C_{ox} W}{2L} (V_{GS} - V_T)^2, \qquad V_{GS} - V_T < V_{DS}$$

An alternative single equation to replace the above is proposed as,

$$i_{D2} = \frac{\mu C_{ox} W}{2L} (V_{GS} - V_T)^2 \tanh(\alpha v_{DS}), \qquad V_{GS} - V_T > 0$$

Compare the two model approaches by equating $\partial i_D / \partial v_{DS}$ ($V_{DS} = 0$) to solve for α and plot i_D / I_{D0} as a function of $v_{DS} / (v_{GS} - V_T)$ for values from 0 to 2, where I_{D0} is the value of i_{D1} in the saturation region. What is the worst case error if the error is defined as $(i_{D1} - i_{D2}) / i_{D1}$?

6. Using the values of Tables 3.1-1 and 3.2-1, calculate the values of C_{GB}, C_{GS}, and C_{GD} for a MOS device which has a W of 20 micrometers and an L of 20 micrometers for all three regions of operation

7. Find C_{BX} at $V_{BX} = 0$ V and 0.75 V of Fig. P3.7 assuming the values of Table 3.2-1 apply to the MOS device where $FC = 0.5$ and $PB = 1$ V. Assume the device is NMOS and repeat for a PMOS device.

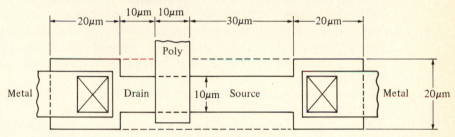

Figure P3.7

8. Calculate the value of C_{GB}, C_{GS}, and C_{GD} for an NMOS device with a length of 25 micrometers and a width of 25 micrometers. Assume $V_D = 1$ V, $V_G = 1.7$ V, and $V_S = 0$ V and let (a) $V_B = -5$ V and (b) $V_B = 0$ V. Assume $V_{T0} = 1$ V, $C_{ox} = 0.45 \times 10^{-15}$ F/μm², $\gamma = 0.5$ V$^{1/2}$, $\phi = 0.6$ V, and an overlap of 0.8 micrometers.

9. Find the bulk ohmic source and drain resistances of the MOS device shown in Fig. P3.9 assuming the diffusion has a sheet resistance of 100 ohms/square. Find C_{GS}, C_{GB}, C_{GD}, C_{BS}, and C_{BD} assuming that the device is in saturation and that $V_{BS} = 0$ V, $V_{BD} = -5$ V, $PB = 1$ V, and $FC = 0.5$. Use the data of Tables 3.1-2 where appropriate.

10. Repeat Examples 3.3-1 and 3.3-2 if the W/L ratio is 100 μm/10 μm.

11. Find the complete small-signal model for an NMOS transistor with the

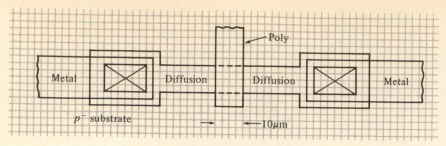

Figure P3.9

drain at 5 V, gate at 3 V, source at 0 V, and the bulk at -5 V. Assume the model parameters of $V_{T0} = 1$ V, $K' = 25$ $\mu A/V^2$, $\gamma = 1$ $V^{1/2}$, $PB = 0.8$ V, $MJ = 0.5$, $\phi = 0.6$ V, $LD = 1$ μm, $C_{ox} = 0.5$ $fF/\mu m^2$, $\lambda = 0.04$ V^{-1}, $C_{BS0} = C_{BD0} = 100$ fF, and $W/L = 100$ $\mu m/10$ μm.

12. Develop an expression for g_m and g_{ds} of a MOS transistor if this transistor is modeled by i_{D2} of Prob. 5.

13. In the devlopment of the simple MOS model of Sec. 3.1, the depletion region under the channel was assumed to be constant width (see Fig. 2.3-3). Rederive the expression for drain current of the MOS device assuming the width of the depletion region linearly causes the channel to have zero width at the drain end of the channel. Assume that the field in the direction perpendicular to the silicon surface is much greater than the field in the direction of current flow so that a one-dimensional MOS analysis is valid. Compare the results with Eq. (1) of Sec. 3.4.

14. Show how one could modify the model of Fig. 3.2–1 to include the effects of breakdown between the drain-source of a device.

15. If $n = 1$ and $V_T = 1$ V, find the value of V_{ON} for n MOS transistor in weak inversion.

16. Develop an expression for the small signal transconductance of a MOS device operating in weak inversion using the large signal expression of Eq. (7) of Sec. 3.5.

17. Develop expressions for g_m, g_{ds}, and g_{mbs} for the small-signal model of a MOS transistor operating in the weak inversion region using the large signal expression of Eq. (4) of Sec. 3.5.

18. If $V_{GS} = V_T$, find the current that flows in an NMOS transistor if $V_{DS} = 5$ V and $V_{BS} = 0$ V. Repeat if $V_{BS} = -5$ V. Assume the parameters of Table 3.1-2 where pertinent.

19. Use SPICE2 to perform the following analyses on the circuit shown in Fig. P3.19: (a) Plot v_{OUT} versus v_{IN} for the nominal parameter set shown. (b) Separately, vary each parameter $+10\%$ and repeat part (a).

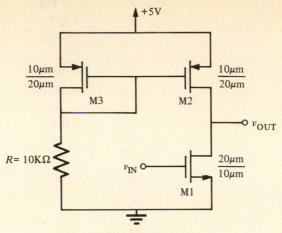

Figure P3.19

Parameter	NMOS	PMOS	Units
V_T	0.7	−0.7	V
K'	14	4	$\mu A/V^2$
λ	0.01	0.01	V^{-1}

20. Use SPICE2 to plot i_2 as a function of v_2 when i_1 has values of 10, 20, 30, 40, 50, 60, and 70 microamperes for Fig. P3.20. The maximum value of v_2 is 10 volts. Use the model parameters of $V_T = 0.7$ volts and $K' = 14$ microamperes/volt^2 and $\lambda = 0.01$ volts^{-1}. Repeat with $\lambda = 0.05$ volts^{-1}.

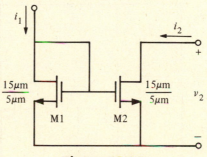

Figure P3.20

21. Use SPICE2 to plot i_D as a function of v_{DS} for values of $V_{GS} = 1, 2, 3, 4$ and 5 volts for an NMOS transistor with $V_T = 1$ volt, $K' = 25$ $\mu A/$Volt^2, and $\lambda = 0.02$ volts^{-1}. Show how SPICE2 can be used to generate and plot these curves simultaneously as illustrated by Fig. 3.1-3.

22. Repeat Example 3.6-1 if R1 = 200 KΩ.
23. Repeat Example 3.6-2 if the transistor of Fig. 3.6-5 is a PMOS having the model parameters given in Table 3.1-2.
24. Repeat Examples 3.6-3 through 3.6-5 for the circuit of Fig. 3.6-2 if R1 = 200 KΩ.
25. The circuit shown in Fig. P3.25 is to be analyzed by SPICE2 using the input file shown below. List the errors you can find which are syntax errors or SPICE errors.

    Problem 3.25 Debugging a SPICE input file.
    VSWP 6 1 DC
    VIC 2 6 DC 0.0
    VID 4 6 DC 0.0
    VBG 3 1 DC
    VT 3 4 DC 2.0
    Q1 2 3 1 NPM1
    M1 5 4 1 1 NMOS1 W = 100 L = 10
    .MODEL NPN (BJ = 245 BR = 0.96 IS = 0.736 VAF = 100)
    .MODEL NMOS VT0 = 1 KP = 25 LAMBDA = 0.02 GAMMA = 1
    .PRINT DC I(V9C) I(VID) V(6) V(9)
    .PLOT DC I(VIC) I(VID) V(6) V(4)
    .AS DEC 10 1 10M
    .PRINT AC VDB(2) VP(2) VDB(5) VP(5)
    .TRAN 0.05U 10U 0 10N
    .PRINT TRAN V(2) V(5)
    .END

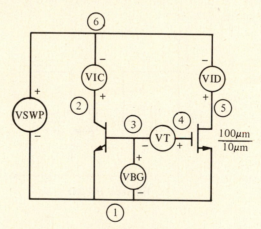

Figure P3.25

26. Problem 21 of Chapter 2 asked for an approximate current-voltage characteristic of a pinched resistor. The objective of this problem is to use SPICE to calculate this characteristic. Pinched resistors can be

thought of as JFETs with floating gates. The parameters for a PMOS device can be expressed as:

$$V_{T0} \cong \frac{-qN_Aa^2}{2\epsilon_{si}} \text{ Volts} \quad \text{BETA} \cong \frac{-2Wq\mu_pN_Aa}{V_{T0}} \text{ Amps/Volt}^2$$

where a = half of the channel depth. If $N_A = 10^{15}$ cm^{-3}, $\mu_p = 477$ cm^2/V·sec, $2a$ = p-well active junction depth = 0.7 μm, $\lambda = 0.01$, and $W/L = 50\mu$m/10μm, use the p-channel JFET model of SPICE2 with a floating gate and plot the drain current versus the drain-source voltage. The floating gate can be achieved by connecting it to a zero-current current source.

REFERENCES

1. Y. Tsividis, "Problems With Modeling of Analog MOS LSI," *IEDM*, (1982) pp. 274–277.
2. C.T. Sah, "Characteristics of the Metal-Oxide-Semiconductor Transistor," *IEEE Transactions on Electron Devices*, ED-11, No. 7(July 1964) pp. 324–345.
3. H. Shichman and D. Hodges, "Modelling and Simulation of Insulated-Gate Field-Effect Transistor Switching Circuits," *IEEE Journal Solid State Circuits*, Vol. SC-3, No. 3 (September 1968) pp. 285–289.
4. A. Vladimirescu and S. Liu, "The Simulation of MOS Integrated Circuits using SPICE2," Memorandum No. UCB/ERL M80/7, October 1980, (Electronics Research Laboratory, College of Engineering, University of California, Berkeley, CA 94720).
5. D.R. Alexander, R.J. Antinone, and G.W. Brown, "SPICE Modelling Handbook," Report BDM/A-77-071-TR, (BDM Corporation, 2600 Yale Blvd., Albuquerque, NM 87106).
6. P.R. Gray and R.G. Meyer, "Analysis and Design of Analog Integrated Circuits," Second Ed., (New York: John Wiley & Sons, 1984), p. 646.
7. P.E. Allen and E. Sanchez-Sinencio, "Switched Capacitor Circuits," (New York: Van Nostrand Reinhold, 1984), p. 589.
8. R.D. Jolly and R.H. McCharles, "A Low-Noise Amplifier for Switched Capacitor Filters," *IEEE Journal of Solid-State Circuits*, Vol. SC-17, No. 6, (December 1982), pp. 1192–1194.
9. Y.P. Tsividis, "Relation Between Incremental Intrinsic Capacitances and Transconductances in MOS Transistors," *IEEE Transactions on Electron Devices*, Vol. ED-27, No. 5 (May 1980) pp. 946–948.
10. S.K. Tewksbury, "N-Channel Enhancement-Mode MOSFET Characteristics from 10°K to 300°K," *IEEE Transactions on Electron Devices*, Vol. ED-28, No. 12 (December 1981) pp. 1519–1529.
11. S. Liu and L.W. Nagel, "Small-Signal MOSFET Models for Analog Circuit Design," *IEEE Journal of Solid-State Circuits*, Vol. SC-17, No. 6 (December 1982) pp. 983–998.
12. F.H. Gaensslen and R.C. Jaeger, "Temperature Dependent Threshold Behavior

of Depletion Mode MOSFET's," *Solid-State Electronics,* Vol. 22, No. 4 (1979) pp. 423–430.

13. D.E. Ward and R.W. Dutton, "A Charge-Oriented Model for MOS Transistor Capacitances," *IEEE Journal of Solid-State Circuits,* Vol. SC-13, No. 5 (October 1978).

14. R. Swanson and J.D. Meindl, "Ion-Implanted Complementary MOS Transistors in Low-Voltage Circuits," *IEEE Journal of Solid-State Circuits,* Vol. SC-7, No. 2 (April 1972) pp. 146–153.

15. Y. Tsividis, "Moderate Inversion In MOS Devices," *Solid State Electronics,* Vol. 25, No. 11 (1982) pp. 1099–1104.

16. P. Antognetti, D.D. Caviglia, and E. Profumo, "CAD Model for Threshold and Subthreshold Conduction in MOSFET's," *IEEE Journal of Solid-State Circuits,* Vol. SC-17, No. 2 (June 1982) pp. 454–458.

17. E. Vittoz and J. Fellrath, "CMOS Analog Integrated Circuits Based on Weak Inversion Operation," *IEEE Journal of Solid-State Circuits,* Vol. SC-12, No. 3 (June 1977) pp. 231–244.

18. M.G. DeGrauwe, J. Rigmenants, E. Vittoz, and H.J. DeMan, "Adaptive Biasing CMOS Amplifiers," *IEEE Journal of Solid-State Circuits,* Vol. SC-17, No. 3 (June 1982) pp. 522–528.

19. W. Steinhagen and W.L. Engl, "Design of Integrated Analog CMOS Circuits— A Multichannel Telemetry Transmitter," *IEEE Journal of Solid-State Circuits,* Vol. SC-13, No. 6 (December 1978) pp. 799–805.

20. Y. Tsividis and R. Ulmer, "A CMOS voltage Reference," *IEEE Journal of Solid-State Circuits,* Vol. SC-13, No. 6 (December 1978) pp. 774–778.

21. S.M. Sze, *Physics of Semiconductor Devices,* Second ed. (New York: John Wiley and Sons, 1981).

22. L.W. Nagel, *SPICE2: A Computer Program to Simulate Semiconductor Circuits,* ERL Memo No. ERL-M520, (Electronics Research Laboratory, University of California, Berkeley, CA, May 1975).

23. A Vladimerescu, A.R. Newton, and D.O. Pederson, *SPICE Version 2G.0 User's Guide,* September 1980 (University of California, Berkeley).

OTHER REFERENCES

24. J. Vlach and K. Singhal, "Computer Methods for Circuit Analysis and Design," (New York: Van Nostrand Reinhold, 1983).

25. P. Antognetti, D.O. Pederson, and H. DeMan, Eds., "Computer Design Aids for VLSI Circuits," (Groningen, the Netherlands: Sijthoff & Noordhoff, 1980).

26. W.J. McCalla and D.O. Pederson, "Elements of Computer-Aided Circuit Analysis," *IEEE Transactions on Circuit Theory,* Vol. CT-18, No. 1, (Jan. 1971), pp. 14–26.

27. G.S. Forsythe and C.B. Moler, "Computer Solution of Linear Algebraic Equations," (Englewood Cliffs, N. J.: Prentice-Hall, 1967).

28. D.O. Pederson, "A Historical Review of Circuit Simulation," *IEEE Transactions on Circuits and Systems,* Vol. CAS-31, No. 1, (Jan. 1984), pp. 103–111.

29. Paul Richman, *MOS Field-Effect Transistors and Integrated Circuits,* (New York: John Wiley and Sons, 1973).

chapter 4

CMOS Device Characterization

In the previous chapter, we presented two strong inversion MOS transistor models that describe the behavior of the transistor over a range of terminal conditions (*S*, *G*, *D*, *B*). These models can be useful for hand calculations and computer simulations using either the simple model or more complex models. However, before the models can be used, proper model parameters that describe the particular characteristics of a given device must be supplied. Generally, extensive device model parameters are not available from wafer manufacturers. Therefore, before a design can begin, devices must be characterized to obtain suitable model parameters. It is prudent practice for the designer to obtain a sampling of the components from the integrated-circuit vendor and characterize them extensively in order to obtain the desired model parameters. The characterization process is the subject of this chapter.

Using the simple model, graphical and numerical techniques will be developed to extract the model parameters. Attention will be given to the geometrical aspects of a good test structure. Some of the techniques and results will be extended to the complex model, where further work will be done to capture some of the second-order parameters associated with this model. Other areas that must be characterized are transistor noise and passive-component parameters.

4.1   *Geometrical Aspects of Characterization*

One uniquely important aspect of integrated-circuit design is the effect of the geometry of a device upon its electrical characteristics. One obvious example of this is the control of transistor current by its *W/L* ratio. Similarly, the channel length has an effect on the transistor's output conductance via the channel-length modulation effect. Unfortunately, the transistor models presented in this text are only approximate (as are all models) and achieve varying degrees of accuracy as a function of the particular device geometry used. The electrical performance of passive components, too, is governed by their geometry. For these reasons, one must consider various components on an integrated circuit from a geometrical as well as an electrical point of view in order to understand their expected electrical behavior.

It would be ideal if the analog designer could design a test chip that contained all of the devices he wanted to characterize. He could then tailor the geometries so that the parameters extracted would be those most applicable to analog circuits. A good example of the importance of this is found in the extraction of the channel-length modulation parameter λ, which is very important for analog circuits. Because λ varies inversely with the length of the device, best accuracy is achieved when devices are characterized at the length that will be most often used in the design. For example, one might decide beforehand to use 12 μm length devices wherever possible throughout an analog design (this might be done as a tradeoff between low output conductance and size or bandwidth). A few of these devices of various widths could be implemented on a test chip to achieve optimum characterization results.

A good example of what not do to in terms of transistor characterization is shown in Fig. 4.1-1. In this case it is difficult to predict the effect of the bends and the true electrical width and length of the device is not clear. Therefore, the results obtained characterizing this device would only be applicable to an identical (or very similar) device. They would not apply to the more common straight device geometry with no bends.

Geometry is also very important when specifying the value of a resistor's sheet resistance. Because out-diffusion can be a significant portion of

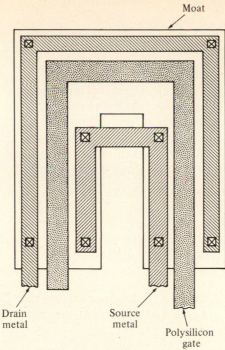

Moat

Drain
metal

Source
metal

Polysilicon
gate

Figure 4.1-1   Example of a structure difficult to characterize because of corner effects on the *W* and *L* values.

the width of a resistor (especially in a lightly doped resistor made using p-well diffusion) and the impurity concentration can vary across the width of the resistor, it is critical that its value be characterized by width. The result will be that a very wide resistor (e.g. 25 micrometers) will likely have a different value of sheet resistance than a narrow resistor (e.g. 5 micrometers). The corner and contact termination effects discussed in Section 2.6 will also be different for resistors with different widths.

4.2   *Characterization of the Simple Transistor Model*

The equations that model a MOS transistor in strong-inversion saturated and nonsaturated regions were given in Section 3.1 and are repeated here for convenience.

$$i_D = K'\left(\frac{W_{\text{eff}}}{2L_{\text{eff}}}\right)(v_{GS} - V_T)^2(1 + \lambda v_{DS}) \qquad (1)$$

$$i_D = K'\left(\frac{W_{\text{eff}}}{L_{\text{eff}}}\right)\left[(v_{GS} - V_T)v_{DS} - \frac{v_{DS}^2}{2}\right](1 + \lambda v_{DS}) \qquad (2)$$

where

$$V_T = V_{T0} + \gamma[\sqrt{2|\phi_F| + v_{SB}} - \sqrt{2|\phi_F|}] \tag{3}$$

The primary parameters of interest are $V_{T0}(V_{SB} = 0)$, K', γ, and λ. In this section, we will concentrate on techniques for determining these parameters. It was noted earlier that the parameter K' is different for the saturation and nonsaturation region and thus must be characterized differently for each case. In order to keep the terminology simple, we adopt (in this chapter) the terms K'_S for the saturation region and K'_L for the nonsaturation region. Using this terminology, Eqs. (1) and (2) can be rewritten

$$i_D = K'_S\left(\frac{W_{eff}}{2L_{eff}}\right)(v_{GS} - V_T)^2(1 + \lambda v_{DS}) \tag{4}$$

$$i_D = K'_L\left(\frac{W_{eff}}{L_{eff}}\right)\left[(v_{GS} - V_T)v_{DS} - \frac{v_{DS}^2}{2}\right](1 + \lambda v_{DS}) \tag{5}$$

First assume that v_{DS} is chosen such that the λv_{DS} term in Eq. (4) is much less than one and v_{SB} is zero, so that $V_T = V_{T0}$. Therefore, Eq. (4) simplifies to

$$i_D = K'_S\left(\frac{W_{eff}}{2L_{eff}}\right)(v_{GS} - V_{T0})^2 \tag{6}$$

This equation can be manipulated algebraically to obtain the following

$$i_D^{1/2} = \left(\frac{K'_S W_{eff}}{2L_{eff}}\right)^{1/2} v_{GS} - \left(\frac{K'_S W_{eff}}{2L_{eff}}\right)^{1/2} V_{T0} \tag{7}$$

which has the form

$$y = mx + b \tag{8}$$

This equation is easily recognized as the equation for a straight line with m as the slope and b as the y-intercept. Comparing Eq. (7) to Eq. (8) gives

$$y = i_D^{1/2} \tag{9}$$

$$x = v_{GS} \tag{10}$$

$$m = \left(\frac{K'_S W_{eff}}{2L_{eff}}\right)^{1/2} \tag{11}$$

and

$$b = -\left(\frac{K'_S W_{eff}}{2 L_{eff}}\right)^{1/2} V_{TO} \tag{12}$$

It can be seen from these equations that if $i_D^{1/2}$ versus v_{GS} is plotted and the slope of the straight line portion of the curve is measured, then K'_S can be easily extracted assuming that W_{eff} and L_{eff} are known. Fig. 4.2-1(a) demonstrates these techniques. It is very important in choosing a device to characterize, that the W and L are chosen large so that L_{eff} and W_{eff} are close as possible to the drawn values resulting in good accuracy when extracting K'_S.

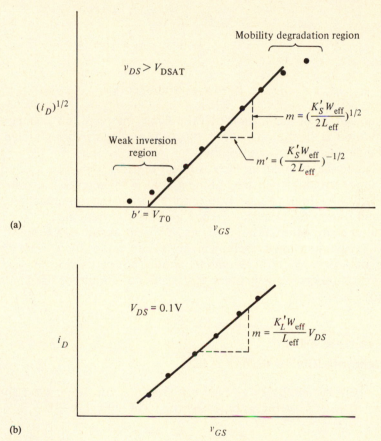

(a)

(b)

Figure 4.2-1　(a) $i_D^{1/2}$ versus v_{GS} plot used to determine V_{TO} and K'_S. (b) i_D versus v_{GS} plot used to determine K'_L.

It is worthwhile to rewrite Eq. (4) a different way to make it convenient to extract the threshold voltage parameter, V_{T0}. This is done below.

$$V_{GS} = \left(\frac{K'_S W_{eff}}{2L_{eff}}\right)^{-1/2} i_D^{1/2} + V_{T0} \tag{13}$$

Equation (13) has the form of

$$x = m'y + b' \tag{14}$$

where

$$x = V_{GS} \tag{15}$$
$$y = i_D^{1/2} \tag{16}$$
$$m' = \frac{\Delta X}{\Delta Y} = \left(\frac{K'_S W_{eff}}{2L_{eff}}\right)^{-1/2} \tag{17}$$

and

$$b' = V_{T0} \text{ (x-intercept)} \tag{18}$$

The advantage of this form is that the x-axis intercept is V_{T0}. Once the data are plotted as illustrated in Fig. 4.2-1(a), K'_S and V_{T0} can be extracted directly and independently of one another.

Numerical techniques such as linear regression may be used to extract the above parameters instead of the graphical techniques illustrated in Fig. 4.2-1(a). A word of caution is in order before using numerical techniques. Second-order effects such as weak-inversion current near $v_{GS} = V_{T0}$ and mobility degradation at large values of v_{GS} can result in a best fit line that is somewhat perturbed, thereby giving wrong values for V_{T0} and K'_S. Therefore, data that do not fit the model should not be used in the parameter-extraction procedure. Any numerical technique should be used with this in mind.

Example 4.2-1

Determination of V_{T0} and $K'_S W/L$

Given the following transistor data shown in Table 4.2-1 and linear regression formulas based on the form, $y = mx + b$

$$m = \frac{\Sigma x_i y_i - (\Sigma x_i \Sigma y_i)/n}{\Sigma x_i^2 - (\Sigma x_i)^2/n} \tag{19}$$

Table 4.2-1

Data for Example 4.2-1[1].

V_{GS} (volts)	I_D (μA)	$\sqrt{I_D}$ (μA)$^{1/2}$	V_{SB} (volts)
1.000	0.70	0.837	0.000
1.200	2.00	1.414	0.000
1.500	8.00	2.828	0.000
1.700	13.95	3.735	0.000
1.900	22.1	4.701	0.000

[1]Data consistent with a 3½ digit DVM scales of 2.000, 20.00 and 200.0.

and

$$b = \bar{y} - m\bar{x} \tag{20}$$

determine V_{T0} and $K'_s W/2L$. The data in Table 4.2-1 also give $I_D^{1/2}$ as a function of V_{GS}.

The data must be checked for linearity before linear regression is applied. Checking slopes between data points is a simple numerical technique for determining linearity. Using the formula that

$$\text{Slope} = m' = \frac{\Delta x}{\Delta y} = \frac{V_{GS2} - V_{GS1}}{\sqrt{I_{D2}} - \sqrt{I_{D1}}}$$

gives

$$m'_1 = \frac{0.2}{1.414 - 0.837} = 0.3466$$

$$m'_2 = \frac{0.3}{2.828 - 1.414} = 0.2122$$

$$m'_3 = \frac{0.2}{3.735 - 2.828} = 0.2205$$

$$m'_4 = \frac{0.2}{4.701 - 3.735} = 0.2070$$

These results indicate that the first (lowest value of V_{GS}) data point is either bad, or at a point where the transistor is in weak inversion. This data point will not be included in subsequent analysis. Performing the linear regression yields the following results.

$$V_{T0} = 0.898 \text{ volts}$$

and

$$\frac{K_S' W_{eff}}{2L_{eff}} = 21.92 \ \mu A/V^2$$

Next consider the extraction of the parameter K_L' for the nonsaturation region of operation. Assume that v_{DS} is very small (e.g. 0.1 V) so that the channel-length modulation term $(1 + \lambda v_{DS})$ is approximately one. Equation (5) can be rewritten as

$$i_D = K_L'\left(\frac{W_{eff}}{L_{eff}}\right)v_{DS}v_{GS} - K_L'\left(\frac{W_{eff}}{L_{eff}}\right)v_{DS}\left(V_T + \frac{v_{DS}}{2}\right) \tag{21}$$

If i_D is plotted versus v_{GS} as shown in Fig. 4.2-1(b), then the slope is seen to be

$$m = \frac{\Delta i_D}{\Delta v_{GS}} = K_L'\left(\frac{W_{eff}}{L_{eff}}\right)v_{DS} \tag{22}$$

Knowing the slope, the term K_L' is easily determined to be

$$K_L' = m\left(\frac{L_{eff}}{W_{eff}}\right)\left(\frac{1}{v_{DS}}\right) \tag{23}$$

if W_{eff}, L_{eff}, (assuming these are large to reduce the effects of dimensional variations and out-diffusion) and v_{DS} are known. The approximate value of the zero-field mobility parameter μ_o can be extracted from the value of K_L' using Eq. (8) of Section 3.1. A more accurate technique for determining μ_o will be given in the next section.

At this point, γ is unknown and remains to be determined in the characterization process. Using the same techniques as before, Eq. (3) is written in the linear form where

$$y = V_T \tag{24}$$
$$x = \sqrt{2|\phi_F| + v_{SB}} - \sqrt{2|\phi_F|} \tag{25}$$
$$m = \gamma \tag{26}$$
$$b = V_{T0} \tag{27}$$

The term $2|\phi_F|$ is unknown but is normally in the range of 0.6 to 0.7 volts. Once γ is calculated, then N_{SUB} can be calculated using Eq. (4) of Section

3.1 and $2|\phi_F|$ can be calculated using Eq. (5) of Section 3.1. This value can then be used in with Eqs. (24) through (27) to determine a new value of γ. Iterative procedures can be used to achieve the desired accuracy of γ and $2|\phi_F|$. Generally, an approximate value for $2|\phi_F|$ gives adequate results.

By plotting V_T versus x of Eq. (25) one can measure the slope of the best fit line from which the parameter γ can be extracted. In order to do this, V_T must be determined at various values of v_{SB} using the technique previously described. Fig. 4.2-2 illustrates the procedure. Each V_T determined in Fig. 4.2-2 must be plotted against the v_{SB} term. The result is shown in Fig. 4.2-3. The slope m, measured from the best fit line, is the parameter γ.

As before, numerical techniques can be employed to extract γ. Care should be exercised to determine how well the data are correlating before the extracted parameters are accepted as reasonable. An example illustrates the procedure.

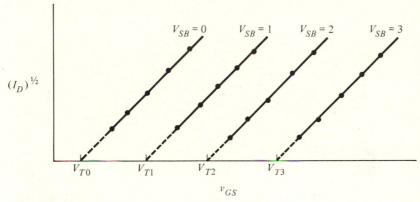

Figure 4.2-2 $(i_D)^{0.5}$ versus v_{GS} plot at different v_{SB} values to determine γ.

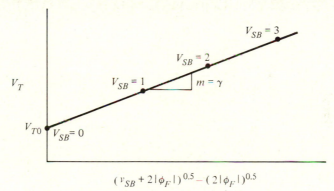

Figure 4.2-3   Plot of V_T versus v_{SB} to determine γ.

Example 4.2-2

Determination of γ

Using the results from Ex. 4.2-1 and the following transistor data, determine the value of γ using linear regression techniques. Assume that $2|\phi_F|$ is 0.6 volts.

Table 4.2-2 shows data for $V_{SB} = 1$ volt and $V_{SB} = 2$ volts. A quick check of the data in this table reveals that $I_D^{1/2}$ versus V_{GS} is linear and thus may be used in the linear regression analysis. Using the same procedure as in Ex. 4.2-1, the following thresholds are determined: $V_{T0} = 0.898$ volts (from Ex. 4.2-1), $V_T = 1.143$ volts ($V_{SB} = 1$ volt), and $V_T = 1.322$ volts ($V_{SB} = 2$ volts). Table 4.2-3 gives the value of V_T as a function of $[(2|\phi_F| + V_{SB})^{1/2} - (2|\phi_F|)^{1/2}]$ for the three values of V_{SB}. With these data, linear regression must be performed on the data of V_T versus $[(2|\phi_F| + V_{SB})^{1/2} - (2|\phi_F|)^{1/2}]$. The regression parameters of Eq. (19) are

$$\Sigma x_i y_i = 1.668$$

$$\Sigma x_i \Sigma y_i = 4.466$$

$$\Sigma x_i^2 = 0.9423$$

$$(\Sigma x_i)^2 = 1.764$$

These values result in $m = 0.506 = \gamma$.

The three major parameters of the simple model have been determined and all that is left to do is extract the three remaining parameters λ, ΔL,

Table 4.2-2

Data for Example 4.2-2.[1]

V_{SB} (volts)	V_{GS} (volts)	I_D (μA)
1.000	1.400	1.431
1.000	1.600	4.55
1.000	1.800	9.44
1.000	2.000	15.95
2.000	1.700	3.15
2.000	1.900	7.43
2.000	2.10	13.41
2.000	2.30	21.2

Table 4.2-3

Data for Example 4.2-2'.

V_{SB} (volts)	V_T (volts)	$[\sqrt{2|\phi_F| + V_{SB}} - \sqrt{2|\phi_F|}]$ (volts)$^{1/2}$
0.000	0.898	0.000
1.000	1.143	0.490
2.000	1.322	0.838

and ΔW. The channel length modulation parameter λ should be determined for all device lengths that might be used. For the sake of simplicity, Eq. (4) is rewritten as

$$i_D = i'_D \lambda v_{DS} + i'_D \tag{28}$$

which is in the familiar linear form where

$$y = i_D \qquad \text{(Eq. (4))} \tag{29}$$
$$x = v_{DS} \tag{30}$$
$$m = \lambda i'_D \tag{31}$$
$$b = i'_D \qquad \text{(Eq. (4) with } \lambda = 0) \tag{32}$$

By plotting i_D versus v_{DS}, measuring the slope of the data in the saturation region, and dividing that value by the y-intercept, λ can be determined. Figure 4.2-4 illustrates the procedure. The following example demonstrates the procedure.

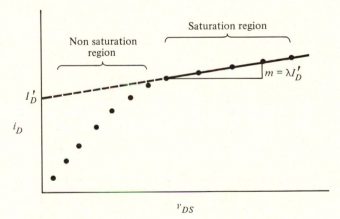

Figure 4.2-4   Plot of i_D versus v_{DS} to determine λ.

Example 4.2-3

Determination of λ

Given the data of I_D versus V_{DS} in Table 4.2-4, determine the parameter λ. We note that the data of Table 4.2-4 covers both the saturation and nonsaturation regions of operation. A quick check shows that saturation is reached near $V_{DS} = 2.0$ volts. To calculate λ, we shall use the data for V_{DS} greater than or equal to 2.5 volts. The parameters of the linear regression are

$$\Sigma x_i y_i = 1277.85$$

$$\Sigma x_i \Sigma y_i = 5096.00$$

$$\Sigma x_i^2 = 43.5$$

$$(\Sigma x_i)^2 = 169$$

These values result in $m = \lambda I'_D = 3.08$ and $b = I'_D = 88$, giving $\lambda = 0.035$ volts^{-1}.

The slope in the saturation region is typically very small, making it necessary to be careful that two data points taken with low resolution are not subtracted (to obtain the slope) resulting in a number that is of the same order of magnitude as the resolution of the data point measured. If this occurs, then the value obtained will have significant and unacceptable error. The fallacy of doing this will be demonstrated by the following example.

Table 4.2-4

Data for Example 4.2-3[1].

I_D (μA)	V_{DS} (volts)
39.2	0.500
68.2	1.000
86.8	1.500
94.2	2.000
95.7	2.50
97.2	3.00
98.8	3.50
100.3	4.00

Example 4.2-4

Measurement Errors Resulting from Small Differences

Suppose that the following two points were measured using a 3-½ digit digital voltmeter (DVM).

$$V_{DS1} = \underline{\ 5.0\ 0} \qquad I_{D1} = \underline{\ 3.2\ 7}$$
$$V_{DS2} = \underline{\ 2.0\ 0} \qquad I_{D2} = \underline{\ 3.1\ 8}$$

Typical accuracy of a DVM might be 0.25% Full Scale (FS) $\pm$ 1 Least Significant Digit (LSD). The readings given above could lie anywhere in the following limits:

$$4.94 < V_{DS1} < 5.06 \text{ volts}$$

$$3.21 < I_{D1} < 3.33 \text{ } \mu A$$

$$1.94 < V_{DS2} < 2.06 \text{ volts}$$

$$3.12 < I_{D2} < 3.24 \text{ } \mu A$$

This range of values translate into the following range of values for λ.

$$0.00 < \lambda < 0.02$$

It is obvious that one must be very careful in determining λ since very small changes must be measured accurately.

In the equations presented thus far, L_{eff} and W_{eff} have been used to describe the transistor's length and width, respectively. This terminology has been used because these dimensions, as they result in fabricated form, differ from the drawn values (due to out-diffusion, oxide encroachment, mask tolerance, and so on). The following analysis will determine the difference between the effective and drawn values.

Consider two transistors, with the same widths but different lengths, operating in the nonsaturation region with the same v_{DS}. The widths of the transistors are assumed to be very large so that $W \cong W_{eff}$. The large signal model is given as

$$i_D = \frac{K_L' W_{eff}}{L_{eff}} \left[(v_{GS} - V_{T0}) v_{DS} - \left(\frac{v_{DS}^2}{2} \right) \right] \qquad \textbf{(33)}$$

and

$$\frac{\partial i_D}{\partial v_{GS}} = g_m = \left(\frac{K'_L W_{\text{eff}}}{L_{\text{eff}}}\right) V_{DS} \tag{34}$$

The aspect ratios (W/L) for the two transistors are

$$\frac{W_1}{L_1 + \Delta L} \tag{35}$$

and

$$\frac{W_2}{L_2 + \Delta L} \tag{36}$$

Implicit in Eqs. (35) and (36) is that ΔL is assumed to be the same for both transistors. Combining Eq. (34) with Eqs. (35) and (36) gives

$$g_{m1} = \frac{K'_L W}{L_1 + \Delta L} V_{DS} \tag{37}$$

and

$$g_{m2} = \frac{K'_L W}{L_2 + \Delta L} V_{DS} \tag{38}$$

where $W_1 = W_2 = W$ (and are assumed to equal the effective width). With further algebraic manipulation of Eqs. (37) and (38), one can show that,

$$\frac{g_{m1}}{g_{m1} - g_{m2}} = \frac{L_2 + \Delta L}{L_2 - L_1} \tag{39}$$

which further yields

$$L_2 + \Delta L = L_{\text{eff}} = \frac{(L_2 - L_1)g_{m1}}{g_{m1} - g_{m2}} \tag{40}$$

The values of L_2 and L_1 are known and the small signal parameters g_{m1} and g_{m2} can be measured so that L_{eff} (or ΔL) can be calculated. Similar analysis can be performed to obtain W_{eff} yielding the following result.

$$W_2 + \Delta W = W_{\text{eff}} = \frac{(W_1 - W_2)g_{m2}}{g_{m1} - g_{m2}} \tag{41}$$

Eq. (41) is valid when two transistors have the same length but different widths.

One must be careful in determining ΔL (or ΔW) to make the lengths (or widths) sufficiently different in order to avoid the numerical error due to subtracting large numbers, and small enough that the transistor model chosen is still valid for both transistors. The following example demonstrates the determination of ΔL.

Example 4.2-5

Determination of ΔL

Given two transistors with the same widths and different lengths, determine L_{eff} and ΔL based upon the following data.

$W_1/L_1 = 20\ \mu m/10\ \mu m$ (drawn dimensions)

$W_2/L_2 = 20\ \mu m/20\ \mu m$ (drawn dimensions)

$g_{m1} = 6.65\ \mu S$ at $V_{DS} = 0.1$ volt and $I_D = 6.5\ \mu A$

$g_{m2} = 2.99\ \mu S$ at $V_{DS} = 0.1$ volt and $I_D = 2.75\ \mu A$

Using Eq. (40), L_{eff} can be determined as follows.
$L_{eff} = L_2 + \Delta L = (20 - 10)(6.65)/(6.65 - 2.99) = 18.17\ \mu m$.
Therefore, $\Delta L = -1.83\ \mu m$. From this result, we can estimate the lateral diffusion as
Lateral diffusion $(LD) = |\Delta L|/2 = 0.915\ \mu m$

In this section, we have shown how to characterize the simple model parameters V_{T0}, K_S', K_L', γ, $2|\phi_F|$, λ, ΔL, and ΔW. We have assumed that $2|\phi_F|$ could be determined by iteration if N_{SUB} were known. (If N_{SUB} is not known, then N_{SUB} must be measured by other means, e.g., bulk resistance.) Also, one must remember that these model parameters, with the exception of λ, ΔL, and ΔW, are dependent upon temperature.

In conclusion, it is important to realize that the extracted parameters may not be the end result of the characterization process. The goal of a model is to simulate the experimental response. The next step is to put the extracted parameters into the simulation models and compare the simulation results with the experimental results. If the results do not compare well, then it is necessary to modify the appropriate model parameters until a good match is obtained. The result often is that the actual value of the model parameter no longer agrees with the value expected from the derivation based on the physical relationships of the transistor. This has led to the use of purely empirical models, which tend to have a better match between the simulated and experimental performance.

4.3 *Transistor Characterization for the Extended Model*

Thus far, techniques have been introduced that extract values for V_{T0}, K'_S, K'_L, γ, $2|\phi_F|$, λ, ΔL, and ΔW using the simple MOS model. These results are satisfactory when all calculations or simulations will use the simple model (in one or the other of the saturation and nonsaturation regions) to get approximate results. To use a more accurate model (in this case, the extended model presented in Chapter 3), its particular model parameters must be extracted.

Equations (1) and (2) represent a simplified version of the extended model for a relatively wide MOS transistor operating in the nonsaturation, strong-inversion region with $V_{SB} = 0$.

$$i_D = \frac{\mu_s C_{ox} W}{L} \left\{ (V_{GS} - V_T) V_{DS} - \left(\frac{v_{DS}^2}{2} \right) + \gamma v_{DS} \sqrt{2|\phi_F|} \right. $$
$$\left. - \left(\frac{2\gamma}{3} \right) [(V_{DS} + 2|\phi_F|)^{1.5} - (2|\phi_F|)^{1.5}] \right\} \qquad (1)$$

where W and L are effective electrical equivalents (dropping the subscript, "eff", for convenience).

$$\mu_s = \mu_o \left[\frac{(UCRIT)\epsilon_{si}}{C_{ox}[V_{GS} - V_T - (UTRA)v_{DS}]} \right]^{UEXP} \qquad (2)$$

Eq. (2) holds when the denominator term in the brackets is less than unity. Otherwise, $\mu_o = \mu_s$. To develop a procedure for extracting μ_o, consider the case where mobility degradation effects are not being experienced, i.e. $\mu_s = \mu_o$, Eq. (1) can be rewritten in general as

$$i_D = \mu_o f(C_{ox}, W, L, v_{GS}, V_T, v_{DS}, \gamma, 2|\phi_F|) \qquad (3)$$

This equation is a linear function of v_{GS} and is in the familiar form of

$$y = mx + b \qquad (4)$$

where $b = 0$. To determine the slope μ_o all that need be done is to plot i_D versus the function, $f(C_{ox}, W, L, v_{GS}, V_T, v_{DS}, \gamma, 2|\phi_F|)$ and measure the slope of the curve. In order to achieve accurate results in the calculation of μ_o, the following constraints on the data must be observed:

1. The data are limited to the nonsaturation region (small v_{DS}).
2. The transistor must be in the strong-inversion region ($v_{GS} > V_T$).
3. The transistor must operate below the critical-mobility point.

Of these three requirements, the most difficult one to achieve is the third, since one does not know beforehand at what value of v_{GS} the critical field point is reached. One should try to keep v_{GS} as low as possible without encroaching on the weak-inversion region of operation.

A good way of testing the data to find where mobility degradation occurs and where weak inversion begins is to plot i_D versus v_{GS} for small values of v_{DS} and note where the curve is linear. This region can be identified as the strong-inversion, constant-mobility region. The data corresponding to higher currents can be identified as the variable mobility region and should be used to calculate UCRIT and UEXP. Fig. 4.3-1 shows a plot of i_D versus v_{GS} for a transistor in the nonsaturation region. This figure illustrates the three regions of interest.

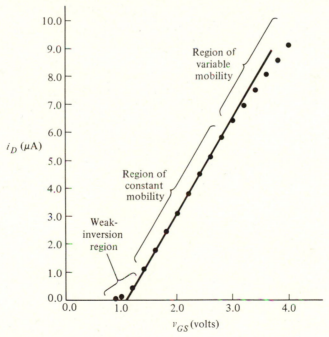

Figure 4.3-1 Plot of i_D versus v_{GS} in the nonsaturation region ($V_{DS} = 0.1$ V).

Once μ_o is determined, there is ample information to determine UCRIT and UEXP. Consider Eqs. (1) and (2) rewritten and combined as follows.

$$i_D = \mu_o[(\text{UCRIT})f_2]^{\text{UEXP}}f_1 \tag{5}$$

where

$$f_1 = \left(\frac{C_{ox}W}{L}\right)\left\{(v_{GS} - v_T)v_{DS} - \left(\frac{v_{DS}^2}{2}\right) + \gamma v_{DS}\sqrt{2|\phi_F|}\right.$$

$$\left. - \left(\frac{2}{3}\right)\gamma[(v_{DS} + 2|\phi_F|)^{1.5} - (2|\phi_F|)^{1.5}]\right\} \tag{6}$$

and

$$f_2 = \frac{\epsilon_{si}}{[v_{GS} - V_T - (\text{UTRA})v_{DS}]C_{ox}} \tag{7}$$

The units of f_1 and f_2 are FV^2/cm^2 and cm/V respectively. Notice that f_2 includes the parameter UTRA, which is an unknown. Typically, UTRA is very small (less than 0.5) so if v_{DS} is chosen to be small for the data taken ($v_{DS} \simeq 0.1$ volts) then the effect of UTRA on the final result is small enough to be insignificant. In the extraction procedure that follows, one should assume UTRA to be 0.1 to 0.5.

Equation (5) can be manipulated algebraically to yield

$$\log\left(\frac{i_D}{f_1}\right) = \log(\mu_o) + \text{UEXP}[\log(\text{UCRIT})] + \text{UEXP}[\log(f_2)] \tag{8}$$

This is in the familiar form of Eq. (4) with

$$x = \log(f_2) \tag{9}$$

$$y = \log\left(\frac{i_D}{f_1}\right) \tag{10}$$

$$m = \text{UEXP} \tag{11}$$

$$b = \log(\mu_o) + \text{UEXP}[\log(\text{UCRIT})] \tag{12}$$

By plotting Eq. (8) and measuring the slope, UEXP can be determined. The y-intercept can be extracted from the plot and UCRIT can be determined by back calculation given UEXP, μ_o, and the intercept, b.

Example 4.3-1

Determination of Mobility, UCRIT, and UEXP

Given the following data, determine approximately where mobility degradation effects begin. Using this information calculate μ_o, UCRIT, and UEXP. Assume that the previous analysis resulted in the following model parameters: $V_{T0} = 1.0$ volt, $\gamma = 1.68$ volts$^{1/2}$, $W_{eff} = 10$ μm, $L_{eff} = 8.4$ μm, and $2|\phi_F| = 0.6$ volts. Further, assume that $t_{ox} = 750$ Angstroms and the dielectric constants of the oxide and silicon are 3.9 and 11.7, respectively.

Data for $V_{DS} = 0.1$ volts:

V_{GS}	I_D (μA)	V_{GS}	I_D (μA)	V_{GS}	I_D (μA)
1.000	0.020	2.00	3.13	3.00	6.40
1.200	0.460	2.20	3.79	3.20	6.96
1.400	1.130	2.40	4.46	3.40	7.50
1.600	1.800	2.60	5.12	3.60	8.04
1.800	2.46	2.80	5.79	3.80	8.58
				4.00	9.10

These data are plotted in Fig. 4.3-1. To determine μ_o, the region between $V_{GS} = 1.8$ and 2.8 volts can be used. Since μ_o is the slope of the curve i_D versus f_1 [where f_1 is given in Eq. (6)], it is convenient to make a table of I_D as a function of f_1. This table is shown below.

V_{GS}	$f_1(\times 10^{-9})$	I_D (μA)
1.800	3.821	2.46
2.000	4.918	3.13
2.20	6.010	3.79
2.40	7.110	4.46
2.60	8.206	5.12
2.80	9.302	5.79

Performing a least-squares fit on the data results in a value for μ_o of 607 cm^2/V·s.

Now that the mobility has been determined, there is enough information to determine UEXP and UCRIT. For these calculations, data must be taken from the variable-mobility region of operation shown on Fig. 4.3-1. The term UEXP is the slope of the curve of log (I_D/f_1) versus log(f_2), where f_1 and f_2 are defined in Eqs. (6) and (7). Again, for convenience, a table will be generated to show the required information.

V_{GS}	$\log(I_D/f_1)$	$\log(f_2)$
3.2	2.782	-4.988
3.4	2.775	-5.026
3.6	2.769	-5.061
3.8	2.764	-5.093
4.0	2.758	-5.123

Using linear regression techniques, the slope and intercept are calculated to be

$$m = 0.175 = UEXP$$

$$b = 3.656$$

UCRIT can be calculated from the following relationship.

$$UCRIT = \log^{-1}\left[\frac{b - \log(\mu_o)}{UEXP}\right]$$

$$UCRIT = \log^{-1}\left[\frac{3.656 - \log(607)}{0.175}\right] = 97,161 \ (V/cm)$$

4.4   1/F Noise

In many applications, good noise performance is a very important requirement for an analog design. Consequently, the noise performance of transistors must be characterized. The equation defining the mean-square noise current in an MOS transistor given in Eq. (12) of Sec. 3.2 is repeated here.

$$\overline{i_N^2} = \left[\frac{8kTg_m(1 + \eta)}{3} + \frac{(KF)I_D}{fC_{ox}L^2}\right]\Delta f \qquad (\text{amperes}^2) \qquad \textbf{(1)}$$

All notation is consistent with that in Sec. 3.2. At high frequencies the first term in Eq. (1) dominates, whereas at low frequencies the second term dominates. Since the second term is the only one with model parameters, it is the only portion of the expression that must be considered for characterization. Equation (2) describes the mean-square noise current at low frequencies as

$$\overline{i_N^2} = \left[\frac{(KF)I_D}{fC_{ox}L^2}\right]\Delta f \qquad \qquad \textbf{(2)}$$

This mean-square noise current is represented as a current source across the drain and source nodes in the small-signal model for the transistor. This is illustrated in Fig. 4.4-1. Since noise is more generally considered at the input rather than the output, the input-referred, noise-current is given by multiplying Eq. (2) by g_m^{-2} to get

$$\overline{v}_N^2 = \left[\frac{(KF)I_D}{g_m^2 f C_{ox} L^2} \right] \Delta f \qquad (\text{volts}^2) \tag{3}$$

Substituting the relationship for g_m in the saturation region

$$g_m = \sqrt{2K_S'(W/L)I_D} \tag{4}$$

into Eq. (3) gives a convenient form for the input-noise voltage

$$\overline{v}_N^2 = \left[\frac{KF}{2K_S'WLfC_{ox}} \right] \Delta f \tag{5}$$

For characterization purposes, assume that the noise voltage is measured at a 1 Hz bandwidth so that the Δf term is unity. Equation (5) can be rewritten as

$$\log[\overline{v}_N^2] = \log\left[\frac{KF}{2K_S'WLC_{ox}} \right] - \log[f] \tag{6}$$

By plotting $\log[f]$ versus $\log[\overline{v}_N^2]$ and measuring the intercept, which is $\log[KF/(2K_S'WLC_{ox})]$, one can extract the parameter KF. This technique is illustrated by the following example.

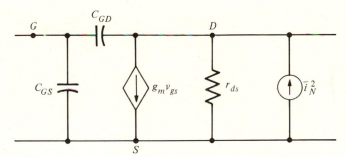

Figure 4.4-1   Small-signal MOS transistor model including noise current.

Example 4.4-1

Calculation of KF

Consider the noise data for a 50 μm/10 μm n-channel device plotted in Fig. 4.4-2[1]. Using device data given in Tables 3.1-1 and 3.1-2 calculate the model parameter *KF*. Assume that t_{ox} = 800 Angstroms. Each measurement was made at a 1 Hz bandwidth.

The first step requires a modified graph of the data. Make another plot, this time plotting $\log[v_N^2]$ as the ordinate variable. This is illustrated in Fig. 4.4-3. In order to determine *KF*, an approximate $1/f$ line must be fit to the data that is considered to be dominated by $1/f$ noise. In this example, the data do not follow the $1/f$ characteristic exactly, so a judgment must be made about the y-axis intercept. For a $1/f$ line intercept of -10.25, *KF* is calculated to be

$$KF = 2K_S'WLC_{ox}(10^{-10.25})$$
$$KF = 2(17 \times 10^{-6})(50 \times 10^{-6})(10 \times 10^{-6})(4.3 \times 10^{-4})(10^{-10.25})$$
$$KF \cong 4 \times 10^{-28}$$

4.5 *Characterization of Other Active Components*

In the previous sections, characterization of most of the more important parameters of the large-signal MOS model has been covered. This section will be devoted to characterization of other components found in a typical CMOS process.

One of the important active components available to the CMOS designer is a substrate bipolar junction transistor (BJT), (see Section 2.4). The collector of this BJT is always common with the substrate of the CMOS process. For example, if the CMOS process is a p-well process, the n-substrate is the collector, the p-well is the base, and the n^+ diffusions in the p-well are the emitter. The substrate BJT is used primarily for two applications. The first is as an output driver. Because the g_m of a BJT is greater than the g_m of an MOS device, the output impedance, which is typically $1/g_m$, is lower for a BJT. The second application is in bandgap voltage-reference circuits, which will be studied in Chapter 11. For these two applications, the parameters of interest are the dc beta, β_{dc}, and the leakage current density J_s. We are also interested in the dependence of β_{dc} upon dc emitter current. These parameters effect device operation as defined by the following equations.

$$v_{BE} = \frac{kT}{q} \ln\left(\frac{i_c}{J_s A_E}\right) \tag{1}$$

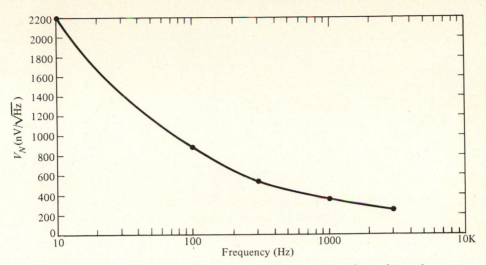

Figure 4.4-2   Noise data for a $W/L = 50\ \mu m/10\ \mu m$ N-channel transistor.

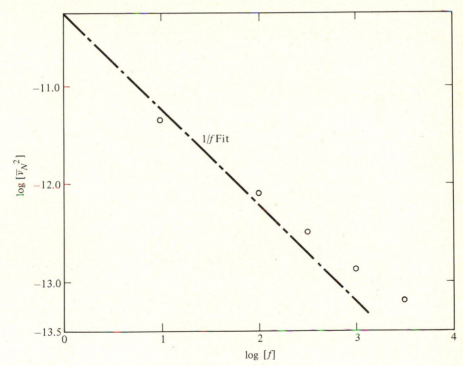

Figure 4.4-3   Plot of $\log [v_N^2]$ versus $\log [f]$ for a 50 $\mu m/10\ \mu m$ N-channel device.

and

$$\beta_{dc} = \frac{i_E}{i_B} - 1 \qquad (2)$$

A_E is the cross-sectional area of the emitter-base junction of the BJT. In order to determine the parameter β_{dc}, Eq. (2) can be rearranged to the form given in Eq. (3) which gives i_E as a function of i_B in a linear equation.

$$i_E = i_B(\beta_{dc} + 1) \qquad (3)$$

The current i_B can be plotted as a function of i_E, and the slope measured to determine β_{dc}. Once β_{dc} is known, then Eq. (1) can be rearranged and modified as follows.

$$V_{BE} = \frac{kT}{q} \ln\left[\frac{i_E \beta_{dc}}{1 + \beta_{dc}}\right] - \frac{kT}{q} \ln(J_s A_E) = \frac{kT}{q} \ln(\alpha_{dc} i_E) - \frac{kT}{q} \ln(J_s A_E) \qquad (4)$$

Plotting $\ln[i_E \beta_{dc}/(1 + \beta_{dc})]$ versus V_{BE} results in a graph where

$$m = \text{slope} = \frac{kT}{q} \qquad (5)$$

and

$$b = y\text{-intercept} = -\left(\frac{kT}{q}\right) \ln(J_s A_E) \qquad (6)$$

Since the emitter area is known, J_s can be directly determined.

Example 4.5-1

Determination of β_{dc} and J_s

Consider a bipolar transistor with a 1000 μm^2 emitter area on which the following data has been taken.

$I_E(\mu A)$	$I_B(\mu A)$	$V_{BE}(V)$
100	0.90	0.540
136	1.26	0.547
144	1.29	0.548
200	1.82	0.558
233	2.11	0.560

Using this data, determine β_{dc} and J_s.

The data for I_E versus I_B can be analyzed using linear regression techniques to determine the slope m. The result is

$$m = 110 = \beta_{dc}$$

with β_{dc} known, the terms of Eq. (4) can be tabulated to obtain the slope and intercept, from which J_s can be calculated. The table below supplies the x and y terms of Eq. (4) for subsequent calculation.

$V_{BE}(V)$	$\ln[I_E\beta_{dc}/(1 + \beta_{dc})]$
0.540	−9.20
0.547	−8.89
0.548	−8.84
0.558	−8.51
0.560	−8.36

From this data, the slope and intercept is calculated to be

$$m = 0.025 \text{ volts} = kT/q$$
$$b = 0.769 = -(kT/q)[\ln(J_sA_E)]$$

from which J_sA_E is determined to be 43.8 fA. With an emitter area of 1000 μm^2, $J_s = 4.38 \times 10^{-17}$ A/μm^2.

4.6   *Characterization of Resistive Components*

Thus far, characterization procedures for the primary active components available to the circuit designer in a typical CMOS process have been developed. This section will deal with passive components and their parasitics. These include resistors, contact resistance, and capacitance.

Consider first the characterization of sheet resistance. While there are a number of ways to do this, the most useful results can be obtained by characterizing resistor geometries exactly as they will be implemented in a design. This is advantageous because (1) sheet resistance is not constant across the width of a resistor, (2) the effects of bends result in inaccuracies, and (3) termination effects are not accurately predictable. Therefore a resistor should be characterized as a function of width and the effects of bends and terminations should be carefully considered.

Figure 4.6-1 illustrates a structure that can be used to determine sheet resistance, and geometry width variation (bias)[2]. By forcing a current into node *A* with node *F* grounded while measuring the voltage drops across *BC*

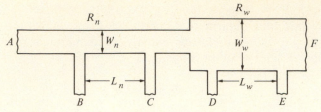

Figure 4.6-1 Sheet resistance and bias monitor.

(V_n) and DE (V_w), the resistors R_n and R_w can be determined as follows

$$R_n = \frac{V_n}{I} \tag{1}$$

$$R_w = \frac{V_w}{I} \tag{2}$$

The sheet resistance can be determined from these to be

$$R_S = R_n\left(\frac{W_n - \text{Bias}}{L_n}\right) \tag{3}$$

$$R_S = R_w\left(\frac{W_w - \text{Bias}}{L_w}\right) \tag{4}$$

where

R_n = resistance of narrow resistor (ohms)

R_w = resistance of wide resistor (ohms)

R_S = sheet resistance of material (polysilicon, diffusion etc. ohms/ square)

L_n = drawn length of narrow resistor

L_w = drawn length of wide resistor

W_n = drawn width of narrow resistor

W_w = drawn width of wide resistor

Bias = difference between drawn width and actual device width

Solving equations (3) and (4) yields

$$\text{Bias} = \frac{W_n - kW_w}{1 - k} \tag{5}$$

where

$$k = \frac{R_w L_n}{R_n L_w} \tag{6}$$

and

$$R_s = R_n\left(\frac{W_n - \text{Bias}}{L_n}\right) = R_w\left(\frac{W_w - \text{Bias}}{L_w}\right) \tag{7}$$

This technique eliminates any effects due to contact resistance since no appreciable current flows through a contact causing a voltage drop. The example that follows illustrates the use of this technique.

Example 4.6-1
Determination of R_s and Bias

Consider the structure shown in Fig. 4.6-1 where the various dimensions are given below

$W_n = 10 \ \mu\text{m}$

$L_n = 40 \ \mu\text{m}$

$W_w = 50 \ \mu\text{m}$

$L_w = 200 \ \mu\text{m}$

A current of 1 mA is forced into node A with node F grounded. The following voltages are measured

$V_n = 133.3 \text{ mV}$

$V_w = 122.5 \text{ mV}$

Therefore R_n and R_w are

$$R_n = \frac{0.1333}{0.001} = 133.3 \text{ ohms}$$

$$R_w = \frac{0.1225}{0.001} = 122.5 \text{ ohms}$$

Determine the sheet resistance R_s and Bias. Using Eq. (6), the value k is calculated to be

$$k = \frac{R_w L_n}{R_n L_w} = \frac{122.5(40)}{133.3(200)} = 0.184$$

Use Eq. (5) with the value calculated for k to determine Bias

$$\text{Bias} = \frac{10 - (0.184)50}{1 - 0.184} = 0.98 \ \mu m$$

Having calculated the bias, the sheet resistance can be determined using Eq. (7)

$$R_s = 133.3\left(\frac{10 - 0.98}{40}\right) = 30.06 \text{ ohms/square}$$

Further work can be done to determine not only the sheet resistivity of the resistor but also the contact resistance. Consider the two resistors shown in Fig. 4.6-2. The values of these two resistors are (assuming contact resistance is the same for both),

$$R_A = R_1 + 2R_c; \qquad R_1 = N_1 R_s \tag{8}$$

and

$$R_B = R_2 + 2R_c; \qquad R_2 = N_2 R_s \tag{9}$$

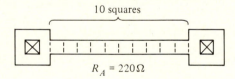

10 squares

$R_A = 220\Omega$

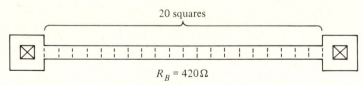

20 squares

$R_B = 420\Omega$

Figure 4.6-2   Two resistors used to determine K_R and R_c.

where N_1 is the number of squares for R_1, R_s is the sheet resistivity in ohms/square, and R_c is the contact resistance. These two equations can be solved simultaneously to get

$$R_S = \frac{R_B - R_A}{N_2 - N_1} \tag{10}$$

and

$$2R_c = R_A - N_1 R_S = R_B - N_2 R_S \tag{11}$$

The application of these equations is illustrated in the following example.

Example 4.6-2

Calculation of Contact Resistance and Sheet Resistivity

Consider the resistors shown in Fig. 4.6-2. Their values can be expressed as

$$R_A = 2R_c + 10R_S$$

and

$$R_B = 2R_c + 20R_S$$

The measured values for R_A and R_B are 220 ohms and 420 ohms, respectively. Using Eqs. (8) and (9), R_s and R_c are determined to be

$$R_S = \frac{R_B - R_A}{N_2 - N_1} = \frac{420 - 220}{20 - 10} = 20 \text{ ohms/square}$$

$$2R_c = R_A - N_1 R_S = 220 - 10(20) = 20 \text{ ohms}$$

$$R_c = 10 \text{ ohms}$$

Lightly-doped resistors, p- or n-well, and pinched resistors, should be characterized considering the back-bias effects of the substrate in which the resistor is fabricated (that is, the effect of the potential difference between a resistor and the substrate which it is diffused into). In order to do this, one should measure the resistance as a function of the voltage between the terminals and the substrate. This is illustrated in Fig. 4.6-3. Depending upon the application, one may prefer to characterize the resis-

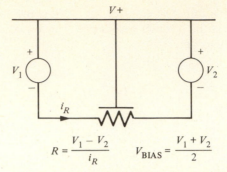

$$R = \frac{V_1 - V_2}{i_R} \qquad V_{BIAS} = \frac{V_1 + V_2}{2}$$

Figure 4.6-3 P-well resistor showing the back-bias dependence.

tor as a function of the back-bias voltage with tabulated values as the result, or model the resistor as a JFET, resulting in model equations similar to those presented for the MOSFET. Fig. 4.6-4 shows the dependence of p-well resistance upon the back bias.

A simpler method for directly determining contact resistance might be more appropriate in some cases. Figure 4.6-5 shows a structure [2] that has two different materials (metal and some other material) contacting together through a single contact. The equivalent circuit for this structure is shown in Fig. 4.6-6. It is easy to see that if current is forced between pads one and two while voltage is measured across pads three and four, the resulting ratio of voltage to current is the contact resistance.

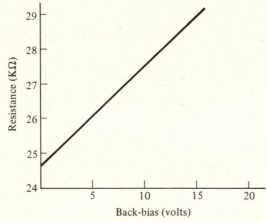

Figure 4.6-4 P-well resistance as a function of back-bias voltage.

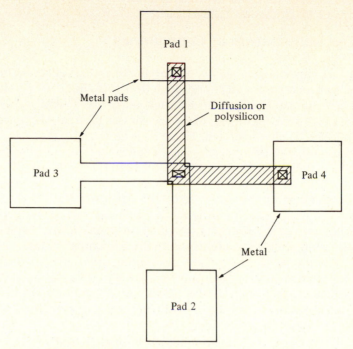

Figure 4.6-5   Contact resistance test structure.

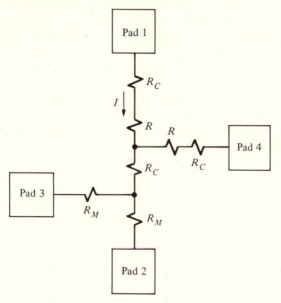

Figure 4.6-6   Electrical equivalent circuit of Fig. 4.6-5.

4.7   *Characterization of Capacitance*

Various capacitances are encountered in a CMOS process. They can be broken into two types: MOS and depletion capacitance. The MOS capacitors include parasitic capacitances such as C_{GS}, C_{GD}, and C_{GB}. The depletion capacitors are C_{DB} and C_{SB}. There are various interconnect capacitances that must also be characterized. These include $C_{\text{Poly}-\text{field}}$, $C_{\text{Metal}-\text{field}}$, and $C_{\text{Metal}-\text{poly}}$. It is desirable to characterize these capacitances for use in simulation models (e.g. the SPICE2 circuit simulator), and for use as parasitic circuit components to be included in circuit simulations. The capacitors that must be characterized for use in the SPICE2 transistor model are C_{GS0}, C_{GD0}, and C_{GB0} (at $V_{GS} = V_{GB} = 0$). Normally SPICE2 calculates C_{DB} and C_{SB} using the areas of the drain and source and the junction (depletion) capacitance, C_J (zero-bias value), that it calculates internally from other model parameters. Two of these model parameters, *MJ* and *MJSW*, are used to calculate the depletion capacitance as a function of voltage across the capacitor. The other parasitic capacitors must be characterized so that interconnect capacitances can be estimated and included in circuit simulations.

Consider the transistor-parasitic capacitances C_{GS0}, C_{GD0}, and C_{GB0}. The first two, C_{GS0} and C_{GD0}, are modeled in SPICE2 as a function of the device width, while the capacitor C_{GB0} is per length of the device. All three capacitances are in units of F/m in the SPICE2 model. The gate/drain overlap capacitances for typical transistor structures are small so that direct measurement, although possible, is difficult. In order to reduce the requirements on the measurement setup, it is desirable to multiply the parasitic capacitance of interest, i.e., measure the C_{GS} of a very wide transistor and divide the result by the width in order to get C_{GS0} (per unit width). An example of a test structure useful for measuring C_{GS} and C_{DS} is given in Fig. 4.7-1[3]. This structure uses multiple, very wide transistors to achieve large, easily measurable capacitances. The metal line connecting to the source and drain that runs parallel to the gate should be sufficiently separated from the gate to minimize the gate to metal capacitance. Figure 4.7-2 shows an experimental setup for measuring the gate to drain/source capacitance. The capacitance measured is

$$C_{\text{meas}} = W(n)(C_{GS0} + C_{GD0}) \tag{1}$$

where

$\quad C_{\text{meas}}$ = total measured capacitance

$\quad W$ = total width of one of the transistors

$\quad n$ = total number of transistors

Gate extension, overhang

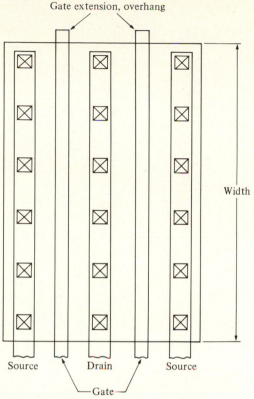

Width

Source   Drain   Source

Gate

Figure 4.7-1   Structure for determining C_{GS} and C_{GD}.

Assuming C_{GS0} and C_{GD0} are equal, they can be determined from the measured data using Eq. (1).

     For very narrow transistors, the capacitance determined using the previous technique will not be very accurate because of fringe field and other edge effects at the edge of the transistor. In order to characterize C_{GS0} and C_{GD0} for these narrow devices, a structure similar to that given in Fig. 4.7-1

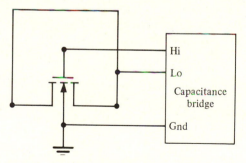

Figure 4.7-2   Experimental setup for measuring C_{GS} & C_{GD}.

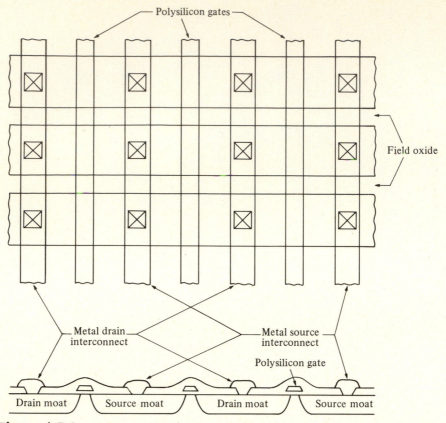

Figure 4.7-3   Structure for measuring C_{GS} and C_{GD}, including fringing effects, for transistors having small L.

can be used, substituting different device sizes. Such a structure is given in Fig. 4.7-3. The equations used to calculate the parasitic capacitances are the same as those given in Eq. (1).

The capacitance C_{GB0} is due to the overhang (see Rule 4D of Table 2.6-1) of the transistor gate required at one end as shown in Fig. 4.7-4. This capacitance is approximated from the interconnect capacitance $C_{poly-field}$ (note that the overhang capacitor is not a true parallel-plate capacitor due to the field-oxide gradient). Therefore, the first step in characterizing C_{GB0} is to obtain a value for $C_{poly-field}$ by measuring the total capacitance of a strip of polysilicon on field (width should be chosen to be the same as the desired device length) and divide this value by the total area as given in the following relationship.

$$C_{poly-field} = \frac{C_{meas}}{L_R W_R} \; (F/m^2)$$

(2a)

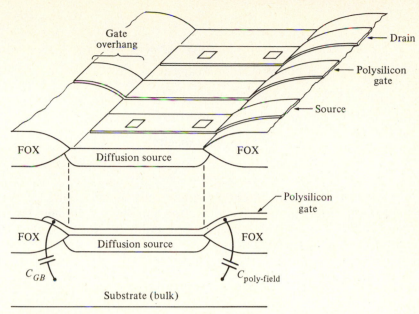

Figure 4.7-4   Illustration of gate-to-bulk and poly-field capacitance.

where

C_{meas} = measured value of the polysilicon strip

L_R = length of the centerline of the polysilicon strip

W_R = width of the polysilicon strip (usually chosen as device length)

Having determined $C_{poly-field}$, C_{GBO} can be approximated as (see Fig. 3.2-6)

$$C_{GBO} \cong 2 \, (C_{poly\text{-}field}) \, (d_{overhang}) = 2C_5 \text{ (F/m)} \tag{2b}$$

where

$d_{overhang}$ = overhang dimension (see Rule 4D, Table 2.6-1)

Next consider the junction capacitances C_{BD} and C_{BS}. As described by Eqs. (5) and (6) of Sec. 3.2, these capacitances are made up of a bottom and a sidewall component. For convenience, the equations are repeated here using the notation of Sec. 3.6.

$$C_J(V_J) = AC_J(0)\left(1 + \frac{V_J}{PB}\right)^{-MJ} + PC_{JSW}(0)\left(1 + \frac{V_J}{PB}\right)^{-MJSW} \tag{3}$$

where

$$V_J = \text{the reverse bias voltage across the junction}$$

$$C_J(V_J) = \text{bottom junction capacitance at } V_J$$

$$C_{JSW}(V_J) = \text{junction capacitance of sidewall at } V_J$$

$$A = \text{area of the (bottom) of the capacitor}$$

$$P = \text{perimeter of the capacitor}$$

$$PB = \text{bulk junction potential}$$

The constants C_J and MJ can be determined by measuring a large rectangular capacitor structure like that shown in Fig. 4.7-5 where the contribution from the sidewall capacitance is minimal[4]. For such a structure, $C_J(V_J)$ can be approximated as

$$C_J(V_J) = AC_J(0)\left(1 + \frac{V_J}{PB}\right)^{-MJ} \tag{4}$$

This equation can be rewritten in a way that is convenient for linear regression.

$$\log[C_J(V_J)] = (-MJ)\log\left[1 + \frac{V_J}{PB}\right] + \log[AC_J(0)] \tag{5}$$

By measuring $C_J(V_J)$ at different voltages and plotting $\log[C_J(VJ)]$ versus $\log[1 + V_J/PB]$ one can determine the slope, $-MJ$, and the Y intercept (where Y is the term on the left), $\text{Log}[AC_J(0)]$. Knowing the area of the capacitor, the calculation of the bottom junction capacitance is straightforward.

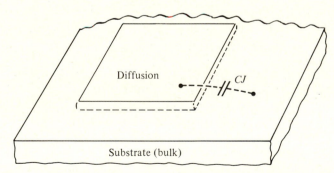

Figure 4.7-5   Rectangular depletion capacitor.

Example 4.7-1

Determination of C_j and MJ

Consider the following data taken on a large (100 μm by 100 μm) junction capacitor at various reverse bias voltages. Assume the sidewall capacitance is negligible and calculate the grading coefficient *MJ* and the bottom junction capacitance, $C_j(0)$. Use *PB* approximately equal to 0.7.

V_j (volts)	C_{meas} (10^{-12}F)
0	3.10
1	1.95
2	1.57
3	1.35
4	1.20
5	1.10

The data should be converted to the form required in Eq. (5) as given below.

$\log[1 + V_j/PB]$	$\log[C_j(VJ)]$
0.0000	-11.51
0.3853	-11.71
0.5863	-11.80
0.7231	-11.87
0.8270	-11.92
0.9108	-11.96

Using linear regression techniques, *MJ* and $C_j(0)$ are determined to be approximately

$$C_j(0) = 3.1 \times 10^{-4} \ (F/m^2)$$
$$MJ = 0.49$$

By using a long narrow structure whose junction capacitance is dominated by the perimeter rather than the area of the bottom, similar techniques as just described can be used to determine C_{JSW} and *MJSW*.

The technique described for obtaining $C_{poly-field}$ can be applied to determining $C_{metal-field}$ and $C_{metal-poly}$ as well. A test structure that can achieve large enough capacitance for easy measurement is required. Once obtained, these (per m^2) capacitances can be used to determine interconnect capacitance.

Other ideas for characterizing circuit capacitances are presented in the literature [5,6]. These have not been addressed in this text for the sake of brevity. The interested reader should refer to the references for further study.

4.8 *Summary*

The primary focus of this chapter has been the determination of model parameters that are implemented in model equations that accurately describe, in a theoretical sense, how a device will operate physically. The ultimate goal of characterization is to obtain accurate models to describe the transistors that will be used to design CMOS analog circuits. This chapter provides the information necessary to model and simulate the circuits. Without this information, the designer has no way of predicting with any degree of confidence how well his design will work.

PROBLEMS — *Chapter 4*

1. Design a resistor test structure that would allow one to predict the resistance of multiple-bend serpentine resistors of various lengths and number of bends. Assume that all bends will be the same, at 90 degrees.

2. What effect on characterization would a device geometry such as that shown in Fig. P4-2 have?

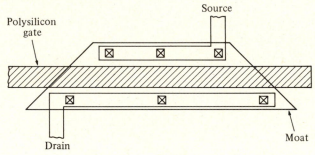

Figure P4-2

3. Given the following measured data taken for $V_{DS} = 3.0$ volts, calculate V_{T0} and $K'_s W_{eff}/L_{eff}$.

V_{GS} (V)	$I_D(\mu A)$
1.5000	76.80
2.0000	203.0

4. Plot the following data and graphically determine V_{T0} and $K'_S W_{eff}/L_{eff}$ if $V_{DS} = 3.0$ volts. Show how these parameters are extracted from the graph.

V_{GS} (V)	$I_D (\mu A)$
1.600	25.00
1.800	49.00
2.000	81.00
2.20	121.0
2.50	169.0

5. Develop a procedure for determining V_{T0} using transistor data in the nonsaturation region.

6. Given the following data, determine $K'_L W_{eff}/L_{eff}$. $V_{DS} = 0.15$ volts.

V_{GS} (V)	$I_D (\mu A)$
1.300	10.73
1.400	14.03
1.500	17.33
1.600	20.6

7. Fig. P4-7 shows a plot of i_D versus v_{GS} for various values of v_{SB}. From this plot, determine V_{T0}, V_{T1}, V_{T2}, V_{T3}, and γ.

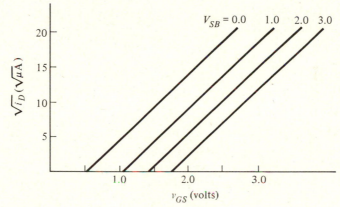

Figure P4-7

8. Given the following data, determine γ and $2|\phi_F|$. Assume $t_{ox} = 800$ Angstroms.

V_{SB} (V)	0.000	1.000	2.000	3.000
V_T (V)	0.900	1.570	2.05	2.45

9. Determine the channel length modulation parameter λ, based on the following data in the saturation region with $V_{T0} = 1$ V, $\gamma = -1.2$ V$^{-0.5}$,

and $V_{GS} = -2.0$ volts.

V_{DS} (V)	-1.500	-2.000	-2.50	-3.00
I_D (μA)	-104.0	-106.0	-107.0	-109.0

10. Develop a technique for determining λ using the test setup shown in Fig. P4-10. Hint: Consider the ratio of the output to the input current.

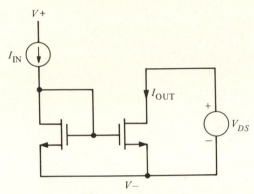

Figure P4-10

11. The following g_m measurements were made on two devices of different widths. In both cases, $V_{DS} = 0.1$ volt. Determine ΔW.

$$I_D = 1.9 \ \mu A \qquad W_1/L_1 = 12\mu m/24\mu m \qquad g_{m1} = 1.35 \ \mu S$$

$$I_D = 4.0 \ \mu A \qquad W_2/L_2 = 24\mu m/24\mu m \qquad g_{m2} = 2.82 \ \mu S$$

12. Develop Eq. (1) of Sec. 4.3 assuming that $v_{SB} = 0$ and $\theta = 1$ in Eq. (1) of Sec. 3.4.

13. The following data were taken on an n-channel transistor whose electrical dimensions were $20\mu m/10\mu m$. Plot I_D versus V_{GS} and determine the various regions of operation as evident on the graph. Calculate the mobility μ_o, UCRIT, and UEXP assuming that UTRA $= 0.2$. Also assume that $V_{TO} = 0.5$ volts, $\gamma = 1.68$ volts$^{1/2}$, $2|\phi_F| = 0.725$ volts, and $t_{ox} = 750$ Angstroms. $V_{DS} = 0.1$ volts for all measurements.

V_{GS} (volts)	I_D
0.500	42.57 nA
0.600	320.6 nA
0.700	941.0 nA
1.000	2.98 μA
1.200	4.33 μA
1.400	5.69 μA

V_{GS} (volts)	I_D
1.800	8.39 μA
2.20	11.09 μA
2.50	13.05 μA
2.80	14.85 μA
3.20	17.20 μA
3.60	19.52 μA
4.00	21.82 μA

14. Verify the units of Eqs. (6) and (7) of Sec. 4.3.

15. Rewrite Eq. (1) of Sec 4.4 in terms of input-referred-noise voltage squared.

16. In Example 4.4-1 it was found that the data taken did not follow the $1/f$ characteristic exactly. How could Eq. (2) of Section 4.4 be modified to account for this slope variation?

17. Using the data given below, determine KF and the approximate transition from the $1/f$ dominated region to the thermal noise dominated region. Effective width and length are 40 μm and 10 μm respectively. The oxide thickness is 800 angstroms. The device is in saturation and K_S' is 20 μA/V^2.

Freq. (Hz)	$\bar{v}_N$ (nV)
10	660
30	220
100	65
300	58
1000	54

18. Suppose a test chip is available that has a number of bipolar transistors with each having a different emitter area. Assume that the β of each of these has already been determined. As part of an automatic test procedure, each of the transistors has been exercised by forcing the same emitter current and measuring the resulting v_{BE}. Develop a characterization procedure that can be used to extract J_S.

19. Given the following data on a bipolar device, determine β and J_S. Also determine the temperature at which the data was taken.

I_B (μA)	I_E (μA)	V_{BE}
1.000	102.0	0.797
1.500	148.0	0.810
2.00	203	0.821
2.50	247	0.830

20. Consider a structure similar to that given in Fig. 4.6-1 with the following dimensions: W_n = 15 μm, L_n = 50 μm, W_w = 60 μm, L_w = 300 μm. A current of 10 mA is forced into node A with node F grounded.

The following voltages are measured: $V_n = 0.714$ V, $V_w = 1.017$ V. Determine the sheet resistance R_s and Bias.

21. Given two resistors identical to those shown in Fig. 4.6-2, determine the sheet resistivity R_s and the contact resistance R_c if $R_A = 466$ ohms and $R_B = 916$ ohms.

22. Fig. P4-22 shows a resistor layout scheme designed to be used to determine contact resistance. If $I_1 = 10$ mA and $V_{TEST} = 100$ mV, determine R_c. Compare the accuracy of this technique to measure R_c with those discussed in Sec. 4.6.

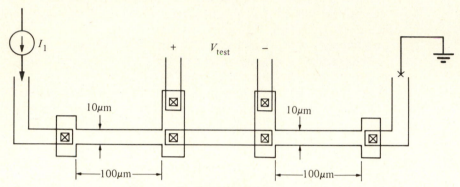

Figure P4-22

23. Consider a structure like that shown in Fig. 4.6-5. If 100 mA is forced into pad 1 with pad 2 grounded, and 500 mV is measured across pads 3 and 4, what is the contact resistance?

24. The following data has been taken on a 150 μm by 150 μm junction capacitor. Assume that the sidewall effects are minimal. Determine the junction grading coefficient MJ, and the junction capacitance C_{j0} given that $PB = 0.7$.

V_j (volts)	$C_{meas}(10^{-12}/\text{F})$
0	3.38
1.000	2.15
2.000	1.70
3.00	1.44
4.00	1.28

25. Assuming that the side view of a diffusion into the background substrate is like that given in Fig. P4-25 where the curve at the edge of the diffusion is described by a circular arc and the junction depth is 1 μm. If the bottom capacitance C_{j0} is 0.35 fF/μm^2, determine the sidewall capacitance C_{jsw}.

26. A serpentine diffusion that is 1 μm wide and has a total periphery of 20,000 μm (total area is approximately 10,000 μm^2) is used to char-

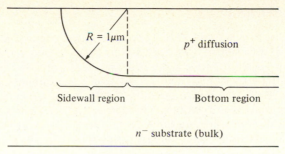

Figure P4-25

acterize sidewall capacitance. The following data has been taken on this device. Assume that $PB = 0.7$ and calculate C_{JSW} and $MJSW$.

V_J (volts)	$CF_{meas}(10^{-12}/F)$
0.000	6.10
1.000	4.65
2.00	4.11
3.00	3.69

Comment about the effects of the bottom junction capacitance on the above results.

REFERENCES

1. D.C. Stone, J.E. Schroeder, et al., "Analog CMOS Building Blocks for Custom and Semicustom Applications," *IEEE Journal of Solid-State Circuits*, Vol. SC-19, (February 1984) pp. 55–61.
2. C. Alcorn, D. Dworak, N. Haddad, W. Henley, and P. Nixon, "Kerf Test Structure Designs for Process and Device Characterization," *Solid State Technology*, Vol. 28, No. 5 (May 1985) pp. 229–235.
3. Petko Vitanov, Ulrich Schwabe, and Iganz Eisele, "Electrical Characterization of Feature Sizes and Parasitic Capacitances Using a Single Test Structure," *IEEE Transactions on Electron Devices*, Vol. ED-31, No. 1 (January 1984) pp. 96–100.
4. A. Vladimirescu and S. Liu, "The simulation of MOS Integrated Circuits using SPICE2," Memorandum No. UCB/ERL M80/7, October 1980, (Electronics Research Laboratory, College of Engineering, University of California, Berkeley, CA 94720).
5. Hiroshi Iwai and Susumu Kohyama, "On-Chip Capacitance Measurement Circuits in VLSI Structures," *IEEE Transactions on Electron Devices*, Vol. ED-29, No. 10 (October 1982) pp. 1622–1626.
6. Morgan J. Thoma and Charles R. Westgate, "A New AC Measurement Technique to Accurately Determine MOSFET Constants," *IEEE Transactions on Electron Devices*, Vol. ED-31, No. 9 (September 1984) pp. 1113–1116.

OTHER REFERENCES

7. Ying-Ren Ma and Kang L. Wang, "A New Method to Electrically Determine Effective MOSFET Channel Width," *IEEE Transactions on Electron Devices*, Vol. ED-29, No. 12 (December 1982) pp. 1825–1827.

8. F.H. De La Moneda and H.N. Kotecha, "Measurement of MOSFET Constants," *IEEE Electron Device Letters*, Vol. EDL-3, No. 1 (January 1982) pp. 10–12.

9. Dezsoe Takacs, Wolfgang Muller, and Ulrich Schwabe, "Electrical Measurement of Feature Sizes in MOS Si^2-Gate VLSI Technology," *IEEE Transactions on Electron Devices*, Vol ED-27, No. 8 (August 1980) pp. 1368–1373.

chapter 5

Analog CMOS Subcircuits

From the viewpoint of Table 1.1-2, the previous three chapters have provided the background for understanding the technology, modeling, and characterization of CMOS devices and components compatible with the CMOS process. The next step toward our objective—methodically developing the subject of CMOS analog-circuit design—is to develop subcircuits. These simple circuits consist of one or more transistors; they are simple; and they generally perform only one function. A subcircuit is typically combined with other simple circuits to generate a more complex circuit function. Consequently, the circuits of this and the next chapter can be considered as building blocks.

The operational amplifier, to be covered in Chapters 8 and 9, is a good example of how simple circuits are combined to perform a complex function. Table 5.0-1 presents a hierarchy showing how an op amp—a complex

Table 5.0-1

Illustration of the Hierarchy of Analog Circuits for an Operational Amplifier.

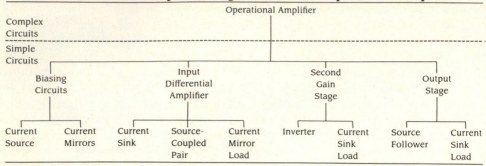

circuit—might be related to various simple circuits. Working our way backward, we note that one of the stages of an op amp is the differential amplifier. The differential amplifier consists of simple circuits that might include a current sink, a current-mirror load, and a source-coupled pair. Another stage of the op amp is a second gain stage, which might consist of an inverter and a current-sink load. If the op amp is to be able to drive a low-impedance load, an output stage is necessary. The output stage might consist of a source follower and a current-sink load. It is also necessary to provide a stabilized bias for each of the previous stages. The biasing stage could consist of a current sink and current mirrors to distribute the bias currents to the other stages.

The subject of basic CMOS analog circuits has been divided into two chapters to avoid one lengthy chapter and yet provide sufficient detail. Chapter 5 covers the simpler subcircuits, including: the MOS switch, active loads, current sinks/sources, current mirrors and current amplifiers, and voltage and current references. Chapter 6 will examine more complex circuits called CMOS amplifiers. That chapter represents a natural extension of the material presented in Chapter 5. Taken together, these two chapters are fundamental for the analog CMOS designer's understanding and capability, as most design will start at this level and progress upward to synthesize the more complex circuits and systems of Table 1.1-2.

5.1   *MOS Switch*

The switch finds many applications in integrated-circuit design. In analog circuits, the switch is used to implement such useful functions as the switched simulation of a resistor [1]. The switch is also useful for multiplexing, modulation, and a number of other applications. The switch is used as a transmission gate in digital circuits and adds a dimension of flexibility not found in standard logic circuits. The objective of this section is to study the characteristics of switches that are compatible with CMOS integrated circuits.

We begin with the characteristics of a voltage-controlled switch. Fig. 5.1-1 shows a model for such a device. The voltage v_c is assumed to control the switch and to determine whether the switch is in the ON or OFF state. The voltage-controlled switch is really a three-terminal network with terminals A and B comprising the switch and terminal C providing the means of applying the control voltage v_c. The most important characteristics of a switch are its ON resistance R_{ON} and its OFF resistance R_{OFF}. Ideally R_{ON} is zero and R_{OFF} is infinite. Also R_{ON} should be linear to avoid harmonic distortion. Most switches have some form of voltage offset which is modelled by V_{OS} of Fig. 5.1-1. V_{OS} represents the small voltage that may exist between terminals A and B when the switch is in the ON state and the current is zero. I_{OFF} represents the leakage current that may flow in the OFF state of the switch. The polarities of the offset sources are not known and have been arbitrarily assigned the directions indicated in Fig. 5.1-1. The parasitic capacitors are an important consideration in the application of analog sampled-data circuits. Capacitors C_A, C_B, and C_{AB} are the parasitic capacitors associated with the switch terminals A and B and ground. Capacitors C_{AC} and C_{BC} are parasitic capacitance that may exist between the voltage-control terminal C and the switch terminals A and B. These capacitors contribute to a problem called feedthrough—where a portion of the control volt-

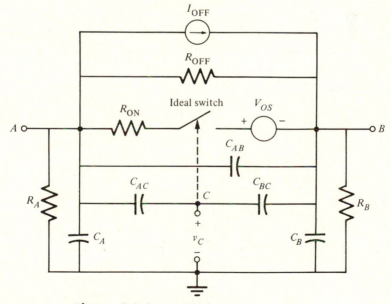

Figure 5.1-1   Model for a nonideal switch.

age appears at the switch terminals A and B. Some other characteristics of switches are not shown in Fig. 5.1-1, one of which is called the commutation rate. The commutation rate of a switch is the time it takes to complete one open-and-close cycle, given in units of cycles per second (Hz). Other switch characteristics of interest are: linearity and maximum analog-signal swing.

One advantage of MOS technology is that it makes a good switch. Figure 5.1-2 shows a MOS transistor that is to be used as a switch. Its performance can be determined by comparing Fig. 5.1-1 with the large-signal model for the MOS transistor. We see that each terminal, A or B, can be either the drain or the source of the MOS transistor. The ON resistance consists of the series combination of r_D, r_S, and whatever channel resistance exists. An expression for the ON resistance can be found as follows. In the ON state of the switch, the voltage across the switch should be small and v_{GS} should be large. Therefore the MOS device is assumed to be in the non-saturation region. Eq. (1) of Sec. 3.1 is used to model this state. If v_{DS} is small, the second-order v_{DS} terms can be neglected to give

$$i_D \cong \frac{\mu C_{ox} W}{L} (v_{GS} - V_T) v_{DS} = \frac{K'W}{L} (v_{GS} - V_T) v_{DS} \tag{1}$$

where v_{DS} is less than $v_{GS} - V_T$ but greater than zero. (v_{GS} becomes v_{GD} if v_{DS} is negative.) Assuming that there is no offset voltage the large-signal channel resistance is

$$R_{ON} = \frac{1}{\partial i_D / \partial v_{DS}} = \frac{L}{\mu_n C_{ox} W (v_{GS} - V_T)} = \frac{L}{K'_n W (v_{GS} - V_T)} \tag{2}$$

Fig. 5.1-3 illustrates Eq. (2) for small values of v_{DS}. When $v_{GS} = V_T$, R_{ON} is infinite. Fig. 5.1-3 is plotted for equal increasing steps of v_{GS} for $W = L = 10 \ \mu m$. Eventually the curves of Fig. 5.1-3 will start to decrease in slope (see Fig. 3.1-2) for increasing v_{DS}. A plot of R_{ON} as a function of v_{GS} is shown in Fig. 5.1-4 for small v_{DS} using the parameters of Table 3.1-2 and for $W/L = 10/10$, $50/10$, $100/10$, and $500/10 \ \mu m/\mu m$. It is seen that a lower value of R_{ON} is achieved for larger values of W/L.

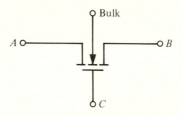

Figure 5.1-2   An NMOS transistor used as a switch.

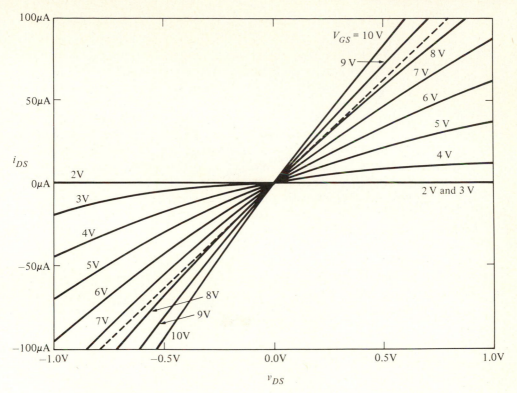

Figure 5.1-3   ON voltage-current characteristics of a MOS switch.

When the switch is OFF, v_{GS} is less than or equal to V_T and the transistor is always in the cutoff region. R_{OFF} is ideally infinite. A typical value is in the range of 10^{12} ohms. Because of this large value, the leakage current is a more important parameter in the OFF state. This leakage current is a combination of the subthreshold current, the surface-leakage current, and the package-leakage current. Typically this leakage current is in the 10 pA range at room temperature and doubles for every 8 °C increase (see Ex. 2.5-1). The resistances designated as R_A and R_B in Fig. 5.1-1 are on the order of 10^{10} ohms due to the surface-leakage effects of the drain or source to the substrate. Unfortunately, R_A and R_B prevent the designer from utilizing the R_{OFF} of 10^{12} ohms.

The offset voltage of the MOS device is zero and does not influence the switch performance. The capacitors C_A, C_B, C_{AC}, and C_{BC} of Fig. 5.1-1 correspond directly to the capacitors C_{BS}, C_{BD}, C_{GS}, and C_{GD} of the MOS transistor (see Fig. 3.2-1). C_{AB} is small for the MOS transistor and is usually negligible. The commutation rate of the MOS switch is determined primarily by the capacitors of Fig. 5.1-1 and the external resistances. Commutation rates for CMOS switches can be as large as 20 MHz.

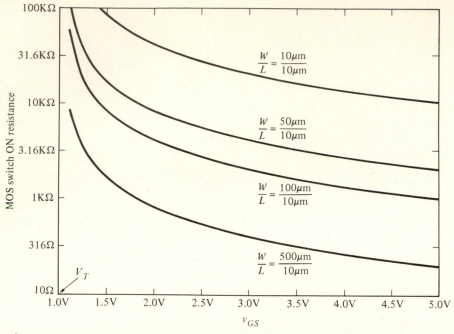

Figure 5.1-4   Illustration of the ON resistance of an NMOS transistor with $W/L = 10\ \mu m/10\ \mu m$, $50\ \mu m/10\ \mu m$, $100\ \mu m/10\ \mu m$, and $500\ \mu m/10\ \mu m$.

One important aspect of the switch is the range of voltages on the switch terminals compared to the control voltage. In the NMOS transistor we see that the gate voltage must be considerably larger than either the drain or source voltage in order to insure that the MOS transistor is ON. (For the PMOS transistor, the gate voltage must be considerably less than either the drain or source voltage.) Typically, the bulk is taken to the most negative potential for the NMOS switch (positive for the PMOS switch). This requirement can be illustrated as follows for the NMOS switch. Suppose that the ON voltage of the gate is the positive power supply V_{DD}. With the bulk to ground this should keep the NMOS switch ON until the signal on the switch terminals (which should be approximately identical at the source and drain) approaches $V_{DD} - V_T$. As the signal approaches $V_{DD} - V_T$ the switch begins to turn OFF. This introduces an undesired nonlinear distortion in the signal. This distortion is typically not serious for $V_{DD} \geq 5$ V and for analog-signal levels of 3 volts or less. For smaller values of V_{DD}, CMOS switches must be used or the control signal that is the clock must have a value greater than V_{DD}. Typical voltages used for an NMOS switch are shown in Fig. 5.1-5 where the switch is connected between the two networks shown.

As an illustration of the influence of the switches on the circuit, con-

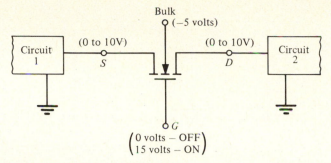

Figure 5.1-5 Application of an NMOS transistor as a switch with typical terminal voltages indicated.

sider the use of a switch to charge a capacitor that is shown in Fig. 5.1-6. M1 is an NMOS transistor used as a switch and ϕ_1 is the control voltage applied to the gate, sometimes called the clock. The ON resistance of the switches can become important during the charge transfer phase of this circuit. For example, when ϕ_1 goes high, M1 connects C_1 to the voltage source v_{in}. The equivalent circuit at this time is shown in Fig. 5.1-7. It can be seen that C_1 will charge to v_{in} with the time constant of $R_{ON}C_1$. For successful operation $R_{ON}C_1 \ll T$ where T is the time ϕ_1 is high.

It is of interest to determine the value of R_{ON} since this will determine the size of M1. Typical values of C_1 are less than 20 pF because the area required to implement larger capacitors would be too large to be practical. If the time ϕ_1 is high is $T = 10 \ \mu s$ and if $C_1 = 20$ pF, then R_{ON} must be less than 0.1 MΩ if charge transfer occurs in 5 time constants. Since small capacitors are used, switches with a large R_{ON} can still perform satisfactorily. As a result the MOS devices used for switching typically use minimum geometries. For a clock of 5 volts, the MOS device of Fig. 5.1-4 with $W = L$ gives R_{ON} of approximately 10 KΩ, which is sufficiently small to transfer the charge in the desired time. The minimum-size switches will also help to reduce parasitic capacitances.

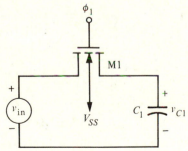

Figure 5.1-6 An application of an MOS switch.

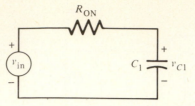

Figure 5.1-7   Model for the ON state of the switch in Fig. 5.1-6.

The OFF state of the switch has little influence upon the performance of the circuit in Fig. 5.1-6 except for the leakage current. Fig. 5.1-8 shows two cases where the leakage current can create serious problems. Fig. 5.1-8a is a sample-and-hold circuit. If C_H is not large enough, then in the hold mode where the MOS switch is OFF the leakage current can charge or discharge C_H a significant amount. Fig. 5.1-8b shows an integrator. The leakage current can cause the circuit to integrate in a continuous mode, which can lead to large values of dc offset unless there is an external feedback path from v_{out} back to the input of the operational amplifier. This dc path may go through other circuits before returning to the input.

One of the most serious limitations of monolithic switches is the clock feedthrough. Fig. 5.1-9 shows all the parasitic capacitors associated with the circuit of Fig. 5.1-6. Since the clock signal is making very large transitions, it can easily couple from gate to source or drain through C_{GS} or C_{GD},

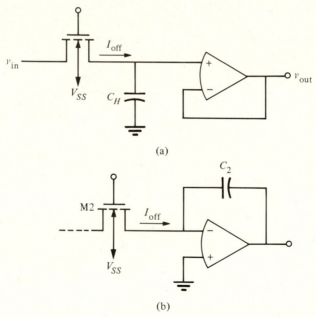

(a)

(b)

Figure 5.1-8   Examples of the influence of I_{OFF}. (a) Sample-and-hold circuit. (b) Integrator.

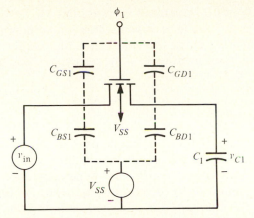

Figure 5.1-9 Illustration of the parasitics associated with the switch of Fig. 5.1-6.

respectively. An example of the effects of clock feedthrough is shown in Fig. 5.1-10, when the ϕ_1 clock is high. The 0.02 pF capacitors represent the C_{GD} and C_{GS} capacitances. Let us assume that the clock varies from 0 to 15 volts. During the rising edge of the ϕ_1 phase, the gate starts at 0 volts and heads toward 15 volts. During the 0 to $v_{in} + V_T$ transition, M1 is off. Consequently, this portion of the clock waveform can couple to C_1 via C_{GD}. The remainder of the rising clock waveform is not coupled to C_1 because M1 turns on and connects C_1 to the low-impedance voltage source v_{in}. As a result,

$$\Delta v_{C1}^+ = \left(\frac{C_{GD}}{C_1 + C_{GD}}\right) \min[(v_{in} + V_T), (v_{C1} + V_T)] \simeq 0.02 v_{in} \tag{3}$$

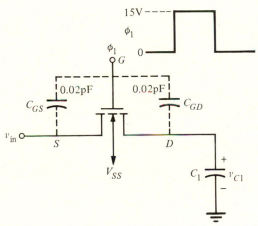

Figure 5.1-10 Illustration of offset caused by clock feedthrough.

is coupled onto C_1 which has been assumed to be 1 pF. However, when M1 turns on, C_1 is charged to v_{in} and it makes no difference what was previously coupled onto C_1. When the clock turns off, feedthrough occurs once more as the clock goes from $v_{in} + V_T$ to zero volts. This feedthrough is given as

$$\Delta v_{C1}^- = -\left(\frac{C_{GD}}{C_1 + C_{GD}}\right)(v_{in} + V_T) \simeq -0.02v_{in} \qquad (4)$$

If v_{in} is 5 volts, then a feedthrough of -100 mV to C_1 occurs during the ϕ_1 phase period. This feedthrough results in an offset, which can be a serious problem. Furthermore, the offset is dependent upon the signal level during the off transition of the ϕ_1 clock. In general, the feedthrough will depend on the switch configuration and the size of the capacitors in the circuit.

It is possible to partially cancel some of the feedthrough effects using the technique illustrated in Fig. 5.1-11. Here a dummy MOS transistor *MD* (with source and drain both attached to the signal line and the gate attached to the inverse clock) is used to apply an opposing clock feedthrough due to M1. The area of *MD* can be designed to provide minimum clock feedthrough. Unfortunately, this method never completely removes the feedthrough and in some cases may worsen it. Also it is necessary to generate an inverted clock, which is applied to the dummy switch. Often the dummy switch is an inverter with the gate attached to the source (or drain) of M1 and the source and drain of the dummy switch connected to the inverse clock. This avoids charge pumping of the substrate which can defeat the purpose of the dummy switch. Clock feedthrough can be reduced by using the largest capacitors possible, using minimum-geometry switches, and keeping the clock swings as small as possible. Typically, these solutions will create problems in other areas, requiring some compromises.

Many of the limitations associated with single-channel MOS switches can be avoided with the CMOS switch shown in Fig. 5.1-12. Using CMOS technology, a switch is usually constructed by paralleling p-channel and n-channel enhancement transistors and thus when ϕ is low both transistors are off, creating an effective open circuit. When ϕ is high both transistors are on, giving a low-impedance state. The ON resistance of the CMOS

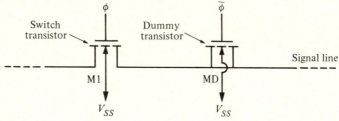

Figure 5.1-11 The use of a dummy transistor to cancel clock feedthrough.

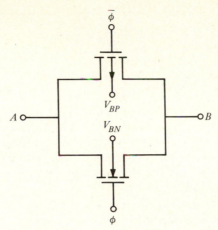

Figure 5.1-12   A CMOS switch.

switch can be lower than 1 KΩ while the OFF leakage current is in the ten picoampere range. The bulk potentials of the p-channel (V_{BP}) and the n-channel (V_{BN}) are taken to the highest and lowest potentials, respectively.

The CMOS switch has two advantages over the single-channel MOS switch. The first advantage is that the dynamic analog-signal range in the ON state is greatly increased. The second is that since the n- and p-channel devices are in parallel and require opposing clock signals, the feedthrough due to the clock will in some cases be diminished through cancellation.

The increased dynamic range of the analog signal can be seen to be a direct result of using complementary devices. When one of the transistors is being turned OFF because of a large analog signal on the drain and source, this same signal will be causing the other transistor to be fully ON. As a consequence, both transistors of Fig. 5.1-12 are ON for analog-signal amplitudes less than the clock magnitude minus V_T and at least one of the transistors is ON for analog-signal amplitudes equal to the magnitude of the clock signal minus V_T.

In some applications, when V_{DD} is low enough and if the clocks cannot swing beyond V_{DD}, the switch will not stay on. This has become a serious problem in the trend of using lower-voltage power supplies for analog CMOS circuits. Depending upon whether the CMOS technology has both p- and n-well tubs or not, several alternatives are available. The effects of V_{DD} on the switch ON resistance can be illustrated with Fig. 5.1-12. Assume that the gate and bulk of the PMOS are taken to ground and V_{DD}, respectively, and the gate and bulk of the NMOS are taken to V_{DD} and ground, respectively. Because the switch is on, a voltage applied to A is effectively applied to B. As this voltage ($V_{A,B}$) is varied from 0 to V_{DD}, the ON resistance of the switch is illustrated by Fig. 5.1-13 when V_{DD} is 5, 4.5, and 4 volts. The peak in the middle of the curves is caused by the variation of the ON resistance of each switch as $V_{A,B}$ varies. At approximately the middle of the $V_{A,B}$ sweep,

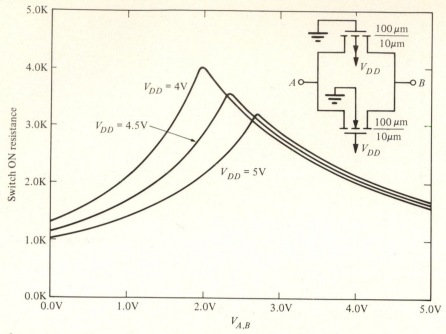

Figure 5.1-13 R_{ON} of Fig. 5.1-12 as a function of the voltage $V_{A,B}$ for various values of V_{DD}.

the increasing NMOS resistance becomes larger than the decreasing PMOS resistance, resulting in a peak. Also, the bulk-bias effect is at its maximum, lowering the thresholds and further increasing the resistance. For a typical CMOS process with $W/L = 100\ \mu m/10\ \mu m$ for both the n- and p-channel transistors, $R_{ON}(max)$ is 4 Kilohms. In general, if $4 < V_{DD} < 5$ volts, one needs a twin-well technology so that both tubs can be switched as will be shown. If $3 < V_{DD} < 4$, it becomes necessary not only to switch tubs but also to generate a gate overdrive (a gate voltage larger than V_{DD}). If V_{DD} is less than 3 volts, switches are not used unless a gate overdrive is used.

A method of generating a gate overdrive in a CMOS technology is shown in Fig. 5.1-14. The operation of this circuit is straightforward. The two capacitors required in this circuit are generally not integrated. To more easily understand the basic principle of the circuit, first ignore the existence of M4, M5, M7, and M8. Consider the other transistors as simple switches. During the phase when ϕ_A is low and ϕ_B is high, C_{pump} is charged with the full supply voltage. When ϕ_A goes high and ϕ_B goes low, M2 turns on along with M6, dumping a portion of the charge in C_{pump} into C_{hold}, causing v_{DBL} to go negative. If both capacitors are equal and the charge is transferred ideally, then after the first pump, v_{DBL} will be equal to $-0.5(V_{DD} - V_{SS})$. After many cycles, v_{DBL} will be at $-V_{DD}$ with respect to V_{SS} (assuming no current drain on the v_{DBL} node).

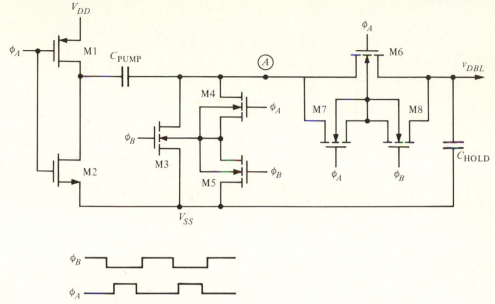

Figure 5.1-14   Voltage doubler used to provide negative gate overdrive.

In order to avoid the real-world problem of turning on substrate diodes, additional transistors are required. Their function is to switch the p-wells of M3 and M6 so that they are always at the most negative supply. For example, when ϕ_B is high and M3 is on, then node A is at V_{SS}. At the same time, v_{DBL} might be at $-V_{DD}$. The p-well of M6 through M8 must be at the most negative potential so M8 is turned on and M7 is turned off. Now consider M3 through M5. In order for M3 to maximize its switching capability, it must have zero back-gate bias. To accomplish this, M5 is turned on. However, when node A goes negative (while ϕ_A is high) M4 must turn on to keep the p-well at the most negative potential.

Another consideration in this implementation is the amount and polarity of charge on C_{hold} at the beginning of the sequence. In order to avoid a potential SCR latch-up problem, v_{DBL} should be at (or near) V_{SS} when the circuit starts its doubling sequence. One must be very careful in the layout of voltage doublers to make sure that any diffusion that has a potential of injecting into the substrate is carefully protected using the guidelines discussed in Sec. 2.5. It is not obvious in Fig. 5.1-14, but the clock signals that drive the switching transistors must derive their negative supply from the v_{DBL} supply being generated. This may seem awkward, but is easily accomplished using standard level translating circuitry.

The feedthrough cancellation of the CMOS switch in Fig. 5.1-12 is not complete for two reasons. One is that the feedthrough capacitances of the n-channel device are not necessarily equal to the feedthrough capacitances

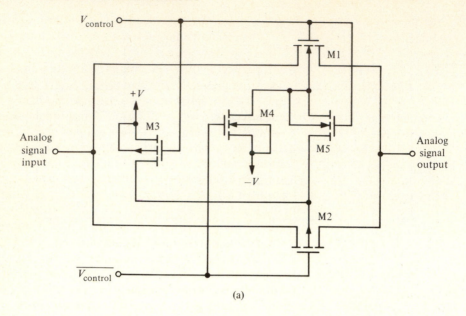

(a)

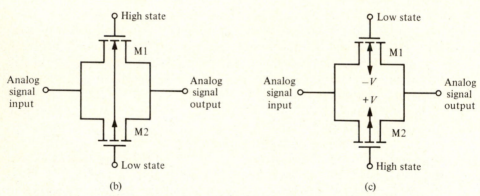

(b)   (c)

Figure 5.1-15 CMOS switch with control circuitry. (a) Circuit. (b) ON state. (c) OFF state.

of the p-channel device. The second reason is that the turn-on delay is not the same for each type of transistor and so the channel conductances do not necessarily track each other during turn-on and turn-off.

CMOS switches are generally used in place of single-channel switches when the technology permits. Although the CMOS switches have larger parasitics than single-channel switches, these parasitics can often be minimized through the use of clever circuit techniques. The clock circuitry for CMOS switches is more complex because of the requirement for a complementary clock. A CMOS switch cell or transmission gate for a twin-tub CMOS technology including the control circuitry is shown in Fig. 5.1-15(a). It is assumed that $V_{control}$ and $\overline{V_{control}}$ are generated from the clock waveform. When $V_{control}$ is HIGH, the transmission gate is ON. The equivalent circuit in this condition is shown in Fig. 5.1-15(b). Here we see that the n-channel gate is taken to the high state and the p-channel gate to the low state. Also, the substrates have been connected together by means of the switches M3, M4, and M5. M3 and M4 are OFF while M5 is ON. This helps to keep the switch ON resistance from being a function of the analog-signal potential. When $V_{control}$ is LOW, the transmission gate is OFF and has the equivalent circuit shown in Fig. 5.1-15(c). The bulks of the n- and p-channel have been taken to $-V$ and $+V$ respectively since M3 and M4 are ON and M5 is OFF. This insures that the OFF state will be maintained, since V_{BS} is strongly reverse biased and causes a large V_T.

In this section we have seen that MOS transistors make one of the best switch realizations available in integrated-circuit form. They require small area, dissipate very little power, and provide reasonable values of R_{ON} and R_{OFF} for most applications. The inclusion of a good realization of a switch into the designer's basic building blocks will produce some interesting and useful circuits and systems which will be studied in the following chapters.

5.2 Active Resistors/Loads

The active resistor is used in place of a polysilicon or diffused resistor to produce a dc voltage drop and/or provide a small signal resistance which is linear over a small range. In many cases the area needed to obtain a small-signal resistance is more important than the linearity. Thus, a small MOS device can simulate the resistance of an equivalent polysilicon or diffused resistor in much less area.

The active resistor is achieved by simply connecting the gate to the drain as shown in Fig. 5.2-1(a) and (b). For the n-channel device the source should be placed at the most negative power supply V_{SS} if possible, so that the body effect is eliminated. Similarly, the source of the p-channel device should be at the most positive power-supply voltage V_{DD}. Since v_{GS} is now v_{DS}, the transconductance curve of Fig. 3.1-4 characterizes the large signal behavior of the active resistor. Fig. 5.2-1(c) illustrates the V-I characteristics of both Fig. 5.2-1(a) or (b). Since the connection of the gate to the drain guarantees operation in the saturation region, the V-I characteristics can be written as

$$i = i_D = \left(\frac{K'W}{2L}\right)[(v_{GS} - V_T)^2] = \frac{\beta}{2}(v_{GS} - V_T)^2 \tag{1}$$

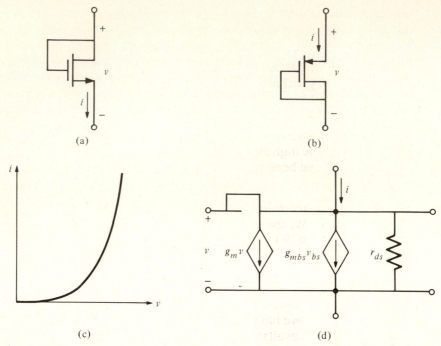

Figure 5.2-1   Active resistor. (a) N-channel. (b) P-channel. (c) *I-V* characteristics. (d) AC model.

or

$$v = v_{GS} = v_{DS} = V_T + \sqrt{2i_D/\beta} \tag{2}$$

where β has been defined in Eq. (8) of Sec. 3.1. If v or i is defined, then the remaining variable can be designed using either Eq. (1) or Eq. (2) and solving for the value of β.

Connecting the gate to the drain means that v_{DS} controls i_D and therefore the channel transconductance becomes a channel conductance. The small-signal model of Fig. 3.3-1 for either the n-channel or p-channel active resistor is shown in Fig. 5.2-1(d). It is easily seen that the small-signal resistance of either of these circuits is

$$r_{out} = \frac{1}{g_m + g_{mbs} + g_{ds}} \cong \frac{1}{g_m} \tag{3}$$

where g_m is greater than g_{mbs} or g_{ds}. Eqs. (6) through (9) of Sec. 3.3 show how this small-signal resistance becomes a function of voltage and current.

An illustration of the application of the active resistor is shown in Fig. 5.2-2 where a voltage V_{OUT} has been derived from V_{DD} and V_{SS}. Both sources are connected so that the body effect has no influence. Since the currents through both transistors must be the same, V_{DS1} can be related to V_{DS2} as follows.

$$V_{DS1} = (\beta_2/\beta_1)^{1/2}[|V_{DS2}| - |V_{T2}|] + V_{T1} \qquad (4)$$

where it is recalled that $v_{GS1} = v_{DS1}$ and $v_{GS2} = v_{DS2}$. A typical situation might be where V_{DD}, V_{SS}, V_{OUT}, and I (the current through the devices) are specified. In that case, only β_1 or β_2 can be found from Eq. (1) and the other from Eq. (4).

Example 5.2-1

Design of a Voltage Divider Using Active Resistors

If $V_{DD} = 5$ V, $V_{SS} = -5$ V, $V_{OUT} = 0$ V, and $I = 8$ microamperes, find the W/L ratios of the transistors of Fig. 5.2-2. Using the model parameters for the MOS transistor of Table 3.1-2, we get $\beta_1 = \beta_2 = 1$ $\mu A/V^2$. This gives a W/L ratio of 1/17 for M1 and 1/8 for M2.

The area available to implement active resistors is often limited, a limitation that influences their application. To illustrate, consider the voltage divider of Ex. 5.2-1. In order to find the active area required for the implementation, set the smaller of W or L to a unit length and make the other some value larger than this unit length to achieve the required W/L ratio.

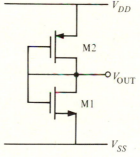

Figure 5.2-2   Voltage division using active resistors.

For example, W_1/L_1 for M1 of Ex. 5.2-1 was 1/17. If we choose W_1 as a unit length, then L_1 is 17 units. Therefore, the active area requires 17 square units. Likewise for M2, set $W_2 = 1$ unit length and $L_2 = 8$ unit lengths giving an active area of 8 square units. The following example will illustrate how increasing the number of devices in an active resistor can reduce the total required area.

Example 5.2-2

Reduction of the Area Required for Active Resistors

Fig. 5.2-2 is to be replaced by Fig. 5.2-3 where M1A and M1B are identical and M2A and M2B are also identical. Neglect the bulk threshold effects on M1A and M2A. Find the W/L ratios of each transistor and the resulting active area to implement Example 5.2-1.

Since 5 volts is to be dropped equally across the upper transistors, each gate-source voltage is 2.5 volts. The gate-source voltage of the lower transistors is also 2.5 volts. Thus, $\beta_1 = \beta_2 = 64/9 \ \mu A/V^2$. This gives a W_1/L_1 of 64/153 and a W_2/L_2 of 8/9. Choosing $W_1 = 1$ gives $L_1 = 153/64$ resulting in an active area of $2 \times (153/64) = 4.781$ squared units. Choosing $W_2 = 1$ gives $L_2 = 9/8$ resulting in an active area of $2 \times (9/8) = 2.25$ square units. The total active area for all four devices is 7.03 square units.

The above example demonstrates the influence of the square of the gate-source voltage upon the current and how this characteristic can be used to reduce the area requirements. Of course, the extra space for contacts has not been considered. This will probably increase the resulting area of Example 5.2-2 but the result will still require less area.

An alternate form of the active resistor of Fig. 5.2-1(a) or (b) can be obtained by connecting the gate of the MOS device to a supply voltage or a bias voltage source rather than to the drain. Fig. 5.2-4 shows a configuration where the transistor's drain and source form the two ends of a "floating" resistor. The region of operation is not guaranteed for all terminal conditions for this type of active resistor. In fact the active floating resistor will act like the switch of the previous section having the *I-V* characteristics given by Fig. 5.1-3. Consequently, the range of resistance values is large but nonlinear. When the transistor is operated in the nonsaturation region, the resistance can be calculated from Eq. (2) of Sec. 5.1, where v_{DS} is assumed small. Note that this equation ignores the effect of λ, so the accuracy decreases near the transition into the saturation region.

If the value of v_{DS} is large enough, the transistor will be in the satura-

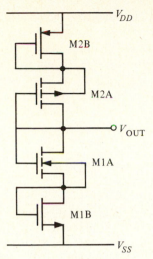

Figure 5.2-3 Implementation of Fig. 5.2-2 that requires less active area.

tion region and very large values of resistance will be obtained. Using the simple model with $|v_{DS}| \geq |v_{GS} - V_T|$, we have from Eq. (9) of Sec. 3.3,

$$r_{ds} = \frac{1 + \lambda V_{DS}}{\lambda I_D} \tag{5}$$

This value corresponds to the slope of the i_{DS} versus v_{DS} curves in the saturation region shown in Fig. 3.1-3 and varies as a function of $|V_{GS} - V_T|$. The resistance of the transistor used as an active floating resistor corresponds to the slope of the curves (in the nonsaturation region) shown in Fig. 3.1-2. It should be obvious that the resistance in most of this region is much less than that in the saturation region.

An important area of operation for an active floating resistor is where v_{DS} is very near, or equal to zero. Many times the active resistor is used in compensation schemes where it is in series with a capacitor. Therefore, no dc current flows and the dc value of v_{DS} is zero. When v_{DS} varies in a small-

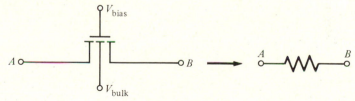

Figure 5.2-4 Floating active resistor using a single MOS transistor.

signal fashion (v_{ds}), the average value of a resistor in such a configuration is

$$r_{ds} = \frac{L}{K'W(V_{GS} - V_T)} \tag{6}$$

Example 5.2-3

Calculation of the Resistance of an Active Resistor

The floating active resistor of Fig. 5.2-4 is to be used to design an 8 Kilohm resistance with the dc value $V_{AB} = 0$. Use the device parameters in Table 3.1-2 and assume the active resistor is an NMOS transistor with $W = L = 10$ "micrometers" or "μm". The drain and source terminals are approximately 0 V and the bulk is -5 V.

Before using Eq. (2) of Sec. 5.1, it is necessary to calculate the new threshold voltage V_T. From Eq. (2) of Sec. 3.1 the new V_T is found to be 3.016 volts. Equating Eq. (6) to 8000 ohms gives a gate-source voltage of 8.016 volts. The slope of the $V_{GS} = 8$ V curve on Fig. 5.1-3 is very close to the dotted line which corresponds to a slope of 1/8000.

The floating active resistor of Fig. 5.2-4 is seen to have a limited range of about ± 0.5 volts for reasonable linearity. In many cases this is not sufficient. A method of increasing the linearity of the floating active resistor is shown in Fig. 5.2-5(a) [2]. If both M1 and M2 are matched and assumed to be nonsaturated, then we may write

$$i_{D1} = \beta_1 \left[(v_{SD} + V_C - |V_T|)v_{SD} - \frac{v_{SD}^2}{2} \right] \tag{7}$$

and

$$i_{D2} = \beta_2 \left[(V_C - V_T)v_{SD} - \frac{v_{SD}^2}{2} \right] \tag{8}$$

Summing both i_{D1} and i_{D2} to get the current i_{AB} gives

$$i_{AB} = i_{D1} + i_{D2} = \beta \left[v_{SD}^2 + (V_C - V_T)v_{SD} - \frac{v_{SD}^2}{2} + (V_C - V_T)v_{SD} - \frac{v_{SD}^2}{2} \right]$$

$$= 2\beta(V_C - |V_T|)v_{AB} \tag{9}$$

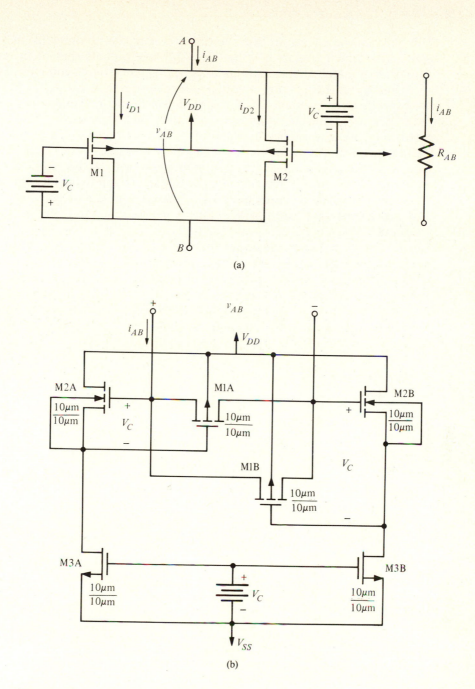

Figure 5.2-5 (a) Extended-range floating active resistor. (b) Practical implementation of Fig. 5.2-5 (a).

The bulk-source effects are the limiting factor in the nonlinearity of Fig. 5.2-4. The principle employed by Fig. 5.2-5 to eliminate this effect is to connect two matched devices in parallel having gate-source voltages of v_{SD} + V_C and V_C. In this manner, the bulk effects introduced by one transistor are cancelled by the other. Fig. 5.2-5(b) shows a practical implementation of Fig. 5.2-5(a). This circuit uses two identical PMOS transistors (M1A and M1B) to create a resistance between points A and B. Fig. 5.2-6 shows the simulation of this implementation using the W/L values indicated and the parameters of Table 3.1-2. These curves show that as the value of V_C increases the linearity of Fig. 5.2-5(b) is extended to approximately ± 1 volt for reasonable linearity. The curves of Fig. 5.2-6 were generated with V_B at 0 V and sweeping V_A from a positive to negative voltage with V_C held constant.

This section has presented realizations for the resistor using active devices resulting in an active resistor. While the area required for such realizations is typically less than that for the resistor, it was observed that the linearity of most active resistors is limited to small voltages. The major sources of nonlinearity are the failure to remain in the nonsaturated region and the influence of the bulk-source voltage on the I-V characteristics.

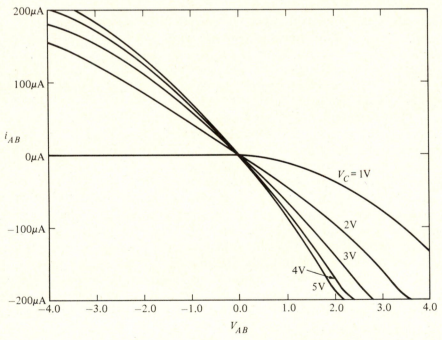

Figure 5.2-6 V-I characteristics of an extended-range active resistor.

5.3   *Current Sinks and Sources*

A current sink and current source are two terminal components whose current at any instant of time is independent of the voltage across their terminals. The current of a current sink or source flows from the positive node, through the sink or source to the negative node. A current sink typically has the negative node at V_{SS} and the current source has the positive node at V_{DD}. Fig. 5.3-1 (a) shows the MOS implementation of a current sink. The gate is taken to whatever voltage necessary to create the desired value of current. The voltage divider of Fig. 5.2-2 can be used to provide this voltage. We note that in the nonsaturation region the MOS device is not a good current source. In fact the voltage across the current sink must be larger than V_{MIN} in order for the current sink to perform properly. For Fig. 5.3-1(a) this means that

$$v_{OUT} \geqq V_{GG} - V_{T0} - V_{SS} \tag{1}$$

If the gate-source voltage is held constant, then the large-signal characteristics of the MOS transistor are given by the output characteristics of Fig. 3.1-3. An example is shown in Fig. 5.3-1 (b). If the source and bulk are both connected to V_{SS}, then the small-signal output resistance is given by (see Eq. (9) of Sec. 3.3)

$$r_{\text{out}} = \frac{1 + \lambda V_{DS}}{\lambda I_D} \simeq \frac{1}{\lambda I_D} \tag{2}$$

If the source and bulk are not connected to the same potential, the characteristics will not change as long as V_{BS} is a constant. This is an advantage

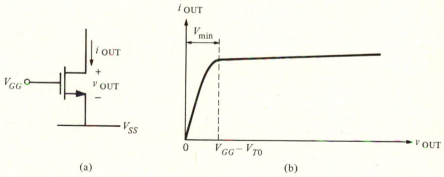

| (a) | (b) |

Figure 5.3-1   (a) Current sink. (b) Current-voltage characteristics of (a).

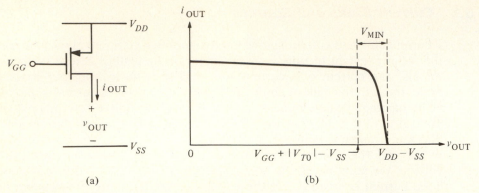

(a)   (b)

Figure 5.3-2   (a) Current source. (b) Current-voltage characteristics of (a).

of CMOS over NMOS or PMOS because the current can always be taken from the drain while the source can be kept at a constant potential.

Fig. 5.3-2 (a) shows an implementation of a current source. Again, the gate is taken to a constant potential as is the source. With the value of V_{GS} held constant, the V-I characteristics of the current source are found from Fig. 3.1-3. With the definition of v_{OUT} and i_{OUT} of the source as shown in Fig. 5.3-2(a), the large-signal V-I characteristic is shown in Fig. 5.3-2(b). The small-signal output resistance of the current source is given by Eq. (2). The source-drain voltage must be larger than V_{MIN} for this current source to work properly. This current source only works for values of v_{OUT} given by

$$v_{OUT} \leqq V_{GG} + |V_{T0}| - V_{SS} \tag{3}$$

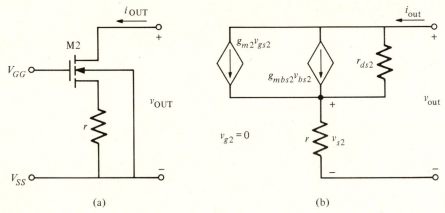

(a)   (b)

Figure 5.3-3   (a) Technique for increasing the output resistance of a resistor r. (b) Small-signal model for the circuit in (a).

The advantage of the current sink and source of Figs. 5.3-1(a) and 5.3-2(a) is their simplicity. However, there are two areas in which their performance may need to be improved for certain applications. One improvement is to increase the small-signal output resistance—resulting in a more constant current over the range of v_{OUT} values. The second is to reduce the value of V_{MIN}, thus allowing a larger range of v_{OUT} over which the current sink/source works properly. We shall illustrate methods to improve both areas of performance. First, the small-signal output resistance can be increased using the principle illustrated in Fig. 5.3-3(a). This principle uses the common-gate configuration to multiply the source resistance r by approximately the voltage gain of the common-gate configuration with an infinite load resistance. The exact small-signal output resistance r_{out} can be calculated from the small-signal model of Fig. 5.3-3(b) as

$$r_{out} = \frac{v_{out}}{i_{out}} = r + r_{ds2} + [(g_{m2} + g_{mbs2})r_{ds2}]r \cong (g_{m2}r_{ds2})r \qquad (4)$$

where $g_{m2}r_{ds2} \gg 1$ and $g_{m2} > g_{mbs2}$.

The above principle is implemented in Fig. 5.3-4(a) where the output resistance (r_{ds1}) of the current sink of Fig. 5.3-1(a) should be increased by approximately the common-gate voltage gain of M2. To verify the principle, the small-signal output resistance of the cascode current sink of Fig. 5.3-4(a) will be calculated using the model of Fig. 5.3-4(b). Since $v_{gs2} = -v_1$ and $v_{gs1} = 0$, summing the currents at the output node gives

$$i_{out} + g_{m2}v_1 + g_{mbs2}v_1 = g_{ds2}(v_{out} - v_1) \qquad (5)$$

Since $v_1 = i_{out}r_{ds1}$, we can solve for r_{out} as

$$\begin{aligned} r_{out} &= \frac{v_{out}}{i_{out}} = r_{ds2}(1 + g_{m2}r_{ds1} + g_{mbs2}r_{ds1} + g_{ds2}r_{ds1}) \\ &= r_{ds1} + r_{ds2} + g_{m2}r_{ds1}r_{ds2}(1 + \eta_2) \end{aligned} \qquad (6)$$

Typically, $g_{m2}r_{ds2}$ is greater than unity so that Eq. (6) simplifies to

$$r_{out} \cong (g_{m2}r_{ds2})r_{ds1} \qquad (7)$$

We see that the small-signal output resistance of the current sink of Fig. 5.3-4(a) is increased by the factor of $g_{m2}r_{ds2}$.

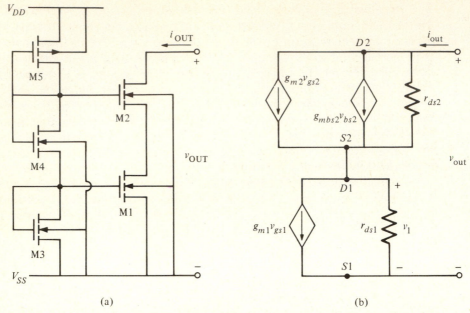

Figure 5.3-4 (a) Circuit for increasing r_{out} of a current sink. (b) Small-signal model for (a).

Example 5.3-1

Calculation of Output Resistance for a Current Sink

Use the model parameters of Table 3.1-2 to calculate: (a) the W/L ratio and the small-signal output resistance for the simple current sink of Fig. 5.3-1(a) if $I_{OUT} = 100\ \mu A$ and $V_{GG1} = -2.5$ V and $V_{SS} = -5$ V; and (b) the small-signal output resistance if the simple current sink of (a) is inserted into the cascode current-sink configuration of Fig. 5.3-4(a). Assume that $W_1/L_1 = W_2/L_2$ and that $V_{GG2} = 0$ V.

(a) Since V_{GS1} is 2.5 volts and we assume that M1 is in saturation, the W_1/L_1 ratio is 5.23. Choosing $L_1 = 10\ \mu m$ gives $W_1 = 52.3\ \mu m$. Using $\lambda = 0.01$ and $I_{OUT} = 100\ \mu A$ gives a small-signal output resistance of 1 MΩ. (b) Assume that if the W/L ratios of M1 and M2 are equal and the currents are equal, then $V_{GS1} \cong V_{GS2}$. This allows one to make an estimate for g_{mbs2} from Eq. (8) of Sec. 3.3 as $0.363 g_{m2}$. Eq. (6) of Sec. 3.3 gives $g_{m1} = g_{m2} = 133.33\ \mu A/V$. Substituting these values into Eq. (6) gives the small-signal output resistance of the cascode current sink as 183 MΩ.

The other performance limitation of the simple current sink/source was the fact that the constant output current could not be obtained for all values of v_{OUT}. This was illustrated in Figs. 5.3-1(b) and 5.3-2(b) by the voltage V_{MIN}. While this problem may not be serious in the simple current sink/source, it becomes more severe in the cascode current–sink/source configuration that was used to increase the small-signal output resistance. It therefore becomes necessary to examine methods of reducing the value of V_{MIN}. Obviously, V_{MIN} can be reduced by increasing the value of W/L and adjusting the gate-source voltage to get the same output current. However, another method which works well for the cascode current–sink/source configuration will be presented [3].

We must introduce an important principle used in biasing MOS devices before showing the method of reducing V_{MIN} of the cascode current sink/source. This principle can be best illustrated by considering two MOS devices, M1 and M2. Assume that the applied dc gate-source voltage V_{GS} can be divided into two parts, given as

$$V_{GS} = \Delta V + V_T \tag{8}$$

where ΔV is that part of V_{GS} which is in excess of the threshold voltage, V_T. This definition allows us to express the minimum value of v_{DS} for which the device will remain in saturation as

$$v_{DS}(\text{sat}) = V_{GS} - V_T = \Delta V \tag{9}$$

Thus, ΔV can be thought of as the minimum drain-source voltage for which the device remains saturated. In saturation, the drain current can be written as

$$i_D = \frac{K'W}{2L} (\Delta V)^2 \tag{10}$$

The principle to be illustrated is based upon Eq. (10). For example, if the currents of two MOS devices are equal (because they are in series), then the following relationship holds.

$$\frac{K'_1 W_1}{L_1} (\Delta V_1)^2 = \frac{K'_2 W_2}{L_2} (\Delta V_2)^2 \tag{11}$$

If both MOS transistors are of the same type, then Eq. (11) reduces to

$$\frac{W_1}{L_1} (\Delta V_1)^2 = \frac{W_2}{L_2} (\Delta V_2)^2 \tag{12}$$

It should be obvious that the W/L ratios can be used to control the values of ΔV or that desired values of ΔV will define the W/L ratios.

The principle above can also be used to define a relationship between the current and W/L ratios. If the gate-source voltages of two similar MOS devices are equal (because they are physically connected), then the ΔV_1 is equal to ΔV_2. From Eq. (10) we can write

$$i_{D1}\left(\frac{W_2}{L_2}\right) = i_{D2}\left(\frac{W_1}{L_1}\right) \tag{13}$$

Eq. (13) is useful even though the gate-source terminals of M1 and M2 may not be physically connected because voltages can be identical without being physically connected as will be seen in later material. Eqs. (12) and (13) represent a very important principle that will be used not only in the material immediately following but throughout this text to determine biasing relationships.

Consider the cascode current sink of Fig. 5.3-5(a) where I_{REF} represents the current provided by M5 in Fig. 5.3-4(a). Our objective is to use the above principle to reduce the value of V_{MIN} $[=V_{OUT}(sat)]$. If we ignore the bulk effects on M2 and M4 and assume that M1, M2, M3, and M4 are all matched with identical W/L ratios, then the gate-source voltage of each transistor can be expressed as $V_T + \Delta V$ as shown in Fig. 5.3-5(a). At the gate of M2 we see that the voltage with respect to the lower power supply is $2V_T + 2 \Delta V$. In order to maintain current–sink/source operation, it will be assumed that M1 and M2 must have at least a voltage of ΔV as given in Eq. (9). In order to find V_{MIN} $[=V_{OUT}(sat)]$ of Fig. 5.3-5(a) we can rewrite Eq.

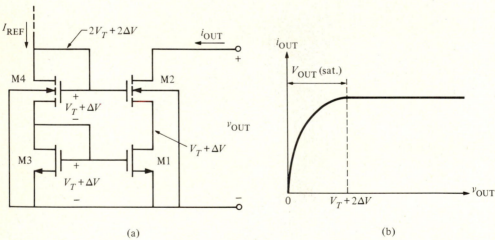

(a)   (b)

Figure 5.3-5   (a) Standard cascode current sink. (b) Output characteristics of (a).

(10) of Sec. 3.1 as

$$v_D \geq v_G - V_T \tag{14}$$

Since $V_{G2} = 2V_T + 2\,\Delta V$, substituting this value into Eq. (14) gives

$$V_{D2}(\text{min}) = V_{\text{MIN}} = V_T + 2\,\Delta V \tag{15}$$

The current-voltage characteristics of Fig. 5.3-5(a) are illustrated in Fig. 5.3-5(b) where the value of V_{MIN} of Eq. (15) is shown.

V_{MIN} of Eq. (15) is dropped across both M1 and M2. The drop across M2 is ΔV while the drop across M1 is $V_T + \Delta V$. From the results of Eq. (9), this implies that V_{MIN} of Fig. 5.3-5 could be reduced by V_T and still keep both M1 and M2 in saturation. Fig. 5.3-6(a) shows how this can be accomplished. All devices are matched and we will ignore the bulk effects on M2, M4, and M6. The W/L ratio of M6 is made ¼ of the identical W/L ratios of M1 through M5. This causes the gate-source voltage across M6 to be $V_T + 2\,\Delta V$ rather than $V_T + \Delta V$. Consequently, the voltage at the gate of M2 is now $V_T + 2\,\Delta V$. Substituting this value into Eq. (14) gives

$$V_{D2}(\text{min}) = V_{\text{MIN}} = 2\,\Delta V \tag{16}$$

The resulting current-voltage relationship is shown in Fig. 5.3-6(b). It can be seen that a voltage of $2\,\Delta V$ is across both M1 and M2 giving the lowest value of V_{MIN} and still keeping both M1 and M2 in saturation. Using this approach and increasing the W/L ratios will result in minimum values of V_{MIN}.

Example 5.3-2

Designing the Cascode Current Sink for a Given V_{MIN}

Use the cascode current-sink configuration of Fig. 5.3-6(a) to design a current sink of 100 µA and a V_{MIN} of 1 V. Assume the device parameters of Table 3.1-2. With V_{MIN} of 1 V, choose $\Delta V = 0.5$ V. Using the saturation model, the W/L ratio of M1 through M5 can be found from

$$\frac{W}{L} = \frac{2 i_{\text{OUT}}}{K' \Delta V^2} = \frac{2 \cdot 100 \times 10^{-6}}{17 \times 10^{-6} \cdot 0.25} = 47.06$$

The W/L ratio of M6 will be ¼ this value or 11.76. To make a more exact calculation, one should include the effects of the bulk voltage on the threshold voltages of M2, M4, and M6.

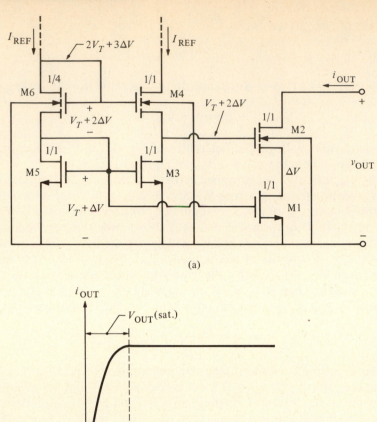

(a)

(b)

Figure 5.3-6 (a) A circuit that reduces the value of V_{OUT} (sat.) of Fig. 5.3-5 (a). Note that $W_4/L_4 = 1/4$ creates twice ΔV of a device with $W/L = 1$. (b) Output characteristics of (a).

The above technique will be useful in maximizing the voltage-signal swings of cascode configurations to be studied later. This section has presented implementations of the current sink/source and has shown how to boost the output resistance of a MOS device. A very important principle that will be used in biasing was based on relationships between the excess gate-source voltage ΔV, the drain current, and the W/L ratios of MOS devices. This principle was applied to reduce the voltage V_{MIN} of the cascode current source.

5.4 *Current Mirrors and Current Amplifiers*

A very useful building block in CMOS analog-circuit design is called the current mirror. This building block has been well developed for bipolar IC technology [4]. The current mirror uses the principle that if the gate-source potential of two identical MOS transistors are equal, the channel currents should be equal. Fig. 5.4-1(a) shows the implementation of a simple n-channel current mirror. The current i_I is assumed to be defined by a current source or some other means and i_O is the output or "mirrored" current. M1 is in saturation because $v_{DS1} = v_{GS1}$. Assuming that $v_{DS2} \geq v_{GS} - V_{T2}$ is greater than V_{T2} allows us to use the equations in the saturation region of the MOS transistor. In the most general case, the ratio of i_O to i_I is

$$\frac{i_O}{i_I} = \left(\frac{L_1 W_2}{W_1 L_2}\right)\left(\frac{V_{GS} - V_{T2}}{V_{GS} - V_{T1}}\right)^2\left(\frac{1 + \lambda v_{DS2}}{1 + \lambda v_{DS1}}\right)\left(\frac{\mu_{o2} C_{ox2}}{\mu_{o1} C_{ox1}}\right) \tag{1}$$

Normally, the components of a current mirror are processed on the same integrated circuit and thus all of the physical parameters such as V_T, μ_o, C_{ox}, etc., are identical for both devices. As a result, Eq. (1) simplifies to

$$\frac{i_O}{i_I} = \left(\frac{L_1 W_2}{W_1 L_2}\right)\left(\frac{1 + \lambda v_{DS2}}{1 + \lambda v_{DS1}}\right) \tag{2}$$

If $v_{DS2} = v_{DS1}$ (not always a good assumption), then the ratio of i_O/i_I becomes

$$\frac{i_O}{i_I} = \left(\frac{L_1 W_2}{W_1 L_2}\right) \tag{3}$$

Consequently, i_O/i_I is a function of the aspect ratios that are under the control of the designer.

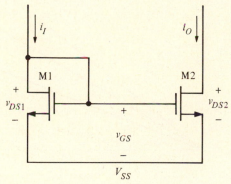

Figure 5.4-1   N-Channel current mirror.

There are three effects that cause the current mirror to be different than the ideal situation of Eq. (3). These effects are: (1) channel-length modulation, (2) threshold offset between the two transistors, and (3) imperfect geometrical matching. Each of these effects will be analyzed separately.

Consider the channel-length modulation effect. Assuming all other aspects of the transistor are ideal and the aspect ratios of the two transistors are both unity, then Eq. (2) simplifies to

$$\frac{i_O}{i_I} = \frac{1 + \lambda v_{DS2}}{1 + \lambda v_{DS1}} \tag{4}$$

with the assumption that λ is the same for both transistors. This equation shows that differences in drain-source voltages of the two transistors can cause a deviation for the ideal unity current gain or current mirroring. Fig. 5.4-2 shows a plot of current ratio error versus $v_{DS2} - v_{DS1}$ for different values of λ with both transistors in the saturation region. Two important facts should be recognized from this plot. The first is that significant ratio error can exist when the mirror transistors do not have the same drain-source voltage and secondly, for a given difference in drain-source voltages, the ratio of the mirror current to the reference current improves as λ becomes smaller (output resistance becomes larger). Thus, a good current mirror or current amplifier should have identical drain-source voltages and a high output resistance.

The second nonideal effect is that of offset between the threshold voltage of the two transistors. For clean silicon-gate CMOS processes, the threshold offset is typically less than 10 mV for transistors that are identical and in close proximity to one another. Consider two transistors in a mirror

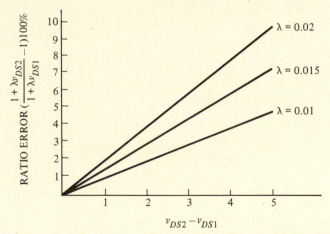

Figure 5.4-2 Plot of ratio error (in %) versus drain voltage difference for the current mirror of Fig. 5.4-1. For this plot, $V_{DS1} = 2.0$ Volts.

configuration where both have the same drain-source voltage and all other aspects of the transistors are identical except V_T. In this case, Eq. (1) simplifies to

$$\frac{i_O}{i_I} = \left(\frac{v_{GS} - V_{T2}}{v_{GS} - V_{T1}}\right)^2 \qquad (5)$$

Fig. 5.4-3 shows a plot of the ratio error versus ΔV_T where $\Delta V_T = V_{T1} - V_{T2}$. It is obvious from this graph that better current-mirror performance is obtained at higher currents, because v_{GS} is higher for higher currents and thus ΔV_T becomes a smaller percentage of v_{GS}.

It is also possible that the transconductance gain K' of the current mirror is also mismatched (due to oxide gradients). A quantitative analysis approach to variations in both K' and V_T is now given. Let us assume that the W/L ratios of the two mirror devices are exactly equal but that K' and V_T may be mismatched. Eq. (5) can be rewritten as

$$\frac{i_O}{i_I} = \frac{K'_2(v_{GS} - V_{T2})^2}{K'_1(v_{GS} - V_{T1})^2} \qquad (6)$$

where $v_{GS1} = v_{GS2} = v_{GS}$. Defining $\Delta K' = K'_2 - K'_1$ and $K' = 0.5(K'_2 + K'_1)$ and $\Delta V_T = V_{T2} - V_{T1}$ and $V_T = 0.5(V_{T2} + V_{T1})$ gives

$$K'_1 = K' - 0.5\Delta K' \qquad (7)$$
$$K'_2 = K' + 0.5\Delta K' \qquad (8)$$

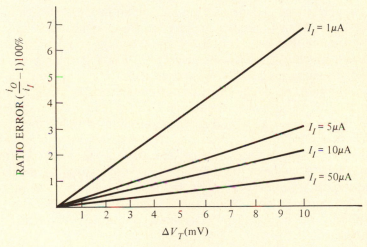

Figure 5.4-3 Plot of ratio error versus offset voltage for the current mirror of Fig. 5.4-1. $V_{T1} = 1.0$ V, $K'(W/L) = 24$ $\mu A/V^2$.

$$V_{T1} = V_T - 0.5\Delta V_T \tag{9}$$
$$V_{T2} = V_T + 0.5\Delta V_T \tag{10}$$

Substituting Eqs. (7) through (10) into Eq. (6) gives

$$\frac{i_O}{i_I} = \frac{(K' + 0.5\Delta K')(v_{GS} - V_T - 0.5\Delta V_T)^2}{(K' - 0.5\Delta K')(v_{GS} - V_T + 0.5\Delta V_T)^2} \tag{11}$$

Factoring out K' and $(v_{GS} - V_T)$ gives

$$\frac{I_O}{i_I} = \frac{\left[1 + \dfrac{\Delta K'}{2K}\right]\left[1 - \dfrac{\Delta V_T}{2(v_{GS} - V_T)}\right]^2}{\left[1 - \dfrac{\Delta K'}{2K}\right]\left[1 + \dfrac{\Delta V_T}{2(v_{GS} - V_T)}\right]^2} \tag{12}$$

Assuming that the quantities in Eq. (12) following the "1" are small, Eq. (12) can be approximated as

$$\frac{i_O}{i_I} \cong \left[1 + \frac{\Delta K'}{2K'}\right]\left[1 + \frac{\Delta K'}{2K'}\right]\left[1 - \frac{\Delta V_T}{2(v_{GS} - V_T)}\right]^2\left[1 - \frac{\Delta V_T}{2(v_{GS} - V_T)}\right]^2 \tag{13}$$

Retaining only first order products gives

$$\frac{i_O}{i_I} \cong 1 + \frac{\Delta K'}{K'} - \frac{2\Delta V_T}{(v_{GS} - V_T)} \tag{14}$$

If the percentage change of K' and V_T are known, Eq. (14) can be used on a worst-case basis to predict the error in the current-mirror gain. For example, assume that $\Delta K'/K' = \pm 5\%$ and $\Delta V_T/(v_{GS} - V_T) = \pm 10\%$. Then the current-mirror gain would be given as $i_O/i_I \cong 1 \pm 0.05 \pm (-0.20)$ or $1 \pm (-0.15)$ amounting to a 15% error in gain.

The third nonideal effect of current mirrors is the error in the aspect ratio of the two devices. We saw in Chapter 3 that there are differences in the drawn values of W and L. These are due to mask, photolithographic, etch, and out-diffusion variations. These variations can be different even for two transistors placed side by side. One way to avoid the effects of these variations is to make the dimensions of the transistors much larger than the typical variation one might see. For transistors of identical size with W and L greater than 10 μm, the errors due to geometrical mismatch will generally be insignificant compared to offset-voltage and v_{DS}-induced errors.

In some applications, the current mirror is used to multiply current and function as a current amplifier. In this case, the aspect ratio of the multiplier transistor (M2) is much greater than the aspect ratio of the reference transistor (M1). To obtain the best performance, the geometrical aspects must be considered. An example will illustrate this concept.

Example 5.4-1

Aspect Ratio Errors in Current Amplifiers

Fig. 5.4-4 shows the layout of a one-to-five current amplifier. Assume that the lengths are identical ($L_1 = L_2$) and find the ratio error if $W_1 = 10 \pm 0.2\ \mu m$. The actual widths of the two transistors are

$$W_1 = 10 \pm 0.2\ \mu m$$

and

$$W_2 = 50 \pm 0.2\ \mu m$$

We note that the tolerance is not multiplied by the nominal gain factor of 5. The ratio of W_2 to W_1 and consequently the gain of the current amplifier is

$$\frac{i_O}{i_I} = \frac{W_2}{W_1} = \frac{50 \pm 0.2}{10 \pm 0.2} = 5 \pm 0.08 \qquad \textbf{(15)}$$

where we have assumed that the variations would both have the same sign. It is seen that this ratio error is almost 2% of the desired current ratio or gain.

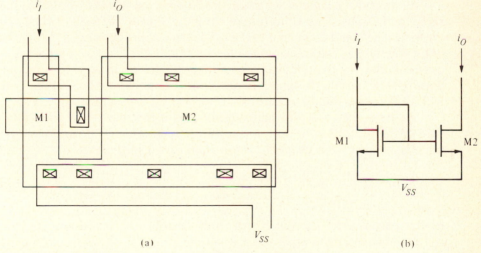

Figure 5.4-4   Layout of current mirror without ΔW correction techniques.

The error noted above would be valid if every other aspect of the transistor were matched perfectly. A solution to this problem can be achieved by using proper layout techniques. The correct one-to-five ratio should be implemented using five duplicates of the transistor M1. In this way, the tolerance on W_2 is multiplied by the nominal current gain. Let us reconsider the above example using this approach.

Example 5.4-2

Reduction of the Aspect Ratio Error in Current Amplifiers

Use the layout technique illustrated in Fig. 5.4-5 and calculate the ratio error of a current amplifier having the specifications of the previous example.

The actual widths of M1 and M2 are

$$W_1 = 10 \pm 0.2 \ \mu\text{m}$$

and

$$W_2 = 5(10 \pm 0.2) \ \mu\text{m}$$

The ratio of W_2 to W_1 and consequently the current gain is seen to be

$$\frac{i_o}{i_I} = \frac{5(10 \pm 0.2)}{10 \pm 0.2} = 5 \tag{16}$$

In the above examples we made the assumption that ΔW should be the same for all transistors. Unfortunately this is not true, but the ΔW matching errors will be small compared to the other error contributions. If the widths of two transistors are equal but the lengths differ, the scaling approach discussed above for the width is also applicable to the length. Usually one does not try to scale the length because the tolerances are greater than the width tolerances due to out diffusion under the polysilicon gate.

We have seen that the small-signal output resistance is a good measure of the perfection of the current mirror or amplifier. The output resistance of the simple n-channel mirror of Fig. 5.4-1 is given as

$$r_{\text{out}} = \frac{1}{g_{ds}} \simeq \frac{1}{\lambda I_D} \tag{17}$$

Higher-performance current mirrors will attempt to increase the value of r_{out}. Eq. (17) will be the point of comparison.

Up to this point we have discussed aspects of and improvements on

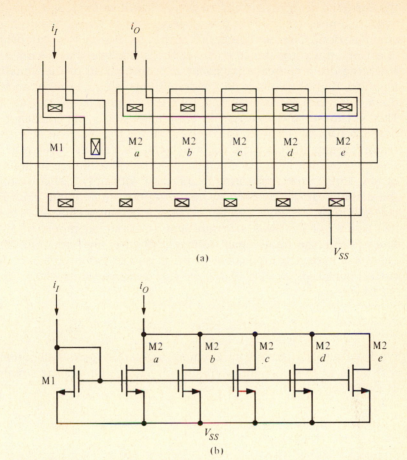

(a)

(b)

Figure 5.4-5 Layout of current mirror with ΔW correction techniques.

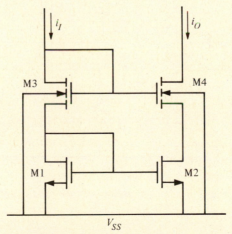

Figure 5.4-6 Cascode current mirror.

233

the current mirror or current amplifier shown in Fig. 5.4-1, but there are ways of improving current-mirror performance by modifying the circuit configuration. One configuration, shown in Fig. 5.4-6, desensitizes the effect of output resistance. (This effect is the one that becomes evident when the drain-source voltages are different.) If all of the transistors are identical, then the voltage on the drain of M1 is equal to that on M2. If the drain voltage of M4 is increased, then the current in M4 tries to increase. This will increase the current in M2 (decreasing v_{GS4}). Transistor M4 will begin to turn off in order to compensate for the increase in current. The result is a small decrease in v_{GS4} which causes a small increase in v_{DS2}. The resulting change in v_{DS2} is much smaller than the change in drain voltage of M4. Thus current mirroring is achieved with only a slight error due to the output-resistance effect. By analyzing the small-signal equivalent circuit, the improvement in output resistance becomes clear.

Figure 5.4-7 shows an equivalent small-signal model of Fig. 5.4-6. Since $i_i = 0$, the small-signal voltages v_1 and v_3 are both zero. Therefore, Fig. 5.4-7 is exactly equivalent to the circuit of Example 5.3-1. Using the correct subscripts for Fig. 5.4-7, we can use the results of Eq. (6) of Sec. 5.3 to write

$$r_{out} = r_{ds2} + r_{ds4} + g_{m4}r_{ds2}r_{ds4}(1 + \eta_4) \qquad (18)$$

We have already seen from Example 5.3-1 that the small-signal output resistance of this configuration is much larger than for the simple mirror of Eq. (17).

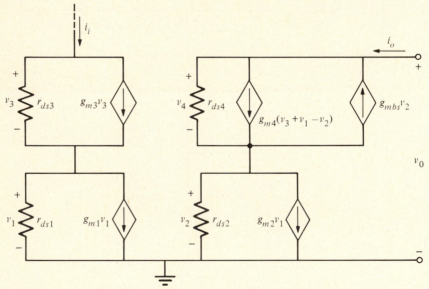

Figure 5.4-7   Small-signal model of the current mirror of Fig. 5.4-6.

Another current mirror is shown in Fig. 5.4-8. This circuit is an n-channel implementation of the well-known Wilson current mirror [5]. The output resistance of the Wilson current mirror is increased through the use of negative, current feedback. If i_o increases, then the current through M2 also increases. However, the mirroring action of M1 and M2 causes the current in M1 to increase. If i_i is constant and if we assume there is some resistance from the gate of M3 (drain of M1) to ground, then the gate voltage of M3 is decreased if the current i_o increases. The loop gain is essentially the product of g_{m1} and the small signal resistance seen from the drain of M1 to ground.

It can be shown that the small-signal output resistance of the Wilson current source of Fig. 5.4-8 is

$$r_{\text{out}} = r_{ds3} + r_{ds2} \left[\frac{1 + r_{ds3}g_{m3}(1 + \eta_3) + g_{m1}r_{ds1}g_{m3}r_{ds3}}{1 + g_{m2}r_{ds2}} \right] \tag{19}$$

The output resistance of Fig. 5.4-8 is seen to be comparable with that of Fig. 5.4-6.

Unfortunately, the behavior described above for the current mirrors or amplifier requires a nonzero voltage at the input and output before it is achieved. Consider the cascode current mirror of Fig. 5.4-6 from a large-signal viewpoint. This voltage at the input, designated as $V_I(\text{min})$, can be shown to depend upon the value of i_I as follows. Since $v_{DG} = 0$ for both M1 and M3, these devices are always in saturation. Therefore we may express $V_I(\text{min})$ as

$$V_I(\text{min}) = \left(\frac{2i_I}{K'} \right)^{1/2} \left[\left(\frac{L_1}{W_1} \right)^{1/2} + \left(\frac{L_3}{W_3} \right)^{1/2} \right] + (V_{T1} + V_{T3}) \tag{20}$$

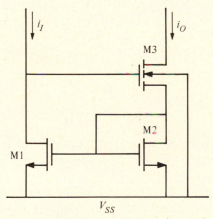

Figure 5.4-8   N-channel version of the Wilson current mirror.

It is seen that for a given i, the only way to decrease $V_i(\min)$ is to increase the W/L ratios of both M1 and M3. One must also remember that V_{T3} will be larger due to the back gate bias on M3. The techniques used to reduce V_{MIN} of the cascode current–sink/source in Sec. 5.3 are not applicable here.

We are also interested in the voltage, $V_{OUT}(\text{sat})$, where M4 makes the transition from the nonsaturated region to the saturated region. This voltage can be found from the relationship

$$V_{DS4} \gtrsim (V_{GS4} - V_{T4}) \tag{21}$$

or

$$V_{D4} \gtrsim V_{G4} - V_{T4} \tag{22}$$

which is when M4 is on the threshold between the two regions. Equation (22) can be used to obtain the value of $V_{OUT}(\text{sat})$ as

$$
\begin{aligned}
V_{OUT}(\text{sat}) &= V_I - V_{T4} \\
&= \left(\frac{2I_I}{K'}\right)^{1/2}\left[\left(\frac{L_1}{W_1}\right)^{1/2} + \left(\frac{L_3}{W_3}\right)^{1/2}\right] + (V_{T1} + V_{T3} - V_{T4})
\end{aligned} \tag{23}
$$

For voltages above $V_{OUT}(\text{sat})$, the transistor M4 is in saturation and the output resistance should be that calculated in Eq. (18). Since the value of voltage across M2 is greater than necessary for saturation, the technique used to decrease V_{MIN} in Sec. 5.3 can be used to decrease $V_{OUT}(\text{sat})$. Unfortunately, the value of $V_i(\min)$ will be increased.

Similar relationships can be developed for the Wilson current mirror or amplifier. If M3 is saturated, then $V_i(\min)$ is expressed as

$$V_i(\min) = \left(\frac{2I_O}{K'}\right)^{1/2}\left[\left(\frac{L_2}{W_2}\right)^{1/2} + \left(\frac{L_3}{W_3}\right)^{1/2}\right] + (V_{T2} + V_{T3}) \tag{24}$$

For M3 to be saturated, v_{OUT} must be greater than $V_{OUT}(\text{sat})$ given as

$$V_{OUT}(\text{sat}) = V_I - V_{T3} = \left(\frac{2I_O}{K'}\right)^{1/2}\left[\left(\frac{L_2}{W_2}\right)^{1/2} + \left(\frac{L_3}{W_3}\right)^{1/2}\right] + V_{T2} \tag{25}$$

It is seen that both of these circuits require at least $2V_T$ across the input before they behave as described above. Larger W/L ratios will decrease $V_i(\min)$ and $V_{OUT}(\text{sat})$.

Experimental results for each of the three types of current mirrors or amplifiers discussed above are presented in Figs. 5.4-9 through 5.4-11. The vertical scale of each of these photographs is i_o and the horizontal scale is

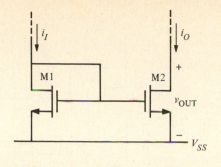

(a)

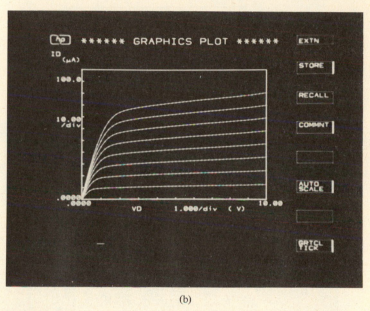

(b)

Figure 5.4-9   (a) Simple current mirror. (b) Experimental results of (a) with $W_2/L_2 = W_1/L_1 = 7.5 \ \mu m/7.5 \ \mu m$. Vertical scale is 10 μA/div. and horizontal scale is 1 volt/div. $V_{BS} = 0$ and step size is 10 μA/step.

v_O. The curves are parameterized with equal and increasing values of i_I. For ideal behavior, the curve should be flat and have a value of I_O equal to the applied value of I_I. The improvement of the cascode mirror and the Wilson mirror are clearly indicated in these photographs.

Each of the current mirrors discussed above can be implemented using p-channel devices. The circuits perform in an identical manner and exhibit the same small-signal output resistance. The use of n-channel and p-channel current mirrors will be useful in dc biasing of CMOS circuits.

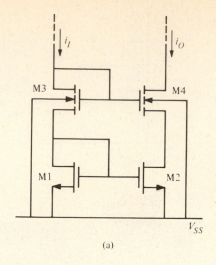

(a)

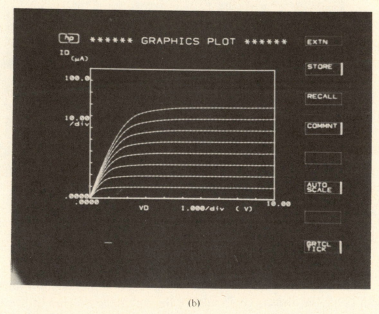

(b)

Figure 5.4-10 (a) An improved current mirror using n-channel MOS transistors. (b) Input-output characteristics of (a) with $V_{BS} = 0$ volts and $W_1/L_1 = W_2/L^2 = 12.5\ \mu\text{m}/12.5\ \mu\text{m}$ and $W_3/L_3 = W_4/L_4 = 7.5\ \mu\text{m}/12.5\ \mu\text{m}$. Vertical scale is 20 μA/div. and the horizontal scale is 1 volt/div. Step size is 10 μA/step.

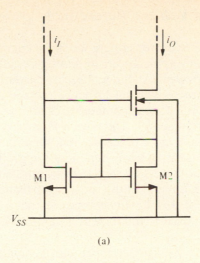

(a)

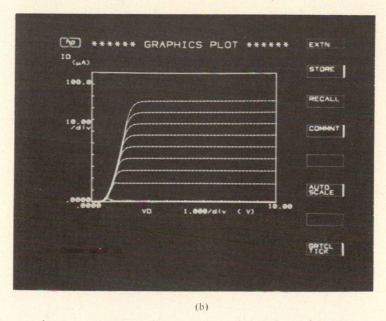

(b)

Figure 5.4-11   (a) Wilson current mirror using n-channel MOS transistors. (b) Experimental characteristics of (a) with $V_{BS} = 0$ volts, and $W_1/L_1 = W_2/L_2 = 25$ μm/7.5 μm and $W_3/L_3 = 125$ μm/12.5 μm. Vertical scale is 10 μA/div. Horizontal scale is 1 volt/div. Step size is 10 μA/step.

239

5.5 *Current and Voltage References*

An ideal voltage or current reference is independent of power supply and temperature. Many applications in analog circuits require such a building block, which provides a stable voltage or current. The large-signal voltage and current characteristics of an ideal voltage and current reference are shown in Fig. 5.5-1. These characteristics are identical to those of the ideal voltage and current source. The term *reference* is used when the voltage or current values have more precision and stability than ordinarily found in a source. A reference is typically dependent upon the load connected to it. It will always be possible to use a buffer amplifier to isolate the reference from the load and maintain the high performance of the reference. In the discussion that follows, it will be assumed that a high-performance voltage reference can be used to implement a high-performance current reference and vice versa.

A very crude voltage reference can be made from a voltage divider between the power supplies. Passive or active components can be used as the divider elements. Fig. 5.5-2(a) and (b) shows an example of each. Unfortunately, the value of V_{REF} is directly proportional to the power supply. Let us quantify this relationship by introducing the concept of *sensitivity* S. The sensitivity of V_{REF} of Fig. 5.5-2(a) to V_{DD} can be expressed as

$$S_{V_{DD}}^{V_{REF}} = \frac{(\partial V_{REF}/V_{REF})}{(\partial V_{DD}/V_{DD})} = \frac{V_{DD}}{V_{REF}}\left(\frac{\partial V_{REF}}{\partial V_{DD}}\right) \tag{1}$$

Eq. (1) can be interpreted as: a 10% change in V_{DD} will result in a 10% change in V_{REF} (which is undesirable for a voltage reference). It may also be shown that the sensitivity of V_{REF} of Fig. 5.5-2(b) with respect to V_{DD} is unity (see Problem 5.24).

A simple way of obtaining a voltage reference is to use an active device

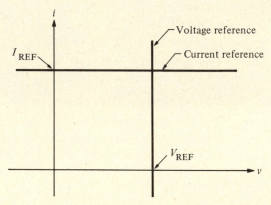

Figure 5.5-1   *V-I* characteristics of ideal voltage and current references.

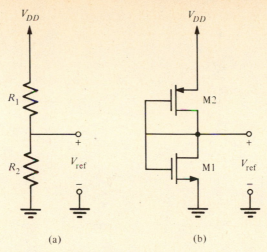

(a)   (b)

Figure 5.5-2   Voltage references using voltage division. (a) Resistor implementation. (b) Active device implementation.

as shown in Fig. 5.5-3(a) and (b). In Fig. 5.5-3(a), the substrate BJT has been connected to power supply through a resistance R. The voltage across the pn junction is given as

$$V_{REF} = V_{EB} = \frac{kT}{q} \ln \left(\frac{I}{I_s} \right) \tag{2}$$

where I_s is the junction-saturation current defined in Eq. (4) of Sec. 2.5. If V_{DD} is much greater than V_{EB}, then the current I is given as

$$I = \frac{V_{DD} - V_{EB}}{R} \simeq \frac{V_{DD}}{R} \tag{3}$$

Thus the reference voltage of this circuit is given as

$$V_{REF} \cong \frac{kT}{q} \ln \left(\frac{V_{DD}}{RI_s} \right) \tag{4}$$

The sensitivity of V_{REF} of Fig. 5.5-3(a) to V_{DD} is shown to be

$$S_{V_{DD}}^{V_{REF}} = \frac{1}{\ln [V_{DD}/(RI_s)]} = \frac{1}{\ln (I/I_s)} \tag{5}$$

Interestingly enough, since I is normally greater than I_s, the sensitivity of V_{REF} of Fig. 5.5-3(a) is less than unity. For example, if $I = 1$ mA and $I_s = 10^{-15}$ amperes, then Eq. (5) becomes 0.0362. Thus, a 10% change in V_{DD}

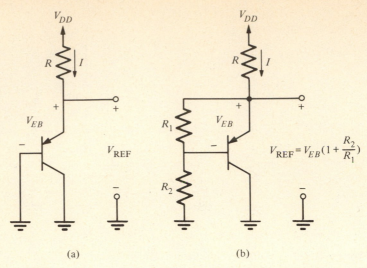

(a)   (b)

Figure 5.5-3   (a) PN junction voltage reference. (b) Increasing V_{REF} of (a).

creates only a 0.362% change in V_{REF}. Fig. 5.5-3(b) shows a method of increasing the value of V_{REF} in Fig. 5.5-3(a). The reference voltage of Fig. 5.5-3(b) can be written as

$$V_{REF} \cong V_{EB} \left(\frac{R_1 + R_2}{R_1} \right) \tag{6}$$

In order to find the value of V_{EB}, it is necessary to assume that the transistor beta is large and/or the resistance $R_1 + R_2$ is large. The larger V_{REF} becomes in Fig. 5.5-3(b), the more the current I becomes a function of V_{REF} and eventually an iterative solution is necessary.

The BJT of Fig. 5.5-3(a) may be replaced with an MOS enhancement device to achieve a voltage which is less dependent on V_{DD} than Fig. 5.5-2(a). V_{REF} can be found from Eq. (2) of Sec. 5.2, which gives V_{GS} as

$$V_{GS} = V_T + \sqrt{2I/\beta} \tag{7}$$

Assuming that V_{DD} is much greater than V_{GS} gives

$$V_{REF} = V_{GS} = V_T + \left(\frac{2(V_{DD} - V_{REF})}{R\beta} \right)^{1/2} \cong V_T + \left(\frac{2V_{DD}}{R\beta} \right)^{1/2} \tag{8}$$

If V_{DD} is not greater than V_{GS}, then an iterative solution can be used to solve for V_{REF}. If $V_{DD} = 10$ volts, $W/L = 10$, and R is 100 kilohms, the values of Table 3.1-2 give a reference voltage of 2.085 volts (assuming V_{DD} is greater than V_{REF}). An iterative solution gives 1.972 volts for V_{REF}. The sensitivity of Fig. 5.5-4(a) can be found as

$$\underset{V_{DD}}{\overset{V_{REF}}{S}} = \frac{0.5}{1 + V_T\sqrt{(R\beta)/2V_{DD}}} \tag{9}$$

Using the previous values gives a sensitivity of V_{REF} to V_{DD} of 0.260. This sensitivity is not as good as the BJT because the logarithmic function is much less sensitive to its argument than the square root. The value of V_{REF} of Fig. 5.5-4(a) can be increased using the technique employed for the BJT reference of Fig. 5.5-3(b), with the result shown in Fig. 5.5-4(b), where the reference voltage is given as

$$V_{REF} = V_{GS}\left(1 + \frac{R_2}{R_1}\right) \tag{10}$$

In the types of voltage references illustrated in Fig. 5.5-3 and Fig. 5.5-4, the designer can use geometry to adjust the value of V_{REF}. In the BJT reference the geometric-dependent parameter is I_s and for the MOS reference it is W/L. The small-signal output resistance of these references is a measure of how dependent the reference will be on the load (see Problem 5.28).

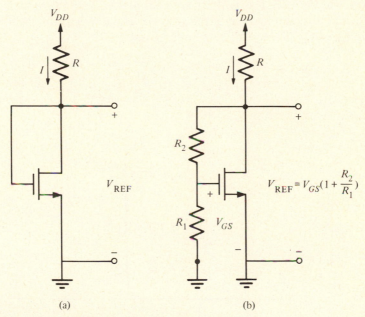

(a)   (b)

Figure 5.5-4   (a) MOS equivalent of the pn junction voltage reference. (b) Amplification of V_{REF} of (a).

A voltage reference can be implemented using the breakdown phenomenon that occurs in the reverse-bias condition of a heavily-doped pn junction discussed in Sec. 2.2. The symbol and current-voltage characteristics of the breakdown diode are shown in Fig. 5.5-5. The breakdown in the reverse direction (v and i are defined for reverse bias in Fig. 5.5-5) occurs at a voltage BV. BV falls in the range of 6 to 8 volts, depending on the doping concentrations of the n^+ and p^+ regions. The knee of the curve depends upon the material parameters and should be very sharp. The small-signal output resistance in the breakdown region is low, typically 30 to 100 ohms, which makes an excellent voltage reference or voltage source. The temperature coefficient of the breakdown diode will vary with the value of breakdown voltage BV as seen in Fig. 5.5-6. Breakdown by the Zener mechanism has a negative temperature coefficient while the avalanche breakdown has a positive temperature coefficient. The breakdown voltage for typical CMOS technologies is around 6.5 to 7.5 volts which gives a temperature coefficient around +3 mV/°C.

The breakdown diode can be used as a voltage reference by simply connecting it in series with a voltage-dropping element (resistor or active device) to V_{DD} or V_{SS} as illustrated in Fig. 5.5-7(a). The dotted load line on Fig. 5.5-5 illustrates the operation of the breakdown-diode voltage reference. If V_{DD} or R should vary, little change in BV will result because of the steepness of the curve in the breakdown region. The sensitivity of the breakdown-diode voltage reference can easily be found by replacing the circuit in Fig. 5.5-7(a) with its small-signal equivalent model. The resistor r_z is equal to the inverse of the slope of Fig. 5.5-5 at the point Q. The sensitivity of V_{REF} to V_{DD} can be expressed as

$$S_{V_{DD}}^{V_{REF}} = \left(\frac{\partial V_{REF}}{\partial V_{DD}}\right)\left(\frac{V_{DD}}{V_{REF}}\right) \simeq \left(\frac{v_{ref}}{v_{dd}}\right)\left(\frac{V_{DD}}{BV}\right) = \left(\frac{r_z}{r_z + R}\right)\left(\frac{V_{DD}}{BV}\right) \qquad (11)$$

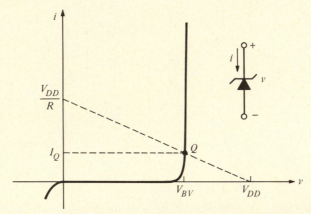

Figure 5.5-5   Current-voltage characteristics of a breakdown diode.

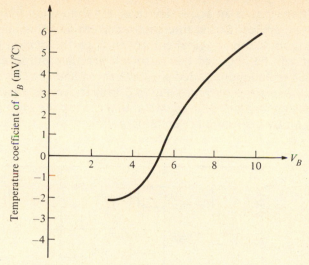

Figure 5.5-6   Variation of the temperature coefficient of the breakdown diode as a function of the breakdown voltage BV [5]. (By permission from John Wiley & Sons, Inc.)

Assume that V_{DD} = 10 volts, BV = 6.5 volts, r_z = 100 ohms, and R = 35 kilohms. Eq. (11) gives the sensitivity of this breakdown-diode voltage reference as 0.0044. Thus a 10% change in V_{DD} would cause only a 0.044% change in V_{REF}. Other configurations of a voltage reference that uses the breakdown diode are considered in the problems.

We have noted in Fig. 5.5-3(a) and Fig. 5.5-4(a) that the sensitivity of

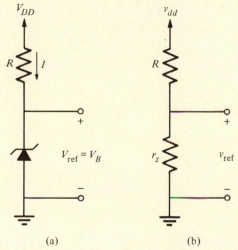

Figure 5.5-7   (a) Breakdown diode voltage reference. (b) Small-signal model of (a).

the voltage across an active device is less than unity. If the voltage across the active device is used to create a current and this current is somehow used to provide the original current through the device, then a current or voltage will be obtained that is for all practical purposes independent of V_{DD}. This technique is called a V_T *referenced source*. This technique is also called a *bootstrap reference*. Fig. 5.5-8(a) shows an example of this technique using all MOS devices. M3 and M4 cause the currents I_1 and I_2 to be equal. I_1 flows through M1 creating a voltage V_{GS1}. I_2 flows through R cre-

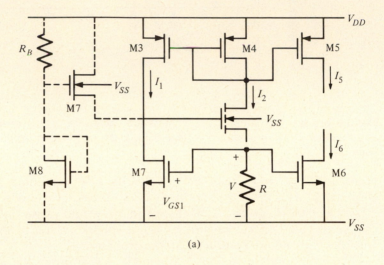

(a)

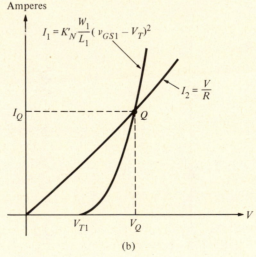

(b)

Figure 5.5-8   (a) Threshold-referenced circuit. (b) Establishment of equilibrium in (a).

ating a voltage I_2R. Because these two voltages are connected together, an equilibrium point is established. Fig. 5.5-8(b) illustrates how the equilibrium point is achieved. On this curve, I_1 and I_2 are plotted as a function of V. The intersection of these curves defines the equilibrium point indicated by Q. The equation describing this equilibrium point is given as

$$I_2R = V_{T1} + \left(\frac{2I_1L_1}{K'_NW_1}\right)^{1/2} \tag{12}$$

This equation can be solved iteratively for $I_1 = I_2 = I_Q$ or alternately, one can assume that V_{GS1} is approximately equal to V_{T1} so that

$$I_Q = I_2 = \frac{V_{T1}}{R} \tag{13}$$

Since I_1 or I_2 does not change as a function of V_{DD}, the sensitivity of I_Q to V_{DD} is essentially zero. A voltage reference can be achieved by mirroring I_2 ($= I_Q$) through M5 or M6 and using a resistor.

   Unfortunately, there are two possible equilibrium points on Fig. 5.5-8(b). One is at Q and the other is at the origin. In order to prevent the circuit from choosing the wrong equilibrium point, a start-up circuit is necessary. The dotted circuit in Fig. 5.5-8(a) functions as a start-up circuit. If the circuit is at the undesired equilibrium point, then I_1 and I_2 are zero. However, M7 will provide a current in M1 that will cause the circuit to move to the equilibrium point at Q. As the circuit approaches the point Q, the source voltage of M7 increases causing the current through M7 to decrease. At Q the current through M1 is essentially the current through M3.

   An alternate version of Fig. 5.5-8(a) that uses V_{BE} to reference the voltage or current is shown in Fig. 5.5-9. It can be shown that the equilibrium point is defined by the relationship

$$I_2R = V_{BE1} = V_T \ln\left(\frac{I_1}{I_s}\right) \tag{14}$$

This reference circuit also has two equilibrium points and a start-up circuit similar to Fig. 5.5-8(a) is necessary. The reference circuits in Fig. 5.5-8(a) and Fig. 5.5-9 represent a very good method of implementing power supply independent references. Either circuit can be operated in the weak-threshold inversion in order to develop a low-power, low-supply voltage reference.

   Unfortunately, supply-independent references are not necessarily temperature independent because the pn junction and gate-source voltage drops are temperature dependent as noted in Sec. 2.5. The concept of *fractional temperature coefficient* (TC_F), defined in Eq. (7) of Sec. 2.5 will be

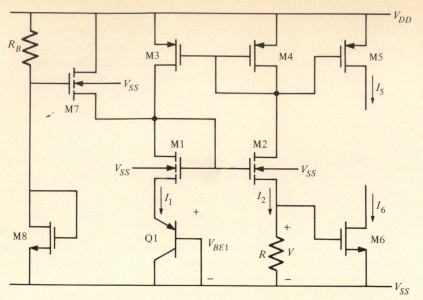

Figure 5.5-9   Base-emitter voltage-referenced circuit.

used to characterize the temperature dependence of voltage and current references. We see that TC_F is related to the sensitivity as defined in Eq. (1)

$$TC_F = \frac{1}{T}\left(\overset{x}{\underset{T}{S}}\right) \tag{15}$$

Let us now consider the temperature characteristics of the simple pn junction of Fig. 5.5-3(a). If we assume that V_{DD} is much greater than V_{REF}, then Eq. (4) describes the reference voltage. Although V_{DD} is independent of temperature, R is not and must be considered. The fractional temperature coefficient of this voltage reference can be expressed using the results of Eq. (17) of Sec. 2.5 as

$$TC_F = \frac{1}{V_{REF}}\frac{dV_{REF}}{dT} \cong \frac{V_{REF} - V_{GO}}{V_{REF}T} - \frac{kT}{V_{REF}q}\left(\frac{dR}{R\ dT}\right) \tag{16}$$

if $v_E = V_{REF}$. Assuming a V_{REF} of 0.6 volts, room temperature, and that R is a polysilicon resistor (see Table 2.4-1), the TC_F of the simple pn voltage reference is -3426 ppm/°C.

Fig. 5.5.4(a) is the MOS equivalent of the simple pn junction voltage reference. The temperature dependence of V_{REF} of this circuit can be written as

$$\frac{dV_{REF}}{dT} = \frac{-\alpha + \sqrt{\dfrac{V_{DD} - V_{REF}}{2\beta R}\left(\dfrac{1.5}{T} - \dfrac{1}{R}\dfrac{dR}{dT}\right)}}{1 + \dfrac{1}{\sqrt{2\beta R}\,(V_{DD} - V_{REF})}} \tag{17}$$

This expression can be divided by V_{REF} to get the TC_F of Fig. 5.5-4(a). If $W = L$, then for $V_{DD} = 5$ volts, $R = 200$ kilohms, and for the parameters of Table 3.1-2 an iterative solution gives $V_{REF} = 2.268$ volts. If R is a polysilicon resistor, then the fractional temperature coefficient of V_{REF} is -29.2 ppm/°C. The improvement over the TC_F of Fig. 5.5-3(a) comes about because V_{REF} is larger and the temperature dependence of the mobility cancels with the temperature dependence of V_T and R. Unfortunately, the TC_F of this example is not realistic because the values of α and the TC_F of the resistor do not have the implied accuracy.

The temperature characteristics of the breakdown diode were illustrated in Fig. 5.5-6. Typically, the temperature coefficient of the breakdown diode is positive. If the breakdown diode can be suitably combined with a negative temperature coefficient, then the possibility of temperature independence exists. Unfortunately, the temperature coefficient depends upon the processing parameters and cannot be well defined, so this approach is not attractive.

The threshold-referenced circuit of Fig. 5.5-8(a) has its current I_2 given by Eq. (13) if V_{GS1} is approximately equal to V_{T1}. Therefore the TC_F of the threshold-referenced circuit can be written as

$$TC_F = \frac{1}{V_T}\frac{dV_T}{dT} - \frac{1}{R}\frac{dR}{dT} = \frac{-\alpha}{V_T} - \frac{1}{R}\frac{dR}{dT} \tag{18}$$

Assuming that R is a polysilicon resistor, the fractional temperature coefficient is -800 ppm/°C. The influence of the TC_F of the resistors on the temperature performance of the above references should be noted. For example, if R was an implanted resistor with a TC_F of $+400$ ppm/°C, the TC_F of the reference circuit would be -1900 ppm/°C. In some cases, combinations of resistors having different TC_Fs can achieve a lower TC_F of the voltage reference.

The temperature behavior of the base-emitter–referenced circuit of Fig. 5.5-9 is identical to that of the threshold-referenced circuit of Fig. 5.5-8(a). Eq. (14) showed that I_2 is equal to V_{BE1} divided by R. Thus, Eq. (18) above expresses the TC_F of this reference if V_T is replaced by V_{BE} as follows.

$$TC_F = \frac{1}{V_{BE}} \frac{dV_{BE}}{dT} - \frac{1}{R} \frac{dR}{dT} \tag{19}$$

Assuming V_{BE} of 0.6 volts gives a TC_F −2333 ppm/°C.

A good reference circuit that is temperature independent should have a fractional temperature coefficient of ±100 ppm/°C or smaller. So far, the references developed for supply independence have not met this criterion. One of the problems is that the temperature dependences of the active devices and resistors have not been compatible. Consider a simple CMOS voltage reference made up of two devices connected in the configuration shown in Fig. 5.5-10. The reference voltage V_{REF} with respect to ground can be expressed as

$$V_{REF} = \frac{V_{SS} + V_{TN} + \sqrt{\beta_2/\beta_1}\,(V_{DD} - |V_{TP}|)}{1 + \sqrt{\beta_2/\beta_1}} \tag{20}$$

If we assume that the temperature dependence of β_1/β_2 is negligible, then the temperature dependence is due to V_{TP} and V_{TN}. Differentiating Eq. (20) with respect to temperature results in

$$\frac{dV_{REF}}{dT} = \frac{1}{1 + \sqrt{\beta_2/\beta_1}}\left[\frac{dV_{TN}}{dT} - \left(\frac{\beta_2}{\beta_1}\right)^{1/2}\frac{d|V_{TP}|}{dT}\right]$$

$$= \frac{1}{1 + \sqrt{\beta_2/\beta_1}}\left[-\alpha_N + \left(\frac{\beta_2}{\beta_1}\right)^{1/2}\alpha_P\right] \tag{21}$$

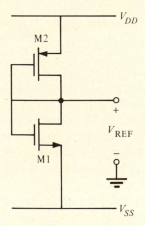

Figure 5.5-10   A simple CMOS voltage reference.

For zero temperature dependence, the following condition must be true.

$$\alpha_N = \left(\frac{\beta_2}{\beta_1}\right)^{1/2} \alpha_P \tag{22}$$

Thus V_{REF} becomes,

$$V_{REF} = \frac{V_{SS} + V_{TN} + (\alpha_N/\alpha_P)(V_{DD} - |V_{TP}|)}{1 + (\alpha_N/\alpha_P)} \tag{23}$$

Unfortunately, the voltage V_{REF} is not a free variable under the conditions of Eq. (22). Normally α_N/α_P is close to unity so that the reference voltage would be approximately half of the rail-to-rail power supply plus half of the difference between the magnitude of the threshold voltages. It should be noted that this voltage reference is not independent of power supplies.

The voltage and current references presented in this section have the objective of providing a stable value of current with respect to changes in power supply and temperature. It was seen that while power-supply independence could thus be obtained, satisfactory temperature performance could not. References that offer better temperature performance will be considered in Chapter 11.

5.6 *Summary*

This chapter has introduced CMOS subcircuits, including the switch, active resistors, current sinks/sources, current mirrors or amplifiers, and voltage and current references. The general principles of each circuit were covered as was their large-signal and small-signal performance. Remember that the circuits presented in this chapter are rarely used by themselves, rather they are joined with other such circuits to implement a desired analog function.

The approach used in each case was to present a general understanding of the circuit and how it works. This presentation was followed by analysis of large-signal performance, typically a voltage-transfer function or a voltage-current characteristic. Limitations such as signal swing or nonlinearity were identified and characterized. This was followed by the analysis of small-signal performance. The important parameters of small-signal performance include ac resistance, voltage gain, and bandwidth.

The subject matter presented in this chapter will be continued and extended in the next chapter. A good understanding of the circuits in this and the next chapter will provide a firm foundation for the later chapters and subject material.

PROBLEMS — *Chapter 5*

1. What is the ON resistance of an enhancement, n-channel switch using the values $V_{TO} = 1$ volt and $K' = 2.5 \times 10^{-5}$ amperes/volt^2 if $V_S = 0$ volts, $W/L = 1$ and if (a) $V_G = 10$ volts and (b) $V_G = 0$ volts? Assume that V_D is greater than or equal to V_S.

2. An n-channel, enhancement MOS switch is connected as shown in Fig. P5.2. If the gate is taken from 0 volts to 10 volts for 10 microseconds and $v_{c1} = 5V$, what is the W/L value necessary to discharge C_1 to within 5% of its initial charge? The parameters of Table 3.1-2 are valid for this device. Assume that v_{DS} is approximately zero. If C_2 was initially discharged, what will be the value of the voltage at the output of the op amp at the end of the 10 microseconds?

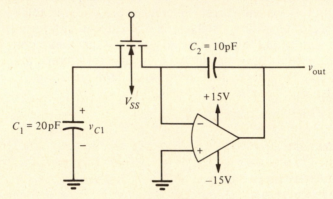

Figure P5.2

3. Calculate the feedthrough of the clock in Fig. 5.1-10 if (a) C_1 is 5 pF and $C_{GS} = C_{GD} = 0.02$ pF and (b) if C_1 is 1 pF and $C_{GS} = C_{GD} = 0.05$ pF.

4. If the range of voltages across C_1 of Fig. P5.2 is 0 to +5 volts, what is the minimum value of clock (gate) voltage that will insure the switch is ON and the maximum value of clock voltage that will insure the switch is OFF? Assume the MOS device has the parameters of Table 3.1-2.

5. Repeat the previous problem if the range of voltages across C_1 is -5 to 0 volts.

6. Explain why Fig. 5.1-13 is offset from the center of the V_{AB} axis.

7. Use the SPICE2 computer program to obtain the curves of Fig. 5.1-13.

8. Describe two advantages of a CMOS switch as compared to a single-channel MOS switch. List two disadvantages.

9.   If $V_{DD} = -V_{SS} = 5$ V, $I = 75$ μA, $W_1/L_1 = 14.8$, and $W_2/L_2 = 0.124$ of Fig .5.2-2, find β_1, β_2, V_{DS1}, V_{DS2}, and V_{OUT}.

10. Assume the gate of M1 in Fig. 5.2-2 is connected to a dc voltage V_{GG}. Design the W/L ratio of M1 and the value of V_{GG} necessary to produce an output voltage of -4 V if $W_2/L_2 = 30$ μm/20 μm, $V_{DD} = 3$ V and $V_{SS} = -6$ V. Use the model parameters of Table 3.1-2 and assume $\lambda = 0$. What is the drain current under these conditions?

11. Assume that the area required for a transistor is $W \times L$ in μm^2 plus 100 μm^2 for minimum source/drain area. Plot the area required to drop 5 volts using a 20 μA current versus the number of transistors. Ignore the bulk effects and use NMOS devices assuming the parameters of Table 3.1-2. Assume a minimum value for W or L of 10 μm.

12. Assume that the active floating resistor of Fig. 5.2-4 is an n-channel transistor used as a resistor in series with a capacitor and that the gate is connected to $+5$ volts. Plot the small-signal resistance of this resistor as the source/drain potential is increased from 0 volts to 5 volts.

13. Use the SPICE computer program to obtain the results of Fig. 5.2-6 for the range of values for V_C shown when the W/L ratios of M1A and M1B are (a) 10 μm/20 μm and (b) 20 μm/10 μm. What is the approximate range of linearity for both cases?

14. Solve for the small-signal output resistance of Fig. 5.3-3(a) by solving for the large-signal output current as a function of v_{OUT} and differentiating this result.

15. Evaluate the output resistance of Example 5.3-1 using the parameters of Table 3.1-2 for (a) $I_D = 10$ microamperes and (b) $I_D = 200$ microamperes.

16. Assume that an NMOS device is used as a current sink and is biased at the edge of saturation with $I_D = 100$ μA. If $\beta = 20$ μA/V^2 and $V_{T0} = 2$ V, find the small-signal output resistance.

17. If r of Fig. 5.3-3(a) is 50 KΩ, find the W/L ratio of M2 that will increase this resistance to 5 MΩ assuming the parameters of Table 3.1-2 apply to the devices and the dc value of i_{OUT} is 100 μA. What is the value of V_{GG} necessary if $V_{SS} = 0$ V?

18. Design the W/L ratios of Fig. 5.3-6(a) if $I_{REF} = 50$ μA and V_{OUT}(sat) (V_{MIN}) is 1 V. What is the minimum voltage across M5 and M6 and M3 and M4 which is necessary to keep these devices in saturation?

19. Assume that M1 and M2 of Fig. P5.19 are matched and the parameters of Table 3.1-2 correspond to the devices shown. Find i_{OUT}. State and justify any assumptions that are necessary.

20. For the resistor-biased current mirror shown in Fig. P5.20, find i_{OUT} when $R_1 = 20$ Kilohms and $R_2 = 30$ Kilohms. Assume both transistors are matched, W/L are equal, and both are in the saturated region. Use the parameters of Table 3.1-2.

21. Verify Eq. (19) of Sec. 5.4.

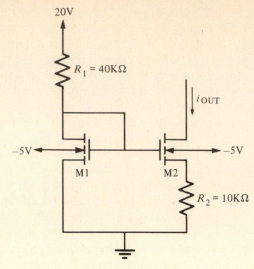

Figure P5.19

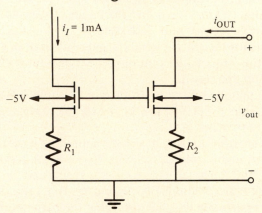

Figure P5.20

22. (a) For the cascode current mirror of Fig. 5.4-6, what is the necessary value of voltage across M1 and M3 for $i_I = 100$ μA? Assume that $W_1/L_1 = W_3/L_3 = 2$ and $V_{T1} = V_{T2} = 1$ V. (b) If the threshold voltages are unchanged and the maximum value of voltage across M1 and M3 is constrained to 5 V, how can the same current for i_I be obtained?

23. An improved Wilson current source is shown in Fig. P5.23. Draw the small-signal model and find the expression for the small-signal output resistance.

24. Show that the sensitivity as defined in Eq. (1) of Sec. 5.5 for Fig. 5.5-2(b) is unity.

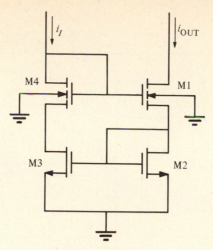

Figure P5.23

25. Derive Eq. (5) of Sec. 5.5.
26. Assume for Fig. 5.5-3(b) that $V_{DD} = 10$ V, $R_2 = 2R_1 = 2$ Megohms, $R = 100$ Kilohms, and $I_s = 10^{-15}$ A and find V_{REF}.
27. Repeat the previous problem for Fig. 5.5-4(b) using Table 3.1-2.
28. Find the small-signal output resistance of Fig. 5.5-3(b) and Fig. 5.5-4(b).
29. Replace R in Fig. 5.5-7(a) with a PMOS transistor with the gate and the drain connected to the breakdown diode and the source to V_{DD}. If V_{DD} is 10 V and $BV = 6.5$ V, design the W/L ratio to give a current of 100 μA using the device parameters of Table 3.1-2. If r_z at 100 μA is 100 ohms, what is the sensitivity of this voltage reference with respect to V_{DD}?
30. Show how to create a current reference from Fig. 5.5-7(a). Derive an expression for the sensitivity of your current reference in terms of changes in V_{DD}.
31. A threshold-referenced current reference is shown in Fig. P5.31 along with the W/L values of the transistors. Find the W/L value of M1 to give an output current of 100 μA using the following device parameters.

	V_T	K'	γ	ϕ	λ
NMOS	0.79V	23.6 μA/V^2	0.53V$^{0.5}$	0.590V	0.02V^{-1}
PMOS	−0.52V	5.8 μA/V^2	0.67V$^{0.5}$	0.600V	0.012V^{-1}

Develop an expression relating the current in M1 (I_1) to V_{DD}, including second-order effects, and find the sensitivity of I_{OUT} to V_{DD} when I_{OUT} is 100 μA. Use SPICE to plot I_{OUT} versus V_{DD} over the range of V_{DD} from 0 to 10 volts. Compare your theoretical and simulated results.

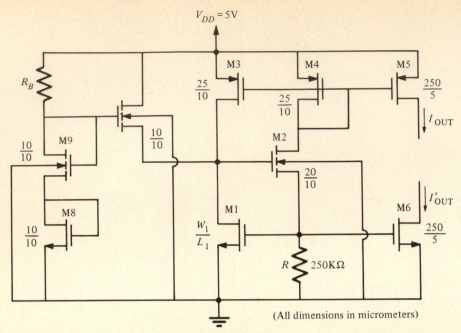

Figure P5.31

32. Develop an expression for the TC_F of Fig. 5.5-3(a) if the assumption that V_{DD} is greater than V_{REF} is no longer true.

33. Find the TC_F of Fig. 5.5-3(b) if $R_2 = 2R_1 = 1$ Megohm and both are polysilicon resistors. Assume that $V_{BE} = 0.6$ V.

34. Use the simple CMOS voltage-reference configuration of Fig. 5.5-10 along with the parameters of Table 3.1-2 to find the value of V_{REF} which will have zero temperature coefficient if $V_{DD} = -V_{SS} = 5$ V. Evaluate the sensitivity of this voltage with respect to V_{DD}.

REFERENCES

1. P.E. Allen and E. Sanchez-Sinencio, *Switched Capacitor Circuits,* (New York: Van Nostrand Reinhold, 1984) Chapter 8.
2. M. Banu and Y. Tsividis, "Floating Voltage-Controlled Resistors in CMOS Technology," *Electronic Letters,* Vol. 18, No. 15 (July 1982) pp. 678–679.
3. T.C. Choi, R.T. Kaneshiro, R.W. Brodersen, P.R. Gray, W.B. Jett, and M. Wilcox, "High-Frequency CMOS Switched-Capacitor Filters for Communications Applications," *IEEE Journal of Solid-State Circuits,* Vol. SC-18, No. 6 (December 1983) p. 661.
4. P.R. Gray and R.J. Meyer, *Analysis and Design of Analog Integrated Circuits,* (New York: John Wiley & Sons, 1977) Chapter 4.
5. G.R. Wilson, "A Monolithic Junction FET-npn Operational Amplifier," *IEEE J. of Solid-State Circuits,* Vol. SC-3, No. 5 (December 1968) pp. 341–348.

chapter 6

CMOS Amplifiers

This chapter uses the basic subcircuits of the last chapter to develop various forms of CMOS amplifiers. We begin by examining the inverter—the most basic of all amplifiers. The order of presentation that follows has been arranged around the stages that might be put together to form a high-gain amplifier. The first circuit will be the differential amplifier. This serves as an excellent input stage. The next circuit will be the cascode amplifier, which is similar to the inverter, but has higher overall performance and more control over the small-signal performance. It makes an excellent gain stage and provides a means of compensation. The output stage comes next. The objective of the output stage is to drive an external load without deteriorating the performance of the high-gain amplifier. The final section of the chapter will examine how these circuits can be combined to achieve a given high-gain amplifier requirement.

The approach used will be the same as that of Chapter 5, namely to present a general understanding of the circuit and how it works, followed by a large-signal analysis and a small-signal analysis. In this chapter we begin the transition in Table 5.0-1 from the lower to the upper level of simple circuits. At the end of this chapter, we will be in a position to consider complex analog CMOS circuits. The section on architectures for high-gain amplifiers will lead directly to the comparator and op amp.

As the circuits we study become more complex, we will have an opportunity to employ some of the analysis techniques detailed in Appendix A. We shall also introduce new techniques where pertinent in developing the subject of CMOS analog integrated-circuit design. Examples of such techniques are the dominant pole approximation used for solving the roots of a second-order polynomial with algebraic coefficients and the use of an NMOS-PMOS analog concept for simplifying the calculations of circuits that are identical but use opposite type MOSFETs.

6.1 *Inverters*

The inverter is the basic gain stage for CMOS circuits. Typically, the inverter uses the common-source configuration with either an active resistor for a load or a current sink/source as a load resistor. There are a number of ways in which the active load can be configured. Only a few of the basic configurations will be described here, leaving the others to the imagination and innovation of the reader.

Many times a low-gain inverting stage is desired that has highly predictable small- and large-signal characteristics. One configuration that meets this need is shown in Fig. 6.1-1(a). This circuit uses a common-source, n-channel transistor with a p-channel transistor connected as an

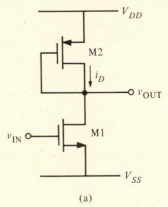

(a)

Figure 6.1-1 (a) Inverter with an active load. (b) Output characteristics of the inverter of (a). (c) Transfer characteristics of the inverter of (a).

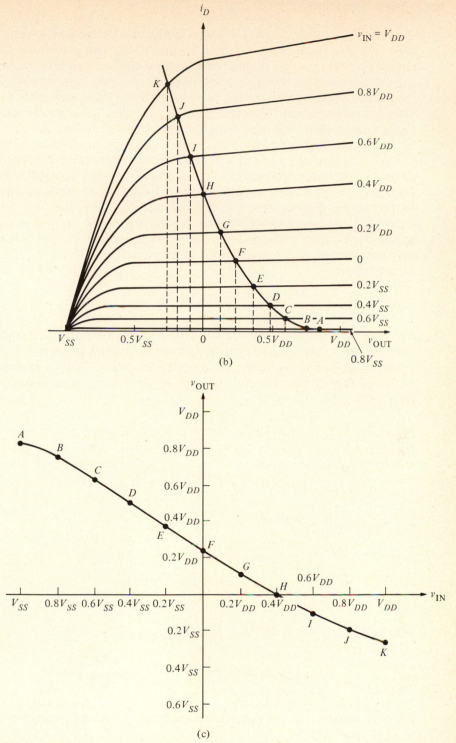

Figure 6.1-1 (Continued)

active resistor for the load. The large-signal characteristics are best illustrated graphically in Fig. 6.1-1(b). This plot shows the i_D versus v_{DS} characteristics of M1 plotted on the same graph with the "load line" (i_D versus v_{DS}) characteristic of the p-channel, diode-connected, transistor, M2. The "load line" of the active resistor, M2, is simply the transconductance characteristic reversed and subtracted from V_{DD}. It is apparent that the output-signal swing will experience a limitation for negative swings. A plot of v_{OUT} versus v_{IN} can be obtained from Fig. 6.1-1(b) by plotting the points marked A, B, etc., on the curve of Fig. 6.1-1(c). This curve is called a large-signal voltage-transfer function curve. It is obvious that this type of inverting amplifier has limited output voltage range and low gain (gain is determined by the slope of the v_{OUT} versus v_{IN} curve).

It is of interest to consider the large-signal swing limitations of the active-resistor load inverter. From Fig. 6.1-1(b) we can see that the maximum output voltage, $v_{OUT}(max)$, is approximately equal to $V_{DD} - V_{TP}$ (point A). Therefore,

$$v_{OUT}(max) \cong V_{DD} - |V_{TP}| \tag{1}$$

This limit is determined by the fact that at this value of v_{OUT} there is no current flow through M2 to further increase the output voltage.

In order to find $v_{OUT}(min)$ we first assume that M1 will be in the non-saturated region and that $V_{T1} = |V_{T2}| = V_T$. We can write the current through M1 as

$$i_D = \beta_1 \left[(V_{GS1} - V_T)v_{DS1} - \frac{v_{DS1}^2}{2} \right]$$

$$= \beta_1 \left[(V_{DD} - V_{SS} - V_T)(v_{OUT} - V_{SS}) - \frac{(v_{OUT} - V_{SS})^2}{2} \right] \tag{2}$$

and the current through M2 as

$$i_D = \frac{\beta_2}{2} (V_{SG2} - V_T)^2 = \frac{\beta_2}{2} (V_{DD} - V_{OUT} - V_T)^2$$

$$= \frac{\beta_2}{2} (v_{OUT} + V_T - V_{DD})^2 \tag{3}$$

Equating Eq. (2) to (3) and solving for v_{OUT} gives

$$v_{OUT}(min) = V_{DD} - V_T - \frac{V_{DD} - V_{SS} - V_T}{\sqrt{1 + (\beta_2/\beta_1)}} \tag{4}$$

We have assumed in developing this expression that the maximum value of v_{IN} is equal to V_{DD}.

The small-signal voltage gain of the inverter with an active-resistor load inverter can be found from Fig. 6.1-2. This gain can be expressed as

$$\frac{V_{out}}{V_{in}} = \frac{-g_{m1}}{g_{ds1} + g_{ds2} + g_{m2}} \simeq -\frac{g_{m1}}{g_{m2}}$$

$$= -\left[\frac{K_N' W_1 L_2}{K_P' L_1 W_2}\right]^{1/2} \tag{5}$$

The small-signal output resistance can also be found from Fig. 6.1-2 as

$$r_{out} = \frac{1}{g_{ds1} + g_{ds2} + g_{m2}} \simeq \frac{1}{g_{m2}} \tag{6}$$

The output resistance of the active-resistor load inverter will be low because of the low resistance of the diode-connected transistor M2. The resulting low output resistance can be very useful in situations where a large bandwidth is required from an inverting gain stage.

The small-signal frequency response of the active-resistor load inverter will be examined next. Figure 6.1-3(a) shows a general inverter configuration and the important capacitors. The point x is connected to V_{out} for the case of Fig. 6.6-1(a). C_{gd1} and C_{gd2} represent the overlap capacitances, C_{bd1} and C_{bd2} are the bulk capacitances, C_{gs2} is the overlap plus gate capacitance, and C_L is the load capacitance seen by the inverter, which can consist of the next gate(s) and any parasitics associated with the connections. Fig. 6.1-3(b) illustrates the resulting small-signal model assuming that V_{in} is a voltage source. (The case when V_{in} has a high source resistance will be examined in the third section of this chapter, which deals with the cascode amplifier.) The frequency response of this circuit is

$$\frac{V_{out}(s)}{V_{in}(s)} = \frac{-g_m R}{(s/\omega_1) + 1} \tag{7}$$

where

$$g_m = g_{m1} \tag{8}$$

$$\omega_1 = \frac{1}{RC} \tag{9}$$

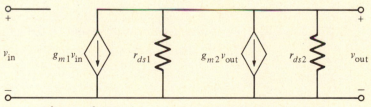

Figure 6.1-2 Small-signal model of Fig. 6.1-1(a).

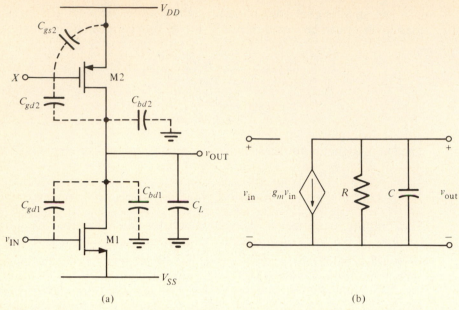

Figure 6.1-3 (a) General configuration of an inverter illustrating parasitic capacitances. (b) Small-signal model of (a).

and

$$R = [g_{ds1} + g_{ds2} + g_{m2}]^{-1} \simeq g_{m2}^{-1} \tag{10}$$
$$C = C_{gd1} + C_{bd1} + C_{bd2} + C_{gs2} + C_L \tag{11}$$

Eq. (10) shows that the −3 dB frequency of the active resistor-load inverter is approximately proportional to the square root of the drain current. As the drain current increases, the bandwidth will also increase because R will decrease.

Example 6.1-1

Performance of an Active Resistor-Load Inverter

Calculate the output-voltage swing limits for $V_{DD} = 5$ volts and $V_{SS} = -5$ volts, the small-signal gain, the output resistance, and the −3 dB frequency of Fig. 6.1-1 if (W_1/L_1) is 50 μm/10 μm and $W_2/L_2 = 10$ μm/40 μm, $C_{gd1} = 0.02$ pF, $C_{bd1} = 0.1$ pF, $C_{bd2} = 0.01$ pF, $C_{gs2} = 0.12$ pF, $C_L = 1$ pF, and $I_{D1} = I_{D2} = 50$ microamperes, using the parameters in Table 3.1-2.

We find that the $v_{OUT}(max) = 4$ volts and $v_{OUT}(min) = -4.9$ volts. Using Eq. (5) we find that the small-signal voltage gain is -6.52. Using Eq. (6) of Sec. 3.3 and Eq. (6) above, we get an output resistance of 66.5 KΩ if we include g_{ds1} and g_{ds2} (λ_2 assumed to be equal to 0.008) and 70.7 KΩ if we consider only g_{m2}. Finally, the -3 dB frequency is 1.8 MHz.

Although the output resistance of Example 6.1-1 is large, it is considered low for resistance levels associated with MOS circuits.

Often an inverting amplifier is required that has gain higher than that achievable by the active-resistor load inverter stage. A second inverter configuration, which has higher gain, is shown in Fig. 6.1-4. Instead of an active resistor as the load, a current-source load is used. The current source is a common gate configuration using a p-channel transistor with the gate tied to a dc bias voltage. The large-signal characteristics can best be determined graphically. Fig. 6.1-5 shows a plot of i_D versus v_{OUT}. On this voltage-current characteristic the output characteristics of M2 are plotted. Since v_{IN} is the same as v_{GS1}, the curves have been labeled accordingly. Superimposed upon these characteristics are the output characteristics of M2 with $v_{OUT} = V_{DD} - v_{SD2}$. The large-signal voltage-transfer function curve can be obtained in a manner similar to Fig. 6.1-1 for the active-resistor load inverter. Transferring the points A, B, etc., from Fig. 6.1-5 for a given value of V_{SG2}, to Fig. 6.1-6 results in the large-signal voltage-transfer function curve shown.

The limits of the large-signal output-voltage swing can be found by a method similar to that used for the active resistor inverter. $v_{OUT}(max)$ is equal to V_{DD} since when M1 is off, M2 can pull the output voltage up to V_{DD}

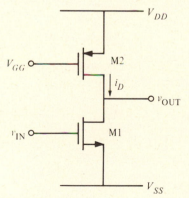

Figure 6.1-4   Inverter with a current source load.

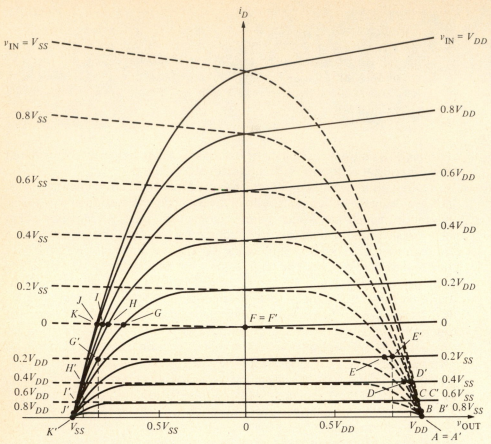

Figure 6.1-5 Large-signal output characteristics of Fig. 6.1-4 with $V_{GG} = 0$.

providing there is not a low resistance dc load. Thus, the maximum positive output voltage is

$$v_{OUT}(\max) \cong V_{DD} \tag{12}$$

The lower limit can be found by assuming that M1 will be in the nonsaturation region. $v_{OUT}(\min)$ can be given as

$$v_{OUT}(\min) = (V_{DD} - V_{SS} - V_{T1})$$
$$\left\{1 - \left[1 - \left(\frac{\beta_2}{\beta_1}\right)\left(\frac{V_{DD} - V_{GG} - |V_{T2}|}{V_{DD} - V_{SS} - V_{T1}}\right)^2\right]^{1/2}\right\} + V_{SS} \tag{13}$$

This result assumes that v_{IN} is taken to V_{DD}.

The small-signal performance can be found using the model of Fig. 6.1-2 with $g_{m2}v_{out} = 0$ (this is to account for the fact that the gate of M2 is

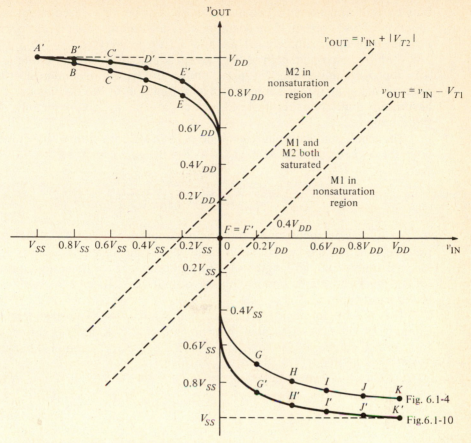

Figure 6.1-6 Transfer characteristics of CMOS inverters ($V_{T1} = |V_{T2}| = 0.2 V_{DD}$)

on ac ground). The small-signal voltage gain is given as

$$\frac{V_{out}}{V_{in}} = \frac{-g_{m1}}{g_{ds1} + g_{ds2}} = \left(\frac{2K_N'W_1}{L_1 I_D}\right)^{1/2} \left(\frac{-1}{\lambda_1 + \lambda_2}\right) \quad (14)$$

This is a significant result in that the gain increases as the dc current decreases. This occurs because the output conductance is proportional to the bias current whereas the transconductance is proportional to the square root of the bias current. This of course assumes that the simple relationship for the output conductance expresssed by Eq. (9) of Sec. 3.3 is valid. The increase of gain as I_D decreases holds true until this current reaches the subthreshold region of operation, where weak inversion occurs. At this point the transconductance becomes proportional to the bias current and the small-signal voltage gain becomes a constant as a function of bias current. If we assume that the subthreshold current occurs at a level of approximately 0.1 μA and if $(W/L)_1 = (W/L)_2 = 10 \ \mu m/10 \ \mu m$, then the maximum gain of the current-load CMOS inverter of Fig. 6.1-4 is approximately -600. Fig. 6.1-7 shows the typical dependence of the inverter using a current-source load as a function of the dc bias current, assuming that the subthreshold effects occur at approximately 0.1 microampere.

The small-signal output resistance of the CMOS inverter with a current source load can be found from Fig. 6.1-2 (with $g_{m2}V_{out} = 0$) as

$$r_{out} = \frac{1}{g_{ds1} + g_{ds2}} \cong \frac{1}{I_D(\lambda_1 + \lambda_2)} \tag{15}$$

If $I_D = 50$ microamperes and using the parameters of Table 3.1-2, the output resistance of the current source CMOS inverter is approximately 0.67 MΩ, assuming channel lengths of 10 μm. Compared to the active resistor CMOS inverter, this is a much higher output resistance. The result is a lower bandwidth.

The -3 dB frequency of the current source CMOS inverter can be found from Fig. 6.1-3 assuming that the gate of M2 (point X) is connected to a voltage source, V_{GG}. In this case, R and C of Eqs. (10) and (11) become

$$R = [g_{ds1} + g_{ds2}]^{-1} \tag{16}$$
$$C = C_{gd1} + C_{gd2} + C_{bd1} + C_{bd2} + C_L \tag{17}$$

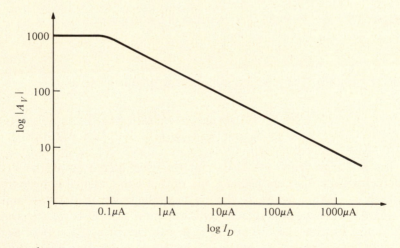

Figure 6.1-7   Illustration of the magnitude of small-signal voltage gain dependence upon the biasing current of the CMOS inverter.

Therefore, the -3 dB frequency response can be expressed as

$$\omega_1 = \frac{g_{ds1} + g_{ds2}}{C_{gd1} + C_{gd2} + C_{bd1} + C_{bd2} + C_L} \tag{18}$$

If the current-load inverter has a dc current of 50 microamperes and if the capacitors have the values given in Example 6.1-1 (with $C_{gd1} = C_{gd2}$), we find that the -3 dB frequency is 208 KHz (assuming channel lengths of 10 μm). The difference between this frequency and that found in Ex. 6.1-1 is due to the difference in the output resistance.

Example 6.1-2

Performance of a Current-Sink Inverter

    The performance of a current-sink CMOS inverter is to be examined. The current-sink inverter is shown in Fig. 6.1-8. Assume that $W_1 = 20$ μm, $L_1 = 10$ μm, $W_2 = 10$ μm, $L_2 = 10$ μm, $V_{DD} = 5$ volts, $V_{SS} = -5$ volts, $V_{GG} = -3$ volts, and the parameters of Table 3.1-2 describe M1 and M2. Use the capacitor values of Example 6.1-1 ($C_{gd1} = C_{gd2}$). Calculate the output-swing limits and the small-signal performance.

    Since the current-sink inverter is inverted with respect to the current-source inverter of Fig. 6.1-4(a), the easiest way to work this example is to use Fig. 6.1-4(a) with the PMOS parameters for M1 and the NMOS parameters for M2. This technique will be called *NMOS-PMOS analog* and is illustrated in Fig. 6.1-9. Any CMOS circuit can be converted to its NMOS-PMOS analog by inverting the power-supply busses and replacing each NMOS transistor by a PMOS transistor having the W/L and model values of the NMOS transistor and vice versa. The primed notation in Fig. 6.1-9 is to designate that the NMOS-PMOS analog concept is being used. This concept will allow us to develop relationships for one type of CMOS circuit and apply it to the opposite type.

    Applying the output signal-swing limitations developed for Fig. 6.1-4 (Eqs. 12 and 13) to the middle circuit of Fig. 6.1-9 gives $v_{OUT}(max)' = 5$ V and $v_{OUT}(min)' = -4.94$ V. Converting these results back to the circuit of Fig. 6.1-8 gives $v_{OUT}(max) = 4.94$ V and $v_{OUT}(min) = -5$ V. In applying the NMOS-PMOS analog, it is important to be sure to use the NMOS parameters for the PMOS parameters and vice versa. The small-signal gain can be calculated from Eq. (14) as -64.7. The output resistance and the -3 dB frequency are 1.96 MΩ and 70.6 KHz, respectively.

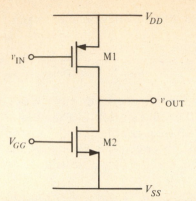

Figure 6.1-8   Current-sink CMOS inverter.

One of the disadvantages of the current-source inverter is that it requires the bias voltage V_{GG}. If the gate of M2 in Fig .6.1-4(a) is taken to the gate of M1, the push-pull CMOS inverter of Fig. 6.1-10 is achieved. The large-signal voltage-transfer function plot for the push-pull inverter can be found in a similar manner as the plot for the current-source inverter. In this case the points, A', B', etc., describe the load line of the push-pull inverter on Fig. 6.1-5. The large-signal voltage-transfer function characteristic is found by projecting these points down to the horizontal axis and plotting the results on Fig. 6.1-6. In comparing the large-signal voltage-transfer function characteristics between the current-source and push-pull inverters, it is seen that the push-pull inverter will have a higher gain assuming identical transistors. This is due to the fact that both transistors are being driven by v_{IN}. Another advantage of the push-pull inverter is that

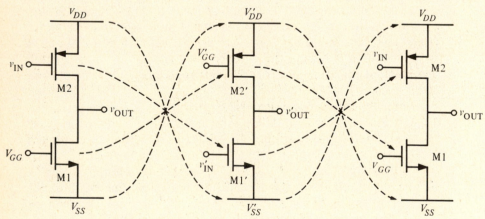

Figure 6.1-9   Illustration of the NMOS-PMOS analog concept.

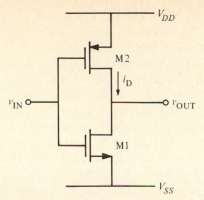

Figure 6.1-10   Push-pull CMOS inverter.

the output swing is capable of operation from V_{DD} to V_{SS} where we have shown that a current-source inverter cannot drive all the way to V_{SS}. The operating regions of M1 and M2 are shown by the dashed lines on Fig. 6.1-6.

The small-signal performance can be analyzed with the aid of Fig. 6.1-11. The small-signal voltage gain is

$$\frac{V_{out}}{V_{in}} = \frac{-(g_{m1} + g_{m2})}{g_{ds1} + g_{ds2}} = -\sqrt{(2/I_D)}\left[\frac{\sqrt{K'_N(W_1/L_1)} + \sqrt{K'_P(W_2/L_2)}}{\lambda_1 + \lambda_2}\right] \quad \textbf{(19)}$$

We note the same dependence of the gain upon the dc current that was observed for the current–source/sink inverters. If I_D is 0.1 microamperes and $W_1/L_1 = W_2/L_2 = 1$, then using the parameters of Table 3.1-2, the maximum small-signal voltage gain is -1036. The output resistance and the -3 dB frequency response of the push-pull inverter are identical to those of the current-source inverter given in Eqs. (15) through (18).

Fig. 6.1-12(a), (b), and (c) show the simulated and experimental responses for the voltage-transfer curve for (a) the active-resistor inverter, (b) the current-source inverter, and (c) the push-pull inverter. These inverters were implemented using a 5 μm, double-poly, n-well CMOS technology.

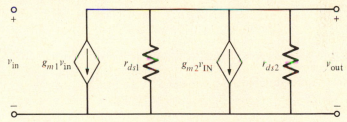

Figure 6.1-11   Small-signal model for the CMOS inverter of Fig. 6.1-10.

The W/L of the p-channel devices is 30 μm/10 μm and 10 μm/10 μm for the n-channel. The n-channel devices have large-signal parameters $V_T = 0.81$ volts, $K'_N = 16.5$ microamperes/volt^2, $\lambda = 0.02$ volts^{-1}, $\gamma = 0.45$ volts$^{1/2}$, and $2|\phi_F| = 0.45$ volts. The p-channel devices have large-signal parameters $V_T = -0.5$ volts, $K'_P = 8.5$ microamperes/volt^2, $\lambda = 0.01$ volts^{-1}, $\gamma = 0.8$ volts$^{1/2}$, and $2|\phi_F| = 0.5$ volts.

It is of interest to analyze the inverters of this section in terms of their noise performance. First consider the active load inverter of Fig. 6.1-1. Our approach will be to assume a mean-square input-voltage-noise spectral density $\overline{e}_n^2$ in series with each gate of each device and then to calculate the output-voltage-noise spectral density $\overline{e}_{out}^2$. In this calculation, all sources are assumed to be additive. Dividing $\overline{e}_{out}^2$ by the square of the voltage gain of the inverter will give the equivalent input-voltage-noise spectral density,

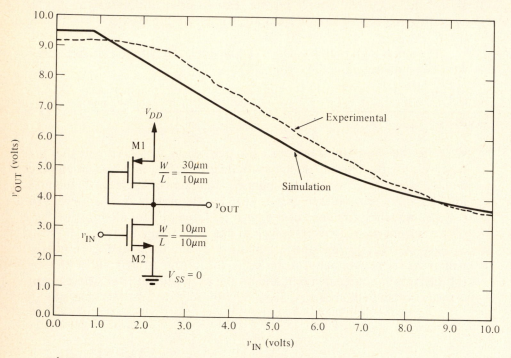

Figure 6.1-12   Experimental and simulated inverter transfer curves. (a) CMOS active-resistor inverter. (b) CMOS current-source inverter. (c) CMOS (push-pull) inverter. The conditions and parameters for each case are $V_{DD} = 10$V, $V_{SS} = 0$V, $K_N = 16.5$ μA/V^2, $K_p = 8.5$ μA/V^2, $V_{TN} = 0.81$V, $V_{TP} = -0.5$ V, and $\lambda_N = 2\lambda_p = 0.02$ V^{-1}.

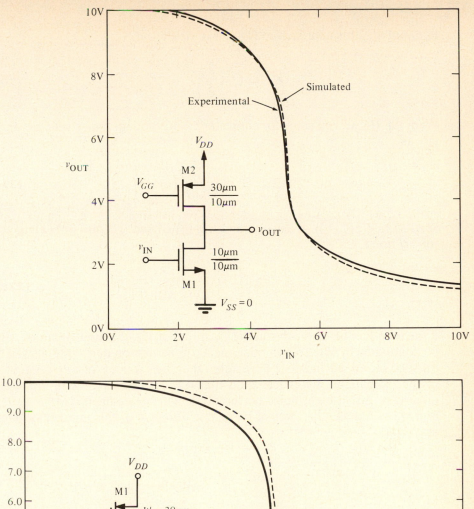

Figure 6.1-12 (Continued)

$\bar{e}_{eq}^2$. Applying this approach to Fig. 6.1-1 yields,

$$\bar{e}_{out}^2 = \bar{e}_{n1}^2 \left(\frac{g_{m1}}{g_{m2}}\right)^2 + \bar{e}_{n2}^2 \tag{20}$$

From Eq. (5) we can solve for equivalent input-voltage-noise spectral density as

$$\bar{e}_{eq} = \bar{e}_{n1} \left[1 + \left(\frac{g_{m2}}{g_{m1}}\right)^2 \left(\frac{\bar{e}_{n2}}{\bar{e}_{n1}}\right)^2 \right]^{1/2} \tag{21}$$

Substituting Eq. (15) of Sec. 3.2 and Eq. (6) of Sec. 3.3 into Eq. (21) gives for $1/f$ noise,

$$\bar{e}_{eq(1/f)} = \left[\frac{B_1}{fW_1L_1} \right]^{1/2} \left[1 + \left(\frac{K_2'B_2}{K_1'B_1}\right)\left(\frac{L_1}{L_2}\right)^2 \right]^{1/2} \tag{22a}$$

If the length of M1 is much smaller than that of M2, the input $1/f$ noise will be dominated by M1. To minimize the $1/f$ contribution due to M1, its width must be increased. In Sec. 2.5 we noted that the $1/f$ noise of a PMOS device is approximately 5 times less than that of a NMOS device with the same area. Obviously, lower noise performance will be obtained if M1 is PMOS and M2 is NMOS. The thermal-noise performance of this inverter is given as

$$\bar{e}_{eq(th)} = \left[\left(\frac{8kT(1 + \eta_1)}{3[2K_1'(W/L)_1I_1]^{1/2}} \right) [1 + \left(\frac{W_2L_1K_2'}{L_2W_1K_1'}\right)^{1/2} \left(\frac{1 + \eta_2}{1 + \eta_1}\right)] \right]^{1/2} \tag{22b}$$

The output-voltage-noise spectral density of the current-source load inverter of Fig. 6.1-4 can be written as

$$\bar{e}_{out}^2 = (g_{m1}r_{out})^2 \bar{e}_{n1}^2 + (g_{m2}r_{out})^2 \bar{e}_{n2}^2 \tag{23}$$

Dividing Eq. (23) by the square of the gain of this inverter as given by Eqs. (14) and (15), and taking the square root results in an expression similar to Eq. (21). Thus the noise performance of the two circuits are equivalent although the small-signal voltage gain is significantly different.

The output-voltage-noise spectral density of the push-pull inverter of Fig. 6.1-10 is given by Eq. (23). Dividing this expression by the square of the gain gives the equivalent input-voltage-noise spectral density of the push-pull inverter as

$$\bar{e}_{eq} = \left[\left(\frac{g_{m1}\bar{e}_{n1}}{g_{m1} + g_{m2}} \right)^2 = \left(\frac{g_{m2}\bar{e}_{n2}}{g_{m1} + g_{m2}} \right)^2 \right]^{1/2}$$

If the transconductances are balanced ($g_{m1} = g_{m2}$), then the noise contribution of each device is divided by two. The total noise contribution can be reduced only by reducing the noise contributed by each device individually. The calculation of thermal and $1/f$ noise in terms of device dimensions and currents is left as an exercise for the reader.

The inverter is one of the basic amplifiers in analog circuit design. Three different configurations of the CMOS inverter have been presented in this section. If the inverter is driven from a voltage source, then the frequency response consists of a single dominant pole at the output of the inverter. The small-signal gain of the inverters with current–sink/source loads was found to be inversely proportional to the square root of the current, which led to high gains. However, the high gain of current–source/sink and push-pull inverters can present a problem when one is trying to establish dc biasing points. High gain stages such as these will require the assistance of a dc negative feedback path in order to stabilize the biasing point. In other words, one should not expect to find the dc output voltage well defined if the input dc voltage is defined.

6.2 *Differential Amplifiers*

The differential amplifier is one of the more versatile circuits in analog circuit design. It is also very compatible with integrated-circuit technology. The objective of the differential amplifier is to amplify only the difference between two different potentials regardless of the common-mode value. Thus, a differential amplifier can be characterized by its *common-mode rejection ratio* (CMRR) which is the ratio of the differential gain to the common-mode gain. In addition, the *input common-mode range* (CMR) specifies over what range of common-mode values the differential amplifier continues to sense and amplify the difference signal with the same gain. Another characteristic affecting performance of the differential amplifier is offset. In CMOS differential amplifiers, the most serious offset is the voltage offset. If the input terminals of the differential amplifiers are connected together, the output offset voltage is the voltage which appears at the output of the differential amplifier. If this voltage is divided by the differential voltage gain of the differential amplifier, then the offset voltage is called the *input voltage offset* (V_{OS}). Typically, the input voltage offset of a CMOS differential amplifier is 5 to 20 millivolts.

Let us consider the large- and small-signal characteristics of the CMOS differential amplifier. Fig. 6.2-1 shows a CMOS differential amplifier that uses n-channel MOS devices, M1 and M2, as the differential pair. I_{SS} is a current source, which can be any of the types previously considered. The loads for M1 and M2 are obtained from a simple p-channel current mirror. If M3 and M4 are matched, then the current of M1 will determine the current in M3. This current will be mirrored in M4. If $v_{GS1} = v_{GS2}$, then the currents in M1 and M2 are equal. Thus the current that M4 sources to M2 should be equal to the current that M2 requires, causing i_{OUT} to be zero– provided that the load is negligible. If $v_{GS1} > v_{GS2}$, then i_{D1} increases with respect to i_{D2} since $I_{SS} = i_{D1} + i_{D2}$. This increase in i_{D1} implies an increase in i_{D3} and i_{D4}. However, i_{D2} decreased when v_{GS1} became greater than v_{GS2}. Therefore, the only way to establish circuit equilibrium is for i_{OUT} to become positive. It can be seen that if $v_{GS1} < v_{GS2}$ then i_{OUT} becomes negative. This is a simple way in which the differential output signal of the differential amplifier can be converted back to a single-ended signal, i.e., one referenced to ac ground.

Fig. 6.2-2 shows a CMOS differential amplifier that uses p-channel MOS devices, M1 and M2, as the differential pair. The circuit operation is identical to that of Fig. 6.2-1, but there is a difference in the bulk effect. If we assume that the CMOS technology uses an n-substrate then the bulks of the p-channel devices can only be taken to the most positive potential, V_{DD}. However, the n-channel bulks need not necessarily be connected to the

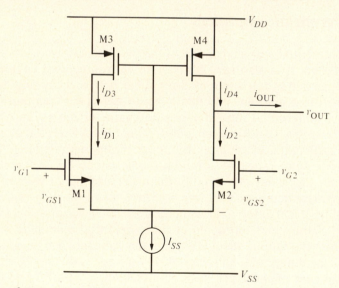

Figure 6.2-1   CMOS differential amplifier using n-channel input devices.

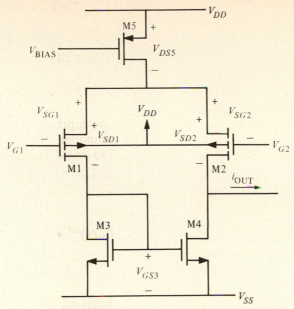

Figure 6.2-2   CMOS differential amplifier using p-channel input devices.

most negative potential, V_{SS}. The differential amplifier of Fig. 6.2-1 can have the sources of M1 and M2 connected to their bulks. Thus, M1 and M2 are made in a p-tub which is allowed to float. Therefore, the circuit in Fig. 6.2-1 is insensitive to the bulk effects, whereas that in Fig. 6.2-2 is not. Of course this is reversed if the CMOS technology uses a p-substrate.

The large-signal characteristics can be developed by assuming that M1 and M2, the differential pair, are always in saturation. This condition is reasonable in most cases and illustrates the behavior even when this assumption is not valid. The pertinent relationships describing large-signal behavior are given as

$$V_{ID} = V_{GS1} - V_{GS2} = \left(\frac{2i_{D1}}{\beta}\right)^{1/2} - \left(\frac{2i_{D2}}{\beta}\right)^{1/2} \tag{1}$$

and

$$I_{SS} = i_{D1} + i_{D2} \tag{2}$$

where it has been assumed that M1 and M2 are matched. Substituting Eq. (2) into Eq. (1) and forming a quadratic allows the solution for i_{D1} and i_{D2} as

$$i_{D1} = \frac{I_{SS}}{2} + \frac{I_{SS}}{2}\left(\frac{\beta v_{ID}^2}{I_{SS}} - \frac{\beta^2 v_{ID}^4}{4I_{SS}^2}\right)^{1/2} \tag{3}$$

and

$$i_{D2} = \frac{I_{SS}}{2} - \frac{I_{SS}}{2} \left(\frac{\beta v_{ID}^2}{I_{SS}} - \frac{\beta^2 v_{ID}^4}{4I_{SS}^2} \right)^{1/2} \tag{4}$$

where these relationships are valid only for $v_{ID} < (2I_{SS}/\beta)^{1/2}$. Fig. 6.2-3 shows a plot of the normalized drain current of M1 versus the normalized differential input voltage. The dotted portions of the curves are meaningless and are ignored.

The above analysis has resulted in i_{D1} or i_{D2} in terms of the differential input voltage, v_{ID}. Although this is a voltage-to-current transfer function, it turns out that this function is as useful as the voltage-transfer function characteristic. Under this circumstance, the appropriate term for Fig. 6.2-1 would be differential transconductance amplifier. It is of interest to determine the slope of this curve which leads to one definition of transconductance for the differential amplifier. Differentiating Eq. (3) with respect to v_{ID} and setting $V_{ID} = 0$ gives the differential transconductance of the differential amplifier as

$$g_m = \partial i_{D1} / \partial v_{ID} (V_{ID} = 0) = (\beta I_{SS}/4)^{1/2} = \left(\frac{K_1' I_{SS} W_1}{4L_1} \right)^{1/2} \tag{5}$$

We note in comparing this result to Eq. (6) of Sec. 3.3 with $I_{SS}/2 = I_D$ that a difference of two exists. The reason for this difference is that only half of v_{ID} is being applied to M1. It is also interesting to note that as I_{SS} is increased the transconductance also increases. The important property—that small-

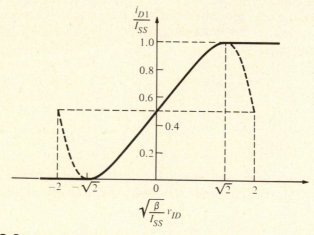

Figure 6.2-3 Large-signal transfer characteristics of a CMOS differential amplifier.

signal performance can be controlled by a dc parameter—is illustrated yet again.

If we assume that the currents in the current mirror are identical, then i_{OUT} can be found by subtracting i_{D2} from i_{D1} for the n–channel differential amplifier of Fig. 6.2-1 and i_{D1} from i_{D2} for the p-channel differential amplifier of Fig. 6.2-2. Since i_{OUT} is a differential output current, we distinguish this transconductance from that of Eq. (5) by using the notation g_{md}. The differential-in, differential-out transconductance is twice g_m and can be written as

$$g_{md} = \frac{\partial i_{OUT}}{\partial v_{ID}}; (V_{ID} = 0) = \left(\frac{K_1' I_{SS} W_1}{L_1}\right)^{1/2} \tag{6}$$

which is exactly equal to the transconductance of the common-source FET if $I_D = I_{SS}/2$.

The output voltage v_{OUT} of the differential amplifier can be found by assuming that a load resistance R_L is connected from the output of the differential amplifier to ground. Therefore, for the n-channel differential amplifier we have

$$v_{OUT} = (i_{D1} - i_{D2})R_L = I_{SS}\left(\frac{\beta v_{ID}^2}{I_{SS}} - \frac{\beta^2 v_{ID}^4}{4I_{SS}^2}\right)^{1/2} R_L \tag{7}$$

The voltage-transfer curve is similar to Fig. 6.2-3 except the vertical scale is multiplied by R_L and shifted by $R_L I_{SS}/2$.

Fig. 6.2-4(a) shows a CMOS, n-channel input, differential amplifier and Fig. 6.2-4(b) shows the simulated voltage-transfer curves of this differential amplifier. These curves were made under the conditions that $V_{DD} = 5$ volts, $V_{SS} = -5$ volts, the input was applied to the gate of M1 (V^+), and the voltage applied to the gate of M2 (V^-) was -1, 0, and $+1$ volt. We immediately note two important characteristics of the differential amplifier. The first is that the assumption that both devices are in saturation only holds for the linear part of the curves, where the magnitude of v_{ID} is very small. Second, the common-mode input voltage has a significant effect on the transfer function, particularly the output-signal swing. Fig. 6.2-5(a) and (b) show the similar results for a p-channel input, differential amplifier.

The small-signal voltage gain of the differential amplifier can be found by differentiating Eq. (7) with respect to v_{ID} and setting $V_{ID} = 0$, giving

$$A_v = \frac{\partial v_{OUT}}{\partial v_{ID}} = \sqrt{\beta I_{SS}} R_L = \left(\frac{K_1' I_{SS} W_1}{L_1}\right)^{1/2} R_L \tag{8}$$

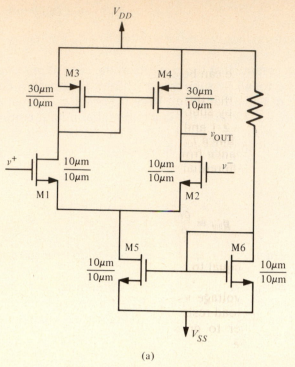

V_{DD}

M3 M4

$\frac{30\mu m}{10\mu m}$ $\frac{30\mu m}{10\mu m}$

v^+ $\frac{10\mu m}{10\mu m}$ $\frac{10\mu m}{10\mu m}$ v_{OUT}

M1 v^-

M2

M5 M6

$\frac{10\mu m}{10\mu m}$ $\frac{10\mu m}{10\mu m}$

V_{SS}

(a)

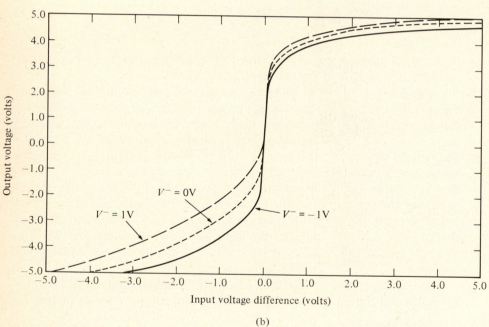

(b)

Figure 6.2-4 (a) N-channel input differential amplifier. (b) Simulation of voltage transfer curve for $V^- = -1, 0,$ and 1 volt. (V_{DD} = 5 volts, V_{SS} = -5 volts, K'_N = $2K'_P$ = 28 $\mu A/volts^2$, V_T = ± 0.7 volts, and λ_N = λ_P = 0.01 volts^{-1})

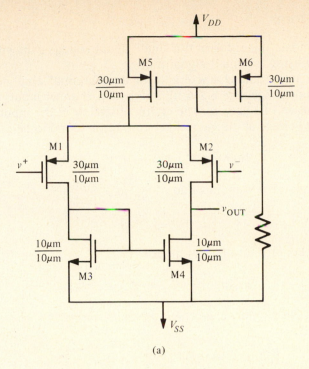

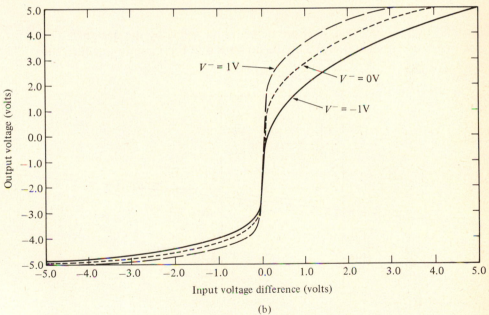

Figure 6.2-5 (a) P-channel input differential amplifier. (b) Simulation of the voltage transfer curve for $V^- = 1, 0,$ and 1 volt. ($V_{DD} = 5$ volts, $V_{SS} = -5$ volts, $K_N' = 2K_P' = 28\ \mu A/volts^2$, $V_T = \pm 0.7$ volts, and $\lambda_N = \lambda_P = 0.01\ volts^{-1}$)

We shall compare this value of the small-signal voltage gain with that calculated later using the small-signal model.

Another important characteristic of a differential amplifier is input common-mode range. Fig. 6.2-6 illustrates a p-channel input, differential amplifier that will be used for the characterization of the input CMR. The lowest input voltage at the gate of M1 (or M2) is found to be

$$V_{G1}(\min) = V_{SS} + V_{GS3} + V_{SD1} - V_{SG1} \tag{9}$$

For saturation, the minimum value of V_{SD1} is

$$V_{SD1} = V_{SG1} - |V_{T1}| \tag{10}$$

Substituting Eq. (10) into Eq. (9) gives

$$V_{G1}(\min) = V_{SS} + V_{GS3} - |V_{T1}| \tag{11}$$

Replacing V_{GS3} in Eq. (11) gives the final result, namely

$$V_{G1}(\min) = V_{SS} + \left(\frac{I_{SS}}{\beta_3}\right)^{1/2} + V_{T03} - |V_{T1}| \tag{12}$$

The first two terms in Eq. (12) are determined solely by the designer. The last two terms are determined by the process and the manner in which the

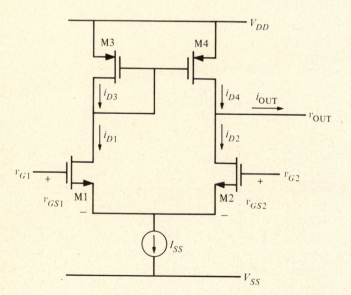

Figure 6.2-6   Circuit for common-mode voltage input range consideration.

substrate is connected for M1. Assume for the moment that an n-well CMOS process is used and that the sources of M1 and M2 are connected to the well. Eq. (12) now becomes

$$V_{G1}(\text{min}) = V_{SS} + \left(\frac{I_{SS}}{\beta_3}\right)^{1/2} + V_{T03} - |V_{T01}| \qquad (13)$$

To design the input stage to meet a specified negative common-mode range, the designer must consider the worst case V_T spread (specified by the process) and adjust I_{SS} and β_3 to meet the requirements. The worst case V_T spread affecting negative common-mode range for the configuration of Fig. 6.2-6 is a high n-channel threshold (V_{T03}) and a low (absolute value) p-channel threshold (V_{T01}).

An improvement can be obtained when the substrates of the input devices are connected to V_{DD}. This connection results in negative feedback to the sources of the input devices. For example, as the common-source node moves negative, the substrate bias increases, resulting in an increase in the threshold voltages (V_{T1} and V_{T2}). Eq. (12) shows that the negative common-mode range will increase as the magnitude of V_{T1} increases.

A similar analysis can be used to determine the highest common-mode voltage possible for the p-channel input differential amplifier of Fig. 6.2-6. The maximum gate voltage on M1 can be shown to be

$$V_{G1}(\text{max}) = V_{DD} - V_{SD5} - V_{SG1} = V_{DD} - V_{SD5} - \left(\frac{I_{SS}}{\beta_1}\right)^{1/2} - |V_{T1}| \qquad (14)$$

The relationships expressed by Eqs. (12) or (13) and (14) allow the designer to achieve maximum common-mode range by making the W/L of the transistors M1 (and M2) and M3 (and M4) as large as possible and minimizing V_{SD5}. Furthermore, the smaller the value of I_{SS}, the greater the input CMR. The above results hold for the n-channel input differential amplifier if formulas for $V_{G1}(\text{max})$ and $V_{G1}(\text{min})$ are interchanged.

Example 6.2-1

Calculation of the Worst-Case Input Common-Mode Range of the p-Channel Input, Differential Amplifier

Assume that V_{DD} varies from 8 to 12 volts and that $V_{SS} = 0$, and use the values of Table 3.1-2 under worst-case conditions to calculate the input common-mode range of Fig. 6.2-6. Assume that I_{SS} is 100 μA, $W_1/L_1 = W_2/L_2 = 5$, $W_3/L_3 = W_4/L_4 = 1$, and $V_{SD5} = 0.2$ V. Include worst-case variation in K' in calculations.

If V_{DD} varies 10 ± 2 volts, then Eq. (14) gives

$$V_{G1}(\text{max}) = 8 - 0.2 - 1.58 - 1.2 = 5.02 \text{ volts}$$

and Eq. (13) gives

$$V_{G1}(\text{min}) = 0 + 2.42 + 1.2 - .8 = 2.83 \text{ volts}$$

which gives a worst-case input common-mode range of 2.19 volts with a nominal 10-volt power supply.

Reducing V_{DD} by several volts more will result in a worst-case common-mode range of zero. In the calculation of $V_{G1}(\text{max})$, we have assumed that V_{SD5} is approximately 0. These results are slightly in error due to the dependence of the threshold voltages upon the bulk-source voltage which has been ignored in this example.

The small-signal analysis of the differential amplifier of Fig. 6.2-1(a) can be accomplished with the assistance of the model shown in Fig. 6.2-7(a). This model can be simplified to that shown in Fig. 6.2-7(b) and is only appropriate for differential analysis when both sides of the amplifier are assumed to be perfectly matched. If this condition is satisfied, then the point where the two sources of M1 and M2 are connected can be considered at ac ground. If we assume that the differential stage is unloaded, then with the output shorted to ac ground, the differential-transconductance gain can be expressed as

$$i_{\text{out}} = \frac{g_{m1}g_{m4}r_{p1}}{1 + g_{m3}r_{p1}} v_{gs1} - g_{m2}v_{gs2} \tag{15}$$

or

$$i'_{\text{out}} \simeq g_{m1}v_{gs1} - g_{m2}v_{gs2} = g_{md}v_{id} \tag{16}$$

where $g_{m1} = g_{m2} = g_{md}$, $r_{p1} = r_{ds1} \| r_{ds3}$ and i'_{out} designates the output current into a short circuit. The unloaded differential voltage gain can be determined by finding the small-signal output resistance of the differential amplifier. It is easy to see that r_{out} is

$$r_{\text{out}} = \frac{1}{g_{ds2} + g_{ds4}} \tag{17}$$

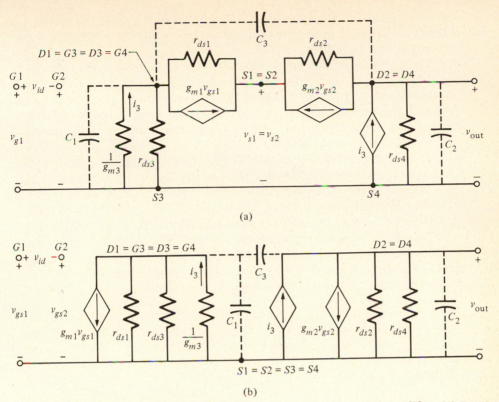

Figure 6.2-7   Small-signal model for the CMOS differential amplifier. (a) Exact model. (b) Simplified equivalent model.

Therefore, the voltage gain is given as the product of g_{md} and r_{out}

$$A_v = \frac{V_{out}}{V_{id}} = \frac{g_{md}}{g_{ds2} + g_{ds4}} \tag{18}$$

Substituting Eqs. (6) of Sec. 6.2 and (9) of Sec. 3.3 in Eq. (18) gives

$$A_v = \frac{V_{out}}{V_{id}} = \frac{(K_1' I_{SS} W_1 / L_1)^{1/2}}{(\lambda_2 + \lambda_4)(I_{SS}/2)} = \frac{2}{\lambda_2 + \lambda_4}\left(\frac{K_1' W_1}{I_{SS} L_1}\right)^{1/2} \tag{19}$$

If we consider r_{out} of Eq. (17) to be equal to R_L of Eq. (8), then Eqs. (8) and (19) are identical. Again we note the dependence of the small-signal gain on the inverse of $I_{SS}^{1/2}$ similar to that of the inverter. This relationship is in fact valid until I_{SS} approaches subthreshold values. Assuming that $W_1/L_1 = 10\ \mu m/10\ \mu m$ and that $I_{SS} = 10\ \mu A$, the small-signal voltage gain of the n-channel differential amplifier is 87. The small-signal gain of the p-channel differential amplifier under the same conditions is 60. This difference is due to the different mobilities.

The common-mode gain of the CMOS differential amplifiers shown in Fig. 6.2-1 is ideally zero. This is because the current-mirror load rejects any common-mode signal. The fact that a common-mode response might exist is due to the mismatches in the differential amplifier. These mismatches consist of a nonunity current gain in the current mirror and geometrical mismatches between M1 and M2 (see Sec. 5.4).

The frequency response of the CMOS differential amplifier is due to the various parasitic capacitors at each node of the circuit. The parasitic capacitors associated with the CMOS differential amplifier are shown as the dotted capacitors in Fig. 6.2-7(b). C_1 consists of C_{gd1}, C_{bd1}, C_{bd3}, C_{gs3}, and C_{gs4}. C_2 consists of C_{bd2}, C_{bd4}, C_{gd2}, and any load capacitance C_L. C_3 consists only of C_{gd4}. In order to simplify the analysis, we shall assume that C_3 is approximately zero. In most applications of the differential amplifier, this assumption turns out to be valid. With C_3 approximately zero, the analysis of Fig. 6.2-7(b) is straightforward. The voltage-transfer function can be written as

$$V_{out}(s) \simeq \frac{g_{m1}}{g_{ds2} + g_{ds4}} \left[\left(\frac{g_{m3}}{g_{m3} + sC_1} \right) V_{gs1}(s) - V_{gs2}(s) \right] \left(\frac{\omega_2}{s + \omega_2} \right) \qquad (20)$$

where ω_2 is given as

$$\omega_2 = \frac{g_{ds2} + g_{ds4}}{C_2} \qquad (21)$$

If we further assume that

$$\frac{g_{m3}}{C_1} \gg \frac{g_{ds2} + g_{ds4}}{C_2} \qquad (22)$$

then the frequency response of the differential amplifier reduces to

$$\frac{V_{out}(s)}{V_{id}(s)} \simeq \left(\frac{g_{m1}}{g_{ds2} + g_{ds4}} \right) \left(\frac{\omega_2}{s + \omega_2} \right) \qquad (23)$$

Thus, the first-order analysis of the frequency response of the differential amplifier consists of a single pole at the output given by $(g_{ds2} + g_{ds4})/C_2$. We shall consider the frequency response of the differential amplifier in more detail later.

The slew-rate performance of the CMOS differential amplifier depends upon the value of I_{ss} and the capacitance from the output node to ac ground. *Slew rate* (SR) is defined as the maximum output-voltage rate, either positive or negative. Since the slew rate in the CMOS differential amplifier is determined by the amount of current that can be sourced or sunk into the output/compensating capacitor, we find that the slew rates of the CMOS differential amplifiers of Fig. 6.2-1 are given by

$$\text{Slew rate} = I_{SS}/C \tag{24}$$

where C is the total capacitance connected to the output node. For example, if $I_{ss} = 10 \mu A$ and $C = 5$ pF the slew rate is found to be 2 volts/micro-seconds. The value of I_{ss} must be increased to increase the slew-rate capability of the differential amplifier.

The noise performance of the CMOS differential amplifier can be due to both thermal and $1/f$ noise. Depending upon the frequency range of interest, one source can be neglected in favor of the other. At low frequencies, $1/f$ noise is important whereas at high frequencies/low currents thermal noise is important. Fig. 6.2-8(a) shows the p-channel differential amplifier with equivalent-noise voltage sources shown at the input of each device. The equivalent-noise voltage sources are those given in Eq. (13) of Sec. 3.2 with the noise of I_{DD} ignored. In this case we solve for the total output-noise current i_{to}^2 at the output of the circuit. Further, let us assume that the output is shorted to ground to simplify calculations. The total output-noise current is found by summing each of the noise-current contributions to get

$$\overline{i}_{to}^2 = g_{m1}^2 v_{n1}^2 + g_{m2}^2 v_{n2}^2 + g_{m3}^2 v_{n3}^2 + g_{m4}^2 v_{n4}^2 \tag{25}$$

Since the equivalent output-noise current is expressed in terms of the equivalent input-noise voltage, we may use

$$\overline{i}_{to}^2 = g_{m1}^2 v_{eq}^2 \tag{26}$$

to get

$$\overline{v}_{eq}^2 = v_{n1}^2 + v_{n2}^2 + (g_{m3}/g_{m1})^2 [v_{n3}^2 + v_{n4}^2] \tag{27}$$

We assume that $g_{m1} = g_{m2}$ and $g_{m3} = g_{m4}$ in the above. Eq. (27) can be expressed in terms of voltage spectral-noise densities and is given as

$$\overline{e}_{eq}^2 = \overline{e}_{n1}^2 + \overline{e}_{n2}^2 + \left(\frac{g_{m3}}{g_{m1}}\right)^2 [\overline{e}_{n3}^2 + \overline{e}_{n4}^2] \tag{28}$$

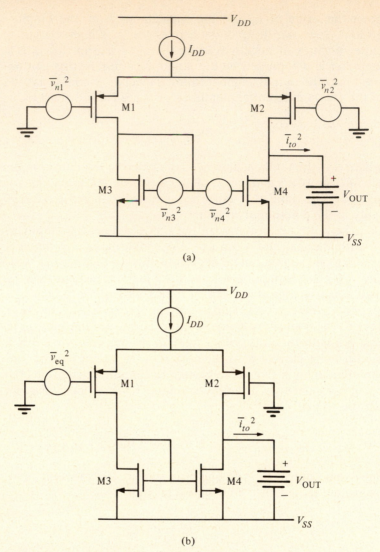

Figure 6.2-8 Noise analysis of a CMOS differential amplifier. (a) Noise model of each device. (b) Equivalent noise model of (a).

Assuming that $\bar{e}_{n1} = \bar{e}_{n2}$ and $\bar{e}_{n3} = \bar{e}_{n4}$, and substituting Eq. (15) of Sec. 3.2 into Eq. (28) results in

$$\bar{e}_{eq(1/f)} = \left[\frac{2B_P}{fW_1L_1}\right]^{1/2} \left[1 + \frac{K'_N B_N}{K'_P B_P}\left(\frac{L_1}{L_3}\right)^2\right]^{1/2} \tag{29}$$

which is the equivalent-input $1/f$ noise for the differential amplifier. By substituting the thermal-noise relationship into Eq. (28) the equivalent-input thermal noise is seen to be

$$\overline{e}_{eq(th)} = \frac{16kT(1 + \eta_1)}{3(2K_p'I_1(W/L)_1]^{1/2}} \left[1 + \frac{K_N'(W/L)_3}{K_P'(W/L)_1} \right] \tag{30}$$

If the load device length is much larger than that of the gain device then the input-referred $1/f$ noise is determined primarily by the contribution of the input devices. Making the aspect ratio of the input device much larger than that of the load device insures that the total thermal-noise contribution is dominated by the input devices.

There are many configurations of the CMOS differential amplifier other than that presented above. One variation of interest uses active-resistor or current–source/sink loads. This variation allows access to the single-ended voltages at each of the drains of M1 and M2. Unfortunately, if a single-ended output voltage is desired, then other circuitry is necessary to accomplish what is automatically done in Figs. 6.2-1 and 6.2-2. Also, the use of the cascode or Wilson current mirror in place of the simple mirror of the differential amplifier studied will increase both the small-signal gain and the output resistance of the differential amplifier. Some of these variations in the basic CMOS differential amplifier will be studied later when considering the design of op amps.

6.3 Cascode Amplifier

The cascode amplifier has two distinct advantages over the inverter amplifier. It provides a higher output impedance similar to the cascode current sink of Fig. 5.3-4 and the cascode current mirror of Fig. 5.4-6. It also reduces the effect of the Miller capacitance on the input of the amplifier, which will be very important in designing the frequency behavior of the op amp. Fig. 6.3-1 shows a simple cascode amplifier consisting of transistors M1, M2, and M3. Except for M2, the cascode amplifier is identical to the current-source CMOS inverter of Sec. 6.1. The primary function of M2 is to keep the small-signal resistance at the drain of M1 low (approximately $1/g_{m2}$ in this case). For the reasons illustrated in the cascode current sink of Fig. 5.3-4, the small-signal resistance looking back into the drain of M2 will be larger than that of M1 without M2. The small-signal gain of the cascode amplifier is approximately the same as the inverter with all of it obtained from the common-gate configuration of M2.

The large-signal voltage-transfer curve of Fig. 6.3-1 is shown in Fig. 6.3-2. The solid line is simulated using the parameters $V_T = \pm 0.7$ volts, $K_N' = 28 \ \mu A/\text{volt}^2$, $K_P' = 8 \ \mu A/\text{volt}^2$, and a channel length modulation factor of 0.01 volts^{-1} for both n- and p-channel devices. The dotted line represents the experimental results when $V_{DD} = 10$ volts, $V_{GG3} = 7.5$ volts, V_{GG2}

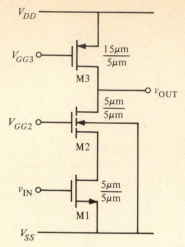

Figure 6.3-1 Simple cascode amplifier.

= 5 volts, and V_{SS} = 0 volts. The graphing of this curve follows the same principles used for the inverters of Sec. 6.1 after the output characteristics of the cascode have been obtained. Fig. 6.3-3 shows the technique for obtaining the voltage-current output characteristics of the cascode configuration consisting of M1 and M2 of Fig. 6.3-1.

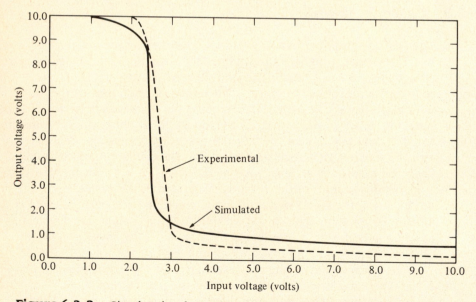

Figure 6.3-2 Simulated and experimental results for Fig. 6.3-1. (V_{DD} = 10 volts, V_{GG3} = 7.5 volts, V_{GG2} = 5 volts, V_{SS} = −0 volts, K'_N = 28 μA/volts^2, K'_P = 8 μA/volts^2, V_T = ±0.7 volts, and λ_N = λ_P = 0.01 volts^{-1})

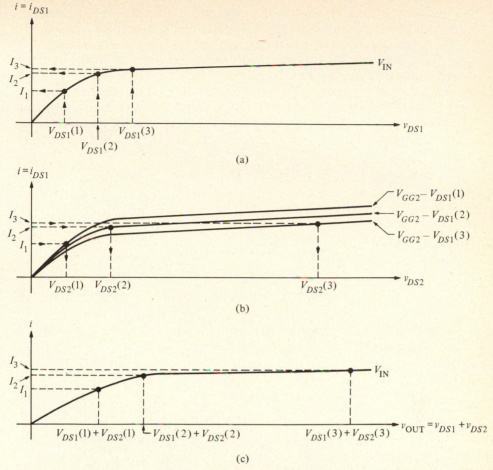

$$i = i_{DS1}$$

$V_{DS1}(1)$ $V_{DS1}(3)$
$V_{DS1}(2)$

v_{DS1}

V_{IN}

(a)

$$i = i_{DS1}$$

$V_{GG2} - V_{DS1}(1)$
$V_{GG2} - V_{DS1}(2)$
$V_{GG2} - V_{DS1}(3)$

$V_{DS2}(1)$ $V_{DS2}(2)$ $V_{DS2}(3)$ v_{DS2}

(b)

i

$V_{DS1}(1) + V_{DS2}(1)$ $V_{DS1}(2) + V_{DS2}(2)$ $V_{DS1}(3) + V_{DS2}(3)$ $v_{OUT} = v_{DS1} + v_{DS2}$

V_{IN}

(c)

Figure 6.3-3 Graphical technique for the construction of the cascode voltage-current output characteristics.

The technique begins with assuming a given value of v_{IN}. This corresponds to the single curve shown in Fig. 6.3-3(a) for the output characteristics of M1. The next step is to select values of v_{DS1}, e.g., $V_{DS1}(1)$, $V_{DS1}(2)$, and $V_{DS1}(3)$ and identify the currents I_1, I_2, and I_3. The next step is to plot the output characteristics of M2 that correspond to the gate voltages $V_{GG2} - V_{DS1}(1)$, $V_{GG2} - V_{DS1}(2)$, and $V_{GG2} - V_{DS1}(3)$. This is illustrated in Fig. 6.3-3(b). Now the currents I_1, I_2, and I_3 are used to identify the voltages $V_{DS2}(1)$, $V_{DS2}(2)$, and $V_{DS2}(3)$. The combined-output characteristic of the cascode configuration corresponding to v_{IN} is found by summing $V_{DS1}(i)$ with $V_{DS2}(i)$ for the current I_i to get one point on the characteristic as shown in Fig. 6.3-3(c). We note that the output characteristic of the cascode configuration has a wider region of nonsaturation behavior and flatter characteristics in the saturation region. This graphing technique is confirmed experimentally by comparing the curves of Fig. 5.4-9(b) with Fig. 5.4-10(b) which are for the output characteristics of the simple and cascode current mirrors.

Fig. 6.3-2 shows that the simple cascode amplifier is capable of swinging to V_{DD}, like the inverter, but cannot reach V_{SS}. The lower limit of v_{OUT}, designated as $v_{OUT}(min)$, can be found as follows. First, assume that both M1 and M2 will be in the nonsaturation region. If we reference all potentials to V_{SS}, we may express the current through each of the devices, M1 through M3, as

$$i_{D1} = \beta_1 \left[(V_{DD} - V_{T1})v_{DS1} - \frac{v_{DS1}^2}{2} \right] \simeq \beta_1 (V_{DD} - V_{T1})v_{DS1} \tag{1}$$

$$i_{D2} = \beta_2 \left[(V_{GG2} - v_{DS1} - V_{T2})(v_{OUT} - v_{DS1}) - \frac{(v_{OUT} - v_{DS1})^2}{2} \right] \tag{2}$$
$$\simeq \beta_2 (V_{GG2} - V_{T2})(v_{OUT} - v_{DS1})$$

and

$$i_{D3} = \frac{\beta_3}{2} (V_{DD} - V_{GG3} - |V_{T3}|)^2 \tag{3}$$

where we have also assumed that both v_{DS1} and v_{OUT} are small, and $v_{IN} = V_{DD}$. We may solve for v_{OUT} by realizing that $i_{D1} = i_{D2} = i_{D3}$ and $\beta_1 = \beta_2$ to get

$$v_{OUT}(min) = \frac{\beta_3}{2\beta_2} (V_{DD} - V_{GG3} - |V_{T3}|)^2 \left[\frac{1}{V_{GG2} - V_{T2}} + \frac{1}{V_{DD} - V_{T1}} \right] \tag{4}$$

Example 6.3-1

Calculation of the Minimum Output Voltage for the Simple Cascode Amplifier

Assume the values and parameters used for the cascode configuration plotted in Fig. 6.3-2 and calculate the value of $v_{OUT}(min)$.

From Eq. (4) we find that $v_{OUT}(min)$ is 0.472 volts. While this matches the simulation results of Fig. 6.3-2, it is approximately a factor of two larger than the experimental results. One of the problems contributing to this discrepancy is that the simulation model parameters are usually measured in the saturation region (see Sec. 4.2), and consequently are not necessarily accurate for M1 and M2 operating in the nonsaturated region.

The small-signal performance of the simple cascode amplifier of Fig. 6.3-1 can be analyzed using the small-signal model of Fig. 6.3-4(a) which has been simplified in Fig. 6.3-4(b). The simplification uses the current-source rearrangement and substitution principles described in Appendix A. Using nodal analysis, we may write

$$[g_{ds1} + g_{ds2} + g_{m2} + g_{mbs2}]v_1 - g_{ds2}v_{out} = -g_{m1}v_{in} \tag{5}$$

$$-[g_{ds2} + g_{m2} + g_{mbs2}]v_1 + (g_{ds2} + g_{ds3})v_{out} = 0 \tag{6}$$

Solving for v_{out}/v_{in} yields

$$\frac{v_{out}}{v_{in}} = \frac{-g_{m1}(g_{ds2} + g_{m2} + g_{mbs2})}{g_{ds1}g_{ds2} + g_{ds1}g_{ds3} + g_{ds2}g_{ds3} + g_{ds3}(g_{m2} + g_{mbs2})}$$

$$\simeq \frac{-g_{m1}}{g_{ds3}} = -\left(\frac{2K'_1 W_1}{L_1 I_D \lambda_3^2}\right)^{1/2} \tag{7}$$

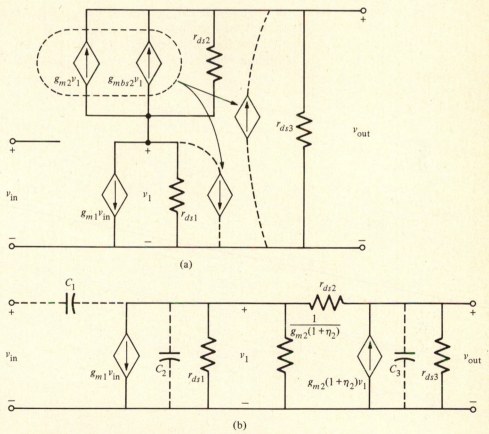

(a)

(b)

Figure 6.3-4 (a) Small-signal model of Fig. 6.3-1. (b) Simplified equivalent model of Fig. 6.3-4(a).

Eq. (7) should be compared with Eq. (14) of Sec. 6.1. The primary difference is that the cascode configuration has made the output resistance of M2 negligible compared with r_{ds3}. We further note the dependence of small-signal voltage gain on the bias current as before. It is also of interest to calculate the small-signal voltage gain from the input v_{in} to the drain of M1 (v_1). From Eqs. (5) and (6) we may write

$$\frac{v_1}{v_{in}} = \frac{-g_{mi}(g_{ds2} + g_{ds3})}{g_{ds1}g_{ds2} + g_{ds1}g_{ds3} + g_{ds2}g_{ds3} + g_{ds3}(g_{m2} + g_{mbs2})}$$

$$\simeq \left(\frac{g_{ds2} + g_{ds3}}{g_{ds3}}\right)\left(\frac{-g_{m1}}{g_{m2} + g_{mbs2}}\right) \simeq \frac{-2g_{m1}}{g_{m2} + g_{mbs2}} = -\left(\frac{W_1L_2}{L_1W_2}\right)^{1/2}\frac{2}{1 + \eta_2}$$

$$(8)$$

It is seen that if the W/L ratios of M1 and M2 are identical and $g_{ds2} = g_{ds3}$, then v_1/v_{in} is approximately -2.

The frequency behavior of the cascode can be studied by reanalyzing Fig. 6.3-4 with the capacitors indicated included. C_1 includes only C_{gd1}, while C_2 includes C_{bd1}, C_{bs2}, and C_{gs2} and C_3 includes C_{bd2}, C_{bd3}, C_{gd2}, C_{gd3}, and any load capacitance C_L. Including these capacitors, Eqs. (5) and (6) become

$$(g_{m2} + g_{mbs2} + g_{ds1} + g_{ds2} + sC_1 + sC_2)v_1 - g_{ds2}v_{out}$$
$$= -(g_{m1} - sC_1)v_{in} \qquad (9)$$

and

$$-(g_{ds2} + g_{m2} + g_{mbs2})v_1 + (g_{ds2} + g_{ds3} + sC_3)v_{out} = 0 \qquad (10)$$

Solving for $V_{out}(s)/V_{in}(s)$ gives,

$$\frac{V_{out}(s)}{V_{in}(s)} = \left[\frac{1}{1 + as + bs^2}\right]\left(\frac{-(g_{m1} - sC_1)(g_{ds2} + g_{m2} + g_{mbs2})}{g_{ds1}g_{ds2} + g_{ds3}(g_{m2} + g_{mbs2} + g_{ds1} + g_{ds2})}\right)$$

$$(11)$$

where

$$a = \frac{C_3(g_{ds1} + g_{ds2} + g_{m2} + g_{mbs2}) + C_2(g_{ds2} + g_{ds3}) + C_1(g_{ds2} + g_{ds3})}{g_{ds1}g_{ds2} + g_{ds3}(g_{m2} + g_{mbs2} + g_{ds1} + g_{ds2})}$$

$$(12)$$

and

$$b = \frac{C_3(C_1 + C_2)}{g_{ds1}g_{ds2} + g_{ds3}(g_{m2} + g_{mbs2} + g_{ds1} + g_{ds2})}$$

$$(13)$$

One of the difficulties with straightforward algebraic analysis is that often the answer, while correct, is meaningless. Such is the case with Eqs. (11) through (13). We can observe that if $s = 0$ Eq. (13) reduces to Eq. (7). Fortunately, we can make some simplifications that bring the results of the above analysis back into perspective. We will develop the method here since it will become useful later when considering the compensation of op amps.

Let us assume that a general second-order polynomial can be written as

$$P(s) = 1 + as + bs^2 = \left(1 - \frac{s}{p_1}\right)\left(1 - \frac{s}{p_2}\right)$$

$$= 1 - s\left(\frac{1}{p_1} + \frac{1}{p_2}\right) + \frac{s^2}{p_1 p_2} \tag{14}$$

Now if $|p_2| \gg |p_1|$, then Eq. (14) can be simplified as

$$P(s) = 1 - \frac{s}{p_1} + \frac{s^2}{p_1 p_2} \tag{15}$$

Therefore we may write p_1 and p_2 in terms of a and b as

$$p_1 = \frac{-1}{a} \tag{16}$$

and

$$p_2 = \frac{-a}{b} \tag{17}$$

The key in this technique is the assumption that the magnitude of the root p_2 is greater than the magnitude of the root p_1. Typically, we are interested in the smallest root so that this technique is very useful. Assuming that the roots of the denominator of Eq. (11) are split apart sufficiently, Eq. (16) gives

$$p_1 = \frac{-[g_{ds1}g_{ds2} + g_{ds3}(g_{m2} + g_{mbs2} + g_{ds1} + g_{ds2})]}{C_3(g_{ds1} + g_{ds2} + g_{m2} + g_{mbs2}) + C_2(g_{ds2} + g_{ds3}) + C_1(g_{ds2} + g_{ds3})} \tag{18a}$$

$$p_1 \cong \frac{-g_{ds3}}{C_3} \tag{18b}$$

The nondominant root p_2 is given as

$$p_2 = \frac{-[C_3(g_{ds1} + g_{ds2} + g_{m2} + g_{mbs2}) + C_2(g_{ds2} + g_{ds3}) + C_1(g_{ds2} + g_{ds3})]}{C_3(C_1 + C_2)}$$

(19a)

$$p_2 \cong \frac{-g_{m2}}{C_1 + C_2}$$

(19b)

Assuming that C_1, C_2, and C_3 are the same order of magnitude, and that g_{m2} is greater than g_{ds3}, then $|p_1|$ is smaller than $|p_2|$ (closer to the origin). Therefore the approximation technique is valid. Eqs. (18) and (19) show a typical trend of CMOS circuits. The poles of the frequency response tend to be associated with the inverse product of the resistance and capacitance of a node to ground. For example, the inverse RC product of the output node is approximately g_{ds3}/C_3 whereas the inverse RC product of the node where v_1 is defined is approximately $g_{m2}/(C_1 + C_2)$.

A zero also occurs in the frequency response and has the value of

$$z_1 = \frac{g_{m1}}{C_1}$$

(20)

The intuitive reason for this zero is the result of two paths from the input to the output. One path couples directly through C_1 and the other through the $g_{m1}v_{in}$ controlled source.

The above analysis of Fig. 6.3-1 did not include a very important property of the cascode amplifier because we assumed the circuit was voltage driven. In general, the source resistance in a CMOS circuit is large enough so that it cannot be neglected as above. Fig. 6.3-5(a) shows an inverter consisting of M1 and M2 and it is driven from the drains of M3 and M4. The equivalent-source resistance is $(g_{ds3} + g_{ds4})^{-1}$ and is not negligible. In this case the capacitance indicated as C_{gd1} becomes multiplied by the Miller effect and creates a large capacitance in parallel with R_1. Let us consider the small-signal circuit in Fig. 6.3-5(b). Assuming the input is I_{in}, the nodal equations are

$$[G_1 + s(C_1 + C_2)]V_1 - sC_2V_{out} = I_{in}$$

(21)

and

$$(g_{m1} - sC_2)V_1 + [G_3 + s(C_2 + C_3)]V_{out} = 0$$

(22)

The values of G_1 and G_3 are $g_{ds3} + g_{ds4}$ and $g_{ds1} + g_{ds2}$, respectively. C_1 is C_{gs1}, C_2 is C_{gd1}, and C_3 is the sum of C_{bd1}, C_{bd2}, and C_{gd2}. Solving for $V_{out}(s)/I_{in}(s)$

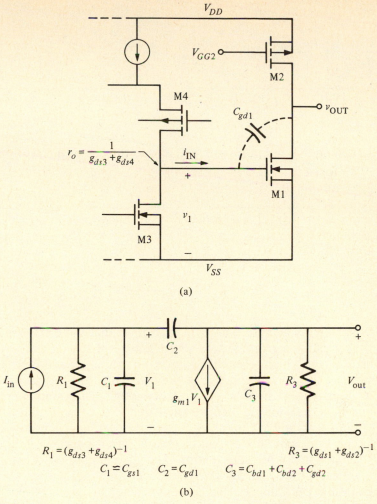

Figure 6.3-5 (a) Situation where inverter M1 — M2 is driven from a high resistance source. (b) Small-signal model.

gives

$$\frac{V_{out}(s)}{I_n(s)} =$$

$$\frac{(sC_2 - g_{m1})}{G_1G_3 + s[G_3(C_1 + C_2) + G_1(C_2 + C_3) + g_{m1}C_2] + (C_1C_2 + C_1C_3 + C_2C_3)s^2}$$

(23)

or

$$\frac{V_{out}(s)}{I_{in}(s)} = \left(\frac{-g_{m1}}{G_1 G_3}\right)$$

$$\frac{[1 - s(C_2/g_{m1})]}{1 + [R_1(C_1 + C_2) + R_3(C_2 + C_3) + g_{m1}R_1R_3C_2]s + (C_1C_2 + C_1C_3 + C_2C_3)R_1R_3s^2}$$

(24)

Assuming that the poles are split allows the use of the previous technique to get,

$$p_1 = \frac{-1}{R_1(C_1 + C_2) + R_3(C_2 + C_3) + g_{m1}R_1R_3C_2} \simeq \frac{-1}{g_{m1}R_1R_3C_2} \quad (25)$$

and

$$p_2 \simeq \frac{-g_{m1}C_2}{C_1C_2 + C_1C_3 + C_2C_3} \quad (26)$$

Obviously, p_1 is more dominant than p_2 so that the technique is valid. Eq. (25) illustrates an important disadvantage of the regular inverter if it is driven from a high resistance source. The Miller effect essentially takes the capacitance, C_2, multiplies it by the low-frequency voltage gain from V_1 to V_{out}, and places it in parallel with R_1, resulting in a dominant pole (see Problem 32). The equivalent capacitance due to C_2 seen at node 1 is called the Miller capacitance. The Miller capacitance can have a negative effect on a circuit from several viewpoints. One is that it creates a dominant pole. A second is that it provides a large capacitive load to the driving circuit.

One of the advantages of the cascode amplifier is that it greatly reduces the Miller capacitance. This is accomplished by keeping the low-frequency voltage gain across M1 low so that C_2 is not multiplied by a large factor. Unfortunately, to repeat the analysis of Fig. 6.3-4(b) with a current-source driver would lead to a third-order denominator polynomial which masks the results. An intuitive approach is to note that the cascode circuit essentially makes the load resistance in the above analysis approximately equal to the reciprocal of the transconductance of the cascode device, M2, in Fig. 6.3-1 (beware that this approximation deteriorates as the load impedance seen by the drain of M2 grows much larger than r_{ds2}). Consequently, R_3 of Eq. (25) becomes $\cong 1/g_m$ of the cascode device. Thus if the two transconductances are approximately equal, the new location of the input pole is

$$p_1 \cong \frac{-1}{R_1(C_1 + C_2) + R_3(C_2 + C_3) + R_1C_2} \quad (27)$$

which is much larger than that of Eq. (25). Eq. (8) also confirms this result in that the gain across M1 is limited to less than 2 so that the Miller effect is minimized. This property of the cascode amplifier—removing a dominant pole at the input—is very useful in controlling the frequency response of an op amp.

The output resistance can be found by combining in parallel the small-signal output resistance of Fig. 5.3-4(a) with r_{ds3} of Fig. 6.3-1. Therefore, the small-signal output resistance of the cascode amplifier is given as

$$r_{out} = [r_{ds1} + r_{ds2} + g_{m2}r_{ds1}r_{ds2}(1 + \eta_2)] \| r_{ds3} \simeq r_{ds3} \qquad (28)$$

We note that although the cascode configuration consisting of M1 and M2 has a high output resistance, that the lower resistance of M3 does not allow the realization of the high output resistance. For this reason, the current source load is often replaced by a cascode current-source load as shown in Fig. 6.3-6(a). The small-signal model of this circuit is shown in Fig. 6.3-6 (b). It is most efficient to consider first the output resistance of this circuit. The small-signal output resistance can be found from Eq. (28) as

$$r_{out} = [r_{ds1} + r_{ds2} + g_{m2}r_{ds1}r_{ds2}(1 + \eta_2)] \| [r_{ds3} + r_{ds4} + g_{m3}r_{ds3}r_{ds4}(1 + \eta_3)]$$
$$\simeq [g_{m2}r_{ds1}r_{ds2}] \| [g_{m3}r_{ds3}r_{ds4}] \qquad (29)$$

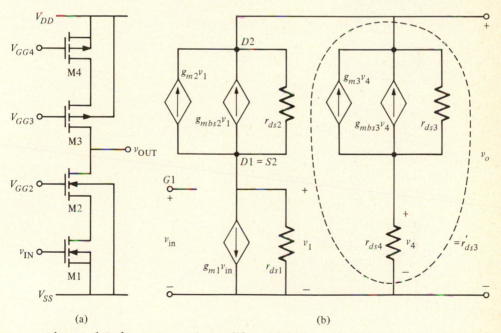

(a)   (b)

Figure 6.3-6 (a) Cascode amplifier with higher gain and output resistance. (b) Small-signal model of (a).

In terms of the large signal model parameters the small signal output resistance is

$$r_{out} \cong \frac{I_D^{-1.5}}{\left(\dfrac{\lambda_1 \lambda_2}{[2K_2'(W/L)_2]^{1/2}}\right) + \left(\dfrac{\lambda_3 \lambda_4}{[2K_3'(W/L)_3]^{1/2}}\right)} \tag{30}$$

Knowing r_{out}, the gain is simply

$$A_v = -g_{m1} r_{out} \cong -g_{m1}\{[g_{m2} r_{ds1} r_{ds2}] \,\|\, [g_{m3} r_{ds3} r_{ds4}]\} \tag{31}$$

$$\cong \frac{\{[2K_1'(W/L)_1]^{1/2}\} I_D^{-1}}{\left(\dfrac{\lambda_1 \lambda_2}{[2K_2'(W/L)_2]^{1/2}}\right) + \left(\dfrac{\lambda_3 \lambda_4}{[2K_3'(W/L)_3]^{1/2}}\right)}$$

Eqs. (30) and (31) are rather surprising in that the voltage gain is proportional to I_D^{-1} and the output resistance varies inversely with the 3/2 power of I_D resistance.

Let us use an example to compare the results obtained so far.

Example 6.3-2

Comparison of the Cascode Amplifier Performance

Calculate the small-signal voltage gain, output resistance, the dominant pole, and the nondominant pole for the cascode amplifier of Fig. 6.3-1 and Fig. 6.3-6 (a). Assume that $I_D = 50$ microamperes, that all W/L ratios are unity with 10 μm channel lengths, and that the parameters of Table 3.1-2 are valid. The capacitors are assumed to be: $C_{gd} = 3.5$ fF, $C_{gs} = 30$ fF, $C_{bsn} = C_{bdn} = 24$ fF, $C_{bsp} = C_{bdp} = 12$ fF, and $C_L = 1$ pF.

The simple cascode amplifier of Fig. 6.3-1 has a small-signal voltage of -82.5 as calculated from the approximate expression in Eq. (7). The output resistance is found from Eq. (28) as 1 MΩ. The dominant pole is found from Eq. (18) as 159 KHz. The nondominant pole is found from Eq. (19) as 77 MHz.

The cascode amplifier of Fig. 6.3-6(a) has a voltage gain of -995 as found from Eq. (31). Eq. (30) gives an output resistance of 24.1 MΩ. The dominant pole is found from the relationship $1/RC$, where R is the output resistance and C is the load capacitance. This calculation yields a dominant pole at 6.6 KHz. There is a nondominant pole associated with the source of M2. This pole can be calculated in the same way as the simple cascode. Assuming the resistance seen at the source of M2 is $1/g_{m2}$, the nondominant pole is approximately 77 MHz.

The small-signal voltage gain of Fig. 6.3-1 [or Fig. 6.3-6(a)] can be increased by increasing the dc current in M1 without changing the current in M2 and the other transistors. This can be done by simply connecting a current source from V_{DD} to the drain of M1 (source of M2). It can be shown that the gain is increased by the square root of the ratio of I_{D1} to I_{D2} (see Problems 33 and 34).

This section has introduced a very useful component in analog integrated-circuit design. The cascode amplifier is very versatile and gives the designer more control over the small-signal performance of the circuit than was possible in the inverter amplifier. In addition, the cascode circuit provides extremely high voltage gains in a single stage with a well defined dominant pole. Both of these characteristics will be used in the more complex circuits yet to be studied.

6.4   *Output Stages*

So far we have considered three types of CMOS amplifiers—the inverter (Sec. 6.1), the differential amplifier (Sec. 6.2), and the cascode amplifier (Sec. 6.3). A common characteristic of all of these is a large output resistance. A large output resistance can be undesirable when the load consists of a small resistance and/or large capacitance. A small load resistance requires a large current in order to provide a large output-voltage swing. A large-load capacitor requires large output currents to supply charging currents needed to meet transient response requirements. In order to provide a sufficient output current on a steady-state or transient basis, it is necessary to use a low-resistance output stage.

The primary objective of the CMOS output stage is to function as a current transformer. Most output stages have a high current gain and a low voltage gain. The specific requirements of an output stage might be: (1) provide sufficient output power in the form of voltage or current, (2) avoid signal distortion, (3) be efficient, and (4) provide protection from abnormal conditions (short circuit, over temperature, and so on). The second requirement results from the fact that the signal swings are large and that nonlinearities normally not encountered in small-signal amplifiers will become important. The third requirement is born out of the need to minimize power dissipation in the driver transistors themselves compared with that dissipated in the load. The fourth requirement is normally met with CMOS output stages since MOS devices are, by nature, self-limiting.

We shall consider several approaches to implementing the output amplifier in this section. These approaches include the Class-A amplifier, source followers, the push-pull amplifier, the use of the substrate bipolar junction transistor (BJT), and the use of negative feedback. Each one of these amplifiers will be considered briefly from the viewpoint of each of the above applicable requirements.

In order to reduce the output resistance and increase the current driving capability, a straightforward approach is simply to increase the bias current in the output stage. Fig. 6.4-1 shows a CMOS inverter with a current-source load. The load of this inverter consists of a resistance R_L and a capacitance C_L. There are several ways to specify the performance of the output amplifier. One is to specify the ac output resistance of the amplifier, which in the case of Fig. 6.4-1 is

$$r_{out} = \frac{1}{g_{ds1} + g_{ds2}} = \frac{1}{(\lambda_1 + \lambda_2)I_D} \tag{1}$$

Another is to specify the output swing V_P for a given R_L. In this case, the maximum current to be sourced or sunk is equal to V_P/R_L. The maximum sinking current of the simple output stage of Fig. 6.4-1 is given as

$$I_{OUT}^- = \frac{K_1' W_1}{2L_1} (V_{DD} - V_{SS} - V_{T1})^2 \tag{2}$$

where it has been assumed that v_{IN} can be taken to V_{DD}. The maximum sourcing current of the simple output stage of Fig. 6.4-1 is given as

$$I_{OUT}^+ = \frac{K_2' W_2}{2L_2} (V_{DD} - V_{GG2} - |V_{T2}|)^2 \tag{3}$$

It can be seen from Eqs. (2) and (3) that the maximum sourcing current will typically be the limit of the output current.

The capacitor C_L of Fig. 6.4-1 also places a requirement on the output current through the slew-rate specification. This limit can be expressed as

$$|I_{OUT}| \cong C_L \left(\frac{dv_{OUT}}{dt}\right) = C_L(SR) \tag{4}$$

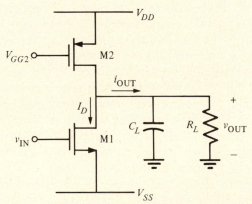

Figure 6.4-1   Simple output amplifier using a class-A, current-source inverter.

This approximation becomes poor when the load resistance shunting C_L is low enough to divert a significant portion of I_{OUT} making it unavailable as charging current. For such cases, the familiar exponential relationship describing the voltage across C_L is needed for accuracy. Therefore, when designing an output stage, it is necessary to consider the effects of both R_L and C_L.

Example 6.4-1

Design of a Simple Class-A Output Stage

Use the values of Table 3.1-2 and design the W/L ratios of M1 and M2 so that a voltage swing of ± 2 volts and a slew rate of $\cong 1$ volt/μs is achieved if $R_L = 20$ KΩ and $C_L = 1000$ pF. Assume that $V_{DD} = -(V_{SS}) = 5$ volts and $V_{GG2} = 2$ volts. Channel lengths are 10 μm.

Let us first consider the effects of R_L. The peak output current must be ± 100 μA. In order to meet the SR requirements a current magnitude of ± 1 mA is needed to charge the load capacitance. Since this current is so much larger than the current needed to meet the voltage specification across R_L, we can safely assume that all of the current supplied by the inverter is available to charge C_L. Using a value of ± 1 mA W_1/L_1 needs to be 15 μm/10 μm and W_2/L_2, 625 μm/10 μm.

The small-signal gain of this amplifier is -7.53. This is low because of the low output resistance in shunt with R_L.

Efficiency is defined as the ratio of the power dissipated in R_L to the power required from the power supplies. For Fig. 6.4-1, the efficiency is given as

$$\text{Efficiency} = \left(\frac{V_{out}(\text{peak})}{V_{DD} + |V_{SS}|} \right)^2 \qquad (5)$$

If $V_{DD} = -(V_{SS})$, the maximum efficiency of the class-A output stage is 25%.

An amplifier's distortion can be characterized by the influence of the amplifier upon a pure sinusoidal signal. Distortion is caused by the nonlinearity of the transfer curve of the amplifier. If a pure sinusoid given as

$$V_{in}(\omega) = V_p \sin(\omega t) \qquad (6)$$

is applied to the input, the output of an amplifier with distortion will be

$$V_{out}(\omega) = a_1 V_p \sin(\omega t) + a_2 V_p \sin(2\omega t) + \cdots + a_n \sin(n\omega t) \qquad (7)$$

Harmonic distortion (*HD*) for the *i*th harmonic can be defined as the ratio of the magnitude of the *i*th harmonic to the magnitude of the fundamental. For example, second-harmonic distortion would be given as

$$HD_2 = \frac{a_2}{a_1} \tag{8}$$

Total harmonic distortion (*THD*) is defined as the square root of the ratio of the sum of all of the second and higher harmonics to the magnitude of the first or fundamental harmonic. Thus, *THD* can be expressed in terms of Eq. (7) as

$$THD = \frac{[a_2^2 + a_3^2 + \cdots + a_n^2]^{1/2}}{a_1} \tag{9}$$

The distortion of Fig. 6.4-1 will not be good because of the nonlinearity of the voltage transfer curve for large-signal swing as shown in Sec. 6.1.

A second approach to implementing an output amplifier is to use the common-drain or source-follower configuration of the MOS transistor. This configuration has both large current gain and low output resistance. Unfortunately, since the source is the output node, the MOS device becomes dependent on the body effect. The body effect causes the threshold voltage V_T to increase as the output voltage is increased, creating a situation where the maximum output is substantially lower than V_{DD}. Fig. 6.4-2 shows a general configuration for the CMOS source follower. It is seen that two n-channel devices are used, rather than a p-channel and an n-channel. It is possible to use a push-pull configuration of the source follower, but we shall not consider that circuit here. Although the gate of M2 can be taken to the output to form an active resistor, we will consider only the source follower which uses a current-source load.

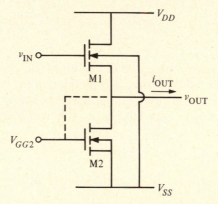

Figure 6.4-2   Common-drain (source-follower) output amplifier.

The large-signal considerations of the source follower will illustrate one of its disadvantages. Fig. 6.4-2 shows that $v_{OUT}(\text{min})$ can be essentially V_{SS} because of the common-source device M2 and the constant gate voltage V_{GG2}. The maximum value of v_{OUT} is given as

$$v_{OUT}(\text{max}) = V_{DD} - V_{T1} \tag{10}$$

assuming that v_{IN} can be taken to V_{DD}. However, V_{T1} is a function of v_{OUT} so that we must substitute Eq. (2) of Sec. 3.1 into Eq. (10) and solve for v_{OUT}. To simplify the mathematics, we approximate Eq. (2) of Sec. 3.1 as

$$V_{T1} \cong V_{T0} + \gamma(v_{SB})^{1/2} = V_{T01} + \gamma_1(v_{out}(\text{max}) - V_{SS})^{1/2} \tag{11}$$

Substituting Eq. (11) into Eq. (10) and solving for v_{OUT} gives

$$v_{OUT}(\text{max}) \cong V_{DD} + \frac{\gamma_1}{2} - V_{T01} - \frac{\gamma_1}{2}[\gamma_1^2 + 4(V_{DD} - V_{SS} - V_{T01})]^{1/2} \tag{12}$$

Using the nominal values of Table 3.1-2 and assuming that $V_{DD} = -V_{SS} = 5$ volts, we find that $v_{OUT}(\text{max})$ is approximately 0.66 volts. This limit could be alleviated in a p-well process by placing M1 in its own p-well and connecting source to bulk.

The maximum output-current sinking and sourcing of the source follower is considered next. The maximum output-current sinking is determined by M2 just like the maximum output-current sourcing of M2 in Fig. 6.4-1. The maximum output-current sourcing is determined by M1 and v_{IN}. If we assume that v_{IN} can be taken to V_{DD}, then the maximum value of I_{OUT} is given by

$$I_{OUT}^{+} = \frac{K_1' W_1}{2L_1}[V_{DD} - v_{OUT}(\text{max})]^2 \tag{13}$$

Assuming a W_1/L_1 of 10 and using the value of $v_{OUT}(\text{max})$ as calculated above, the maximum value of I_{OUT} is 1.60 mA.

The efficiency of the source follower can be shown to be similar to the class-A amplifier of Fig. 6.4-1 (see Problem 39). The distortion of the source follower will be better than the class-A amplifier because of the inherent negative feedback of the source follower.

Fig. 6.4-3 shows the small-signal model for the source follower. The effect of the bulk is implemented by the g_{mbs1} transconductance. The small-signal voltage gain can be found as

$$\frac{V_{out}}{V_{in}} = \frac{g_{m1}}{g_{ds1} + g_{ds2} + g_{m1}(1 + \eta_1)} \simeq \frac{1}{1 + \eta_1} \tag{14}$$

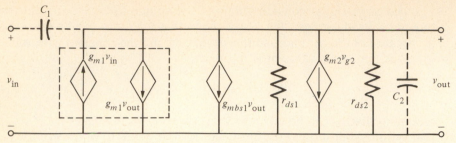

Figure 6.4-3   Small-signal model of Fig. 6.4-2.

If we assume that $V_{DD} = 5$ V, $V_{SS} = -5$ V, $V_{out} = 0$ V, $W_1/L_1 = 100$ μm/10 μm, $W_2/L_2 = 10$ μm/10 μm, and $I_D = 100$ μA, then using the parameters of Table 3.1-2 we find that $\eta_1 = 0.2723$ and the small-signal voltage gain is 0.786. If the bulk effect were not present, $\eta_1 = 0$, the small-signal voltage gain would become 0.951. Because of the bulk effect, the small-signal voltage gain of MOS source followers is less than unity.

The frequency response of the source follower is determined by two capacitances designated as C_1 and C_2 in the small-signal model of Fig. 6.4-3. C_1 consists of capacitances connected between the input and output of the source follower, which are primarily C_{gs1}. C_2 consists of capacitances connected from the output of the source follower to ground. This includes C_{gd2}, C_{bd2}, C_{bs1} and C_L of the next stage. The small-signal frequency response can be found as

$$\frac{V_{out}(s)}{V_{in}(s)} = \frac{(g_{m1} + sC_1)}{g_{ds1} + g_{ds2} + g_{m1}(1 + \eta_1) + s(C_1 + C_2)} \qquad (15)$$

The pole is located at approximately $-g_{m1}/(C_1 + C_2)$ which is much greater than the dominant pole of the inverter, differential amplifier, or the cascode amplifier. The presence of a zero leads to the possibility that in most cases the pole and zero will provide some degree of cancellation leading to a broadband response.

A third method of designing output amplifiers is to use the push-pull amplifier. The push-pull amplifier has the advantage of better efficiency. It is well known that a class-B, push-pull amplifier has a maximum efficiency of 78.5% which means that less quiescent current is needed to meet the output-current demands of the amplifier. Smaller quiescent currents imply smaller values of W/L and smaller area requirements. There are many versions of the push-pull amplifier. For example, Fig. 6.4-1 could be a push-pull amplifier if the gate of M2 is simply connected to v_{IN}. This configuration has the disadvantage that large quiescent current flows when operated in the high-gain region (i.e. class AB operation). If a voltage V_{TR} is inserted between the gates as shown in Fig. 6.4-4, then higher efficiency can be

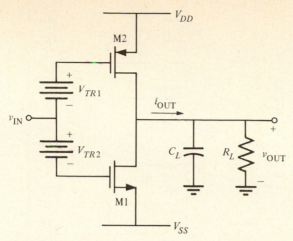

Figure 6.4-4 Push-pull, inverting CMOS amplifier.

obtained. The usefulness of this circuit is shown by considering the case when v_{IN} is precisely at the threshold of the n-channel device and V_{TR} is such that the p-channel device is also operating at the verge of turn-on. If v_{IN} is perturbed in the positive direction, then the p-channel device turns off and all of the current in the n-channel sinks the load current. Similar action occurs when v_{IN} is perturbed negatively, with the result that all of the load current is sourced by the p-channel device. One can easily see that no current is wasted because all of the current supplied flows into (or out of) the load. This configuration does not reduce the output resistance although the distortion will be slightly improved because of the symmetry of the voltage-transfer curve around the midpoint.

An example of how Fig. 6.4-4 might be implemented is illustrated in Fig. 6.4-5. The operation (class AB or B) can be determined by the voltages at the gates of M3 and M4. When the input is taken positive, the current in M1 increases and the current in M2 decreases. If the operation is class B, then M2 turns off. As the current in M1 increases, it is mirrored as an increasing current in M8, which provides the sinking capability for the output current. When v_{IN} is decreased, M6 can source output current.

The push-pull concept can also be applied to the source follower as shown in Fig. 6.4-6(a). Again, the value of V_{TR} determines whether the operation is class AB or class A. This circuit has all of the desired properties of an output stage except for signal swing. Fig. 6.4-6(b) shows how Fig. 6.4-6(a) could be implemented. In this case, we note that the gate of M2 will not reach V_{DD} for the maximum positive output swing.

In order to reduce the output resistance and decrease the area of output stages, the substrate bipolar junction transistor available in the standard CMOS process has been used. For example, in a p-well process, a sub-

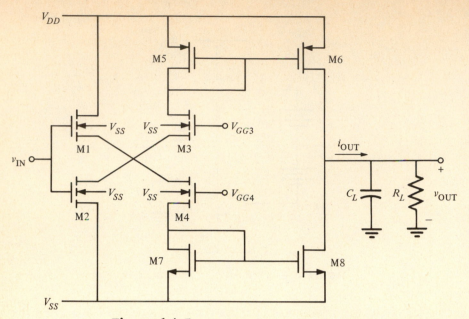

Figure 6.4-5 Implementation of Fig. 6.4-4.

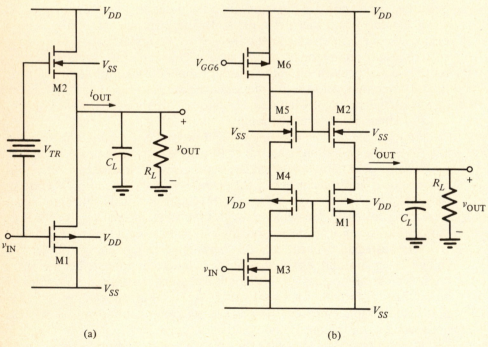

(a)

(b)

Figure 6.4-6 (a) Push-pull source follower. (b) Implementation of (a).

strate NPN BJT is available (see Sec. 2.5). Since the collector must be tied to V_{DD}, the push-pull follower configuration is suitable for the substrate BJT. The advantage of using the BJT is that the output resistance is approximately $1/g_m$, which for a BJT can be in the 100 Ω range. The disadvantage of the BJT is that the positive and negative parts of the voltage transfer curve are not symmetrical and therefore large distortion is encountered. Another disadvantage of the BJT is that as it begins to source more current, more base current is required. It is difficult for the driver to provide this base current when the base is approaching V_{DD}.

A technique which has proven very useful in lowering the output resistance of a CMOS output stage and maintaining its other desirable properties is the use of current negative feedback. The push-pull, inverting CMOS amplifier of Fig. 6.4-4 is very attractive from all viewpoints but output resistance. Fig. 6.4-7 shows a configuration using two differential-error amplifiers to sample the output and input and apply negative feedback to the gate of the common-source MOS transistors. The error amplifiers must be designed to turn on M1 (or M2) to avoid crossover distortion and yet maximize efficiency. The output resistance will be approximately equal to that of Fig. 6.4-4 divided by the loop gain. Since the circuit has sufficient gain, the error amplifiers can be replaced by a resistive feedback network as shown in Fig. 6.4-8. The resistors could be polysilicon or could be MOS transistors properly biased. If the resistors were equal, the output resistance of Fig. 6.4-4 would be divided by approximately $g_{m1}R_L/2$ (see Problem 45).

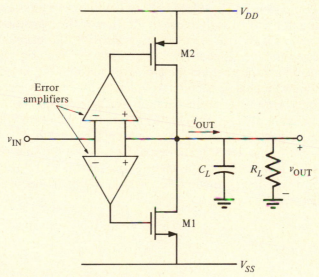

Figure 6.4-7   Use of negative feedback to reduce the output resistance of Fig. 6.4-4.

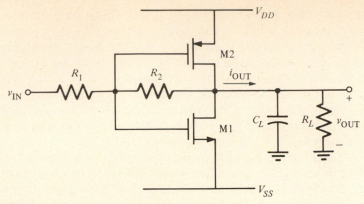

Figure 6.4-8 Use of resistive feedback to decrease the output resistance of Fig. 6.4-4.

We have not discussed how to protect the output amplifier from abnormal conditions. This subject and others not included in this section are best left to specific applications in the following chapters. The general principles of output amplifiers have been identified and illustrated.

6.5 High-Gain Amplifier Architectures

This section starts the transition from the simple circuits of Table 1.1-2 to more complex circuits. While not all complex circuits are amplifiers, the subject is general enough to use in illustrating the transition. The high-gain amplifier is a widely used circuit in analog-circuit design and will serve as the step to the next higher level of complexity—analog systems.

The philosophy behind the high-gain amplifier is based on the concept of feedback. In analog circuits, we must be able to precisely define transfer functions. A familiar representation of this concept is illustrated by the block diagram of Fig. 6.5-1. In this diagram, X may be a voltage or current, A is the high-gain amplifier, F is the feedback network, and the feedback signal X_f is subtracted from the input signal X_s in the summer. If we assume the signal flow is unidirectional as shown and A and F are independent of the source or load resistance (not shown), then the overall gain of the amplifier can be written as

$$A_f = \frac{X_o}{X_s} = \frac{A}{1 + AF} \qquad (1)$$

The principle of the high-gain amplifier can be seen from Eq. (1). If A is sufficiently large, then even though the gain through the feedback network

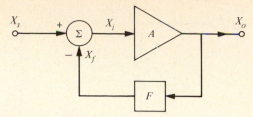

Figure 6.5-1   A general, signal-loop, negative feedback circuit.

may be less than unity, the magnitude of *AF* is much greater than unity. Consequently, Eq. (1) reduces to

$$A_f = \frac{X_o}{X_s} \simeq \frac{1}{F} \tag{2}$$

In order to precisely define A_f we need only define F, if A is sufficiently large. Typically, F is implemented with passive components such as resistors or capacitors.

The high-gain amplifier is defined in terms of Fig. 6.5-1 as

$$A = \frac{X_o}{X_i} \tag{3}$$

Because X can be voltage or current there are four different types of high-gain amplifiers that will be examined in this section. The first type of high-gain amplifier to be considered is called *voltage-controlled, current-source* (VCCS). Fig. 6.5-2(a) illustrates the use of a VCCS, which is represented by the circuit within the dotted box. R_i is the input resistance, G_m is the gain, and R_o is the output resistance. R_s represents the resistance of the external source V_s and R_L represents the load resistance. The loaded VCCS gain can be expressed as

$$G_M = \frac{G_m R_o R_i}{(R_i + R_s)(R_o + R_L)} \tag{4}$$

For an ideal VCCS, R_i and R_o should be infinite so that G_m approaches G_M. The VCCS is sometimes called the *operational transconductance amplifier* (OTA).

The architecture that will implement the VCCS must have a high input resistance, large transconductance gain, and a high output resistance. In most practical implementations of Fig. 6.5-1, the summing junction is incorporated within the amplifier A. In this case, a differential input is

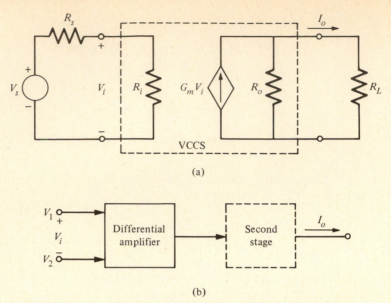

Figure 6.5-2   (a) VCCS circuit. (b) Possible architecture of (a).

desired. The differential input is also desirable from the viewpoint of performance in an integrated-circuit implementation. With this insight and the material presented in the earlier sections of this chapter and the previous chapter, we can propose the architecture of Fig. 6.5-2(b) as an implementation of Fig. 6.5-2(a). The input stage will be a differential amplifier similar to that studied in Sec. 6.2. Since the output resistance of the differential amplifier is reasonably high, a simple differential amplifier may suffice to implement the VCCS. If more gain is required, a second stage consisting of an inverter can be added. If a higher output resistance is required, the differential amplifier transistors M1 through M4 can be replaced with cascode equivalents. If both higher output resistance and more gain is required, then the second stage could be a cascode with a cascode load [Fig. 6.3-6(a)]. The choices depend upon the specifications of the VCCS and the judgment of the designer.

Fig. 6.5-3(a) shows the case where X_o and X_i of A are both voltage. This amplifier is called *voltage-controlled, voltage-source* (VCVS). R_s and R_L are the source and load resistances, respectively. The loaded gain of the VCVS can be expressed as

$$A_V = \frac{A_v R_i R_L}{(R_s + R_i)(R_o + R_L)}$$ (5)

It is seen that the ideal VCVS requires R_i to be infinite and R_o to be zero, so that A_V approaches A_v. The architecture of a VCVS begins with a differential

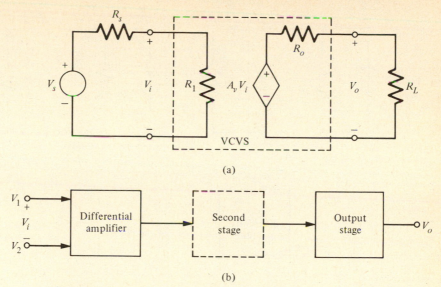

Figure 6.5-3   (a) VCVS circuit. (b) Possible architecture of (a).

amplifier as the input stage, as shown in Fig. 6.5-3(b). If sufficient gain is available from the differential amplifier, then only an output stage is needed in order to reduce the output resistance. If more gain is required, then a second stage consisting of an inverter or cascode is needed, as illustrated in Fig. 6.5-3(b).

Fig. 6.5-4(a) shows the case where X_o and X_i of A are both current. This amplifier is called *current-controlled, current-source* (CCCS). R_s and R_L are the source and load resistances, respectively. The loaded gain of the CCCS can be expressed as

$$A_I = \frac{A_i R_s R_o}{(R_s + R_i)(R_o + R_L)} \tag{6}$$

It is seen that the ideal CCCS requires R_i to be zero and R_o to be infinite, so that A_I approaches A_i. The architecture of a CCCS has a problem with the input in terms of the circuits we have studied. The input should be current-driven and have low resistance. Let us propose the architecture shown in Fig. 6.5-4(b), which consists of a current, differential amplifier and a second stage if needed. A possible implementation of the current, differential amplifier is shown in Fig. 6.5-4(c). It consists of two current mirrors with the output current of the first subtracting from the input of the second. As a consequence, the current in M4, assuming ideal current mirrors, is $I_1 - I_2$. This current can be taken to a current-source load to generate a voltage

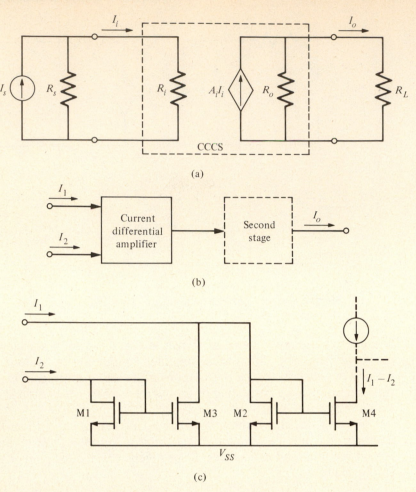

Figure 6.5-4   (a) CCCS circuit. (b) Possible architecture of (a). (c) Implementation of a current differential amplifier.

proportional to the difference between I_1 and I_2. The input resistance is $1/g_m$ for both I_1 and I_2—which is the lowest resistance possible for the MOS device without using feedback. Note that input CMR and CMRR are now in terms of current and not voltage. A second stage is necessary in the CCCS architecture because the current, differential input stage will have a low current gain. If the output resistance must be high, a cascode amplifier can be used as the second stage as was done in the VCCS.

The last type of amplifier is shown in Fig. 6.5-5(a) where X_o is voltage and X_i is current. This amplifier is called *current-controlled, voltage-source*

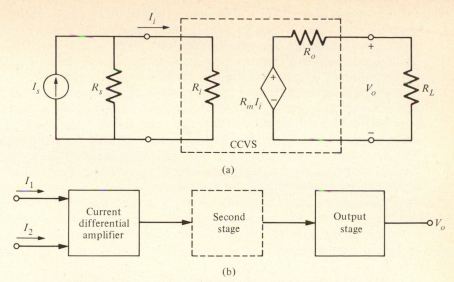

Figure 6.5-5 (a) CCVS circuit. (b) Possible architecture of (a).

(CCVS). The loaded amplifier gain can be expressed as

$$R_M = \frac{R_m R_s R_L}{(R_i + R_s)(R_o + R_L)} \tag{7}$$

The ideal CCVS has R_s infinite and R_o zero so that R_M becomes equal to R_m. Using the current, differential input stage of Fig. 6.5-4(c) allows us to propose the architecture shown in Fig. 6.5-5(b) for implementing the CCVS. If the output stage has high input resistance, a second gain stage may not be necessary if the gain is sufficient.

The architecture proposed for the high-gain amplifier in this section should be considered only as a starting point in designing such an amplifier. In most cases, the boundaries between the stages will become fuzzy as the designer begins to optimize performance. The advantage of the architectural viewpoint is that it identifies the generic elements of the design. Thus designs that are completely different in the final version can start from a common basis using generic elements.

6.6 *Summary*

This chapter has introduced the important subject of basic CMOS amplifiers. The inverter, differential, cascode, and output amplifiers were presented. We then saw how the various stages could be assembled to implement a high-gain amplifier and identified the generic building blocks at the simple circuit level. Although we have now covered the primary building blocks, many different implementations have yet to be considered. Specific implementations will be given as needed in the use of simple circuits to realize complex circuits and complex circuits to realize systems.

The principles introduced in this chapter include the method of characterizing an amplifier from both a large-signal and small-signal basis. The dependence of small-signal performance upon the dc or large-signal conditions continued to be evident and is a very important principle to grasp. Another important problem is that the analysis of a circuit can quickly be complicated beyond the designer's ability to interpret the results. It is always necessary to simplify the analysis and concepts in design as much as possible. Otherwise the intuitive aspects of the circuit can be lost to the designer. A computer can always be used to perform a more detailed and extensive analysis of the design, but the computer is not yet capable of making design decisions and performing the synthesis of complex analog circuits.

With the background now assimilated, we shall move into the study of more complex analog circuits. Chapters 7 through 9 will cover comparators and operational amplifiers.

PROBLEMS — *Chapter 6*

1. Develop a large-signal expression for v_{OUT} of Fig. 6.1-1 as a function of v_{IN} and verify Eq. (5) of Sec. 6.1 by differentiating v_{OUT} with respect to v_{IN} of this expression.
2. Using the specifications of Example 6.1-1, find the small-signal voltage gain and -3 dB frequency in Hertz of the active-resistor, current-load, and push-pull CMOS inverters if $I_D = 10$ microamperes and $W_1 = L_1 = W_2 = L_2 = 10\ \mu m$.
3. M1 and M2 of the active-resistor load inverter have the following parameter values: $W_1/L_1 = W_2/L_2 = 5$, $C_{ox} = 0.5 \times 10^{-7}$ F/cm², $\mu_{oN} = 550$ cm²/V, $\mu_{oP} = 225$ cm²/V, and $|V_T| = 1.5$ V. Find I_D and V_{OUT} if $V_{IN} = 2.5$ V when $V_{DD} = 10$ V and $V_{SS} = 0$ V. What is the small-signal voltage gain of this circuit?
4. Assume that $V_{DD} = -V_{SS} = 5$ V and $V_{TN} = -V_{TP} = 1$ V and plot $v_{OUT}(min)$ of Eq. (4) of Sec. 6.1 as a function of the ac voltage gain (v_{out}/v_{in}) for the active-resistor load inverter for ac voltage gains from -1 to -10.
5. What is the small-signal voltage gain of a current-sink inverter with $W_1 = L_1 = W_2 = L_2 = 10\ \mu m$ at $I_D = 0.1$, 5 and 100 μA? Assume that the parameters of the devices are given by Table 3.1-2.
6. Develop an expression for v_{OUT} of Fig. P6.6 and use this result to

develop an expression for the small signal voltage gain. Assume that M1, M2, and M3 are in the saturation region and that the channel-modulation effect can be neglected. If $\beta_2/\beta_1 = 50$, what is the small-signal voltage gain?

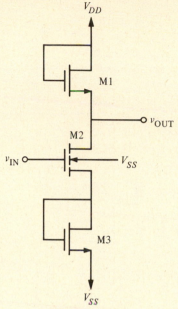

Figure P6.6

7. Find the midband gain of (a) a CMOS push-pull inverter and (b) an inverter with an active load using the parameters of Table 3.1-2 when $V_{DD} = -V_{SS} = 5$ V and the input has a bias of $+1$ V. Assume the W/L ratios of both transistors are 10 μm/10 μm.

8. Derive the expressions given in Eqs. (19) and (15) of Sec. 6.1 for the CMOS push-pull inverter of Fig. 6.1-10. If $C_{gd1} = C_{gd2} = C_{bd1} = C_{bd2} = 0.2$ pF, $C_L = 10$ pF, and $I_D = 200$ μA, find the small-signal voltage gain and the -3 dB frequency if $W_1/L_1 = W_2/L_2 = 5$.

9. For the active-resistor load inverter, the current-source load inverter, and the push-pull inverter compare the active channel area if the gain is to be -1000 at a current of $I_D = 0.1$ μA and the PMOS transistor has a W/L of 1.

10. Use the SPICE2 computer program to obtain a voltage-transfer curve for Fig. 6.1-10 using the parameters of Table 3.1-2 if $V_{DD} = 5$ V and $V_{SS} = 0$ V. Assume that $W_1/L_1 = 15$ μm/10 μm and $W_2/L_2 = 30$ μm/10 μm.

11. Derive Eqs. (3) and (4) of Sec. 6.2 for the CMOS differential amplifier of Fig. 6.2-1.

12.   A CMOS differential amplifier is shown in Fig. P6.12. Assume that the transistor parameters are those of Table 3.1-2. (a) Find the value of the unloaded differential-transconductance gain, g_{md}, and the unloaded differential-voltage gain, A_v, when $I_{ss} = 10$ microamperes. (b) Repeat when $I_{ss} = 1$ microampere.

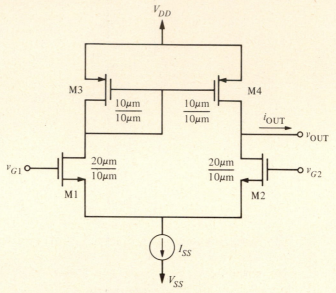

Figure P6.12

13.   What is the slew rate of the differential amplifier in the previous problem if a 100 pF capacitor is attached to the output?

14.   Verify Eq. (12) of Sec. 6.2 for the minimum output-voltage swing for the current-source inverter.

15.   Use the parameters of Table 3.1-2 to calculate the small-signal, differential-in, differential-out transconductance g_{md} and voltage gain A_v for the n-channel input, differential amplifier when $I_{ss} = 100$ μA and $W_1/L_1 = W_2/L_2 = W_3/L_3 = W_4/L_4 = 1$. Repeat if $W_1/L_1 = W_2/L_2 = 10W_3/L_3 = 10W_4/L_4 = 1$.

16.   Repeat the previous problem for the p-channel input, differential amplifier.

17.   Find the expressions for the maximum and minimum input voltages, $v_{G1}(\max)$ and $v_{G1}(\min)$ for the n-channel differential amplifier with enhancement loads shown in Fig. P6.17.

18.   Repeat Example 6.2-1 for an n-channel input, differential amplifier.

19.   (a) For the n-channel differential amplifier find r_{out}, g_{md}, and A_v assuming that $I_{ss} = 20$ μA and $W_1/L_1 = 1$. (b) Find the value of W_1/L_1 needed

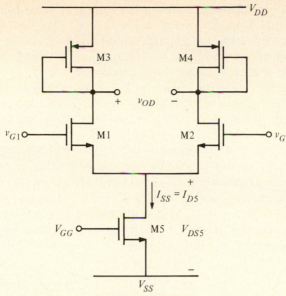

Figure P6.17

for a p-channel differential amplifier to have the same A_v as in (a). Assume that $I_{SS} = 20 \ \mu A$.

20. Use the parameters of Table 3.1-2 to calculate the differential-in-to-single-ended-output voltage gain of Fig. P6.17. Assume that I_{SS} is 50 microamperes.

21. If all the devices in the differential amplifier of Fig. P6.17 are saturated, find the worst-case input offset voltage, V_{OS}, if $|V_{Ti}| = 1 \pm 0.01$ volts and $\beta_i = 10^{-5} \pm 5 \times 10^{-7}$ amperes/volt². Assume that

$$\beta_1 = \beta_2 = 10\beta_3 = 10\beta_4$$

and

$$\frac{\Delta \beta_1}{\beta_1} = \frac{\Delta \beta_2}{\beta_2} = \frac{\Delta \beta_3}{\beta_3} = \frac{\Delta \beta_4}{\beta_4}$$

Carefully state any assumptions that you make in working this problem.

22. If all the devices in the differential amplifier shown in Fig. 6.2-1 are saturated, find the worst-case input-offset voltage V_{OS} using the parameters of Table 3.1-2. Assume that $10(W_4/L_4 = 10(W_3/L_3) = W_2/L_2 = W_1/L_1 = 10 \ \mu m/10 \ \mu m$. State and justify any assumptions used in working this problem.

23. Verify Eqs. (15) and (16) of Sec. 6.2.
24. Verify Eq. (20) of Sec. 6.2.

25. If the equivalent input-noise voltage of each transistor of the differential amplifier of Fig. 6.2-1 is $1 nV/\sqrt{Hz}$, find the equivalent input noise voltage for the differential amplifier of Fig. 6.2-1 if $W_1/L_1 = 20$ $\mu m/10\ \mu m$ and $W_3/L_3 = 10\ \mu m/20\ \mu m$. What is the equivalent output noise current under these conditions?

26. Assume that the curves of Fig. 5.4-10(b) represent the output characteristics of M1 and M2 of Fig. 6.3-1. Plot the voltage-transfer function of Fig. 6.3-1 if $V_{TP} = -1$ volt, $K_P' = 8\ \mu A/volt^2$, and $\lambda = 0.02$ volts^{-1} where $V_{GS3} = 2.5$ volts.

27. For the CMOS cascode amplifier shown in Fig. P6.27, (a) calculate V_1, V_2, V_3, I_1, and I_2. Assume that the W/L ratios of M1, M2, M3, M5, and M6 and the W/L ratios of M4 and M7 are equal. All NMOS transistors are identical with $\beta = 40\ \mu A/V^2$ while β of the PMOS transistors is 10 $\mu A/V^2$. $V_{TN} = -V_{TP} = 2$ V. (b) Derive an expression for the small-signal voltage gain, v_{out}/v_{in} and evaluate. (c) Find the small-signal output resistance, r_{out} and evaluate.

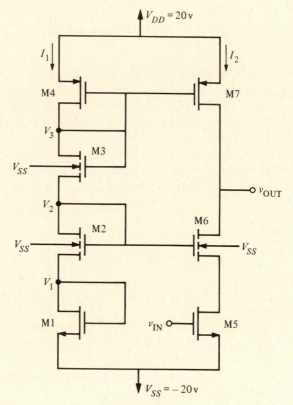

Figure P6.27

28. Derive the minimum output-voltage expression for the cascode amplifier given in Eq. (4) of Sec. 6.3.

29. Consider the current-source load inverter of Fig. 6.1-4 and the simple cascode amplifier of Fig. 6.3-1. If the W/L ratio for M1 is 5 μm/5 μm and for M2 is 15 μm/5 μm of Fig. 6.1-4, compare the minimum output-voltage swing, v_{OUT}(min) of both amplifiers if $V_{GG2} = 0$ V and $V_{GG3} = 2.5$ V when $V_{DD} = -V_{SS} = 5$ V.

30. Find the small-signal output resistance for the cascode amplifier of Fig. 6.3-1.

31. Use nodal analysis techniques on the cascode amplifier of Fig. 6.3-6(a) to find v_{out}/v_{in}. Verify the result with Eq. (30) of Sec. 6.3.

32. Use the Miller simplification technique described in Appendix A on the capacitor C_2 of Fig. 6.3-5(b) and derive an expression for p_1 assuming that the reactance of C_2 at the frequencies of interest is greater than R_3. Compare your result with Eq. (25) of Sec. 6.3.

33. Assume that a dc current source is connected to the drain of M1 in Fig. 6.3-1. Derive an expression similar to Eq. (7) in terms of I_{D1} and I_{D3} for the small-signal voltage gain of the simple cascode amplifier. If $I_{D2} = 10$ μA, what value for this current source would increase the voltage gain by a factor of 10? How is the output resistance affected?

34. Show how adding a dc current source to the drain of M1 in Fig. 6.3-6(a) could be used to increase the small-signal voltage gain. Derive an expression similar to that of Eq. (30) in terms of I_{D1} and I_{D4}. How is the output resistance affected?

35. Find the numerical value of the small signal voltage gain, v_{out}/v_{in}, for the circuit of Fig. P6.35. Assume that all devices are saturated and that $K_N' = 25$ μA/V^2, $K_P' = 10$ μA/V^2, $g_{mbs}/g_m = 0.1$, and $\lambda = 0.1$ V^{-1}.

36. Use the values of Table 3.1-2 and design the W/L ratios of M1 and M2 of Fig. 6.4-1 so that a voltage swing of ± 3 volts and a slew rate of 5 volts/μs is achieved if $R_L = 10$ KΩ and $C_L = 1$ nF. Assume that $V_{DD} = -V_{SS} = 5$ volts and $V_{GG2} = 2$ volts.

37. Find the W/L for the source follower of Fig. 6.4-2 that will source 1 mA of output current with the gate of M2 connected to its drain.

38. Find the small-signal voltage gain and output resistance of the source follower of Fig. 6.4-2 when the gate is connected to V_{GG2}. Assume that $V_{DD} = -V_{SS} = 5$ V, $V_{OUT} = 1$ V, $I_D = 50$ μA, and the W/L ratios of both M1 and M2 are 20 μm/10 μm. Use the parameters of Table 3.1-2 where pertinent.

39. Develop an expression for the efficiency of the source follower of Fig. 6.4-2 in terms of the maximum symmetrical peak-output voltage swing. Ignore the effects of the bulk-source voltage. What is the maximum possible efficiency?

40. Find the pole and zero location of the source follower of Fig. 6.4-2 if

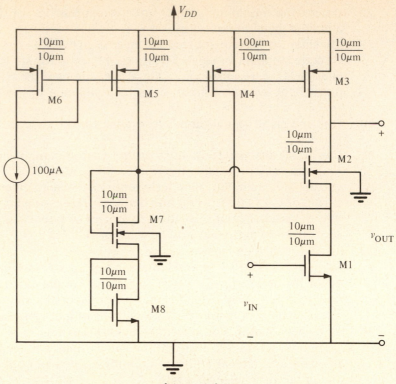

Figure P6.35

$C_{gs1} = C_{gd2} = 0.2$ pF and $C_{bs1} = C_{bd2} = 0.5$ pF and $C_L = 1$ pF. Assume the device parameters of Table 3.1-2, $I_D = 100$ μA, and $V_{SB} = 5$ volts.

41. Show that a class B, push-pull amplifier has a maximum efficiency of 78.5% for a sinusoidal signal.

42. Assume the parameters of Table 3.1-2 are valid for the devices of Fig. 6.4-4. Design $V_{TR1} = V_{TR2} = V_{TR}$ so that M1 and M2 are working in class-B operation, i.e., M1 starts to turn on when M2 starts to turn off.

43. Repeat the previous problem for Fig. 6.4-6(a).

44. Given the push-pull inverting CMOS amplifier shown in Fig. P6.44, show how short-circuit protection can be added to this amplifier. Note that R_1 could be replaced with an active load if desired.

45. If $R_1 = R_2$ of Fig. 6.4-8, find an expression for the small-signal output resistance not including R_L or C_L.

46. Develop a table that expresses the dependence of the small-signal voltage gain, output resistance, and the dominant pole as a function of dc drain current for the differential amplifier of Fig. 6.2-1, the cascode amplifier of Fig. 6.3-1, the high-output-resistance cascode of Fig. 6.3-6, the inverter of Fig. 6.4-1, and the source follower of Fig. 6.4-2.

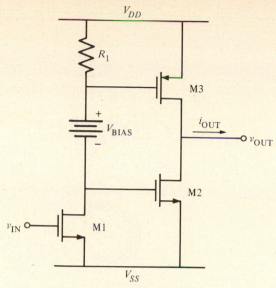

Figure P6.44

For the following problems use appropriate circuits from Sections 6.1 through 6.4 to propose implementations of the amplifier architectures of Section 6.5. Do not make any dc or ac calculations. Give a circuit schematic that would be the starting point of a more detailed design.

47. Propose an implementation of the VCCS of Fig. 6.5-2(b).
48. Propose an implementation of the VCVS of Fig. 6.5-3(b).
49. Propose an implementation of the CCCS of Fig. 6.5-4(b).
50. Propose an implementation of the CCVS of Fig. 6.5-5(b).

REFERENCES

1. P.R. Gray, "Basic MOS Operational Amplifier Design—An Overview," *Analog MOS Integrated Circuits,* A.B. Grebene, ed., (New York: IEEE Press, 1980) pp. 28–49.
2. D.A. Hodges and H.G. Jackson, *Analysis and Design of Digital Integrated Circuits,* (New York: McGraw-Hill, 1983).
3. P.R. Gray and R.G. Meyer, *Analysis and Design of Analog Integrated Circuits,* Second Ed., (New York: John Wiley & Sons, 1984).
4. Y.P. Tsividis, "Design Considerations in Single-Channel MOS Analog Integrated Circuits—A Tutorial," *IEEE Journal of Solid-State Circuits,* Vol. SC-13, no. 3 (June 1978) pp. 383–391.

chapter 7

Comparators

Referring to Table 1.1-2, we see that the previous two chapters dealt with basic building blocks of CMOS circuits. This chapter and those following will build on the basic circuits developed in chapters 5 and 6 to make complex circuits and systems that perform various functions. In this chapter, we will develop CMOS implementations of the comparator.

In recent years, scaled-down processes (see Fig. 1.1-3) have encouraged digital designers to use the tremendous amount of computing power available on single integrated circuits. In order for the analog world to take advantage of this computing power, there must be some means of bridging the analog-digital gap. Analog-digital (A/D), and digital-analog (D/A) converters (introduced in Chapter 10) perform this function. A most important component in A/D converters is the comparator.

The comparator is a circuit that compares an analog signal with another analog signal, and outputs a binary signal based upon the comparison. What is meant here by an analog signal is one that can have any of a continuum of amplitude values at a given point in time (see Sec. 1.1). In the strictest sense a binary signal can have only one of two given values at any point in time, but this concept of a binary signal is too ideal for real-world situations, where there is a transition region between the two binary states.

7.1 Model of a Comparator

Figure 7.1-1 shows the circuit symbol for the comparator that will be used throughout this book. This symbol is similar to that for an operational amplifier, because a comparator has many of the same characteristics as a high-gain amplifier. A comparator was defined in the introduction as a circuit that has a binary output whose value is based upon a comparison of two analog inputs. This is illustrated in Fig. 7.1-2. As shown in this figure, the output of the comparator is high (V_{OH}) when the difference between the non-inverting and inverting inputs is positive, and low (V_{OL}) when this difference is negative. Even though this type of behavior is impossible in a real-world situation, it can be modeled with ideal circuit elements with mathematical descriptions. One such circuit model is shown in Fig. 7.1-3. It comprises a voltage-controlled voltage source (VCVS) whose characteristics are described by the mathematical formulation given on the figure.

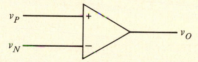

Figure 7.1-1   Circuit symbol for a comparator.

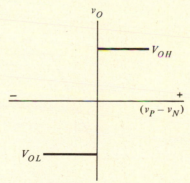

Figure 7.1-2   Zero-order transfer curve of a comparator.

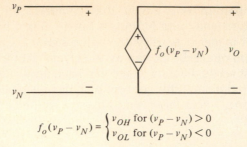

$$f_o(v_P - v_N) = \begin{cases} v_{OH} \text{ for } (v_P - v_N) > 0 \\ v_{OL} \text{ for } (v_P - v_N) < 0 \end{cases}$$

Figure 7.1-3   Zero-order model for a comparator.

The ideal aspect of this model is the way in which the output makes a transition between V_{OL} and V_{OH}. The output changes states for an input change of ΔV, where ΔV approaches zero. This implies a gain of infinity, as shown below.

$$\text{Gain} = A_v = \lim_{\Delta V \to 0} \frac{V_{OH} - V_{OL}}{\Delta V} \tag{1}$$

Figure 7.1-4 shows the dc transfer curve of a first-order model that is an approximation to a realizable comparator circuit. The difference between this model and the previous one is the gain, which can be expressed as

$$A_v = \frac{V_{OH} - V_{OL}}{V_{IH} - V_{IL}} \tag{2}$$

where V_{IH} and V_{IL} represent the input-voltage difference $v_P - v_N$ needed to just saturate the output at its upper and lower limit respectively. Gain is a very important characteristic describing comparator operation, for it

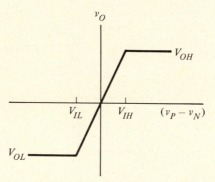

Figure 7.1-4   First-order transfer curve of a comparator.

defines the minimum amount of input change necessary to make the output swing between the two binary states. These two output states are usually defined by the input requirements of the digital circuitry driven by the comparator output. The voltages V_{OH} and V_{OL} must be adequate to meet the V_{IH} and V_{IL} requirements of the following digital stage. For CMOS technology, these values are usually 70% and 30%, respectively, of the rail-to-rail supply voltage.

The transfer curve of Fig. 7.1-4 is modeled by the circuit of Fig. 7.1-5. This model looks similar to the one for the zero-order model, the only difference being the functions f_1 and f_0.

The second nonideal effect seen in comparator circuits is input-offset voltage (see Sec. 6.2 for a definition of V_{OS}). In Fig. 7.1-2 the output changes as the input difference crosses zero. If the output did not change until the input difference reached a value $+V_{OS}$, then this difference would be

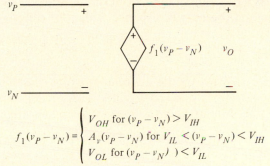

$$f_1(v_P - v_N) = \begin{cases} V_{OH} \text{ for } (v_P - v_N) > V_{IH} \\ A_v(v_P - v_N) \text{ for } V_{IL} < (v_P - v_N) < V_{IH} \\ V_{OL} \text{ for } (v_P - v_N)) < V_{IL} \end{cases}$$

Figure 7.1-5   First-order model for a comparator.

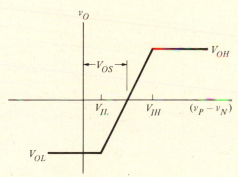

Figure 7.1-6   First-order transfer curve including offset.

defined as the offset voltage. This would not be a problem if the offset could be predicted, but it varies randomly from circuit to circuit [1] for a given design. Figure 7.1-6 illustrates offset in the transfer curve of the first-order model of the comparator, with the circuit model including an offset generator shown in Fig. 7.1-7. The ± sign of the offset voltage accounts for the fact that V_{os} is unknown in polarity.

Up to this point in our discussion we have considered only the dc model for a comparator. This model included the parameters of gain, amplitude saturation, and offset voltage. The comparator must also be modeled in the time domain. Figs. 7.1-2 and 7.1-4 show that the output of a comparator changes from one state to another for a given amount of differential input. We do not know, at this point, how long it takes for the comparator to respond to the given differential input. The characteristic delay between input excitation and output transition is the time response of the comparator. Figure 7.1-8 illustrates the response of a comparator to an input as a function of time. Notice that there is a delay between the input excitation and the output response. This time difference is called the *prop-*

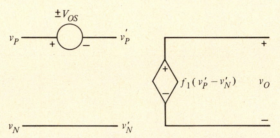

Figure 7.1-7   First-order model of a comparator including offset.

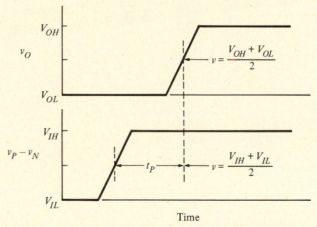

Figure 7.1-8   Time response of a noninverting first-order comparator.

agation delay of the comparator. It is a very important parameter since it is often the speed limitation in the conversion rate of an A/D converter. The propagation delay in comparators generally varies as a function of the amplitude of the input. A larger input will result in a smaller delay time. There is an upper limit at which a further increase in the input voltage will no longer effect the delay.

7.2 Development of a CMOS Comparator

With a firm understanding of what comparators are and what they should do, we are ready to study practical implementation of CMOS comparator circuits. One of the simplest circuits available (in CMOS technology) that can function as a comparator is the current-sink inverter first described in Sec. 6.1 (Fig. 6.1-8). A similar circuit is shown in Fig. 7.2-1. The transfer curve of this inverter is shown in Fig. 7.2-2 with the trip voltage V_{TRP}

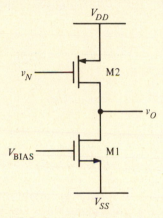

Figure 7.2-1   Simple inverting comparator.

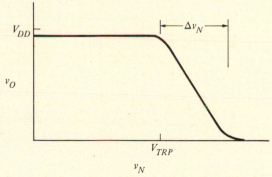

Figure 7.2-2   DC transfer curve of a simple comparator.

described by the following equations.

$$V_{TRP} \cong V_{DD} - \left[\frac{2I_B}{K_2'(W_2/L_2)\left[1 + \lambda_2\left(\frac{V_{DD} - V_{SS}}{2}\right)\right]} \right]^{1/2} - |V_{T2}| \qquad (1)$$

where

$$I_B \cong (K_1'W_1/2L_1)(V_{BIAS} - V_{SS} - V_{T1})^2\left[1 + \lambda_2\left(\frac{V_{DD} - V_{SS}}{2}\right)\right] \qquad (2)$$

The equation for V_{TRP} is derived from the approximation that v_o begins going high when the current in M2 (in saturation) equals the bias current I_B. It is obvious from these relationships that the trip voltage is a function of supply voltages and model parameters that describe the behavior of the transistors. This is illustrated by the following example.

Example 7.2-1

Calculation of Comparator Trip Voltage

Using the parameters given in Table 3.1-2, calculate the variation in trip voltage as a function of supply voltage and threshold voltage for the circuit shown in Fig. 7.2-1. $V_{DD} = 9 \pm 1$ volt, $V_{BIAS} = 2$ volts, $(W_1/L_1) = 1$, $(W_2/L_2) = 10$ (effective device length is 10 μm), and $V_{SS} = 0$ V. The maximum trip voltage occurs when V_{DD} is maximum, $|V_{T2}|$ is minimum, and I_B is minimum (thus V_{T1} is maximum and K_1' is minimum). The minimum trip voltage occurs when V_{DD} is minimum, $|V_{T2}|$ is maximum, and I_B is maximum (thus V_{T1} is minimum and K_1' is maximum).

$$I_B(\text{min}) = [(15.3)(1)/2](2 - 1.2)^2[1 + 5(.01)] = 5.14 \ \mu\text{A}$$

$$V_{TRP}(\text{max}) = 10 - \left[\frac{2(5.14)}{(8.8)(10)[1 + 5(0.02)]} \right]^{1/2} - 0.8$$

$$V_{TRP}(\text{max}) = 8.87 \text{ volts}$$

$$I_B(\text{max}) = [(18.7)(1)/2](2 - 0.8)^2[1 + 4(.01)] = 14.0 \ \mu\text{A}$$

$$V_{TRP}(\text{min}) = 8 - \left[\frac{2(14.0)}{(7.2)(10)[1 + 4(0.02)]} \right]^{1/2} - 1.2$$

$$V_{TRP}(\text{min}) = 6.20 \text{ volts}$$

The gain of the comparator shown in Fig. 7.2-1 is the same as the small-signal gain of the inverter derived in Sec. 6.1, and is given in Eq. (3) below as

$$A_v = - \left[\frac{2K_2'W_2}{L_2I_B} \right]^{1/2} \frac{1}{\lambda_1 + \lambda_2} \tag{3}$$

Gain is important in determining the drive requirements of the simple comparator. It must be pointed out that the output does not switch from V_{OL} to V_{OH} (or V_{OH} to V_{OL}) at the input voltage V_{TRP}. This could only be the case if the gain were infinite. In fact, there is a range of voltages around V_{TRP} where the output is changing in a roughly linear way as a function of the input, as shown in Fig. 7.2-2. This range of input voltages is approximately determined by the gain of the comparator. The larger the gain, the smaller the input-voltage drive requirements. The following example illustrates the calculation of gain and input drive requirements for the simple comparator.

Example 7.2-2

Input Drive Requirements for a Simple Comparator

The simple comparator circuit shown in Fig. 7.2-1 is designed so that $I_B = 20$ μA. Transistor M2 has an effective channel width of 100 μm and effective channel length of 10 μm. All of the parameters describing transistor operation may be found in Table 3.1-2. Determine the minimum allowable input swing $\Delta V_N(\text{min})$ to achieve CMOS logic levels at the output for a 5 volt supply.

$A_v = -\{[(2)(8.0)(10)/20]^{1/2}\}/(0.01 + 0.02)$

$A_v = -94.3$

$V_{OH} = (0.7)(5) = 3.5$ volts

$V_{OL} = (0.3)(5) = 1.5$ volts

$V_{OH} - V_{OL} = 2$ volts

$\Delta V_N(\text{min}) = 2/94.3 = 21.2$ mV

The gain of this comparator implementation is low, but could be improved substantially using cascode techniques. The major drawback of this circuit is not gain, however, but the fact that the trip point is dependent upon the power supply. In addition, there is a limited range at which the trip point can be placed while still maintaining adequate gain. These problems can be solved by using a different input scheme.

Figure 7.2-3 shows the differential amplifier stage with n-channel input devices and a single-ended output presented in Sec. 6.2. The key attribute of this circuit is its ability to amplify the difference between the

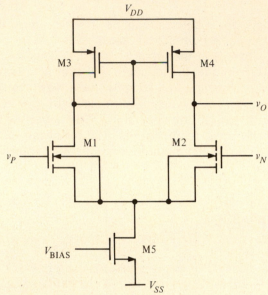

Figure 7.2-3 Differential input comparator.

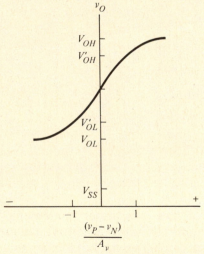

Figure 7.2-4 DC transfer curve of a differential input comparator.

inverting and noninverting inputs. As a result, the trip point of a comparator can be made independent of process and supply variations (to the first order). The transfer curve describing the differential stage has been described in Sec. 6.2 and shown in Figs. 6.2-4(b) and 6.2-5(b). A simplified transfer curve is shown in Fig. 7.2-4. This curve is based upon small-signal differential inputs. The output limits determined using the simplified MOS model are

$$V_{OL} \cong v_N - V_{T2} - \left[\frac{I_5}{K_2'(W_2/L_2)} \right]^{1/2} \tag{4}$$

$$V_{OH} \cong V_{DD} \tag{5}$$

These values are asymptotic limits for the output, and not necessarily the maximum useful output range. High-gain operation is maintained while the output transistors remain in their saturation region of operation. Keeping the output devices in saturation limits the range to

$$V_{OL}' \cong v_N - V_{T2} \tag{6}$$

$$V_{OH}' \cong V_{DD} - \left[\frac{I_5}{K_3'(W_3/L_3)} \right]^{1/2} \tag{7}$$

These values more closely approximate the useful output range. One can see that this range is limited, and would probably not be useful for driving digital circuitry. This is the major drawback of using the differential stage alone as comparator.

While the output transistors are in saturation, the gain is approximately

$$A_v = \frac{\Delta v_O}{\Delta(v_P - v_N)} \cong \frac{g_{m2}}{g_{ds2} + g_{ds4}} = \frac{2(K_2'W_2/L_2)^{1/2}}{I_5^{1/2}(\lambda_2 + \lambda_4)} \tag{8}$$

As in the case of the current-load inverting stage, the gain of the differential amplifier is not very high. There are ways to increase this gain at the expense of the output voltage range. As we shall see in later developments, the limited gain of this single stage is not a major disadvantage.

As mentioned before, the primary advantage of the differential stage is its ability to control the trip voltage very accurately. If there were no non-ideal effects, the trip voltage could be controlled perfectly. The limitation on this is due to process-dependent input-offset voltage, which can be caused by threshold voltage, geometry, and/or temperature mismatch. A further error contribution is the channel-length modulation effect that becomes evident when two matched devices have different drain voltages resulting in current mismatch that is reflected as a voltage mismatch at the input. This effect was introduced in Sec. 5.4 in relation to current mirrors.

Threshold voltage mismatches reflect directly to the input (i.e. a 10 mV threshold difference reflects a 10 mV trip point difference, assuming all else is ideal). Mismatches of this type are unavoidable due to imperfections in the process. Offsets can be minimized by using common centroid geome-

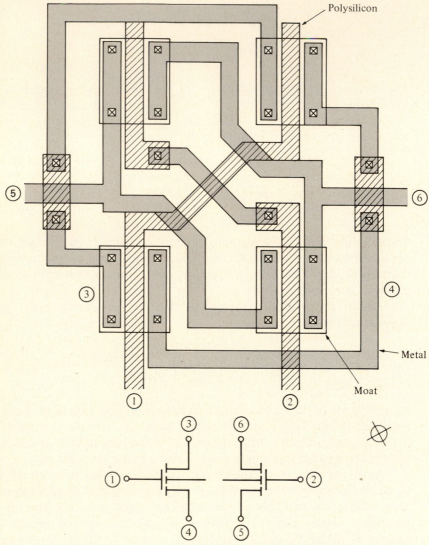

Figure 7.2-5 Layout of a cross-coupled transistor pair.

tries [2]. An example of this is illustrated in Fig. 7.2-5. It is also desirable to keep the number of bends and corners in the transistors to a minimum for two devices that must match.

7.3 Two-Stage Comparator

Thus far, we have presented two circuit techniques to implement the comparator function. Neither of the proposed circuits performs the comparator function satisfactorily by itself. But, though these two circuits do not perform well alone, they can have satisfactory performance if they are combined. Fig. 7.3-1 shows a two-stage comparator that combines the two stages previously described. We will see that the most useful attribute of each of the circuits is realized when they operate together. The poor gain of the differential stage is augmented by the gain of the inverting stage. The output of the differential stage that lies closest to V_{DD} is in the vicinity of the trip point of the inverting stage that follows. Therefore the limited output range, which is a problem for the differential stage by itself, now becomes a good feature in the combination shown. It is now desirable that the output of the differential stage be centered at the trip point of the second stage in balanced quiescent conditions.

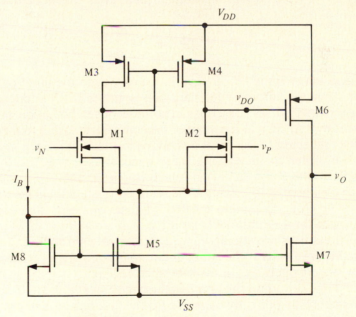

Figure 7.3-1   Two-stage comparator.

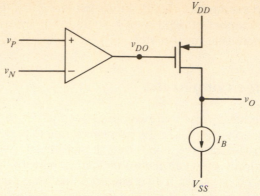

Figure 7.3-2 Equivalent circut for two-stage comparator.

It has been pointed out that one of the problems with the current-sink inverter is the poorly-predicted trip-point voltage. Consider Fig. 7.3-2, in which the complete comparator is drawn in simplified form with an ideal differential input. Assume that the trip voltage at the input of the second stage varies by an amount ΔV_{TRP2} because of process and other variations. This variation is reflected back to the input of the differential stage by dividing by the differential stage voltage gain A_{v1} as shown below

$$\Delta V_{TRP1} = \frac{\Delta V_{TRP2}}{A_{v1}} \tag{1}$$

where ΔV_{TRP1} is the reflected variation of V_{TRP2} to the input of the comparator. Thus the uncontrolled variation is substantially reduced in the two-stage configuration described. The total offset (variation in the nominal value of V_{TRP}) seen at the input of the comparator is due to the second-stage offset reflected to the input and added to the offset in the first stage.

In order to understand these concepts fully, we must first develop some of the basic dc design relationships for the two-stage comparator. Our goal here is to establish the required relationships between transistor sizes in order to achieve a balanced condition with a quiescent balanced input. In this discussion "balance" means a condition in which all n-channel transistors are attempting to sink precisely the same amount of current as their respective p-channel counterparts are trying to source when all devices are operating in the saturation region.

Consider again the comparator circuit in Fig. 7.3-1. It is desired that this circuit be perfectly balanced (to the first order), and that any input of v_P be transmitted to the output the same way (except for polarity) as an input of v_N. Using these guidelines, we can establish design rules for transistor sizes in this circuit. Based upon the desire to match the two paths to the output, it should be obvious that transistors M1 and M2 must be matched, just as M3 and M4 must also be matched. If the input is balanced—that is, v_P and v_N are both equal—then the current that flows in M5 is split equally through M1 and M2. This qualitative analysis leads to the following relationships

$$W_1/L_1 = W_2/L_2 \tag{2a}$$
$$W_3/L_3 = W_4/L_4 \tag{2b}$$
$$i_1 = i_2 = i_5/2 \tag{2c}$$

The current that flows in M5 is mirrored to the output by the ratio of the sizes of M7 to M5. Likewise the current in M4 is ratioed to the output by the ratio of M6 to M4. This results from our assumption of perfectly balanced conditions, which means that the drain voltages of M4 and M3 are equal. Since the gate and drain of M3 are tied together, the drain voltage of M4 is essentially the gate voltage of M3, thus the current i_4 is mirrored to i_6 by the ratios of the sizes of M6 to M4. The following equations describe these relationships

$$i_7 = i_5[(W_7/L_7)/(W_5/L_5)] \tag{3}$$
$$i_6 = i_4[(W_6/L_6)/(W_4/L_4)] \tag{4}$$

For balanced conditions, it is desired that i_6 and i_7 be equal, thus

$$i_7 = i_6 \tag{5}$$
$$[i_5/i_4][(W_7/L_7)/(W_5/L_5)] = (W_6/L_6)/(W_4/L_4) \tag{6}$$

From previous discussions, we know that

$$i_5/i_4 = 2 \tag{7}$$

Therefore

$$(W_6/L_6)/(W_4/L_4) = 2(W_7/L_7)/(W_5/L_5) \tag{8}$$

Equations (2) and (8) completely describe the relationships between transistors to achieve the desired balanced condition. Channel-length modulation effects have not been taken into account in the development of the design equations. As a result, there will be some error in the output resulting from current mismatches. This error at the output is then reflected back to the input as a systematic offset. This is illustrated in the following example.

Example 7.3-1

Calculation of Systematic Offset

A two-stage comparator with the following device sizes has been designed using the balance equations, Eqs. (2) and (8). Assume that $V_{DD} = 10$ volts, $V_{SS} = 0$ volts, $v_O = 5$ volts, $V_{DS5} = 3$ volts, and $V_{DS3} = V_{DS4} = 2$ volts. The effective device sizes are:

$$W_1 = W_2 = W_3 = W_4 = 20 \ \mu m; \ W_5 = W_7 = 10 \ \mu m; \ W_6 = 40 \ \mu m$$
$$L_{1-7} = 10 \ \mu m$$

Pertinent process information can be found in Table 3.1-2. The bias current I_5 is 20 μA.

$$i_7/i_5 = \frac{1 + \lambda_7 v_{DS7}}{1 + \lambda_5 v_{DS5}} \cdot \frac{W_7/L_7}{W_5/L_5} = \frac{1 + (0.01)(5)}{1 + (0.01)(3)} \ (1) = 1.019$$

$$i_6/i_4 = \frac{1 + \lambda_6 v_{DS6}}{1 + \lambda_4 v_{DS4}} \cdot \frac{W_6/L_6}{W_4/L_4} = \frac{1 + (0.02)(5)}{1 + (0.02)(2)} \ (2) = 2.115$$

$$i_5 = 2(i_4)$$

then

$$i_7 = (1.019)(2)(i_4) = 2.038(i_4)$$
$$i_6 = 2.115(i_4)$$

We can see that i_6 is greater than i_7, thus the current in i_6 must be reduced. Determine how much v_{GS6} must be reduced to change i_6 to equal i_7.

$$\Delta v_{GS6} = [\sqrt{i_a} - \sqrt{i_b}][(K_6'/2)(W_6/L_6)]^{-1/2}$$
$$i_a = 2.115(i_4) = 21.15 \ \mu A$$
$$i_b = 2.038(i_4) = 20.38 \ \mu A$$
$$\Delta v_{GS6} = 21.1 \ mV$$

From Eq. (1), we can determine the input referred offset by dividing ΔV_{GS6} by the differential-stage gain. We must first calculate this gain.

$$A_v = \left[\frac{2}{\lambda_2 + \lambda_4}\right] \left[\frac{K_2'(W_2/L_2)}{I_5}\right]^{1/2} = 86.9$$
$$A_v = 86.9$$

The resulting input referred to offset is

$$V_{OS} = \frac{\Delta V_{GS6}}{A_v} = \frac{21.1 \text{ mV}}{86.9} = 0.243 \text{ mV}$$

$$V_{OS} = 0.243 \text{ mV} = \text{systematic offset}$$

Common-mode input range is another important consideration in the design of comparators. The subject has been covered already in Sec. 6.2. The design equations below are given based upon n-channel input transistors.

$$V_{G1}(\text{min}) = V_{SS} + V_{DS5} + (I_5/\beta_1)^{1/2} + V_{T1}(\text{max}) \tag{9}$$

$$V_{G1}(\text{max}) = V_{DD} - (I_5/\beta_3)^{1/2} - |V_{T03}|(\text{max}) + V_{T1}(\text{min}) \tag{10}$$

where

$$\beta = K'(W/L) \tag{11}$$

When designing the input stage for a specific common-mode input range, the procedure is to design the size of M3 (M4) to meet the maximum input requirement, then design the device sizes of M1 (M2) and M5 to meet the minimum input requirement. This can be illustrated by the following example

Example 7.3-2

Designing for Proper Common-Mode Input Range (CMR)

Using the circuit given in Fig. 7.3-1, design transistors M1 through M4 for a common-mode input range of 1.5 to 8.5 volts, with $V_{DD} = 10$ volts and $V_{SS} = 0$ volts. Pertinent process information can be found in Table 3.1-2, with the following exceptions: $|V_{TN,P}| = 0.4$ to 1.0 volts. $I_5 = 20\ \mu\text{A}$, $V_{DS5} = 0.1$ volt. Using Eq. (9),

$$V_{G1}(\text{min}) = V_{SS} + V_{DS5} + (I_5/\beta_1)^{1/2} + V_{T1}(\text{max})$$

$$1.5 = 0 + 0.1 + (20/\beta_1)^{1/2} + 1$$

$$\beta_1 = K_N' W_1/L_1 = 20/[(0.4)^2] = 125\ (\mu\text{A/V}^2)$$

$$W_1/L_1 = W_2/L_2 = (125)/(17) = 7.35$$

Using eq. (10) to calculate maximum input voltage,

$$V_{G1}(\text{max}) = V_{DD} - (I_5/\beta_3)^{1/2} - |V_{T03}|(\text{max}) + V_{T1}(\text{min})$$
$$8.5 = 10 - (20/\beta_3)^{1/2} - 1 + 0.4$$
$$\beta_3 = K'_P W_3/L_3 = 20/[(0.9)^2] = 24.69 \ (\mu A/V^2)$$
$$W_3/L_3 = W_4/L_4 = (24.69)/8.0 = 3.09$$

As mentioned before, gain is another very important parameter in comparator operation. The gain of the differential stage has already been described in Sec. 6.2, so all we need do to get the total gain is multiply the differential-stage gain by the current-source inverter gain.

$$A_v = \left(\frac{g_{m1}}{g_{ds2} + g_{ds4}}\right)\left(\frac{g_{m6}}{g_{ds6} + g_{ds7}}\right) \tag{12}$$

$$A_v = \frac{2\sqrt{K'_1 K'_6 (W_1/L_1)(W_6/L_6)}}{(\lambda_2 + \lambda_4)(\lambda_6 + \lambda_7)\sqrt{I_1 I_6}} \tag{13}$$

When designing the gain of a comparator, the size and current in M1 (M2) have been fixed by other constraints, so the terms that can be designed to achieve proper gain are I_6 and W_6/L_6.

The final characteristic discussed here is comparator speed. As defined before, propogation delay is the delay from the time an input passes the threshold to the time the output is a valid logic signal. This time is determined by the amount of current available to charge (or discharge) parasitic and circuit capacitances. Figure 7.3-3 shows the comparator with the dominant parasitics indicated. The capacitor C_{L1} consists of depletion capacitance C_D and gate-source capacitance C_{GS}. The capacitor C_D is the total depletion capacitance contributed by the drain diffusions of M2 and M4. These capacitances were described in Sec. 3.2.

It will be assumed in the analysis that follows that the comparator is driven by v_P while v_N provides the reference voltage. It will be further assumed that the input is large enough to warrant large-signal analysis techniques. Although this is a simplification, it illustrates some of the basic considerations required when analyzing comparator performance.

Since the comparator is made up of two stages, the total delay is determined by adding together the propagation delays of each stage. The delay for the first stage is the time it takes for v_{DO} to make the transition from its quiescent state to the trip point V_{TRP2} of the second stage. If it is assumed

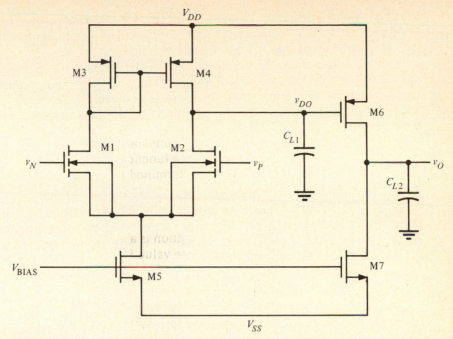

Figure 7.3-3   Simulation of a two-stage comparator with load capacitance.

that the transistors driving the second stage are in the saturation region the majority of the time during transition, then we can make the approximation that a constant current drives the parasitic load capacitors. Once the differential stage is sufficiently unbalanced, the bias current I_5 flows in one side of the stage or the other. Essentially, this means that all of I_5 will be used in the charging or discharging of the parasitic load capacitors. With this in mind, it is easy to approximate the time required for v_{DO} to move from the quiescent state to the trip point. This time is

$$\Delta T_1 = \Delta v_{DO}(C_{L1}/i_5) \tag{14}$$

where

$$\Delta v_{DO} = v_{DO}(t_0) - V_{TRP2} \tag{15}$$

Remember that the assumption is made that the differential stage is unbalanced enough to divert all of I_5 to one of the two paths (through M1 or M2). This requires an amount of differential input voltage sufficient to cause this imbalance.

The output of the first stage has been approximated as a ramp. In order to simplify the calculation of the delay due to the second stage, we will assume that the first-stage output is a step occurring at the end of time ΔT_1 (in some cases, this is a poor approximation). The delay ΔT_2 can be determined based upon the time required for the output to go from either supply to the trip voltage of the following stage. If the output makes the transition from high to low, then the amount of current i_7 delivered to the capacitance at the output C_{L2} is determined by the bias network, and is approximately

$$i_7 \cong [(W_7/L_7)/(W_5/L_5)] i_5 \tag{16}$$

When the output is driven high, M6 sources the current to the load, so i_6 determines the speed of the output. In the unbalanced condition, i_6 is determined by the gate drive on M6, which is a function of the input voltage, v_P. We will assume that the value of i_6 is determined by the minimum value of v_{DO} for the differential stage, which is

$$V_{DO}(\text{min}) \cong v_N - (i_5/\beta_1)^{1/2} - V_{T1} \tag{17}$$

For relatively fast circuits, the above equation is adequate. For slower input stages, $v_N - V_{T1}$ might be a more accurate value for $V_{DO}(\text{min})$. The value of i_6 when M6 is driving the load capacitance is

$$i_6 \cong (K_6'/2)(W_6/L_6)(V_{DD} - V_{DO}(\text{min}) - |V_{T6}|)^2 \tag{18}$$

Assuming the current available to charge the output load capacitance is constant (not a bad assumption), the positive going delay time, ΔT_{2+}, is

$$\Delta T_{2+} \cong C_{L2} \left[\frac{(V_{TRP3} - V_{SS})}{i_6 - i_7} \right] \tag{19}$$

and the negative going delay time is

$$\Delta T_{2-} \cong C_{L2} \left[\frac{V_{DD} - V_{TRP3}}{i_7} \right] \tag{20}$$

where V_{TRP3} is the trip voltage of the third stage. With ΔT_1 and ΔT_2 determined (separately), the total comparator delay is seen to be simply the sum of the delays of each stage.

$$\Delta T = \Delta T_1 + \Delta T_2 \tag{21}$$

The example that follows illustrates how to calculate propagation delays.

Example 7.3-3

Comparator Propagation-Delay Calculation

Consider the comparator shown in Fig. 7.3-3 with the following effective device widths

$$W_1 = W_2 = W_7 = 20 \ \mu m, \qquad W_3 = W_4 = W_5 = 10 \ \mu m$$
$$W_6 = 40 \ \mu m$$

All effective device lengths are 10 μm. Assume total load capacitances, C_{L1} and C_{L2} are 0.3 pF and 10 pF respectively. $V_{DD} = +5$ V, $V_{SS} = -5$ V, and $v_N = 0$ V. $I_5 = 20 \ \mu A$. Using device parameters given in Table 3.1-2, determine the positive-going propagation delay due to an input v_P going from -1 to $+1$ volts in 2 ns. The trip point of the third stage (not shown) is 0 volts.

$$\Delta T_1 = v_{DO}(t_0) - V_{TRP2} \ (C_{L1}/i_5)$$
$$v_{DO}(t_0) = + 5 \ V$$

The trip voltage of the second stage is approximately the value required for i_6 to equal i_7 with M6 and M7 in saturation. Approximating $\lambda = 0$ to simplify calculations, we have:

$$i_7 = 2(i_5) = 40 \ \mu A$$
$$V_{GS6} = \left(\frac{2i_7}{K_6'(W/L)_6} \right)^{1/2} + |V_{T6}| = \left(\frac{(2)(40)}{8(4)} \right)^{1/2} + 1 = \left(\frac{10}{4} \right)^{1/2} + 1$$
$$= 2.58$$
$$V_{TRP2} = V_{DD} - V_{GS6} = 5 - 2.58 = 2.42$$
$$\Delta T_1 = (5 - 2.42)(0.3 \ pF/20 \ \mu A) = 38.7 \ ns$$

With the output going high, the charging current available for C_{L2} is the difference between i_6 and i_7.

$$i_6 = \left(\frac{K_6'}{2} \right) \left(\frac{W_6}{L_6} \right) (V_{DD} - V_{DO}(\text{min}) - |V_{T6}|)^2$$
$$V_{DO}(\text{min}) \cong v_N - (i_5/\beta_1)^{1/2} - V_{T1} = 0 - 0.77 - 1 = -1.77$$
$$i_6 = \left(\frac{8 \times 10^{-6}}{2} \right) (4)[5 - (-1.77) - 1]^2 = 533 \ \mu A$$
$$i_7 = 40 \ \mu A \rightarrow i_6 - i_7 = 493 \ \mu A$$
$$\Delta T_{2+} = C_{L2}(\Delta v_2/i) = 10 \ pF(5/493 \ \mu A) = 101 \ ns$$
$$\text{Total } \Delta T = 101 \ ns + 38 \ ns \cong 139 \ ns$$

The actual results of a PSPICE simulation of this circuit are shown in Fig. 7.3-4. The calculated delay of the first stage agrees closely with that of the simulation, but the second-stage calculated delay is off by about 25%, due to the approximation of the first stage output as a step [to $v_{DO}(\min)$] occurring at the end of ΔT_1.

As stated before, the analysis given thus far has assumed the comparator starts in an unbalanced state and is driven to an imbalance of the opposite

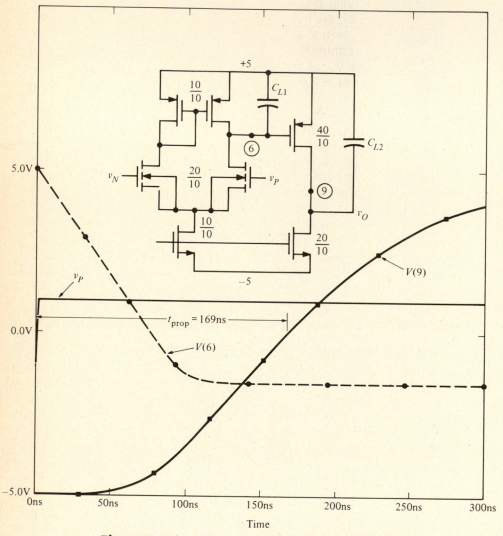

Figure 7.3-4   Differential comparator propagation delay.

polarity. As a result, we have been able to use large-signal analysis only in computing comparator propagation delay.

When the comparator operates in the linear region of the transfer function (resulting from small-signal overdrive), the propagation delay is determined by the small-signal time constants of the comparator circuit. Figure 7.3-5 shows a two-pole small-signal equivalent circuit of the two-stage comparator. In this figure, R_I and R_{II} are small-signal output resistances, and C_I and C_{II} are the load capacitances of the two stages. This circuit can be used to get a feel for the propagation delay. The small-signal, s-domain, transfer function of the circuit in Fig. 7.3-5 is given as

$$V_o(s) = \frac{A_{vo}(\omega_{pI}\omega_{pII})}{(s + \omega_{pI})(s + \omega_{pII})} [V_{in}(s)] \qquad (22)$$

where

$$\omega_{pI} = 1/(R_I C_I) \qquad (23a)$$
$$\omega_{pII} = 1/(R_{II}C_{II}) \qquad (23b)$$
$$A_{vo} = (G_{mI}R_I)(G_{mII}R_{II}) \qquad (23c)$$

Of interest here is not the frequency domain response, but the time domain response. An analytical expression that precisely predicts the propagation delay is not readily obtainable for most cases. What is important is to understand that the small-signal propagation delay is inversely proportional to the location of the two poles p_I and p_{II}. The larger these poles are, the shorter the delay. Since these poles are generally under the control of the designer, the propagation delay is also under his control, up to a limit. For a given design, load capacitance will be fixed based upon transistor sizes. With fixed capacitance, the propagation delay can only be controlled by the output resistances of each stage. A reduction in output resistance causes a corresponding reduction in small-signal propagation delay. Decreasing the output resistance (with all else constant) also decreases the

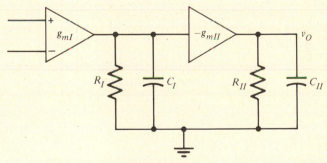

Figure 7.3-5   Small-signal equivalent circuit for comparator.

gain of the comparator. Design tradeoffs must be made between gain and speed.

Thus far in this chapter, a great deal of effort has been put into analyzing the comparator. The remainder of this section will deal with design. The following design procedure can be used to design the comparator of Fig. 7.3-1. It is not the only approach to the design of this comparator. Furthermore, it is important to point out that there is no unique solution to a comparator design problem. Even for the same configuration, there can be many variations in transistor sizes and device currents, due to the fact that the requirements for a comparator design generally do not completely specify the design of all of the transistors. As a result, we are normally unable to present a set of simultaneous equations defining comparator operation with the proper number of unknowns so that all that is required to complete the design is to manipulate these equations mathematically. The design procedure introduced here attempts to minimize the size of the circuit while maintaining required electrical specifications.

At the beginning of the design procedure, all devices are set at their minimum size (all device aspect ratios will be greater than or equal to one). Throughout the procedure, these will be adjusted as required. The first step in the design procedure would be to set the current in M7 so that the output slew-rate (dV/dT) requirements are met for the given load capacitance. Generally, the output slew rate may not be specified. A value of approximately 10 times that of the propagation delay may be chosen as a starting point (e.g. for a 1 μs propagation delay, and a 10 volt supply, a slew-rate of 100 V/μs would be initially chosen). This may be adjusted later. Once I_7 is set, then M6 and M7 are adjusted to achieve proper output voltage swing. Normally it is desired that these transistors be in saturation for some specified amount of the output voltage swing. Therefore, the sizes of M6 and M7 can be calculated based upon the equation for $V_{DS}(\text{sat})$

$$V_{DS}(\text{sat}) = (2I_{DS}/\beta)^{1/2} \tag{24}$$

The sizes calculated based upon a maximum $V_{DS}(\text{sat})$ are minimum sizes to achieve the specification.

Knowing the second-stage current, and minimum sizes for the output transistors, the second-stage gain can be calculated based upon the small-signal equation

$$A_2 = \frac{-g_{m6}}{g_{ds6} + g_{ds7}} \tag{25}$$

After determining the second-stage gain, the requirement for the first-stage gain can be easily calculated.

A minimum current can be chosen for the first stage by remembering that both M4 and M3 are minimum size, and calculating what the resulting current ratio is from I_6 to I_4 (based on transistor ratios). The minimum current calculated must be checked to see that the internal slew rate is at least that of the output. The load capacitance for the slew-rate calculation is the gate capacitance of M6 and the depletion capacitance of M4 and M2 in parallel.

Given the current and gain requirement for the first stage, calculate the minimum device size of M1 (M2) to achieve proper gain. The following small-signal expression can be used.

$$A_1 = -g_{m1}/(g_{ds2} + g_{ds4}) \qquad (26)$$
$$g_{m1} = [2K_1'I_1(W/L)_1]^{1/2} \qquad (27)$$
$$g_{ds} \cong (\lambda I_1) \qquad (28)$$

It is desirable that M5 always remain in saturation, so its device size can be calculated based upon negative input CMR requirements. $V_{DS}(\text{sat})$ can be calculated using Eq. (9) of Sec. 7.3, and then the proper size of M5 can be determined from the value of $V_{DS}(\text{sat})$. At this point, either M5 or M7 must be increased to achieve proper mirroring.

Now M3 (M4) must be designed for proper positive input CMR. Equation (10) of Sec. 7.3 should be used here. Once this is accomplished, M3 (M4) or M6 can be increased for proper mirroring. At this point, all devices have been designed. This work must be checked with computer simulation to see that all design specifications are met and to check the propagation delay. This last parameter must especially be checked since it was not used as a criterion for design. If this delay is too great it must be determined whether the problem is a small- or large-signal one. If it is small-signal, then the size of the output devices could be reduced. This would reduce the loading on the first stage. Another solution might be to increase the current in the first stage thus reducing the output resistance, which would push out the pole associated with this resistance. The above design procedure is summarized in Table 7.3-1. The example that follows shows in detail how this design procedure is used.

Example 7.3-4
Design of a Comparator

The task is to design the comparator shown in Fig. 7.3-1 to meet the following specifications: $A_v > 66$ dB, $P_{\text{diss}} < 10$ mW at 10 V, $C_2 = 2$ pF, $t_{\text{prop}} < 1$ μs, and input CMR = 4–6 V, output voltage swing within 2 volts of either rail. $V_{DD} = 10$ V and $V_{SS} = 0$ V. Pertinent process information is given in Table 3.1-2 with the following exception: $\lambda = 0.07$ for a 5 micrometer device length p-channel device and 0.03 for an n-channel device.

Table 7.3-1

Comparator Design Procedure

1. Set the output current to meet slew rate requirements. This is accomplished by calculating this current based upon the equation

$$I = C(dV/dT)$$

2. Determine minimum sizes for M6 and M7 for proper output voltage swing. To do this, use the equation for $V_{DS}(sat)$

$$V_{DS}(sat) = (2I/\beta)^{1/2}$$

3. Knowing the second stage current, and minimum device size for M6, the second stage gain can be calculated using the following equation

$$A_2 = -g_{m6}/(g_{ds6} + g_{ds7})$$

4. Calculate required first stage gain using the specification, and the previously calculated second stage gain.

5. Determine the current in the first stage based upon proper mirroring, and minimum values calculated for M6 and M7. Verify that P_{diss} specification is met.

6. Given first stage current, and gain requirement of first stage, calculate device size of M1 using the following relationship

$$A_1 = -g_{m1}/(g_{ds1} + g_{ds3})$$
$$g_m = (2K'(W/L)I_1)^{1/2}$$
$$g_{ds} \cong \lambda I_1$$

7. Design minimum device for M5 based upon negative CMR requirement using the following equations

$$V_{G1}(min) = V_{SS} + V_{DS5} + (I_5/\beta_1)^{1/2} + V_{T1}(max)$$
$$V_{DS5} = (2I_5/\beta_5)^{1/2}$$

8. Increase either M5 or M7 for proper mirroring.

9. Design M4 for proper positive CMR using the following equation

$$V_{G1}(max) = V_{DD} - (I_5/\beta_3)^{1/2} - |V_{T03}(max)| + V_{T1}$$

10. Increase M3 or M6 for proper mirroring.

11. Simulate circuit to check to see that all specifications are met. Propagation delay was not used as a parameter for design.

The first step in the design procedure is to set the output current based on slew-rate. For a t_{prop} of 1 μs, we should design for a slew rate ten times faster than 10 V/μs. So, design for 100 V/μs slew rate. For a 2 pF load capacitor, I_7 is calculated to be

$$I_7 = (2 \text{ pF}) \left(\frac{100 \text{ V}}{1 \text{ } \mu s} \right) = 200 \text{ } \mu A$$

Now M6 and M7 must be adjusted so that their V_{DS}(sat) is low enough to meet output-voltage swing requirements.

$$2 > V_{DS7}(\text{sat}) = \left[\frac{2I_7}{\beta_7} \right]^{1/2} = \left[\frac{(2)(200)}{(17.0)(W/L)_7} \right]^{1/2}$$

$$(W/L)_7 > 5.88$$

Similarly for M6

$$2 > V_{DS6}(\text{sat}) = \left[\frac{2I_7}{\beta_6} \right]^{1/2} = \left[\frac{(2)(200)}{(8.0)(W/L)_6} \right]^{1/2}$$

$$(W/L)_6 > 12.5$$

There is enough information to calculate the second-stage gain:

$$A_2 = - \left(\frac{g_{m6}}{g_{ds6} + g_{ds7}} \right) = \frac{-[2K_6'I_6(W/L)_6]^{1/2}}{I_6(\lambda_6 + \lambda_7)} = \frac{-[2(8)200(12.5)]^{1/2}}{200(0.03 + 0.07)}$$

$$A_2 = -10.0$$

Gain in the first stage must be at least 200 to meet the 66 dB requirement.

For a minimum-area design, the first-stage currents can be established by assuming minimum allowable device sizes for balanced conditions. First consider M6 and M4 (M3). For a $(W/L)_4$ of 1.0, I_4 is 200(1.0/12.5) or 16.0 μA. Considering M7 and M5, a unity aspect ratio of M5 results in $I_5 = 34$ μA or $I_4 = 17$ μA, which is larger than the first estimate. The larger of the two must be chosen in order to keep device aspect ratios greater than unity. Therefore set $I_5 = 34$ μA and $I_4 = 17$ μA. The minimum device size of M4 (M3) is now calculated to be

$$(W/L)_4 = (W/L)_6(17/200) = 1.06$$

$$(W/L)_4 > 1.0$$

Power dissipation must be checked at this point.

$$P_{\text{Total}} = (10)(I_7 + I_5) = 2.34 \text{ mW} < 10 \text{ mW}$$

Transistor M1 (M2) can now be adjusted to achieve proper gain

$$|A_1| = \left[\frac{2K_1'(W/L)_1}{I_4}\right]^{1/2} \left(\frac{1}{\lambda_1 + \lambda_4}\right)$$

$$(W/L)_1 = [(\lambda_1 + \lambda_4)A_1]^2 \left(\frac{I_4}{2K_1'}\right) = [(0.1)200]^2 \left(\frac{17 \times 10^{-6}}{2(17.0 \times 10^{-6})}\right)$$

$$(W/L)_1 > 200$$

With the device size of M1 (M2) known, the minimum size of M5 can be adjusted to meet negative input CMR requirements.

$$V_{G1}(\text{min}) = V_{SS} + V_{DS5} + \left[\frac{I_5}{\beta_1}\right]^{1/2} + V_{T1}(\text{max})$$

$$V_{DS5} = 4 - 0 - 1 - \left[\frac{(34 \times 10^{-6})}{(17.0 \times 10^{-6})(200)}\right]^{1/2} = 2.9$$

$$V_{DS5} = \left[\frac{2I_5}{\beta_5}\right]^{1/2} = 2.9$$

$$(W/L)_5 > 0.48$$

To achieve proper mirroring from M5 to M7, M5 must be

$$(W/L)_5 = \left(\frac{I_5}{I_7}\right)(W/L)_7 = \left(\frac{34 \times 10^{-6}}{200 \times 10^{-6}}\right)(5.88)$$

$$(W/L)_5 = 1.0$$

Proper positive input CMR can be designed in the following way

$$V_{G1}(\text{max}) = V_{DD} - \left[\frac{I_5}{\beta_3}\right]^{1/2} - |V_{T03}|(\text{max}) + V_{T1}(\text{min})$$

$$\beta_3 = \frac{I_5}{[V_{DD} - V_{G1}(\text{max}) - |V_{T03}|(\text{max}) + V_{T1}(\text{min})]^2}$$

$$\beta_3 = \frac{34\mu A}{(10 - 6 - 1 + 0.5)^2} = K_p'(W/L)_3$$

$$(W/L)_3 = (W/L)_4 > 0.35$$

Previous calculations set the size of M3 (M4) to be greater than 1.0, so no adjustment is necessary. At this point, the following aspect ratios have been determined.

$$(W/L)_1 = (W/L)_2 = 200$$

$$(W/L)_3 = (W/L)_4 = 1.0$$

$$(W/L)_5 = 1.0$$

$$(W/L)_6 = 12.5$$

$$(W/L)_7 = 5.88$$

Assuming minimum device lengths, and taking into account lateral diffusion, the devices' drawn widths and lengths are (in units of micrometers)

$$(W/L)_1 = (W/L)_2 = 680/5$$

$$(W/L)_3 = (W/L)_4 = 3.4/5 \rightarrow 5/5$$

$$(W/L)_5 = 3.4/5 \rightarrow 5/5$$

$$(W/L)_6 = (8)(W/L)_4 = 40/5$$

$$(W/L)_7 = (4)(W/L)_5 = 20/5$$

Notice that devices have been adjusted to integer values, and proper mirroring observed. The final design is shown in Fig. 7.3-6. This circuit must be simulated using a circuit simulator (e.g. PSPICE) to check for adequate propagation-delay margin.

7.4 Comparator with Hysteresis

Often a comparator is placed in a very noisy environment in which it must detect signal transitions at the threshold point. If the comparator is fast enough (depending upon the frequency of the most prevalent noise) and the amplitude of the noise is great enough, the output will also be noisy. In this situation, a modification on the transfer characteristic of the comparator is desired. Specifically, hysteresis is needed in the comparator.

Hysteresis is the quality of the comparator in which the input threshold changes as a function of the input (or output) level. In particular, when the input passes the threshold, the output changes and the input threshold is subsequently reduced so that the input must return beyond the previous

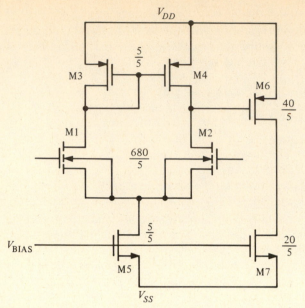

Figure 7.3-6 Two-stage comparator example result (all dimensions in micrometers).

threshold before the comparator's output changes state again. This can be illustrated much more clearly with the diagram shown in Fig. 7.4-1. Notice that as the input starts negative and goes positive, the output does not change until it reaches the positive trip point, V_{TRP+}. Once the output goes high, the effective trip point is changed. When the input returns in the negative direction, the output does not switch until it reaches the negative trip point, V_{TRP-}.

The advantage of hysteresis in a noisy environment can be clearly seen from the illustration given in Fig. 7.4-2. In this figure, a noisy signal is

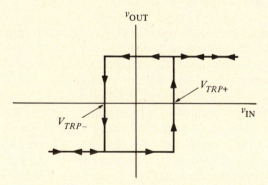

Figure 7.4-1   Comparator hysteresis curve.

shown as the input to a comparator without hysteresis. The intent is to have the comparator output follow the low-frequency signal. Because of noise variations near the threshold points, the comparator output is too noisy. The response of the comparator can be improved by adding hysteresis equal to or greater than the amount of the largest expected noise amplitude. The response of such a comparator is shown in Fig. 7.4-2(b).

There are many ways to accomplish hysteresis in a comparator. All of them involve some form of positive feedback. Consider the differential input stage given in Fig. 7.4-3 [3]. In this circuit there are two paths of feedback. The first is current-series [4, 5] feedback through the common-source node of transistors M1 and M2. This feedback path is negative. The second path is the voltage-shunt feedback through the gate-drain connections of transistors M10 and M11. This path of feedback is positive. If the positive feedback factor is less than the negative feedback factor, then the overall feedback will be negative and no hysteresis will result. If the positive feedback factor becomes greater, the overall feedback will be positive, which will give rise to hysteresis in the voltage transfer curve. As long as the ratio β_{10}/β_3 is less than one, there is no hysteresis in the transfer function. When this ratio is greater than one, hysteresis will result. The follow-

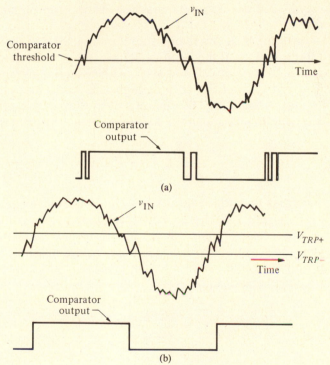

Figure 7.4-2   (a) Comparator response to a noisy input. (b) Comparator response to a noisy input when hysteresis is added.

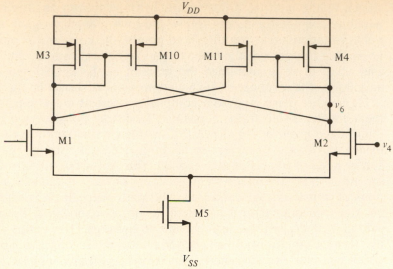

Figure 7.4-3   Comparator with hysteresis applied to the input differential stage only.

ing analysis will develop the equations for the trip points when there is hysteresis.

Assume that plus and minus supplies are used and that the gate of M1 is tied to ground. With the input of M2 much less than zero, M1 is on and M2 off, thus turning on M3 and M10 and turning off M4 and M11. All of i_5 flows through M1 and M3, so v_6 is high. The resulting circuit is shown in Fig. 7.4-4(a). Notice that M2 is shown even though it is off. At this point, M10 is attempting to source the following amount of current

$$i_{10} = \frac{(W/L)_{10}}{(W/L)_3} i_5 \tag{1}$$

As v_{in} increases toward the threshold point (which is not yet known), some of the tail current i_5 begins to flow through M2. This continues until the point where the current through M2 equals the current in M10. Just beyond this point the comparator switches state. To approximately calculate one of the trip points, the circuit must be analyzed right at the point where i_2 equals i_{10}. Mathematically this is

$$i_{10} = \frac{(W/L)_{10}}{(W/L)_3} i_3 \tag{2}$$

$$i_2 = i_{10} \tag{3}$$

$$i_5 = i_2 + i_1 \qquad (i_1 = i_3) \tag{4}$$

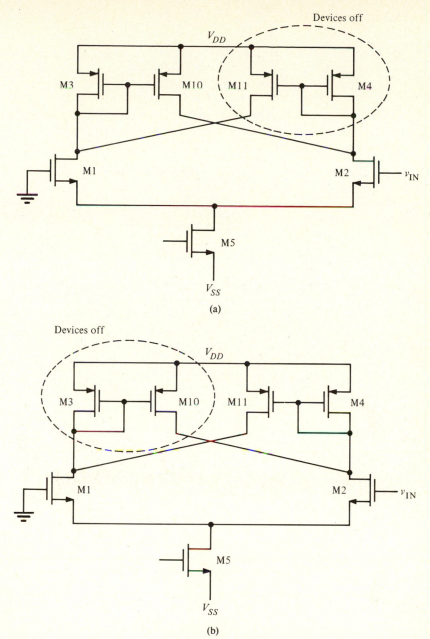

Figure 7.4-4 (a) Comparator in a state where v_{IN} is very negative and increasing toward V_{TRP+}. (b) Comparator in a state where v_{IN} is very positive and decreasing toward V_{TRP-}.

therefore

$$i_3 = \frac{i_5}{1 + [(W/L)_{10}/(W/L)_3]} = i_1 \tag{5}$$

$$i_2 = i_5 - i_1 \tag{6}$$

Knowing the currents in both M1 and M2 it is easy to calculate their respective v_{GS} voltages. Since the gate of M1 is at ground, the difference in their gate-source voltages will yield the positive trip point as given below

$$V_{GS1} = \left(\frac{2i_1}{\beta_1}\right)^{1/2} + V_{T1} \tag{7}$$

$$V_{GS2} = \left(\frac{2i_2}{\beta_2}\right)^{1/2} + V_{T2} \tag{8}$$

$$V_{TRP+} = V_{GS2} - V_{GS1} \tag{9}$$

Once the threshold is reached, the comparator changes state so that the majority of the tail current now flows through M2 and M4. As a result, M11 is also turned on thus turning off M3 and M10 and M1. As in the previous case, as the input returns negative the circuit reaches a point at which the current in M1 increases until it equals the current in M11. The input voltage at this point is the negative trip point V_{TRP-}. The equivalent circuit in this state is shown in Fig. 7.4-4(b). To calculate the trip point, the following equations apply

$$i_{11} = \frac{(W/L)_{11}}{(W/L)_4} i_4 \tag{10}$$

$$i_1 = i_{11} \tag{11}$$

$$i_5 = i_2 + i_1 \tag{12}$$

therefore

$$i_4 = \frac{i_5}{1 + [(W/L)_{11}/(W/L)_4]} = i_2 \tag{13}$$

$$i_1 = i_5 - i_2 \tag{14}$$

Using Eqs. (7) and (8) to calculate v_{GS}, the trip point is

$$V_{TRP-} = V_{GS2} - V_{GS1} \tag{15}$$

These equations do not take into account the effect of channel length modulation. Their use is illustrated in the following example.

Example 7.4-1

Calculation of Trip Voltages for a Comparator with Hysteresis

Consider the circuit shown in Fig. 7.4-3. Using the transistor device parameters given in Table 3.1-2 calculate the positive and negative threshold points. Simulate the results using PSPICE.

The device lengths are all 10 μm and the widths are given as: $W_1 = W_2 = W_{10} = W_{11} = 25$ μm and $W_3 = W_4 = 10$ μm. The gate of M1 is tied to ground and the input is the gate of M2. The current, i_5 = 25 μA. Assume λ for p-channel and n-channel devices is 0.002 and 0.001 in this example.

To calculate the positive trip point, assume that the input has been negative and is heading positive.

$$i_{10} = \frac{(W/L)_{10}}{(W/L)_3} i_3 = (2.5/1)(i_3)$$

$$i_3 = \frac{i_5}{1 + [(W/L)_{10}/(W/L)_3]} = i_1 = \frac{25 \text{ }\mu A}{1 + 2.5} = 7.14 \text{ }\mu A$$

$$i_2 = i_5 - i_1 = 25 - 7.14 = 17.86 \text{ }\mu A$$

$$V_{GS1} = \left(\frac{2i_1}{\beta_1}\right)^{1/2} + V_{T1} = \left(\frac{14.28}{(2.5)17}\right)^{1/2} + 1 = 1.58$$

$$V_{GS2} = \left(\frac{2i_2}{\beta_2}\right)^{1/2} + V_{T2} = \left(\frac{35.72}{(2.5)17}\right)^{1/2} + 1 = 1.92$$

$$V_{TRP+} \cong V_{GS2} - V_{GS1} = 1.92 - 1.58 = 0.34 \text{ volts}$$

Determining the negative trip point, similar analysis yields

$$i_4 = 7.14 \text{ }\mu A$$

$$i_1 = 17.86 \text{ }\mu A$$

$$V_{GS2} = 1.58$$

$$V_{GS1} = 1.92$$

$$V_{TRP-} \cong V_{GS2} - V_{GS1} = 1.58 - 1.92 = -0.34 \text{ volts}$$

PSPICE simulation results of this circuit are shown in Fig. 7.4-5.

The differential stage described thus far is generally not useful alone, and thus requires an output stage to achieve reasonable voltage swings and output resistance. There are a number of ways to implement an output stage for this type of input stage. One of these is given in Fig. 7.4-6. Differ-

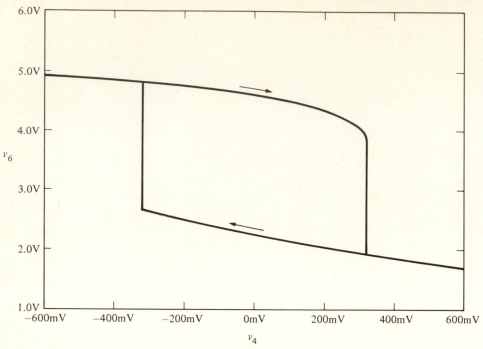

Figure 7.4-5 Simulation of the comparator of Fig. 7.4-3 with hysteresis.

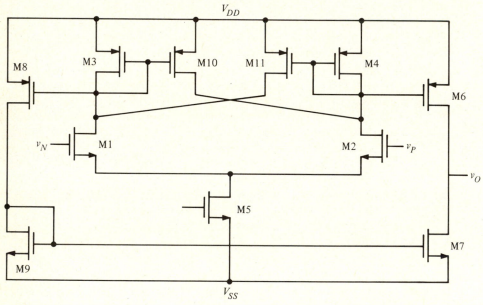

Figure 7.4-6 Complete comparator with hysteresis including output stage.

ential-to-single-ended conversion is accomplished at the output and thus provides a class-AB type of driving capability.

7.5   *Auto Zero Techniques*

Input-offset voltage can be a particularly difficult problem in comparator design. In precision applications, such as high-resolution A/D converters, large input-offset voltages cannot be tolerated. While systematic offset can be nearly eliminated with proper design (though still affected by process variations), random offsets still remain, and are unpredictable. Fortunately, there are techniques in MOS technology to remove a large portion of the input offset using offset-cancellation techniques. These techniques are available in MOS because of the nearly infinite input resistance of MOS transistors. This characteristic allows long-term storage of voltages on the transistor's gate. As a result, offset voltages can be measured, stored on capacitors, and summed with the input so as to cancel the offset.

Figure 7.5-1 shows symbolically an offset-cancellation algorithm. A model of a comparator with an input-offset voltage is shown in Fig. 7.5-1(a). A known polarity is given to the offset voltage for convenience. Neither the value nor the polarity can be predicted in reality. Figure 7.5-1(b) shows the comparator connected in the unity-gain configuration so that the input offset is available at the output. In order for this circuit to work properly, it is necessary that the comparator be stable in the unity-gain configuration. Sec. 8.2 discusses compensation methods applicable to the comparator. Capacitor C_{AZ} measures the offset, storing it across its terminals. In the final operation of the auto-zero algorithm C_{AZ} is placed at the input of the comparator in series with V_{OS}. The voltage across C_{AZ} adds to V_{OS}, resulting in zero volts at the noninverting input of the comparator. Since there is no dc path to discharge the auto-zero capacitor, the voltage across it remains indefinitely (in the ideal case). In reality there are leakage paths in shunt with C_{AZ} that can discharge it over a period of time. The solution to this problem is to repeat the auto-zero cycle periodically.

A practical implementation of an auto-zeroed comparator is shown in Fig. 7.5-2(a). The comparator is modeled with an offset-voltage source as before. Figure 7.5-2(b) shows the state of the circuit during the first phase of the cycle when ϕ_1 is high. The offset is stored across C_{AZ}. Figure 7.5-2(c) shows the circuit in the second phase of the auto-zero cycle when ϕ_2 is high. The offset is cancelled by the addition of V_{OS} to V_{CAZ}. It is during this portion of the cycle that the circuit functions as a comparator.

There are other ways to implement an auto-zeroed comparator other than the one just described. A slight variation of Fig. 7.5-2(a) is shown in Fig. 7.5-3. This is a noninverting version of the previous auto-zeroed comparator. Yet another version is shown in Fig. 7.5-4. This one is simpler in operation since the noninverting input is always connected to ground.

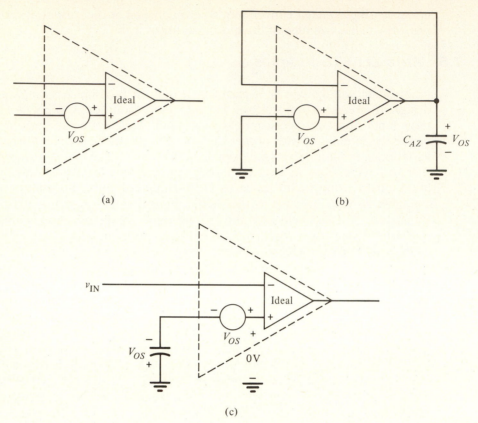

(a)   (b)

(c)

Figure 7.5-1   (a) Simple model of a comparator including offset. (b) Comparator in unity-gain configuration storing offset on auto-zero capacitor C_{AZ}. First half of auto-zero cycle. (c) Comparator in open-loop configuration with offset cancelation achieved at the noninverting input. Second half of auto-zero cycle.

    In all of the implementations of the auto-zeroed comparators, n-channel switches were used. Depending upon the voltage levels at the inputs and outputs of the comparator, it may be necessary to use either p-channel or complementary switches (Fig. 5.1-12). It is also very important to use nonoverlapping clocks to drive the switches so that any given switch turns off before another turns on. For n-channel switches, the clock signals shown in Fig. 7.5-5 are appropriate.

    Auto-zero techniques can be very effective in removing a large amount of a comparator's input offset, but the offset cancellation is not perfect. Charge injection resulting from clock feedthrough (Sec. 5.1) can by itself introduce an offset. This too can be cancelled, but it is usually the cause of the lower limit of the offset voltage being greater than zero.

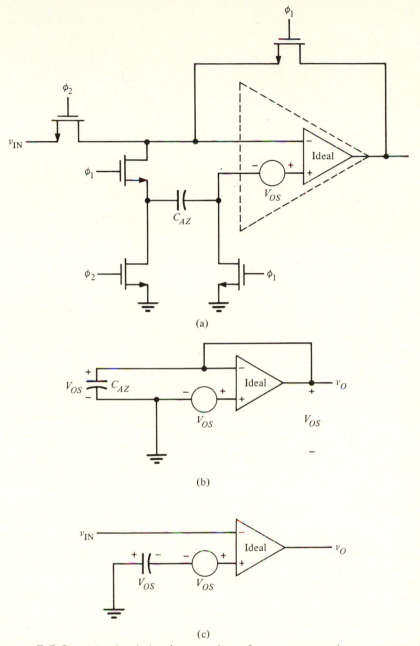

(a)

(b)

(c)

Figure 7.5-2 (a) Circuit implementation of an auto-zeroed comparator. (b) Comparator during ϕ_1 auto-zero state. (c) Comparator during ϕ_2 state.

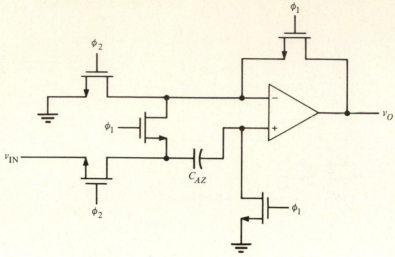

Figure 7.5-3   Noninverting auto-zeroed comparator.

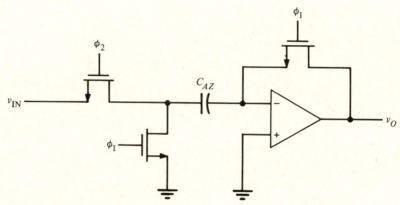

Figure 7.5-4   Inverting auto-zeroed comparator.

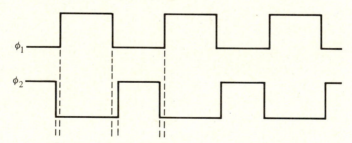

Figure 7.5-5   Nonoverlapping clock phasing.

7.6 Summary

This chapter has introduced the comparator function along with the CMOS implementation of that function. The development of the CMOS comparator relied heavily on previous work covering basic CMOS circuits in chapters 5 and 6. There are many other ways to implement comparators that have not been presented here. One of the primary motivations for presenting the particular topology chosen was to prepare the reader for the material on operational amplifiers that follows.

Using two basic CMOS elements—the current-source load inverter, and the differential-input stage—a very useful comparator circuit has been developed. This circuit has been characterized over the various conditions important in considering the comparison function. Armed with formulas defining the circuit's operation, we have presented a design procedure that allows for a good "first-cut" at the sizes and currents of the comparator. Finally, a few different configurations were presented along with some circuit techniques that correct for offset voltage.

Consistent with the hierarchical approach taken in this book, where circuits build upon earlier circuits, the development of the basic comparator circuit in this chapter will be the foundation for the work on operational amplifiers that follows in the next two chapters. By simply adding compensation to the comparator of Fig. 7.3-1, one can make a stable operational amplifier.

PROBLEMS — *Chapter 7*

1. Draw the first-order dc transfer curves ($v_{O1} - vs - v_{IN}$) for a comparator with a ± 20 mV offset. Illustrate the region of uncertainty resulting from offset variation.

2. Draw the first-order time response of an inverting comparator with a 20 μs propagation delay. The input is described by the following equation

$$\begin{aligned}
v_{in} &= 0 && \text{for t} < 5\ \mu s \\
v_{in} &= 5(t - 5\ \mu s) && \text{for } 5\ \mu s < t < 7\ \mu s \\
v_{in} &= 10 && \text{for t} > 7\ \mu s
\end{aligned}$$

3. Calculate the trip voltage for the comparator shown in Fig. 7.2-1. Use the parameters given in Table 3.1-2. Also, $(W/L)_2 = 100$ and $(W/L)_1 = 10$. $V_{BIAS} = 2.5$ V, $V_{SS} = 0$V, and $V_{DD} = 8$ V.

4. Using Problem 3, compute the worst-case variations of the trip voltage assuming a $\pm 10\%$ variation on V_T, K', V_{DD}, and V_{BIAS}.

5. Calculate the input drive required for the output to swing CMOS logic levels in the circuit of Problem 3.

6. Sketch the output response of the circuit in Problem 3, given a step input that goes from 8 to 5 volts. Assume a 10 pF capacitive load. Also assume the input has been at 8 volts for a very long time. What is the delay time from the step input to when the output changes logical (CMOS) states.

7. Calculate the approximate limits of the output range for the circuit shown in Fig. 7.2-3. Use the device parameters given in Table 3.1-2. Assume $(W/L)_1 = (W/L)_2 = 50$, $(W/L)_3 = (W/L)_4 = 25$, and $I_5 = 20$ μA.

8. Repeat Problem 7, only calculate "useful" output range instead.

9. Calculate the gain of the circuit described in problem 7.

10. Calculate the systematic offset of the comparator shown in Fig. 7.3-1 assuming parametric data given in Table 3.1-2, and the following device sizes: $(W/L)_7 = 50$ μm/10 μm, $(W/L)_5 = 25$ μm/10 μm, $(W/L)_{1,2} = 40$ μm/10μm $(W/L)_{3,4} = 10$ μm/10 μm, $(W/L)_6 = 50$ μm/10 μm. $I_5 = 20$ μA. $\lambda_P = 0.03$, and $\lambda_N = 0.02$. $V_{DD} = 10$ V, $V_{SS} = 0$V, $V_O = 5$ V, and the average value of V_N and V_P is 5 V. What is the most obvious way to improve (reduce) the systematic offset of this circuit?

11. Design transistors M1 through M4 of the circuit in Fig. 7.3-1 for a common mode range of 2 to 8 volts. $V_{DD} = 10$ volts and $V_{SS} = 0$V. Pertinent process information can be found in Table 3.1-2 with the following exceptions: $|V_{TN,P}| = 0.3$ to 1.1 volts, $I_5 = 30$ μA and $(W/L)_5 = 10$.

12. Calculate the common-mode input range for the circuit shown in Fig. P7-12. Use process data given in Table 3.1-2 and V_T variations of 0.4 to 1.0 volts.

13. Calculate the gain of the circuit in Fig. P7-12. Use Table 3.1-2 as required.

14. Calculate the total large-signal propagation delay for the circuit shown in Fig. P7-14. Use Table 3.1-2. Assume that the total depletion capacitance at node A is 0.3 pF. v_{IN} goes from 5 to 6 volts in 10 ns.

15. Design a comparator given the following requirements: $P_{diss} < 5$ mW, $V_{DD} = 10$ V, $V_{SS} = 0$ V $C_{load} = 3$ pF, $t_{prop} < 1$ μs, input CMR $= 3.6 - 6.5$ V, $A_{vo} > 2200$, and output voltage swing within 1.5 volts of either rail. Use Tables 3.1-2 and 3.3-1 with the following exceptions: $\lambda = 0.04$ for a 5 μm device length.

16. Rederive the hysteresis equations in Sec. 7.4, this time including the channel-length modulation effect.

17. Recalculate the trip points of the Example 7.4-1 with the widths of M10 and M11 as 20 μm instead of 25 μm.

18. Would the circuit of Fig. 7.4-3 have any hysteresis if M10 was missing? Calculate the trip point(s) using the device sizes given in Example 7.4-1.

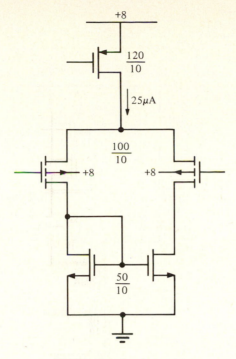

All W/L values in micrometers

Figure P7.12

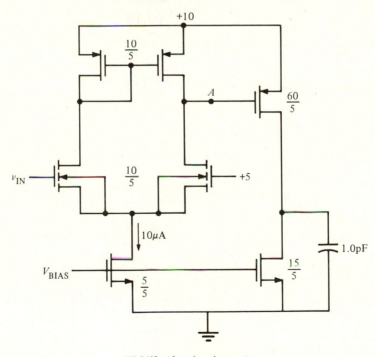

All W/L values in micrometers

Figure P7.14

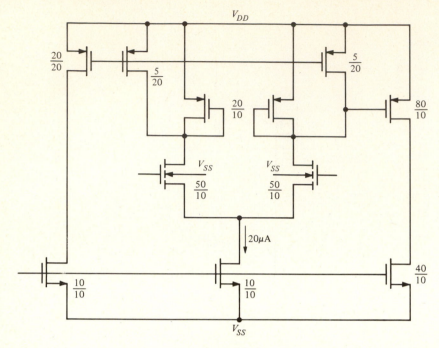

All *W/L* values in micrometers

Figure P7.19

19. Fig. P7.19 shows a circuit called a clamped comparator. Use the parameters of Table 3.1-2 and calculate the gain of this comparator.
20. Repeat the previous problem with the two 5/20 devices removed. Compare the gain of the two configurations of the clamped comparator.

REFERENCES

1. D.C. Stone, et. al., "Analog CMOS Building Blocks for Custom and Semicustom Applications," *IEEE Journal Sold-State Circuits*, Vol. SC-19, no. 1, (February 1984) pp. 55–61.
2. J.L. McCreary and P.R. Gray, "All-MOS Charge Redistribution Analog-to-Digital Conversion Techniques-Part I," *IEEE Journal of Solid State Circuits*, Vol. SC-10, no. 6 (Dec. 1975) pp. 371–379.
3. D.J. Allstot, "A Precision Variable-Supply CMOS Comparator", *IEEE Journal Solid-State Circuits*, Vol. SC-17, no. 6, (December 1982) pp. 1080–1079.
4. Jacob Millman and C.C. Halkias, *Integrated Electronics: Analog and Digital Circuits and Systems* (New York: McGraw-Hill, 1972).
5. A.S. Sedra and K.C. Smith, *Microelectronic Circuits* (New York: Holt, Rinehart and Winston, 1982).

chapter 8

*U*nbuffered CMOS Op Amps

The operational amplifier, which has become one of the most versatile and important building blocks in analog-circuit design, is introduced in this chapter. The op amp fits into the scheme of Table 1.1-2 as an example of a complex circuit. The unbuffered op amps developed in this chapter might be better described as operational-transconductance amplifiers since the output resistance typically will be very high (hence the term "unbuffered"). The term "op amp" has become accepted for such circuits, so it will be used throughout this text. The terms "unbuffered" and "buffered" will be used to distinguish between high output resistance (operational-transconductance amplifiers or OTAs) and low output resistance amplifiers (voltage operational amplifiers). Chapter 9 will consider op amps which have low output resistance (buffered op amps).

Operational amplifiers are amplifiers (controlled sources) that have sufficiently high forward gain so that when negative feedback is applied, the closed-loop transfer function is practically independent of the gain of the op amp (see Fig. 6.5-1). This principle has been exploited to develop many useful analog circuits and systems.

The comparators developed in the previous chapter are legitimate candidates for op amp realizations. In fact, the architectures of some comparators are suited for implementation as op amps once compensation is applied to avoid instability. Compensation is an essential task in the design of good op amps, and will be analyzed in detail. Subsequently, a design procedure for a popular two-stage op amp will be developed. Cascode op amp structures will be introduced as an alternative for better performance in certain cases.

Simulation and measurement of op amp performance will be the final subject of this chapter. Simulation is necessary to verify and refine the design. Experimental measurements are necessary to verify the performance of the op amp with the original design specifications. Typically, the techniques applicable to simulation are also suitable for experimental measurement.

8.1 *Op Amp Design Methodology*

Figure 8.1-1 shows a block diagram that represents the important aspects of an op amp. CMOS op amps are very similar in architecture to their bipolar counterparts. The differential-transconductance stage introduced in Sec. 6.2 forms the input of the op amp and sometimes provides the differ-

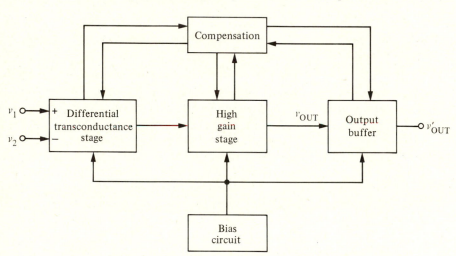

Figure 8.1-1 Block diagram of a two-stage CMOS Op Amp.

ential to single-ended conversion. Normally a good portion of the overall gain is provided by the input-differential stage, which improves noise and offset performance. The second stage is typically an inverter similar to that introduced in Sec. 6.1. When the differential-input stage does not perform the differential-to-single-ended conversion, it is accomplished in the second-stage inverter. If the op amp must drive a low-resistance load, the second stage must be followed by a buffer stage whose objective is to lower the output resistance and maintain a large signal swing. Bias circuits are provided to establish the proper operating point for each transistor in its quiescent state. As noted in the introduction, compensation is required to achieve stable closed-loop performance. Sec. 8.2 will address this important topic.

Ideally, an op amp has infinite differential-voltage gain, infinite input resistance, and zero output resistance. In reality, an op amp only approaches these values. For most applications where unbuffered CMOS op amps are used, an open-loop gain of 5,000 or more is usually sufficient. The symbol for an op amp is shown in Fig. 8.1-2 where, in the nonideal case, the output voltage, V_o, can be expressed as

$$V_o = A_v(V_1 - V_2) \tag{1}$$

A_v is used to designate the open-loop, differential-voltage gain. V_1 and V_2 are the input voltages applied to the noninverting and inverting terminals, respectively. Figure 8.1-3 shows the way an op amp is typically used. With the noninverting input connected to ac ground, the output voltage is described as

$$V_o = -A_v V_i \tag{2}$$

If A_v is assumed to be very high then the external feedback connected from V_o to V_i will cause the input of the op amp to simulate a virtual ground (two terminals having zero potential between them and no current flow into or

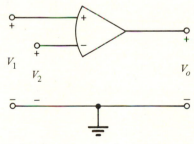

Figure 8.1-2   Symbol for an operational amplifier realized by a voltage-controlled voltage source.

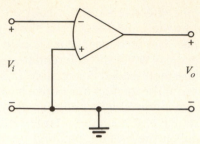

Figure 8.1-3 Typical configuration of the op amp for analog sampled-data circuit applications.

out of the terminals). Therefore, V_i can be assumed to be zero. The current flowing into the op amp is also zero because of the infinite input impedance and the large forward gain.

   In practice, the operational amplifier only approaches the ideal infinite-gain voltage amplifier. Some of its other nonideal characteristics are illustrated in Fig. 8.1-4. The *finite differential-input impedance* is modeled by R_{id} and C_{id}. The *output resistance* is modeled by R_{out}. The *common-mode input resistances* are given as resistances of R_{icm} connected from each of the inputs to ground. V_{os} is the *input-offset voltage* necessary to make the output voltage zero if both of the inputs of the op amp are grounded. I_{os} (not

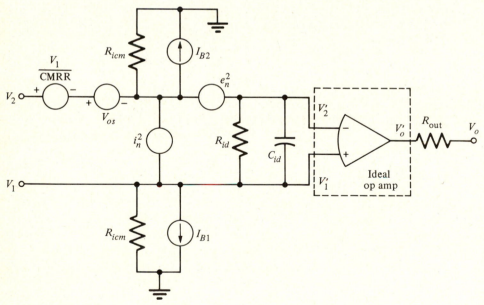

Figure 8.1-4 A model for a nonideal op amp showing some of the nonideal characteristics.

shown) is the *input-offset current*, which is necessary to make the output voltage zero if the op amp is driven from two identical current sources. Therefore, I_{os} is defined as the magnitude of the difference between the two *input-bias currents* I_{B1} and I_{B2}. Since the bias currents for a CMOS op amp are approximately zero, the offset current is also zero.

The *common-mode rejection ratio* (CMRR) is modeled by the voltage-controlled voltage source indicated as V_1/CMRR. This source approximately models the effects of the common-mode input signal on the op amp. The two sources designated as $\overline{e}_n^2$ and $\overline{i}_n^2$ are used to model the op amp noise. These are *voltage- and current-noise spectral densities* with units of mean square volts and mean square amperes, respectively. These noise sources have no polarity and are always assumed to add.

Not all of the nonideal characteristics of the op amp are shown in Fig. 8.1-4. Other pertinent characteristics of the op amp will now be defined. The output voltage of Fig. 8.1-2 can be defined as

$$V_o(s) = A_v(s) \, [V_1(s) - V_2(s)] + A_c(s) \left[\frac{V_1(s) + V_2(s)}{2} \right] \qquad (3)$$

where the first term on the right is the differential portion of $V_o(s)$ and the second term is the common-mode portion of $V_o(s)$. The *differential-frequency response* is given as $A_v(s)$ while the *common-mode frequency response* is given as $A_c(s)$. A typical differential-frequency response of an op amp is given as

$$A_v(s) = \frac{A_{vo}}{(\frac{s}{p_1} - 1) \, (\frac{s}{p_2} - 1) \, (\frac{s}{p_3} - 1) \cdots} \qquad (4)$$

where $p_1, p_2, \ldots$ are poles of the operational amplifier open-loop transfer function. In general, a pole designated as p_i can be expressed as

$$p_i = -\omega_i \qquad (5)$$

where ω_i is the reciprocal time constant or break-frequency of the pole p_i. While the operational amplifier may have zeros, they will be ignored at the present time. A_{vo} or $A_v(0)$ is the gain of the op amp as the frequency approaches zero. Figure 8.1-5 shows a typical frequency response of the magnitude of $A_v(s)$. In this case we see that ω_1 is much lower than the rest of the break-frequencies causing ω_1 to be the dominant influence in the frequency response. The frequency where the -6 dB/oct. slope from the dominant pole intersects with the 0 dB axis is designated as the *unity-gain bandwidth*, abbreviated *GB*, of the op amp. Even if the next higher order poles are smaller than *GB*, we shall still continue to use the unity-gain bandwidth as defined above.

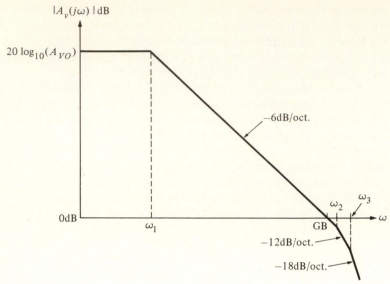

Figure 8.1-5 Typical frequency response of the magnitude $A_v(j\omega)$ for an op amp.

Other nonideal characteristics of the op amp not defined by Fig. 8.1-4 include the *power-supply rejection ratio*, PSRR. The PSRR is defined as the product of the ratio of the change in supply voltage to the change in output voltage of the op amp caused by the change in the power supply and the open-loop gain of the op amp. Thus

$$\text{PSRR} = \frac{\Delta V_{DD}}{\Delta V_{\text{OUT}}} A_v(s) = \frac{V_o/V_{\text{in}} \, (V_{dd} = 0)}{V_o/V_{dd} \, (V_{\text{in}} = 0)} \tag{6}$$

An ideal op amp would have an infinite PSRR. The reader is advised that both this definition for PSRR and its inverse will be found in the literature. The *common-mode input range* is the voltage range over which the input common-mode signal can vary. Typically, this range is several volts less than V_{DD} and several volts more than V_{SS}.

The output of the op amp has several important limits, one of which is the maximum output current sourcing and sinking capability. There is a limited range over which the output voltage can swing while still maintaining high-gain characteristics. The output also has a voltage rate limit called *slew rate*. The slew rate is generally determined by the maximum current available to charge or discharge a capacitance. Normally slew rate is not limited by the output, but by the current–sourcing/sinking capability of the first stage. The last characteristic of importance in analog sampled-data circuit applications is the *settling time*. This is the time needed for the output of the op amp to reach a final value (to within a predetermined tolerance) when excited by a small signal. This is not to be confused with slew rate which is a large-signal phenomenon. Many times, the output response of an op amp is a combination of both large- and small-signal characteristics. Small-signal settling time can be completely determined from the location of the poles and zeros in the small-signal equivalent circuit whereas slew rate is determined from the large-signal conditions of the circuit.

The importance of the settling time to analog sampled-data circuits is illustrated by Fig. 8.1-6. It is necessary to wait until the amplifier has settled to within a few tenths of a percent of its final value in order to avoid errors in the accuracy of processing analog signals. A longer settling time implies that the rate of processing analog signals must be reduced.

Fortunately the CMOS op amp does not suffer all of the nonideal characteristics previously discussed. Because of the extremely high input resistance of the MOS devices, both R_{id} and I_{os} (or I_{B1} and I_{B2}) are of no importance. A typical value of R_{id} is in the range of 10^{14} ohms. Also, R_{icm} is extremely large and can be ignored. If an op amp is used in the configuration of Fig. 8.1-3, then all common-mode characteristics are unimportant.

The design of an op amp can be divided into two distinct design-related activities that are for the most part independent of one another. The first of these activities involves choosing or creating the basic structure of the op amp. A diagram that describes the interconnection of all of the transistors results. In most cases, this structure does not change throughout the remaining portion of the design, but sometimes certain characteristics of the chosen design must be changed by modifying the structure.

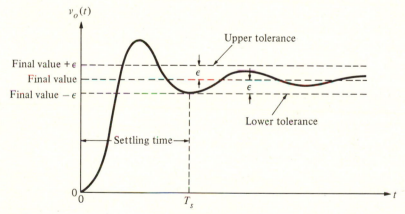

Figure 8.1-6   Transient response of an op amp with negative feedback illustrating settling time T_s. ϵ is the tolerance to the final value used to define the settling time.

Once the structure has been selected, the designer must select dc currents and begin to size the transistors and the compensation components. Most of the work involved in completing a design is associated with this, the second activity of the design process. Devices must be properly scaled in order to meet all of the ac and dc requirements imposed on the op amp. Computer circuit simulations, based on hand calculations, are used extensively to aid the designer in this phase.

Before the actual design of an op amp can begin, though, one must set out all of the requirements and boundary conditions that will be used to guide the design. The following list describes many of the items that must be considered.

Boundary conditions:

1. Process specification (V_T, K', C_{ox}, etc.)
2. Supply voltage and range
3. Supply current and range
4. Operating temperature and range

Requirements:

1. Gain
2. Gain bandwidth
3. Settling time
4. Slew rate
5. Common-mode input range, CMR
6. Common-mode rejection ratio, CMRR
7. Power-supply rejection ratio, PSRR
8. Output-voltage swing
9. Output resistance
10. Offset
11. Noise
12. Layout area

The typical specifications for an unbuffered, CMOS op amp are listed in Table 8.1-1.

The block diagram of Fig. 8.1-1 is useful for guiding the CMOS op amp design process. The compensation method has a large influence on the design of each block. Two basic methods of compensation are suggested by the opposite parallel paths into the compensation block of Fig. 8.1-1. These two methods, feedback and feedforward, are developed in the following section. The method of compensation is greatly dependent upon the number of stages present (differential, second, or buffer stages).

In designing an op amp, one can begin at many points. The design procedure must be iterative, since it is almost impossible to relate all specifi-

Table 8.1-1

Specifications for a Typical Unbuffered CMOS Op Amp.

Boundary Conditions	Requirement
Process Specification	See Tables 3.1-1 and 3.1-2
Supply Voltage	± 5 V $\pm 10\%$
Supply Current	100 μA nominal
Temperature Range	0 to 70°C

Specifications	
Gain	$\geq$ 70 dB
Gainbandwidth	$\geq$ 2 MHz
Settling Time	$\leq$ 1 μsec
Slew Rate	$\geq$ 2 V/μsec
Input CMR	$\geq$ ± 3 V
CMRR	$\geq$ 60 dB
PSRR	$\geq$ 60 dB
Output Swing	$\geq$ ± 4 V
Output Resistance	N/A, capacitive load only
Offset	$\leq$ ± 10 mV
Noise	$\leq$ 100 nV/$\sqrt{Hz}$ at 1 KHz
Layout Area	$\leq$ 120,000 square microns

cations simultaneously. For a typical CMOS op amp design, the following steps may be appropriate.

1. Decide upon a suitable configuration. After examining the specifications in detail, determine the type of configuration required. For example, if extremely low noise and offset is a must, then a configuration that affords high gain in the input stage is required. If there are low-power requirements, then a class-AB-type output stage may be necessary. This in turn will govern the type of input stage that must be used. Often one must create a configuration that meets a specific application.

2. Determine the type of compensation needed to meet the specifications. There are many ways to compensate amplifiers. Some have unique aspects that make them suitable for particular configurations or specifications. For example, an op amp that must drive very large load capacitances might be compensated at the output. If this is the case, then this requirement also dictates the type of input and output stage needed. As this example shows, iteration might be necessary between steps 1 and 2 of the design process.

3. Design device sizes for proper dc, ac, and transient performance. This begins with hand calculations based upon approximate design equations. Compensation components are also sized in this step of the procedure. After each device is sized by hand, a circuit simulator is used to fine tune the design.

One may find during the design process that some specification may be difficult or impossible to meet with a given configuration. At this point the designer has to improve the configuration or search the literature for ideas particularly suited to the requirement. This literature search takes the place of creating a new configuration from scratch. For very critical designs, the hand calculations can achieve about 80% of the complete job in roughly 20% of the total job time. The remaining 20% of the job requires 80% of the time for completion. Sometimes hand calculations can be misleading due to their approximate nature. Nonetheless, they are necessary to give the designer a feel for the sensitivity of the design to parameter variation. There is no other way for the designer to understand how the various design parameters influence performance. Iteration by computer simulation gives the designer very little feeling for the design and is generally not a wise use of computer resources.

In summary, the design process consists of two major steps. The first is the conception of the design and the second is the optimization of the design. The conception of the design is accomplished by proposing an architecture to meet the given specifications. This step is normally accomplished using hand calculations in order to maintain the intuitive viewpoint necessary for choices that must be made. The second step is to take the "first-cut" design and verify and optimize it. This is normally done by using computer simulation and can include such influences as environmental or process variations.

8.2 Compensation of Op Amps

Operational amplifiers are primarily used in a negative-feedback configuration. In this way, the relatively high, inaccurate forward gain can be used with feedback to achieve a very accurate transfer function that is a function of the feedback elements only. Figure 8.2-1 illustrates a general negative-feedback configuration. $A(s)$ is the amplifier gain and will normally be the open-loop, differential voltage gain of the op amp, and $F(s)$ is the transfer function for external feedback from the output of the op amp back to the input (See Sec. 6.5). The loop gain of this system will be defined as

$$\text{Loop Gain} = L(s) = -A(s)F(s) \tag{1}$$

Consider a case where the forward gain from V_{in} to V_{out} is unity. It is easily shown that if the open-loop gain at dc $A(0)$ is between 1000 and 2000, and F equal to 1, the forward gain varies from 0.999 to 0.9995. For very high loop gain (due primarily to a high amplifier gain), the forward transfer function V_{out}/V_{in} is accurately controlled by the feedback network. This is the principle applied in using operational amplifiers.

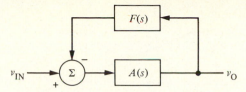

Figure 8.2-1   Feedback system.

It is of primary importance that the signal fed back to the input of the op amp be of such amplitude and phase that it does not continue to regenerate itself around the loop. Should this occur, the result will be either clamping of the output of the amplifier at one of the supply potentials (regeneration at dc), or oscillation (regeneration at some frequency other than dc). The requirement for avoiding this situation can be succinctly stated by the following equation.

$$|A(j\omega_{0°})F(j\omega_{0°})| = |L(j\omega_{0°})| < 1 \tag{2}$$

where $\omega_{0°}$ is defined as

$$\text{Arg}[-A(j\omega_{0°})F(j\omega_{0°})] = \text{Arg}[L(j\omega_{0°})] = 0° \tag{3}$$

Another convenient way to express this requirement is

$$\text{Arg}[-A(j\omega_{0\text{dB}})F(j\omega_{0\text{dB}})] = \text{Arg}[L(j\omega_{0\text{dB}})] > 0° \tag{4}$$

where $\omega_{0\text{dB}}$ is defined as

$$|A(j\omega_{0\text{dB}})F(j\omega_{0\text{dB}})| = |L(j\omega_{0\text{dB}})| = 1 \tag{5}$$

If these conditions are met, the feedback system is said to be stable (i.e., sustained oscillation cannot occur).

This second relationship given in Eq. (4) is best illustrated with the use of Bode diagrams. Figure 8.2-2 shows the response of $|A(j\omega)F(j\omega)|$ and $\text{Arg}[-A(j\omega)F(j\omega)]$ as a function of frequency. The requirement for stability is that the $|A(j\omega)F(j\omega)|$ curve cross the 0 dB point before the $\text{Arg}[-A(j\omega)F(j\omega)]$ reaches 0 degrees. A measure of stability is given by the value of the phase when $|A(j\omega)F(j\omega)|$ is unity. This measure is called *phase margin* and is described by the following relationship

$$\text{Phase margin} = \text{Arg}[-A(j\omega_{0\text{dB}})F(j\omega_{0\text{dB}})] = \text{Arg}[L(j\omega_{0\text{dB}})] \tag{6}$$

The importance of "good stability" obtained with adequate phase margin is best understood by considering the response of the closed-loop sys-

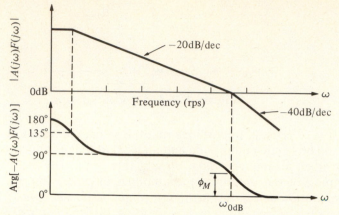

Figure 8.2-2   Frequency and phase response of a second-order system.

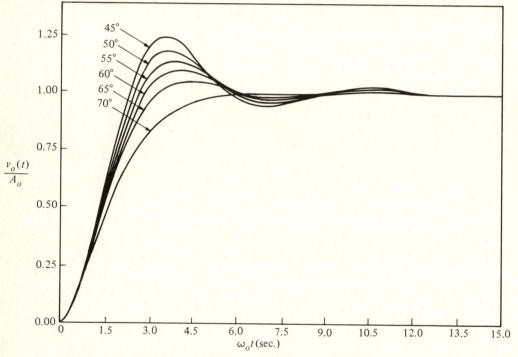

Figure 8.2-3   Response of second-order system with various phase margins.

tem in the time domain. Figure 8.2-3 shows the time response of a second-order closed-loop system with various phase margins. One can see that larger phase margins result in less "ringing" of the output signal. Too much ringing can be undesirable, so it is important to have adequate phase margin keeping the ringing to an acceptable level. It is desirable to have a

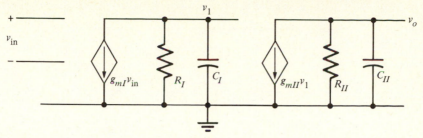

Figure 8.2-4 Second-order small-signal equivalent circuit for a two-stage op amp.

phase margin of at least 45°, with 60° preferable in most situations. Appendix C develops the relationship between phase margin and time domain response for second-order systems.

Now consider the second-order model for an uncompensated op amp shown in Fig. 8.2-4. In order to generalize the results, the components associated with the first stage have the subscript I and those associated with the second stage have the subscript II. The locations for the two poles are given by the following equations

$$p_1' = \frac{-1}{R_I C_I} \tag{7}$$

and

$$p_2' = \frac{-1}{R_{II} C_{II}} \tag{8}$$

where R_I (R_{II}) is the resistance to ground seen from the output of the first (second) stage and C_I (C_{II}) is the capacitance to ground seen from the output of the first (second) stage. In a typical case, these poles are at a high frequency and are relatively close together. This situation is illustrated in the bode diagram of Fig. 8.2-5 where the feedback factor F is assumed equal to 1 (this is worst-case for stability considerations). Notice that the phase margin is significantly less than 45°. This amplifier must be compensated before using it in a closed-loop configuration.

The first compensation method discussed here will be the "Miller" compensation technique. This technique is applied by connecting a capacitor from the output to the input of the second transconductance stage g_{mII}. The resulting small-signal model is illustrated in Fig. 8.2-6. Two results come from adding the compensation capacitor C_c. First, the effective capacitance shunting R_I is increased by the additive amount of approximately $g_{mII}(R_{II})(C_c)$. This moves p_1 down by a significant amount (assuming that the second-stage gain is large). Second, p_2 is moved to a higher frequency resulting from the negative feedback reducing the output resistance of the second stage.

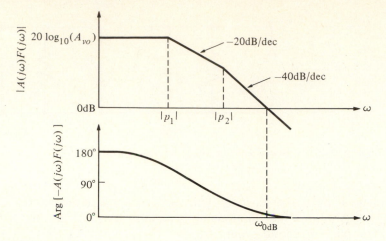

Figure 8.2-5   Uncompensated frequency response of an op amp.

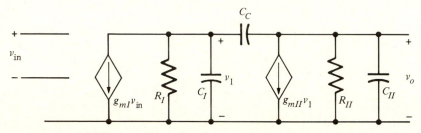

Figure 8.2-6   Miller capacitance applied to two-stage op amp.

The following derivation illustrates this in a rigorous way. The overall transfer function that results from the addition of C_c is

$$\frac{V_o(s)}{V_{in}(s)} = \frac{(g_{mI})(g_{mII})(R_I)(R_{II})(1 - sC_c/g_{mII})}{1 + s[R_I(C_I + C_c) + R_{II}(C_{II} + C_c) + g_{mII}R_IR_{II}C_c] + s^2R_IR_{II}[C_IC_{II} + C_c(C_I + C_{II})]}$$

Using the approach developed in Sec. 6.3 for two widely spaced poles gives the following compensated poles.

$$p_1 \cong \frac{-1}{g_{mII}\,R_I\,R_{II}\,C_c} \tag{10}$$

and

$$p_2 \cong \frac{-g_{mII}C_c}{C_IC_{II} + C_{II}C_c + C_IC_c} \tag{11}$$

If C_{II} is much greater than C_I and C_c is greater than C_I, then Eq. (11) can be approximated by

$$p_2 \cong \frac{-g_{mII}}{C_{II}} \qquad (12)$$

It is of interest to note that a zero occurs in the right-half-plane due to the feed forward path through C_c. The right-half-plane zero is located at

$$z_1 = \frac{g_{mII}}{C_c} \qquad (13)$$

Figure 8.2-7(a) illustrates, with a pole-zero diagram, the movement of the poles from their uncompensated to their compensated positions. Figure 8.2-7(b) shows results of compensation with a Bode diagram. Notice that the second pole does not begin to effect the magnitude until after $|A(j\omega)F(j\omega)|$ is less than unity. The right-half-plane (RHP) zero increases the phase shift (acts like a left-half-plane (LHP) pole) but increases the magnitude (acts like a LHP zero). Consequently, the RHP zero causes the two worst things possible with regard to stability considerations. If either the zero (z_1) or the pole (p_2) moves to a lower frequency, the phase margin will be degraded. The task in compensating an amplifier for closed-loop applications is to move all poles and zeros, except for the dominant pole (p_1), sufficiently beyond the unity-gain bandwidth frequency to result in a phase shift similar to Fig. 8.2-7(c).

Only a second-order (two poles) system has been considered thus far. In practice, there are more than two poles in the transfer function of a CMOS op amp. The rest of this treatment will concentrate on the two most dominant (smaller) poles and the RHP zero. Figure 8.2-8 illustrates a typical CMOS op amp with various parasitic and circuit capacitances shown. The approximate pole and zero locations resulting from these capacitances are given below

$$p_1 \cong \frac{-G_I G_{II}}{g_{mII} C_c} = \frac{-(g_{ds2} + g_{ds4})(g_{ds6} + g_{ds7})}{g_{m6} C_c} \qquad (14)$$

$$p_2 \cong \frac{-g_{mII}}{C_{II}} = \frac{-g_{m6}}{C_2} \qquad (15)$$

and

$$z_1 \cong \frac{g_{mII}}{C_c} = \frac{g_{m6}}{C_c} \qquad (16)$$

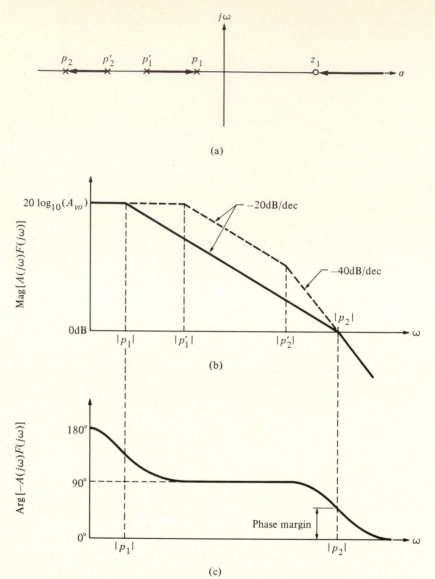

Figure 8.2-7 (a) Root locus movement resulting from Miller compensation as C_c is varied from 0 to the value used to achieve the unprimed roots. (b) Bode magnitude plot showing before and after compensation. (c) Bode phase plot after compensation.

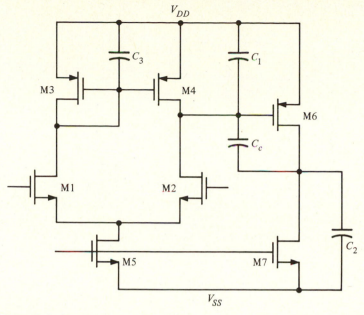

Figure 8.2-8 Op amp with various parasitic and circuit capacitances.

The unity-gain bandwidth as defined in Section 8.1 is easily derived (see Prob. 8.5) and is shown to be approximately

$$GB \cong \frac{g_{ml}}{C_c} = \frac{g_{m2}}{C_c} \qquad (17)$$

As stated before, the goal of the compensation task is to achieve a phase margin greater than 45°. It can be shown (see Prob. 8.6) that if the zero is placed at least ten times higher than the *GB*, then in order to achieve 45° phase margin, the second pole (p_2) must be placed at least 1.22 times higher than *GB*. In order to obtain 60° of phase margin, p_2 must be placed about 2.2 times higher than *GB* (see Prob. 8.7). Assuming that a 60° phase margin is required, the following relationships apply.

$$\frac{g_{m6}}{C_c} > 10\left(\frac{g_{m2}}{C_c}\right) \qquad (18a)$$

Therefore,

$$g_{m6} > 10g_{m2} \qquad (18b)$$

furthermore

$$\left(\frac{g_{m6}}{C_2}\right) > (2.2)\left(\frac{g_{m2}}{C_c}\right) \tag{19}$$

Combining Eqs. (18b) and (19) gives the following requirement

$$C_c > \frac{2.2C_2}{10} = 0.22\ C_2 \tag{20}$$

The RHP zero resulting from the feedforward path through the compensation capacitor tends to limit the GB that might otherwise be achievable if the zero were not present. There are several ways of eliminating the effect of this zero. One approach is to eliminate the feedforward path by placing a unity-gain buffer in the feedback path of the compensation capacitor [1]. This technique is shown in Fig. 8.2-9. Assuming that the output resistance of the unity-gain buffer is small, then the transfer function is given by the following equation

$$\frac{V_o(s)}{V_{in}(s)} = \frac{(g_{mI})(g_{mII})(R_I)(R_{II})}{1 + s[R_IC_I + R_{II}C_{II} + R_IC_c + g_{mII}R_IR_{II}C_c] + s^2[R_IR_{II}C_{II}(C_I + C_c)]} \tag{21}$$

Using the technique as before to approximate p_1 and p_2 results in the following

$$p_1 \cong \frac{-1}{R_IC_I + R_{II}C_{II} + R_IC_c + g_{mII}R_IR_{II}C_c} \cong \frac{-1}{g_{mII}R_IR_{II}C_c} \tag{22}$$

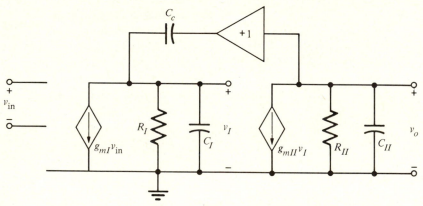

Figure 8.2-9 Miller compensation technique that eliminates the feedforward effect of C_c.

and

$$p_2 \cong \frac{-g_{mII}C_c}{C_{II}(C_I + C_c)} \tag{23}$$

Notice that the poles of the circuit in Fig. 8.2-9 are approximately the same as before, but the zero has been removed. With the zero removed, the pole p_2 must be placed higher than the GB in order to achieve a phase margin of 45°. For a 60° phase margin, p_2 must be placed 1.73 times greater than the GB. Using the type of compensation scheme shown in Fig. 8.2-9 results in greater bandwidth capabilities as a result of eliminating the zero.

The above analysis neglects the output resistance of the buffer amplifier R_o, which can be significant. Taking the output resistance into account, and assuming that it is less than R_I or R_{II}, results in an additional pole p_3 and a LHP zero z_2 given by

$$p_3 \cong \frac{-1}{R_o[C_IC_c/(C_I + C_c)]} \tag{24}$$

$$z_2 \cong \frac{-1}{R_oC_c} \tag{25}$$

Although the LHP zero can be used for compensation, the additional pole makes this method less desirable than the following method.

Another means of eliminating the effect of the right-half-plane zero resulting from feedforward through the compensation capacitor C_c is to insert a nulling resistor in series with C_c. Figure 8.2-10 shows the application of this technique. This circuit has the following node-voltage equations

$$g_{mI}V_{in} + \frac{V_I}{R_I} + sC_IV_I + \left(\frac{sC_c}{1 + sC_cR_z}\right)(V_I - V_o) = 0 \tag{26}$$

$$g_{mII}V_I + \frac{V_o}{R_{II}} + sC_{II}V_o + \left(\frac{sC_c}{1 + sC_cR_z}\right)(V_o - V_I) = 0 \tag{27}$$

These equations can be solved to give

$$\frac{V_o(s)}{V_{in}(s)} = \frac{a\{1 - s[(C_c/g_{mII}) - R_zC_c]\}}{1 + bs + cs^2 + ds^3} \tag{28}$$

where

$$a = g_{mI}g_{mII}R_IR_{II} \tag{29}$$
$$b = (C_{II} + C_c)R_{II} + (C_I + C_c)R_I + g_{mII}R_IR_{II}C_c + R_zC_c \tag{30}$$
$$c = [R_IR_{II}(C_IC_{II} + C_cC_I + C_cC_{II}) + R_zC_c(R_IC_I + R_{II}C_{II})] \tag{31}$$
$$d = R_IR_{II}R_zC_IC_{II}C_c \tag{32}$$

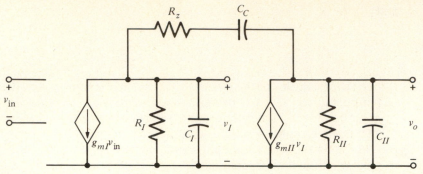

Figure 8.2-10 Compensation technique used to eliminate or relocate the RHP zero.

If R_z is assumed to be less than R_I or R_{II} and the poles widely spaced, then the roots of Eq. (28) can be approximated as

$$p_1 \cong \frac{-1}{(1 + g_{mII}R_{II})R_I C_c} \cong \frac{-1}{g_{mII}R_{II}R_I C_c} \tag{33}$$

$$p_2 \cong \frac{-g_{mII}C_c}{C_I C_{II} + C_c C_I + C_c C_{II}} \cong \frac{-g_{mII}}{C_{II}} \tag{34}$$

$$p_3 = \frac{-1}{R_z C_I} \tag{35}$$

and

$$z_1 = \frac{1}{C_c(1/g_{mII} - R_z)} \tag{36}$$

The resistor R_z allows independent control over the placement of the zero. In order to remove the right-half-plane zero, R_z must be set equal to $1/g_{mII}$. Another option is to move the zero from the RHP to the LHP, and place it on top of p_2. As a result, the pole associated with the output loading capacitance is cancelled. To accomplish this, the following conditions must be satisfied

$$z_1 = p_2 \tag{37}$$

$$\frac{1}{C_c(1/g_{mII} - R_z)} = \frac{-g_{mII}}{C_{mII}} \tag{38}$$

The value of R_z can be found as

$$R_z = \left(\frac{C_c + C_{II}}{C_c}\right)(1/g_{mII}) \tag{39}$$

With p_2 cancelled, the remaining roots are p_1 and p_3. For unity-gain stability, all that is required is that

$$|p_3| > A_v(0)|p_1| = \frac{A_v(0)}{g_{mII}R_{II}R_IC_c} \tag{40}$$

$$(1/R_zC_I) > (g_{mI}/C_c) \tag{41}$$

Substituting Eq. (39) into Eq. (41) and assuming $C_{II} \gg C_I$ results in

$$C_c > \left[\frac{g_{mI}}{g_{mII}} C_IC_{II} \right]^{0.5} \tag{42}$$

Another compensation technique used in CMOS op amps is the feedforward scheme shown in Fig. 8.2-11(a). In this circuit, the buffer is used to break the bidirectional path through the compensation capacitor. Fig. 8.2-11(b) can be used as a model for this circuit. The voltage transfer function $V_o(s)/V_{in}(s)$ can be found to be

$$\frac{V_o(s)}{V_{in}(s)} = \frac{AC_c}{C_c + C_{II}} \left(\frac{s - g_{mII}/AC_c}{s + 1/[R_{II}(C_c + C_{II})]} \right) \tag{43}$$

This scheme results in a slight shift of the original pole $-1/R_{II}C_{II}$ when C_c is zero. However, the zero is located on the positive real axis which causes negative phase shift rather than the desired positive shift. Consequently it is necessary to use an inverting buffer with a gain of $-A$. Alternately, one could feedforward around a noninverting stage. Fig. 8.2-11(c) gives the desired compensation scheme using feedforward techniques. The voltage transfer function is

$$\frac{V_o(s)}{V_{in}(s)} = \frac{AC_c}{C_c + C_{II}} \left(\frac{s + g_{mII}/AC_c}{s + 1/[R_{II}(C_c + C_{II})]} \right) \tag{44}$$

In order to use the circuit in Fig. 8.2-11(c) to achieve compensation, it is necessary to place the zero located at g_{mII}/AC_c above the value of GB so that the boosting of the magnitude will not negate the desired effect of positive phase shift caused by the zero. Fortunately, the phase effects extend over a much broader frequency range than the magnitude effects so that this method will contribute additional phase margin to that provided by the feedback compensation technique. It is quite possible that several zeros can be generated in the transfer function of the op amp. These zeros should all be placed above GB and should be as close to a pole as possible to avoid large settling times caused by a pole-zero doublet in the transient response [2].

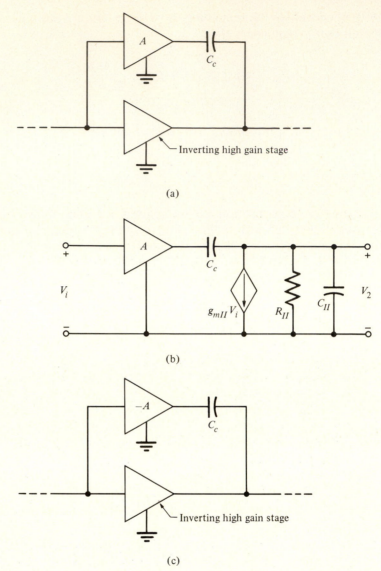

(a)

(b)

(c)

Figure 8.2-11 (a) Feedforward compensation scheme. (b) Model for gain calculation. (c) Proper polarity for feedforward compensation.

8.3 Two-Stage Operational Amplifier Design

The previous two sections described the general approach to op amp design and compensation. In this section, a procedure will be developed that will enable a first-cut design of the op amp shown in Fig. 8.3-1. In order to simplify the notation, it is convenient to define the notation $S_i = W_i/L_i = (W/L)_i$ where S_i is the ratio of W and L of the ith transistor.

Before beginning this task, important relationships describing op amp performance will be summarized from Sec. 8.2 assuming that $g_{m1} = g_{m2} = g_{mI}$, $g_{m6} = g_{mII}$, $g_{ds2} + g_{ds4} = G_I$, and $g_{ds6} + g_{ds7} = G_{II}$. These relationships are based on the circuit shown in Fig. 8.3-1.

Slew rate	$SR = \dfrac{I_5}{C_c}$	(1)
First-stage gain	$A_{v1} = \dfrac{g_{m2}}{g_{ds2} + g_{ds4}} = \dfrac{2g_{m2}}{I_5(\lambda_2 + \lambda_4)}$	(2)
Second-stage gain	$A_{v2} = \dfrac{g_{m6}}{g_{ds6} + g_{ds7}} = \dfrac{g_{m6}}{I_6(\lambda_6 + \lambda_7)}$	(3)
Gain-bandwidth	$GB = \dfrac{g_{m2}}{C_c}$	(4)
Output pole	$p_2 = \dfrac{-g_{m6}}{C_L}$	(5)
RHP zero	$z_1 = \dfrac{g_{m6}}{C_c}$	(6)
Positive CMR	$V_{in(max)} = V_{DD} - \left(\dfrac{I_5}{\beta_3}\right)^{1/2} - |V_{T03}|_{(max)} + V_{T1(min)}$	(7)
Negative CMR	$V_{in(min)} = V_{SS} + \left(\dfrac{I_5}{\beta_1}\right)^{1/2} + V_{T1(max)} + V_{DS5}(\text{sat})$	(8)
Saturation voltage	$V_{DS}(\text{sat}) = \left(\dfrac{2I_{DS}}{\beta}\right)^{1/2}$	(9)

This design procedure assumes that specifications for the following parameters are given

1. Gain at dc, $A_v(0)$
2. Gain-bandwidth, GB
3. Input common-mode range, CMR
4. Load Capacitance, C_L
5. Slew-rate, SR
6. Output voltage swing
7. Power dissipation, P_{diss}

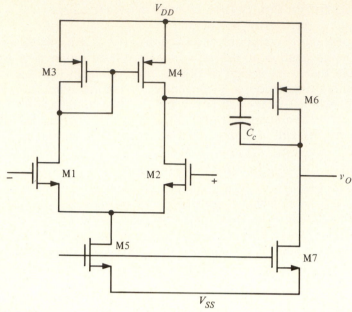

Figure 8.3-1   Schematic of an unbuffered CMOS op amp with n-channel input pair.

The design procedure begins by choosing a device length to be used throughout the circuit. This value will determine the value of the channel-length modulation parameter λ, which will be a necessary parameter in the calculation of amplifier gain. Having chosen the nominal transistor device length, one next establishes the minimum value for the compensation capacitor C_c. It was shown in Sec. 8.2 that placing the loading pole p_2 2.2 times higher than the GB permitted a 60° phase margin (assuming that the RHP zero z_1 is placed at or beyond ten times GB). It was shown in Eq. 20 of Sec. 8.2 that such pole and zero placements result in the following requirement for the minimum value for C_c.

$$C_c > (2.2/10)C_L \qquad \textbf{(10)}$$

Next, determine the minimum value for the tail current I_5, based upon slew-rate requirements. Using Eq. (1), the value for I_5 is determined to be

$$I_5 = SR\ (C_c) \qquad \textbf{(11)}$$

If the slew-rate specification is not given, then one can choose a value based upon settling-time requirements. Determine a value that is roughly ten times faster than the settling-time specification, assuming that the output slews approximately one-half of the supply rail. The value of I_5 resulting from this calculation can be changed later if need be.

The aspect ratio of M3 can now be determined by using the requirement for positive common-mode range. The following design equation for $(W/L)_3$ was derived from Eq. (7).

$$S_3 = (W/L)_3 = \frac{I_5}{(K_3')\,[V_{DD} - V_{in(max)} - |V_{T03}|_{(max)} + V_{T1(min)}]^2} \qquad (12)$$

If the value determined for $(W/L)_3$ is less than one, then it should be increased to a value that minimizes the product of W and L. This minimizes the area of the gate region, which in turn reduces the gate capacitance. This gate capacitance will affect a pole-zero pair which causes a small degradation in phase margin.

Requirements for the transconductance of the input transistors can be determined from knowledge of C_c and GB. The transconductance g_{m2} can be calculated using the following equation

$$g_{m2} = GB(C_c) \qquad (13)$$

The aspect ratio $(W/L)_2$ is directly obtainable from g_{m2} as shown below

$$S_2 = (W/L)_2 = \frac{g_{m2}^2}{(K_2')\,(I_5)} \qquad (14)$$

Enough information is now available to calculate the saturation voltage of transistor M5. Using the negative CMR equation, calculate V_{DS5} using the following relationship derived from Eq. (8).

$$V_{DS5} = V_{in(min)} - V_{SS} - \left(\frac{I_5}{\beta_1}\right)^{1/2} - V_{T1(max)} \qquad (15)$$

If the value for V_{DS5} is less than about 100 mV then the possibility of a rather large $(W/L)_5$ may result. This may not be acceptable. If the value for V_{DS5} is less than zero, then the CMR specification could be too stringent. To solve this problem, I_5 can be reduced or $(W/L)_1$ increased. The effects of these changes must be accounted for in previous design steps. One must iterate until the desired result is achieved. With V_{DS5} determined, $(W/L)_5$ can be extracted using Eq. (9) in the following way

$$S_5 = (W/L)_5 = \frac{2(I_5)}{K_5'(V_{DS5})^2} \qquad (16)$$

For a phase margin of 60°, the location of the loading pole was assumed to be placed at 2.2 times *GB*. Based upon this assumption and the relationship for p_2 in Eq. (5), the transconductance g_{m6} can be determined using the following relationship

$$g_{m6} = 2.2(g_{m2})(C_L/C_c) \tag{17}$$

Combining the defining equation for g_m and V_{DS}(sat) results in an equation relating (W/L), V_{DS}(sat), g_m, and process parameters (see Prob. 8.18). Using this relationship, given below, with the V_{DS}(sat) requirement taken from the output range specification one can determine $(W/L)_6$.

$$S_6 = (W/L)_6 = \frac{g_{m6}}{K_6' V_{DS6}(\text{sat})} \tag{18}$$

Using the following equation, determine a value for I_6

$$I_6 = \frac{g_{m6}^2}{(2)(K_6')(W/L)_6} = \frac{g_{m6}^2}{2K_6' S_6} \tag{19}$$

Recalculate I_6 a different way using the following balance equation

$$I_6 = \frac{(W/L)_6}{(W/L)_3} I_1 = \left(\frac{S_6}{S_3}\right) I_1 \tag{20}$$

Choose the larger of these two values for I_6. If the larger value is found in Eq (19), then $(W/L)_6$ must be increased to satisfy Eq. (20). If the larger value is found in Eq. (20), then no other adjustments must be made.

The device size of M7 can be determined from the balance equation given below

$$S_7 = (W/L)_7 = (W/L)_5 \left(\frac{I_6}{I_5}\right) = S_5 \left(\frac{I_6}{I_5}\right) \tag{21}$$

At this point in the design procedure, the total amplifier gain must be checked against the specifications.

$$A_v = \frac{(2)(g_{m2})(g_{m6})}{I_5(\lambda_2 + \lambda_3)I_6(\lambda_6 + \lambda_7)} \tag{22}$$

If the gain is too low, a number of things can be adjusted. The best way to do this is to use Table 8.3-1, which shows the effects of various device

Table 8.3-1

Dependence of the Performance of Fig. 8.3-1 upon DC Current, W/L Ratios and the Compensating Capacitor.

	Drain Current		M1 and M2		M3 and M4		Inverter	Inverter Load		Comp. Cap.
	I_5	I_7	W/L	L	W	L	(W_6/L_6)	W_7	L_7	C_c
Increase dc Gain	$(\downarrow)^{1/2}$	$(\downarrow)^{1/2}$	$(\uparrow)^{1/2}$	$\uparrow$		$\uparrow$	$(\uparrow)^{1/2}$		$\uparrow$	
Increase GB	$(\uparrow)^{1/2}$		$(\uparrow)^{1/2}$							$\downarrow$
Increase RHP Zero		$(\uparrow)^{1/2}$					$(\uparrow)^{1/2}$			$\downarrow$
Increase Slew Rate	$\uparrow$									$\downarrow$
Increase C_L										$\uparrow$

sizes and currents on the different parameters generally specified. Each adjustment may require another pass through this design procedure in order to insure that all specifications have been met. Table 8.3-2 summarizes the above design procedure.

No attempt has been made to account for noise or PSRR thus far in the design procedure. Now that the preliminary design is complete, these two specifications can be addressed. The input-referred noise voltage results primarily from the load and input transistors. Each of these contribute both thermal and $1/f$ noise. The $1/f$ noise contributed by any transistor can be reduced by increasing device area (e.g., increase WL). Thermal noise contributed by any transistor can be reduced by increasing its g_m. This is accomplished by an increase in W/L, an increase in current, or both. Effective input-noise voltage attributed to the load transistors can be reduced by reducing the g_{m3}/g_{m1} ratio. One must be careful that these adjustments to improve noise performance do not adversely effect some other important performance parameter of the op amp.

The power-supply rejection ratio (PSRR) is to a large degree determined by the configuration used. Some improvement in negative PSRR can be achieved by increasing the output resistance of M5. This is usually accomplished by increasing both W_5 and L_5 proportionately without seriously effecting any other performance. Transistor M7 should be adjusted accordingly for proper matching. A more detailed analysis of the PSRR of the two-stage op amp will be considered in the next section.

At this point, the design should be simulated using SPICE2 or a similar program. During the simulation process, various devices can be adjusted to achieve desired performance. Again, Table 8.3-1 can be extremely useful in deciding what to change to improve any particular specification. The following example illustrates the steps in designing the op amp described.

Table 8.3-2

Unbuffered Op Amp Design Procedure.

This design procedure assumes that the gain at dc (A_v), unity gain bandwidth (GB), input common mode range ($V_{in}(min)$ and $V_{in}(max)$), load capacitance (C_L), slew rate (SR), settling time (T_s), output voltage swing ($V_{out}(max)$ and $V_{out}(min)$), and power dissipation (P_{diss}) are given.

1. Choose the smallest device length which will keep the channel modulation parameter constant and give good matching for current mirrors.

2. Choose the minimum value for C_c based on the following relation.

$$C_c > 0.22C_L \text{ [Gives } 60° \text{ phase margin]}$$

3. Determine the minimum value for the "tail current" (I_5) from the largest of the two values.

$$I_5 = SR \cdot C_c$$

$$I_5 \cong 10\left(\frac{V_{DD} + |V_{SS}|}{2 \cdot T_s}\right)$$

4. Design for S_3 from the maximum input voltage specification.

$$S_3 = \frac{I_5}{K_3'[V_{DD} - V_{in}(max) - |V_{T03}|(max) + V_{T1}(min)]^2} \geq 1$$

5. Verify that the pole of M3 due to C_{gs3} and C_{gs4} ($=0.67W_3L_3C_{ox}$) will not be dominant by assuming it to be greater than 10 GB

$$\frac{g_{m3}}{2C_{gs3}} > 10GB.$$

6. Design for S_2 to achieve the desired GB.

$$g_{m2} = GB \cdot C_c \Rightarrow S_2 = \frac{g_{m2}^2}{K_2'I_5}$$

7. Design for S_5 from the minimum input voltage. First calculate $V_{DS5}(sat)$ then find S_5.

$$V_{DS5}(sat) = V_{in}(min) - V_{SS} - \left[\frac{I_5}{\beta_1}\right]^{0.5} - V_{T1}(max) \geq 100 \text{ mV}$$

$$S_5 = \frac{2I_5}{K_5'[V_{DS5}(sat)]^2}$$

8. Find S_6 by letting the second pole (p_2) be equal to 2.2 times GB. Assume that $V_{DS6} = V_{DS6}(\text{min}) = V_{DS6}(\text{sat}) = V_{DD} - V_{out}(\text{max})$.

$$g_{m6} = 2.2g_{m2}(C_L/C_c)$$

$$S_6 = \frac{g_{m6}}{K_6V_{DS6}(\text{sat.})}$$

9. Calculate I_6 by the following two methods. Choose the larger current and re-adjust S_6 if necessary.

$$\text{Method 1:} \quad I_6 = \frac{g_{m6}^2}{2K_6'S_6}$$

$$\text{Method 2:} \quad I_6 = (S_6/S_3)I_1$$

10. Design S_7 to achieve the desired current ratios between I_5 and I_6.

$$S_7 = (I_6/I_5)S_5$$

11. Check gain and power dissipation specifications.

$$A_v = \frac{2g_{m2}g_{m6}}{I_5(\lambda_2 + \lambda_3)I_6(\lambda_6 + \lambda_7)}$$

$$P_{diss} = (I_5 + I_6)(V_{DD} + |V_{SS}|)$$

12. If the gain specification is not met, then the currents, I_5 and I_6, can be decreased or the W/L ratios of M2 and/or M6 increased. The previous calculations must be rechecked to insure that they have been satisfied. If the power dissipation is too high, then one can only reduce the currents I_5 and I_6. Reduction of currents will probably necessitate increase of some of the W/L ratios in order to satisfy input and output swings.

13. Simulate the circuit to check to see that all specifications are met.

Example 8.3-1

Design of a Two-Stage Op Amp

Using the material and device parameters given in Tables 3.1-1 and 3.1-2, design an amplifier similar to that shown in Fig. 8.3-1 that meets the following specifications

$A_v > 4000$	$C_L = 20\text{pF}$	$V_{SS} = -5$ V
$GB = 1$ MHz	$SR > 2$ V/μs	V_{out} range $= \pm4$ V
$CMR = \pm3$ V	$V_{DD} = +5$ V	$P_{diss} < 10$ mW
		Channel length $= 10$ μm

The first step is to calculate the minimum value of the compensation capacitor C_c, which is

$$C_c > (2.2/10)(20 \text{ pF}) = 4.4 \text{ pF}$$

Using the slew-rate specification and C_c calculate I_5.

$$I_5 = (4.4 \times 10^{-12})(2 \times 10^6) = 8.8 \text{ } \mu\text{A}$$

Next calculate $(W/L)_3$ using CMR requirements. Using Eq. (12) we have

$$(W/L)_3 = \frac{8.8 \times 10^{-6}}{(8 \times 10^{-6})[5 - 3 - 1.2 + 0.8]^2} = 0.43$$

Increase the value to 1.0. Therefore

$$(W/L)_3 = 1.0$$

The next step in the design is to calculate g_{m2} using Eq. (13).

$$g_{m2} = (2)(3.14)(1 \times 10^6)(4.4 \times 10^{-12}) = 27.6 \text{ } \mu\text{S}$$

Therefore, $(W/L)_2$ is

$$(W/L)_2 = \frac{(27.6 \times 10^{-6})^2}{(17 \times 10^{-6})(8.8 \times 10^{-6})} = 5.1$$

Next calculate V_{DS5} using Eq. (15)

$$V_{DS5} = (-3) - (-5) - \left[\frac{8.8 \times 10^{-6}}{(17 \times 10^{-6})(5.1)} \right]^{1/2} - 1.2$$

$$V_{DS5} = 0.48 \text{ V}$$

Using V_{DS5} calculate $(W/L)_5$ from Eq. (16)

$$(W/L)_5 = \frac{2(8.8 \times 10^{-6})}{(17 \times 10^{-6})(0.48)^2} = 4.5$$

Calculate g_{m6} using Eq. (17)

$$g_{m6} = 2.2(27.6 \times 10^{-6})(20/4.4) = 276 \text{ } \mu\text{S}$$

Based upon output range, calculate $(W/L)_6$ using Eq. (18)

$$(W/L)_6 = \frac{276 \times 10^{-6}}{(8 \times 10^{-6})(1)} = 34.5$$

Calculate I_6 using Eq. (19)

$$I_6 = \frac{(276 \times 10^{-6})^2}{(2)(8 \times 10^{-6})(34.5)} = 138 \ \mu A$$

Using balance equations, I_6 is determined to be

$$I_6 = (4.4 \times 10^{-6})(34.5/1) = 152 \ \mu A$$

Choose the larger of these two. Recalculate g_{m6} as

$$g_{m6} = [(2)(8 \times 10^{-6})(34.5)(152 \times 10^{-6})]^{1/2} = 290 \ \mu S$$

Next, calculate $(W/L)_7$ using Eq. (21)

$$(W/L)_7 = 4.5 \left(\frac{152 \times 10^{-6}}{8.8 \times 10^{-6}} \right) = 77.7$$

Now check to see that the gain specification has been met

$$A_v = \frac{(2)(290 \times 10^{-6})(27.6 \times 10^{-6})}{8.8 \times 10^{-6}(.01 + .02)152 \times 10^{-6}(.01 + .02)}$$
$$A_v \cong 13300 \ (\text{within specifications})$$

Power dissipation should be checked to see that it meets the specification

$$P_{diss} = (I_6 + I_5)(V_{DD} - V_{SS}) = (152 + 8.8) \times 10^{-6}(10)$$
$$P_{diss} \cong 1.6 \ mW \ (\text{within specification})$$

   The remaining device sizes are determined based upon matching requirements for proper balance. The final step in the hand design is to establish true electrical widths and lengths based upon ΔL and ΔW variations. In this example ΔL will be due to lateral diffusion only. Unless otherwise noted, ΔW will not be taken into account. All dimensions will be rounded to integer values. Therefore, we have

$$W_1 = W_2 = 5.1(10 - 1.6) = 43 \ \mu m$$
$$W_3 = W_4 = 1.0(10 - 1.6) \cong 8 \ \mu m$$

Since W_3 is so small, increase it to account for ΔW

$$W_3 = W_4 = 10 \ \mu m$$
$$W_5 = 4.5(10 - 1.6) = 38 \ \mu m$$
$$W_6 = 34.5(10 - 1.6) \doteq 290 \ \mu m$$
$$W_7 = 77.7(10 - 1.6) = 653 \ \mu m$$

Readjust W_6 for proper balance.

$$2\frac{W_7}{W_5} = \frac{W_6}{W_4} \text{ then } W_6 = (653/38)(2)(10) = 344 \ \mu m$$

Fig. 8.3-2 shows the results of the first-cut design. The next phase requires simulation.

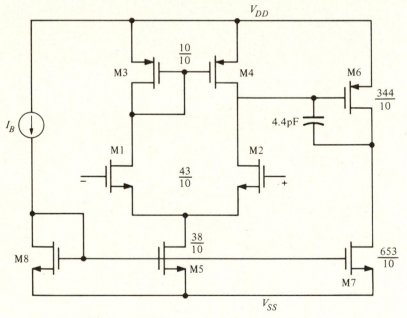

Figure 8.3-2   Results of Example 8.3-1.

The next step in the design of the op amp is the layout. This step cannot be neglected because the performance of the op amp depends on the geometry of the devices. As illustrated in the above example, the designer will typically simulate the op amp after the initial design to confirm and optimize its performance. After the layout is performed, it is necessary to resimulate the op amp because the area and periphery of the sources and drains and the interconnection parasitics can now be identified. Fig. 8.3-3 shows an n-channel input, unbuffered op amp similar to that of Fig. 8.3-2 but different in that the bias has been provided and some of the W/L values are different. Fig. 8.3-4 shows a layout of this op amp on which the various component of Fig. 8.3-3 are identified. This layout uses the design rules of Table 2.6-1. At this point, the designer is able to determine the areas and peripheries of the various devices. Fig. 8.3-5 shows a photomicrograph of Fig. 8.3-4 which has been fabricated using a p-well, double-poly process.

During the simulation process, one might discover that, due to the right-half-plane zero, the phase margin is not adequate for the particular

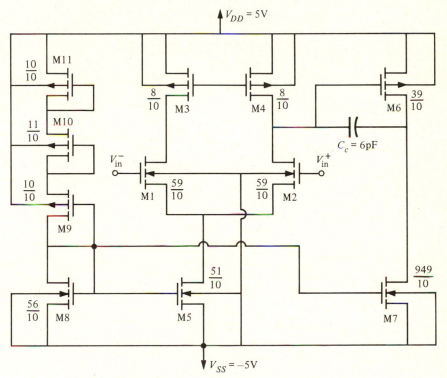

Figure 8.3-3   N-channel input, unbuffered CMOS op amp.

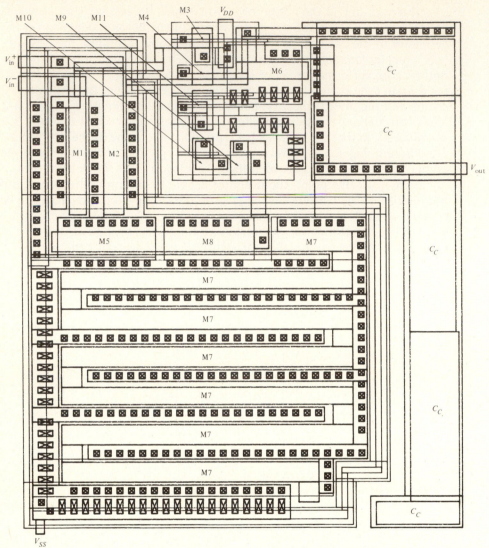

Figure 8.3-4   Layout of Fig. 8.3-3.

application of the op amp. A slight variation in the compensation scheme can be used to our advantage here. Section 8.2 described a technique whereby the RHP zero can be moved to the left-half plane and placed upon the highest non-dominant pole. To accomplish this, a resistor is placed in series with the compensation capacitor. Figure 8.3-6 shows a compensation scheme using a transistor as a resistor. This transistor is controlled by a control voltage V_c that adjusts the resistor so that it maintains the proper value over process variations [2].

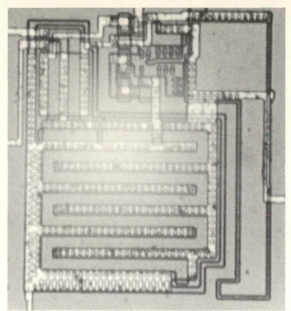

Figure 8.3-5   Microphotograph of Fig. 8.3-3.

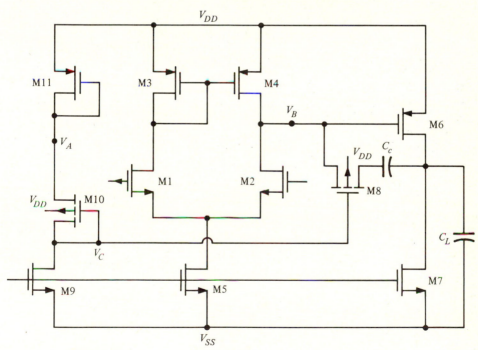

Figure 8.3-6   Op amp using tracking-compensation scheme.

399

With the addition of the resistor in the compensation scheme, the resulting poles and zeros are (see Sec. 8.2, Eqs. 33–36)

$$p_1 = -\frac{g_{m2}}{A_v C_c} = -\frac{g_{m1}}{A_v C_c} \tag{23}$$

$$p_2 = -\frac{g_{m6}}{C_L} \tag{24}$$

$$p_3 = -\frac{1}{R_z C_I} \tag{25}$$

$$z_1 = \frac{-1}{R_z C_c - C_c/g_{m6}} \tag{26}$$

In order to place the zero on top of the second pole, the following relationship must hold

$$R_z = \frac{1}{g_{m6}}\left(\frac{C_L + C_c}{C_c}\right) \tag{27}$$

We know that the value of R_z (as simulated with a transistor) is

$$R_z = \frac{1}{g_{ds8}} \tag{28}$$

Equating Eqs. (27) and (38) results in the following requirement

$$g_{ds8} = g_{m6}\left(\frac{C_c}{C_L + C_c}\right) \tag{29}$$

Transistor M8 is in the nonsaturation region so that

$$g_{ds8} = K_8'(W_8/L_8)(|V_{GS8}| - |V_T|) \tag{30}$$

The bias circuit is designed so that voltage V_A is equal to V_B. As a result,

$$|V_{GS10}| - |V_T| = |V_{GS8}| - |V_T| \tag{31}$$

In the saturation region

$$|V_{GS10}| - |V_T| = \left[\frac{2(I_{10})}{K_{10}'(W_{10}/L_{10})}\right]^{1/2} = |V_{GS8}| - |V_T| \tag{32}$$

substituting into Eq. (30) yields

$$g_{ds8} = K_8' \left(\frac{W_8}{L_8}\right) \left[\frac{2(I_{10})}{K_{10}'(W_{10}/L_{10})}\right]^{1/2} \tag{33}$$

Substituting the relationship for g_{m6} and Eq. (33) into Eq. (29) results in the following design equation

$$\left(\frac{W_8}{L_8}\right) = \left[\left(\frac{W_6}{L_6}\right)\left(\frac{W_{10}}{L_{10}}\right)\left(\frac{I_6}{I_{10}}\right)\right]^{1/2} \left[\frac{C_c}{C_L + C_c}\right] \tag{34}$$

This relationship must be met in order for Eq. (27) to hold. To complete the design of this compensation circuit, M11 must be designed to meet the criteria set forth in Eq. (31). To accomplish this V_{GS11} must be equal to V_{GS6}. Therefore,

$$\left(\frac{W_{11}}{L_{11}}\right) = \left(\frac{I_{10}}{I_6}\right)\left(\frac{W_6}{L_6}\right) \tag{35}$$

The example that follows illustrates the design of this compensation scheme.

Example 8.3-2

RHP Zero Compensation

Use results of Ex. 8.3-1 and design compensation circuitry so that the RHP zero is moved from the RHP to the LHP and placed on top of the output loading pole p_2. Use device data given in Ex. 8.3-1.

The task at hand is the design of transistors M8, M9, M10, M11, and bias current I_{10}. The first step in this design is to establish the bias components. In order to set V_A equal to V_B it is convenient to match M11 to M3 and match their drain currents. As a result the following relationships are valid.

$$(W/L)_{11} = (W/L)_3 = 1.0$$
$$I_{11} = I_{10} = I_3 = 4.4 \ \mu A$$

The aspect ratio of M10 is essentially a free parameter, and will be set equal to $10\mu m/10\mu m$. There must be sufficient supply voltage to support the sum of V_{GS11}, V_{GS10}, and V_{DS9}. The ratio of I_{10}/I_5 determines the (W/L) of M9. This ratio is

$$(W/L)_9 = (I_{10}/I_5)(W/L)_5 = (4.4/8.8)(38/8.4) = 2.26$$
$$W_9 = 2.26[L_9 - 2(LD)] = 2.26(10 - 1.6) \cong 19$$

Now using design Eq. (34), $(W/L)_8$ is determined to be

$$(W/L)_8 = [(344/8.4)(10/8.4)(152/4.4)]^{1/2} [4.4/(4.4 + 20)] = 7.40$$
$$W_8 = 7.40[L_8 - 2(LD)] = 7.40(10 - 1.6) \cong 62$$

It is worthwhile to check that the RHP zero has been moved on top of p_2. To do this, first calculate the value of R_z. V_{GS8} must first be determined. It is equal to V_{GS10}, which is

$$V_{GS10} = \left[\frac{(2)(4.4 \times 10^{-6})}{(8 \times 10^{-6})(10/8.4)} \right]^{1/2} + 1 = 1.96 \text{ V}$$

Next determine g_{ds8}

$$g_{ds8} = (8 \times 10^{-6})(62/8.4)(1.96 - 1) = 56.7 \text{ } \mu\text{S}$$

Therefore

$$R_z = 1/(56.7 \times 10^{-6}) = 17.6 \text{ kilohms}$$

Using Eq. (26), the location of z_1 is calculated to be

$$z_1 = \frac{-1}{(17.6 \times 10^3)(4.4 \times 10^{-12}) - (4.4 \times 10^{-12})/(316 \times 10^{-6})}$$
$$= -1.57 \times 10^7 \text{ rps}$$
$$p_2 = (-316 \times 10^{-6})/(20 \times 10^{-12}) = -1.58 \times 10^7 \text{ rps}$$

Note that the magnitude of p_2 is sufficiently large. The results of this design are summarized below.

$$W_8 = 62 \text{ } \mu\text{m}$$

$$W_9 = 19 \text{ } \mu\text{m}$$

$$W_{10} = 10 \text{ } \mu\text{m}$$

$$W_{11} = 10 \text{ } \mu\text{m}$$

Several aspects of the performance of the two-stage op amp will be considered in later sections. These performance considerations include PSRR (Sec. 8.4) and noise (Sec. 9.5). This section has introduced a design procedure for the two-stage, unbuffered CMOS op amp. We note that except for the compensation and dynamic range considerations, the design procedure is very similar to the two-stage comparator of the previous chapter.

8.4   *Cascode Op Amps*

The previous section introduced the design of a two-stage CMOS op amp. This op amp is probably one of the most widely used CMOS amplifiers to date. Its performance is well understood and experimental results compare closely to the design results. However, there are a number of unbuffered applications in which the performance of the two-stage op amp is not sufficient. Performance limitations of the two-stage op amp include insufficient gain, limited stable bandwidth caused by the inability to control the higher-order poles of the op amp, and a poor power-supply rejection ratio.

In this section we introduce several versions of the cascode CMOS op amp which offer improved performance in the above three areas. Three cascode op amp topologies will be discussed. The first is based on increasing the gain of the op amp, the second on increasing the control of the poles, and the third on increasing the PSRR. The primary difference between these three topologies is where the cascode stage is applied in the block diagram of a general op amp shown in Fig. 8.1-1.

The motivation for using the cascode configuration to increase the gain can be seen by examining how the gain of the two-stage op amp could be increased. There are three ways in which the gain could be increased. They are: (1) add additional gain stages, (2) increase the transconductance of the first or second stage, and (3) increase the output resistance seen by the first or second stage. Due to possible instability, the first approach is not attractive. Of the latter two approaches, the third is the more attractive because the output resistance increases in proportion to a decrease in bias current whereas the transconductance increases as the square root of the increase in bias current. Thus it is generally more efficient to increase r_{out} rather than g_m. Also, r_{out} can be dramatically increased by using special circuit techniques such as the cascode structure introduced in Sections 5.3 and 6.3.

First consider the possibility of increasing the gain of the first stage of Fig. 8.1-1, the differential-input stage, and leaving the second stage alone. Fig. 8.4-1 shows a small-signal model of the five-transistor differential amplifier stage of Sec. 6.2. In order to increase the gain by increasing the output resistances, both r_{o2} and r_{o4} would have to be increased. The resistance associated with the diode-connected transistor M3 does not have to be increased since it only serves as a current mirror. Fig. 8.4-2 shows this

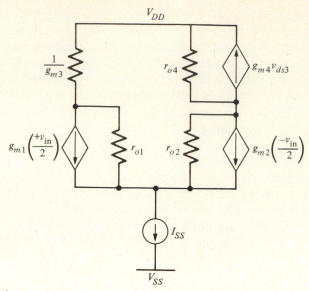

Figure 8.4-1   Small-signal model of differential stage.

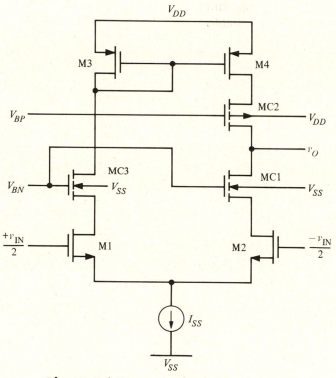

Figure 8.4-2   Cascoded differential stage.

modification of the differential stage using cascoded transistors. Transistors MC1 and MC2 perform the resistance multiplication which was developed in Sec. 5.3 while MC3 is used to keep the drain voltages of the input transistors matched, which helps to reduce the voltage offset. Using the results of Eq. (7) of Sec. 5.3, we may express the output resistance of Fig. 8.4-2 as

$$R_I \cong (g_{mC2} r_{oC2} r_{o4}) \,||\, (g_{mC1} r_{oC1} r_{o2}) \qquad (1)$$

As was shown in Ex. 5.3-1, the output resistance can be increased by over two orders of magnitude.

One of the disadvantages of this design is the requirement for the additional bias voltages V_{BN} and V_{BP}. Furthermore, the common-mode input range is reduced due to the extra voltage drop required by the two cascode devices, MC1 and MC3. In many cases, the CMR limitation is not important since the noninverting input of the op amp will be connected to ground. However, in the buffer configuration of the op amp the cascoded differential-input stage will not be satisfactory. This will be solved by the other cascode configurations to be considered in this section.

In order to prevent common-mode biasing problems, it is necessary to bias MC1 from a voltage which tracks the input common-mode voltage. One way of accomplishing this is shown in Fig. 8.4-3. I_{CM} is a bias current that depends on the common-mode voltage (see Prob. 8.24).

While the objective of increasing the output resistance of the first stage of the two-stage op amp was to achieve a higher gain, Fig. 8.4-2 could be used as an implementation of a single-stage op amp. In many cases a high gain is not needed. The advantage of this single-stage op amp is that it has only one dominant pole, which is at the output of the stage. Consequently, compensation is best accomplished by a small shunt capacitance attached to the output. The voltage gain of Fig. 8.4-2 as a single-stage op amp is

$$A_v = g_{mI} R_I \qquad (2)$$

where R_I is defined in Eq. (1). Assuming a gain-bandwidth of GB implies a dominant pole at GB/A_v. Equating A_v/GB to the product of R_I and a shunt capacitor C_I connected at the output gives the relationship between C_I and the amplifier specifications, namely

$$C_I = \frac{g_{mI}}{GB} \qquad (3)$$

An example will illustrate some of the performance capabilities of Fig. 8.4-2 as a single-stage op amp.

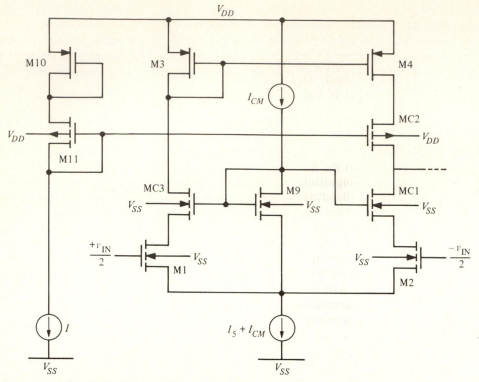

Figure 8.4-3   Differential cascode with common-mode biasing.

Example 8.4-1

Single-Stage, Cascode Op Amp Performance

Assume that $W_1/L_1 = W_2/L_2 = W_{C2}/L_{C2} = W_{C3}/L_{C3} = 100\ \mu m/10$ μm, all other W/L ratios are 10 $\mu m/10\ \mu m$, and that $I_{ss} = 100\ \mu A$ of Fig. 8.4-2. Find the voltage gain of this op amp and the value of C_i if $GB = 10$ MHz. Use the model parameters of Table 3.1-2.

The device transconductances are $g_{m1} = g_{m2} = g_{mi} = 130.4\ \mu S$, $g_{mC1} = 41.23\ \mu S$, and $g_{mC2} = 89.44\ \mu S$. The output resistance of the NMOS and PMOS devices is 2 MΩ and 1 MΩ, respectively. From Eq. (1) we find that $R_i = 58$ MΩ. Therefore the voltage gain is 7562. For a unity-gain bandwidth of 10 MHz, the value of C_i is 2 pF.

The above example shows that the single-stage op amp can have practical values of gain and bandwidth. One other advantage of this configuration is that it is self-compensating as a large capacitive load is attached. If, for example, a 100 pF capacitor is attached to the output, the dominant pole decreases from 1.32 KHz (GB = 10 MHz) to 27 Hz (GB = 0.2 MHz). Of course, the output resistance of the op amp is large (58 MΩ), which is not suitable for low resistive loads.

If higher gain or lower output resistance is required, then Fig. 8.4-2 needs to be cascaded with a second stage. However, we note that the output dc voltage of Fig. 8.4-2 is further away from V_{DD} than that of the two-stage op amp. Driving a common-source PMOS output transistor from this stage would result in a large V_{DS}(sat), which would degrade the output-swing performance. To optimize the output swing of the second stage, it is better to perform a voltage translation before driving the gate of the output PMOS transistor. This is accomplished very easily using the circuit shown in Fig. 8.4-4. Only the output of the first-stage differential amplifier is shown. MT1 and MT2 serve the function of level translation between the first and second stage. MT2 is a current source that biases the source fol-

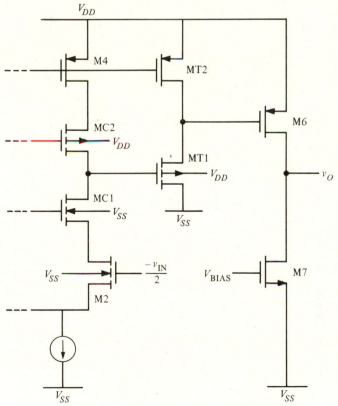

Figure 8.4-4   Level translator for differential cascode.

lower, MT1. The small-signal gain from the output of the differential stage to the output of the voltage translator is close to unity with a small amount of phase shift. The complete cascode input-operational amplifier is shown in Fig. 8.4-5. Compensation can be performed using Miller compensation techniques on the second stage as was illustrated in Sec. 8.2.

One of the disadvantages of the cascode input, operational amplifier of Fig. 8.4-5 is that the compensation is more complicated than the single-stage version examined above. Although the Miller compensation will work satisfactorily, the circuit stability will be diminished because of large capacitive loading at the output. In order to overcome this disadvantage and to solve the common-mode range problem discussed for Fig. 8.4-2, the cascode can be moved to the second stage of the two-stage op amp. In order to increase the gain by increasing the output resistance (of the second stage) it is necessary to use the cascode configuration of Fig. 6.3-6(a). The resulting two-stage op amp is shown in Fig. 8.4-6. The small-signal voltage

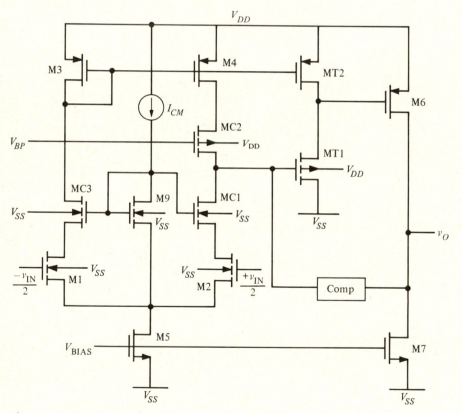

Figure 8.4-5 Op amp using cascode differential stage with level-translated driver for second stage.

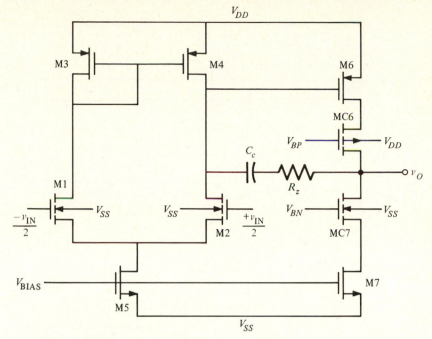

Figure 8.4-6 Two-stage op amp with a cascode second stage.

gain of this circuit is given as

$$A_v = g_{mI}g_{mII}R_IR_{II} \qquad (4)$$

where

$$g_{mI} = g_{m1} = g_{m2} \qquad (5)$$
$$g_{mII} = g_{m6} \qquad (6)$$
$$R_I = \frac{1}{g_{ds2} + g_{ds4}} = \frac{2}{(\lambda_2 + \lambda_4)I_{D5}} \qquad (7)$$

and

$$R_{II} = (g_{mC6}r_{dsC6}r_{ds6}) \,||\, (g_{mC7}r_{dsC7}r_{ds7}) \qquad (8)$$

Comparing with the gain of the normal two-stage op amp, the increase of gain is obtained from R_{II}. Normally, a gain increase of approximately 100 would be possible which will lead to decreased stability. Also, note that the output resistance has increased.

The tradeoff between gain and stability indicated above can be accomplished with the two-stage cascode op amp in Fig. 8.4-7. In this op amp,

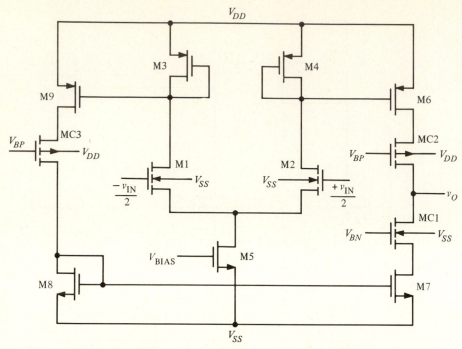

Figure 8.4-7   Op amp using cascode output stage.

the first stage gain has been decreased by using active loads M3 and M4. While the gain is decreased, the pole at the output of the first stage in Fig. 8.4-6 (the junction of M2 and M4) has been increased because of the low resistance of M4 to ac ground. The output signal from the first stage is applied differentially to the cascode output stage. In Fig. 8.4-7, the majority of the gain is obtained in the output stage. The overall gain of the input stage is

$$A_{vI} = g_{m2}/g_{m4} = g_{m1}/g_{m3} \qquad \textbf{(9)}$$

The gain of the second stage is

$$A_{vII} = \left(\frac{g_{m6} + g_{m7}}{2} \right) R_{II} \qquad \textbf{(10)}$$

where R_{II} is given by Eq. (8). Because the only dominant pole is now at the output, the cascode op amp of Fig. 8.4-7 can be compensated by a shunt capacitor at the output in a manner similar to the single-stage op amp of Fig. 8.4-2.

Example 8.4-2
Design of Fig. 8.4-7

Fig. 8.4-7 is a useful alternative to the two-stage op amp. Its design will be illustrated by this example. The pertinent design equations for the op amp of Fig. 8.4-7 are shown in Table 8.4-1. The specifications of the design are as follows:

$V_{DD} = -V_{SS} = 5$ V

Slew rate = 5 V/μs with a 50 pF load

GB = 5 MHz with a 10 pF load

$A_v \geq 5000$

Input CMR = ± 3 V

Output swing = ± 3 V

Use the paramaters of Table 3.1-2 and let all device lengths be 10 μm.

While numerous approaches can be taken, we shall follow one based on the above specifications. The steps will be numbered to help illustrate the procedure.

1.) The first step will be to find the maximum source/sink current. This is found from

$$I_{source}/I_{sink} = C_L \times \text{slew rate} = 50 \text{ pF}(5 \text{ V}/\mu s) = 250 \text{ } \mu A$$

2.) Next some W/L constraints based on the maximum output source/sink current are developed. Under dynamic conditions, all of I_5 will flow in M4; thus we can write

$$\text{Max. } I_{out}(\text{source}) = (S_6/S_4)I_5$$

Table 8.4-1

Pertinent Design Relationships for Fig. 8.4-7.

$$\text{Slew rate} = \frac{I_{out}}{C_{II}} \qquad GB = \frac{g_{m1}(g_{m6} + g_{m7})}{g_{m3}C_{II}}$$

$$A_v = \frac{g_{m1}}{g_{m3}}\left(\frac{g_{m6} + g_{m7}}{2}\right)R_{II}$$

$$V_{in}(\text{max}) = V_{DD} - [I_5/\beta_3]^{1/2} - |V_{TO3}|(\text{max}) + V_{T1}(\text{min})$$

$$V_{in}(\text{min}) = V_{SS} + V_{DS5} + [I_5/\beta_1]^{1/2} + V_{T1}(\text{min})$$

The maximum output sinking current is equal to the maximum output sourcing current if

$$S_3 = S_4, \qquad S_6 = S_9, \qquad \text{and } S_7 = S_8$$

3.) Choose I_5 as 100 μA. Remember that this value can always be changed later on if desirable. This current gives

$$S_9 = S_6 = 2.5S_4 = 2.5S_3$$

4.) Next design for ± 3 V output capability. We shall assume that the output must source or sink the 250 μA at the peak values of output. First consider the negative output peak. Since there is 2 V difference between V_{SS} and the minimum output, let $V_{DSC1}(\text{sat}) = V_{DS7}(\text{sat}) = 1$ V. Under the maximum negative peak assume that $I_{C1} = I_7 = 250$ μA. Therefore

$$1 = \left[\frac{2I_7}{K_N'S_7}\right]^{1/2} = \left[\frac{2I_{C1}}{K_N'S_{C1}}\right]^{1/2} = \left[\frac{500\ \mu A}{(17\ \mu A/V^2)S_7}\right]^{1/2}$$

which gives $S_7 = S_{C1} = 29.4$ and $S_8 = S_7 = 29.4$. Using the same approach for the positive peak gives

$$1 = \left[\frac{2I_6}{K_P'S_6}\right]^{1/2} = \left[\frac{2I_{C2}}{K_P'S_{C2}}\right]^{1/2} = \left[\frac{500\ \mu A}{(8\ \mu A/V^2)S_6}\right]^{1/2}$$

which gives $S_6 = S_{C2} = 62.5$ and $S_3 = S_4 = 25$.

5.) Next the bias voltages V_{BP} and V_{BN} are designed. First, consider V_{BN}. Assume that the current flow in MC1 and M7 is 250 μA. Ignoring the bulk effects for MC1, the saturation condition for MC1 can be written as

$$V_{DSC1}\ (\text{sat.}) = V_{GSC1} - V_{TC1} = 1\ V$$

Solving for V_{GSC1} gives 2 V which implies that V_{BN} is 3 V above V_{SS} or -2 V. Use the same method to design V_{BP}. The saturation condition for MC2 can be written as

$$V_{SDC2}\ (\text{sat.}) = V_{SGC2} - |V_{TC2}| = 1\ V$$

giving $V_{SGC2} = 2$ V and implying that V_{BP} must be 3 V less than V_{DD} or 2 V.

6.)   Now we must consider the possibility of conflict among the specifications. For example, the input CMR will influence S_3 which has already been designed as 25. Using Eq. (7) of Sec. 8.3 we find that S_3 should be at least 4.88. A larger value of S_3 will give a higher value of $V_{in}(\max)$ so that we continue to use $S_3 = 25.\ (V_{in}(\max) = 3.89\ V)$

7.)   Next we find $g_{m1}\ (g_{m2})$. There are two ways of calculating g_{m1}. The first is from the A_v specification. The product of Eqs. (9) and (10) give

$$A_v = (g_{m1}/2g_{m4})(g_{m6} + g_{m7})R_{II}$$

Calculating the various transconductances we get $g_{m4} = 141.4\ \mu S$, $g_{m6} = g_{mC1} = 353.5\ \mu S$, $g_{m7} = g_{mC2} = 353.5\ \mu S$, $r_{ds6} = r_{dC2} = 0.4$ Megohm, and $r_{ds7} = r_{dsC1} = 0.8$ Megohms. Assuming that the gain A_v must be greater than 5000 gives $g_{m1} > 44\ \mu S$. The second method of finding g_{m1} is from the GB specifications. Multiplying the gain by the dominant pole $(1/C_{II}R_{II})$ gives

$$GB = \frac{g_{m1}(g_{m6} + g_{m7})}{g_{m4}C_{II}}$$

Assuming that $C_{II} = 10$ pF and using the specified GB gives $g_{m1} = 62.8\ \mu S$. Knowing I_5 gives $S_1 = S_2 = 2.32$.

8.)   The next step is to check that S_1 and S_2 are large enough to meet the $-3\ V$ input CMR specification. From Eq. (8) of Sec. 8.3, we obtain a negative value of V_{DS5}, which implies that $S_1\ (S_2)$ is not large enough. If we choose $S_5 = 100$, then $V_{DS5}(\text{sat})$ is 0.343 V. This gives $S_1 = S_2 = 28.2$ and $g_{m1} = 219\ \mu S$. This results in $A_v = 24,769$ and $GB = 17.43$ MHz for a 10 pF load. We shall assume that exceeding the specifications in this area is not detrimental to the performance of the op amp.

   Table 8.4-2 summarizes the values of W/L that resulted from this design procedure. The power dissipation for this design is seen to be 3.5 mW. The next step would be to design the transistor widths taking into account lateral diffusion, and then simulate.

   The parasitic capacitance of op amps with a cascode in the output stage such as Fig. 8.4-2, Fig. 8.4-6, or Fig. 8.4-7 can be improved if a double polysilicon process is available [3]. Because the source and drain of a cascode pair are not connected externally, it is possible to treat the cascode

Table 8.4-2

Summary of W/L Ratios for Example 8.4-2.
$S_1 = S_2 = 28.2$
$S_3 = S_4 = 25$
$S_5 = 100$
$S_6 = S_9 = S_{C2} = S_{C3} = 62.5$
$S_7 = S_8 = S_{C1} = 29.4$
$V_{BP} = 2V \qquad V_{BN} = -2V$

pair as a dual-gate MOSFET. This is illustrated by considering the cascode pair shown in Fig. 8.4-8(a). The parasitic capacitance associated with the common drain/source connection can be virtually eliminated, as shown in Fig. 8.4-8(b), if a double polysilicon process is available. Obviously, one wants to minimize the overlap of the polysilicon layers in order to reduce the shunt input capacitance to the cascode pair.

The two-stage, unbuffered op amp of the last section has been used in many commercial products, particularly in the telecommunications area. After the initial successes, it was noted that this op amp suffers from a poor power-supply rejection ratio (PSRR). The ripple on the power supplies contributed too much noise at the output of the op amp. In order to illustrate this problem, which will be solved by the cascode op amp, consider the op amp of Fig. 8.2-8. The definition of PSRR given in Eq. (6) of Sec. 8.1 is stated as the ratio of the differential gain A_v to the gain from the power-supply ripple to the output with the differential input set to zero (A_{dd}). Thus, PSRR can be written as

$$\text{PSRR} = \frac{A_v(V_{dd} = 0)}{A_{dd}(V_{in} = 0)} \qquad \textbf{(11)}$$

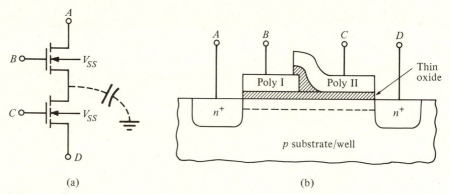

(a)   (b)

Figure 8.4-8 (a) Cascode amplifier with parasitic capacitance. (b) Method of reducing source/drain-to-bulk capacitance.

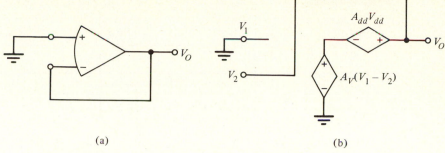

Figure 8.4-9   (a) Method of calculating PSRR. (b) Model of (a).

While one can calculate A_v and A_{dd} and combine the results, it easier to use the unity gain configuration of Fig. 8.4-9(a). Using the model of the op amp shown in Fig. 8.4-9(b) to represent the two gains of Eq. (11), we can show that

$$V_{\text{out}} = \frac{A_{dd}}{1 + A_v} V_{dd} \cong \frac{A_{dd}}{A_v} V_{dd} = \frac{1}{\text{PSRR}^+} V_{dd} \qquad (12)$$

where V_{dd} is the power supply ripple of V_{DD} and PSRR$^+$ is the PSRR for V_{DD}. Therefore, if we connect the op amp in the unity-gain mode and input an ac signal of V_{dd} in series with the V_{DD} power supply, V_o/V_{dd} will be equal to the inverse of PSRR$^+$. This approach will be taken to calculate the PSRR of the two-stage op amp and will then be applied to the cascode op amp to show the improvement afforded by the latter.

The two-stage op amp of Fig. 8.2-8 is shown in Fig. 8.4-10(a) connected in the unity-gain mode with an ac ripple of V_{dd} on the positive power supply. As before, C_I and C_{II} are the parasitic capacitances to ground at the output of the first and second stages, respectively, The unsimplified, small-signal model corresponding to the PSRR$^+$ calculation is shown in Fig. 8.4-10(b). This model has been simplified as shown in Fig. 8.4-10(c), where the $g_{m1}V_5$ and $g_{m2}V_5$ controlled sources have been rerouted and the resulting $1/g_{m1}$ and $1/g_{m2}$ resistances combined in parallel with R_5. The final simplification is shown in Fig. 8.4-10(d) where V_5 is assumed to be zero and the current I_3 flowing through $1/g_{m3}$ is replaced by

$$I_3 = g_{m1}V_o + g_{ds1}\left(V_{dd} - \frac{I_3}{g_{m3}}\right) \cong g_{m1}V_o + g_{ds1}V_{dd} \qquad (13)$$

Even if V_5 is not assumed zero, the contribution of $g_{m1}V_5$ will be subtracted by the current source $g_{m2}V_5$, so that the simplified model of Fig. 8.4-10(d) is further justified. The dotted line represents the portion of Fig. 8.4–10(c) in parallel with V_{dd} and thus is not important to the model.

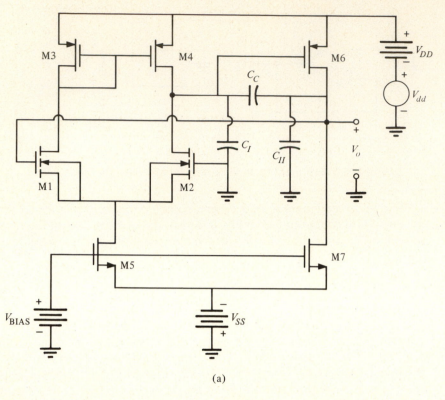

(a)

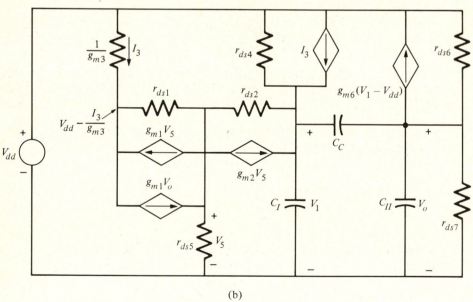

(b)

416

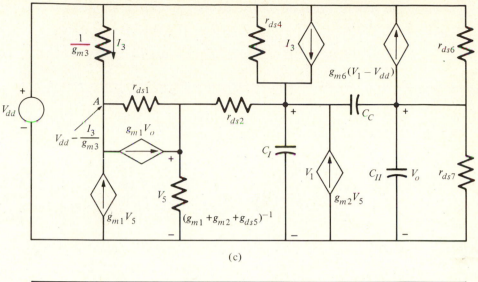

(c)

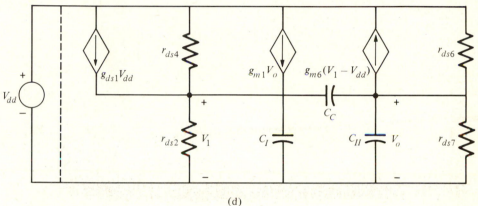

(d)

Figure 8.4-10 Calculation of PSRR$^+$ for the op amp of Ex. 8.3-1. (a) Circuit diagram. (b) Small-signal model. (c) Rearrangement of $g_{m1}V_5$ and $g_{m2}V_5$ controlled sources. (d) Simplified model assuming $V_5 \cong 0$.

The nodal equations for the voltages V_1 and V_o of Fig. 8.4-10(d) can be written as follows. For the node at V_1 we have

$$(g_{ds1} + g_{ds4})V_{dd} = (g_{ds2} + g_{ds4} + sC_c + sC_I)V_1 - (g_{m1} + sC_c)V_o \qquad \textbf{(14)}$$

and for the output node we have

$$(g_{m6} + g_{ds6})V_{dd} = (g_{m6} - sC_c)V_1 + (g_{ds6} + g_{ds7} + sC_c + sC_{II})V_o \qquad \textbf{(15)}$$

Using the generic notation for the two-stage op amp allows Eqs. (14) and (15) to be rewritten as

$$G_I V_{dd} = (G_I + sC_c + sC_I)V_1 - (g_{mI} + sC_c)V_o \tag{16}$$

and

$$(g_{mII} + g_{ds6})V_{dd} = (g_{mII} - sC_c)V_1 + (G_{II} + sC_c + sC_{II})V_o \tag{17}$$

where

$$G_I = g_{ds1} + g_{ds4} = g_{ds2} + g_{ds4} \tag{18}$$
$$G_{II} = g_{ds6} + g_{ds7} \tag{19}$$
$$g_{mI} = g_{m1} = g_{m2} \tag{20}$$

and

$$g_{mII} = g_{m6} \tag{21}$$

Analysis of Fig. 8.4-10(b) gives the following result.

$$\frac{V_{dd}}{V_o} =$$

$$\frac{s^2[C_cC_I + C_IC_{II} + C_{II}C_c] + s[G_I(C_c + C_{II}) + G_{II}(C_c + C_I) + C_c(g_{mII} - g_{mI})] + G_IG_{II} + g_{mI}g_{mII}}{s[C_c(g_{mII} + G_I + g_{ds6}) + C_I(g_{mII} + g_{ds6})] + G_Ig_{ds6}} \tag{22}$$

Using the technique described in Sec. 6.3, we may solve for the approximate roots of Eq. (22) as

$$\text{PSRR}^+ = \frac{V_{dd}}{V_o} \cong \left[\frac{g_{mI}g_{mII}}{G_Ig_{ds6}}\right] \left[\frac{\left(\frac{sC_c}{g_{mI}} + 1\right)\left(\frac{s(C_cC_I + C_IC_{II} + C_cC_{II})}{g_{mII}C_c} + 1\right)}{\left(\frac{sg_{mII}C_c}{G_Ig_{ds6}} + 1\right)}\right] \tag{23}$$

where we have assumed that g_{mII} is greater than g_{mI} and that all transconductances are larger than the channel conductances. For all practical purposes, Eq. (23) reduces to

$$
\text{PSRR}^+ = \frac{V_{dd}}{V_o} = \left[\frac{g_{ml}g_{mll}}{G_Ig_{ds6}} \right] \left[\frac{\left(\dfrac{sC_c}{g_{ml}} + 1 \right)\left(\dfrac{sC_{ll}}{g_{mll}} + 1 \right)}{\dfrac{sg_{mll}C_c}{G_Ig_{ds6}} + 1} \right]
$$

$$
= A_{vo} \left[\frac{\left(\dfrac{s}{GB} + 1 \right)\left(\dfrac{s}{|p_2|} + 1 \right)}{\left(\dfrac{sA_{vo}}{GB} + 1 \right)} \right] \tag{24}
$$

The results of this analysis show that at a frequency of GB/A_{vo} the PSRR$^+$ begins to roll off with a -20 dB/dec slope. Consequently, the PSRR$^+$ becomes degraded at high frequencies, which is a disadvantage of the two-stage, unbuffered op amp. The reason for this poor performance is that as frequency rises, the impedance of the compensation capacitor C_c becomes low and the gate and the drain of M6 begin to track one another. Because M6 is biased by the M7 current source, the gate-source voltage must remain constant. This requirement forces the gate of M6 to track the changes in V_{DD} which are in turn transmitted by C_c to the output of the amplifier. The gain from V_{dd} to V_o becomes approximately unity as frequency increases. Thus the PSRR$^+$ has the approximate frequency response of $A_v(s)$ until the zeros at $-GB$ and $-p_2$ influence the frequency response. For the value of Example 8.3-1, the dc value of PSRR$^+$ is 89 dB and the roots are $z_1 = -1.08$ MHz, $z_2 = -1.59$ MHz, and $p_1 = -56.5$ Hz. Fig. 8.5-17(b) and (c) are the simulated response of the PSRR$^+$ of Ex. 8.3-1 and compare very closely with these results.

Fig. 8.4-11(a) shows the two-stage op amp in the configuration for calculating the negative PSRR (PSRR$^-$). It is important to note that V_{BIAS} is referenced to the V_{ss} rail so that the only source of V_{ss} injection is through r_{ds5} and r_{ds7}. Because common-mode signals injected into the differential amplifier are cancelled, r_{ds7} will be the only effective source of V_{ss} to be considered. The simplified small-signal model of the PSRR$^-$ calculation is shown in Fig. 8.4-11(b). The nodal equations are written as

$$
0 = (G_I + sC_c + sC_I)V_1 - (g_{ml} + sC_c)V_o \tag{25}
$$

and

$$
g_{ds7}V_{ss} = (g_{mll} - sC_c)V_1 + (G_{ll} + sC_c + sC_{ll})V_o \tag{26}
$$

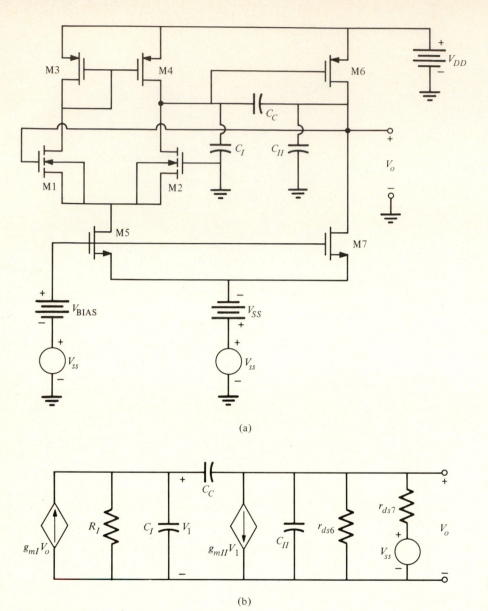

(a)

(b)

Figure 8.4-11 (a) Configuration to calculate the PSRR$^-$ of a two-stage op amp. (b) Small-signal model of (a).

Solving for V_o/V_{ss} and inverting gives

$$\frac{V_{ss}}{V_o} =$$

$$\frac{s^2[C_cC_I + C_IC_{II} + C_{II}C_c] + s[G_I(C_c + C_{II}) + G_{II}(C_c + C_I) + C_c(g_{mII} - g_{mI})] + G_IG_{II} + g_{mI}g_{mII}}{[s(C_c + C_I) + G_I]g_{ds7}}$$

$$(27)$$

Again using the technique described in Sec. 6.3, we may solve for the approximate roots of Eq. (27) as

$$\text{PSRR}^- = \frac{V_{ss}}{V_o} \cong \left[\frac{g_{mI}g_{mII}}{G_Ig_{ds7}}\right]\left[\frac{\left(\dfrac{sC_c}{g_{mI}} + 1\right)\left(\dfrac{s(C_cC_I + C_IC_{II} + C_cC_{II})}{g_{mII}\,C_c} + 1\right)}{\left(\dfrac{s(C_c + C_I)}{G_I} + 1\right)}\right]$$

$$(28)$$

The differences between the PSRR$^+$ and the PSRR$^-$ are in the gain and the pole. Assuming the values of Ex. 8.3-1 gives a gain of 92 dB. The significant difference is that the pole is at -3890 Hz which is larger by a factor of almost 70 over the PSRR$^+$ case. The result is that PSRR$^-$ for the n-channel input, two-stage op amp is always greater than PSRR$^+$. Alternate methods of calculating the PSRR can be found in the references [4, 5].

Cascode op amps offer the possibility of improved PSRR performance. For example, the single-stage cascode op amp of Fig. 8.4-2 will not suffer the degradation of the two-stage op amp because the Miller compensation capacitor is not used. The cascode op amps of Fig. 8.4-5 avoid the PSRR degradation caused by the Miller capacitor by using the buffer MT1, which isolates the signal at the gate of M6 from the compensation capacitor. However, the cascode op amp of Fig. 8.4-6 will have the same PSRR degradation as the two-stage op amp. Fig. 8.4-7 avoids PSRR degradation because compensation is accomplished by a shunt capacitance from the ground.

A cascode op amp which was developed to increase the PSRR and the common-mode input range is shown in Fig. 8.4-12 [5]. This op amp is called the folded cascode op amp. This circuit uses a current-folding circuit technique to permit direct connection of the drains of a p-channel differential amplifier to the sources of the cascode devices. This requires two additional transistors (M10 and M11) operating as current sources to bias the first stage. The current sunk by M10 and M11 equals the sum of the currents flowing from the differential amplifier (M1 and M2) and the cascode mirror (M3-M6). Under balance conditions, the current in M4 is equal

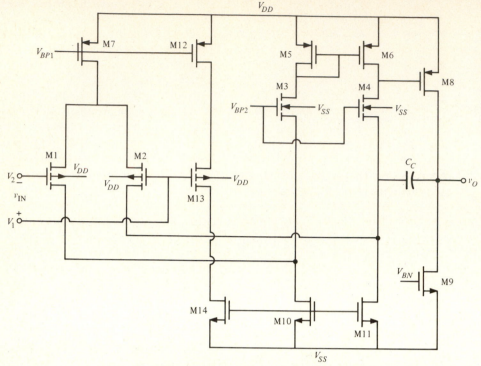

Figure 8.4-12   A folded, cascode op amp.

to the current in M6. If v_{IN} is increased, then I_{D2} increases by ΔI and I_{D1} decreases by ΔI. These changes reflect themselves into a ΔI increase in M3 and a ΔI decrease in M4. The result will be $2\Delta I$ flowing into the resistance at the drains of M4 and M6.

One of the advantages of the folded cascode op amp is an increased common-mode input range due to the opposite polarity of the input pair (M1-M10 and M2-M11). The bias voltage at the gates of M3 and M4 should be just large enough to keep the n-channel current sources M10 and M11 in saturation. This can be accomplished by using the technique illustrated in Sec. 5.3 for reducing the value of V_{out}(sat) across the cascode pairs M3-M10 and M4-M11.

One of the problems with wide common-mode input range op amps is that in the unity-gain configuration for large positive common mode, the output will abruptly spike up to the positive power-supply voltage. This will occur in the circuit of Fig. 8.4-12 if M10 and M11 are biased from a fixed voltage. As the common-mode input is increased, the sources of the input transistors reduce the drain/source voltage across M7, which reduces the bias current for the differential amplifier. If I_{D1} and I_{D2} become too small, they are insignificant in comparison to the fixed-bias current of M10 and M11. As a result, the voltage at the gate of M8 decreases, causing the output to be pulled to V_{DD}. This problem is removed by the biasing technique illustrated by M12, M13, and M14. This arrangement causes the biasing current in M10 and M11 to track the biasing current in M7. Thus, when the common-mode input limit is exceeded, the amplifier experiences soft limiting rather than an output spike.

The small-signal analysis of Fig. 8.4-12 can be accomplished with the small-signal model of Fig. 8.4-13. The two controlled-current sources of value $g_{ml}V_{in}/2$ represent the input-differential amplifier with the right-hand one corresponding to M2 and the left-hand one to M1. G_{o1} represents g_{ds6} and G_{o2} equals the sum of g_{ds8} and g_{ds9}. G_{M2} is equal to $g_{m4} + g_{mbs4}$. The conductance in parallel with the right-hand $g_{ml}V_{in}/2$ source has been neglected because the cascode configuration will cause this conductance to approach G_{M2}. The conductance g_{ds4} has been neglected for simplicity. The capacitor C_c is the compensation capacitor, $C_1 = C_{bd4} + C_{dg4} + C_{bd6} + C_{dg6} + C_{gs8}$, $C_2 = C_{gd8}$, and $C_3 = C_{bd8} + C_{bd9} + C_{gd9}$.

The low-frequency gain of the folded cascode op amp in Fig. 8.4-13 is found as

$$A_v(0) = \frac{V_o(0)}{V_{in}(0)} = [g_{m1}g_{m8}G_{o1}G_{o2}]V_{in}(0) \tag{29}$$

A dominant pole is formed by the Miller capacitance reflected across V_1 of Fig. 8.4-13. The approximate value of this pole is

$$|p_1| \cong \frac{G_{M2}}{(V_o/V_1)C_c} = \frac{G_{M2}}{[G_{M2}g_{m8}/G_{o2}G_{o1}]C_c} = \frac{G_{o1}G_{o2}}{g_{m8}C_c} \tag{30}$$

where V_o/V_1 is the midband gain from V_1 to the output. A more detailed analysis of the larger roots of Fig. 8.4-12 using the model of Fig. 8.4-13 shows both a left-half- and right-half-plane zero and a pair of complex poles [5]. While the nulling resistor method can be used to eliminate the effects of the RHP zero, the designer must be careful to keep the Q of the

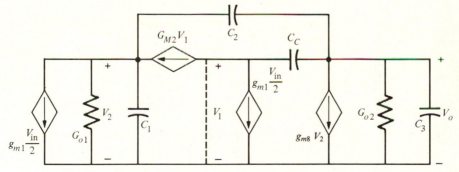

Figure 8.4-13   Small-signal model of the op amp in Fig. 8.4-12.

complex poles small. Otherwise, a gain peaking of the frequency response will occur in the vicinity of *GB*.

The PSRR performance of the folded cascode op amp can be analyzed using the small-signal model of Fig. 8.4-14. It can be shown that the PSRR$^+$ is given as

$$\text{PSRR}^+ \cong \frac{C_c[1 + (g_{m1}/2G_{M2})]\,[s + (g_{m1}/C_c)/1 + (g_{m1}/2G_{M2})]}{C_2[s + (G_{o1}G_{o2}/g_{m8}C_2)]} \qquad (31)$$

Comparing this expression to that of Eq. (24) where $G_{o1} = G_I$, $G_{o2} = G_{II}$, and $g_{m8} = g_{mII}$, we see that the only difference in the pole values is the ratio of C_2 to C_c. From this analysis it is evident that the mechanism causing a zero is the same for either circuit, namely the amount of capacitance coupling the gate of M8 (M6) to the output. However, since C_2 in the folded cascode op amp is approximately C_{gd8} while C_c in the two-stage op amp is the compensating capacitor, a significant improvement can occur. We note that the zero frequency is approximately at *GB* as was the case for the two-stage op amp.

Fig. 8.4-15 shows a version of the folded cascode op amp which has a wide output signal swing [6, 7]. This op amp has been used in telecommunication applications and has a gain of approximately 5000, a PSRR of 60–70 dB, and an input-voltage offset of approximately ±6 mV.

Op amps using the cascode configuration enable the designer to optimize some of the second-order performance specifications not possible with the classical two-stage op amp. This flexibility has allowed the devel-

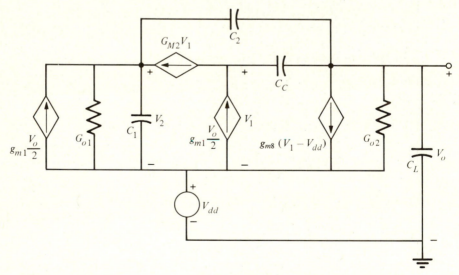

Figure 8.4-14   Model to calculate PSRR$^+$ of Fig. 8.4-12.

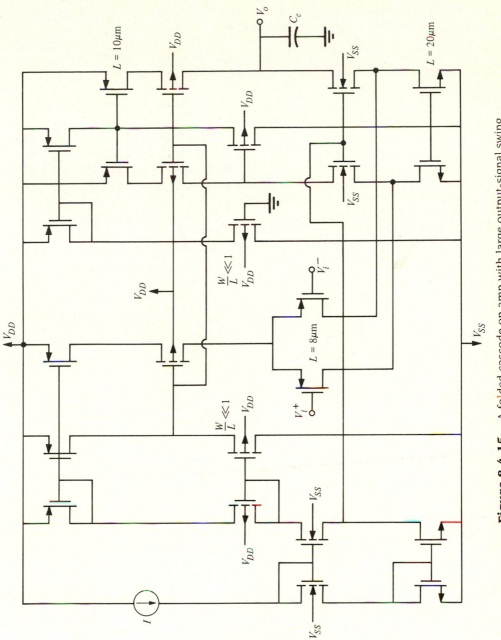

Figure 8.4-15 A folded cascode op amp with large output-signal swing.

opment of high-performance unbuffered op amps suitable for CMOS technology. Such amplifiers are widely used in present-day integrated circuits for telecommunications applications.

8.5   *Simulation and Measurement of Op Amps*

In designing a CMOS op amp, the designer starts with building blocks whose performance can be analyzed to a first-order approximation by hand/calculator methods of analysis. The advantage of this step is the insight it provides to the designer as the design of the circuit develops. However, at some point the designer must turn to a better means of simulation. For the CMOS op amp this is generally a computer-analysis program such as SPICE. With the insight of the first-order analysis and the modeling capability of SPICE, the circuit design can be optimized and many other questions (such as tolerances, stability, and noise) can be examined.

Fabrication follows the simulation and layout of the MOS op amp. After fabrication the MOS op amp must be tested and evaluated. The techniques for testing various parameters of the op amp can be as complex as the design of the op amp itself. Each specification must be verified over a large number of op amps to insure a working op amp in case of process variations.

The objective of this section is to provide the background for simulating and testing a CMOS op amp. We shall consider methods of simulating an op amp that are appropriate to SPICE but the concepts are applicable to other types of computer-simulation programs. Because the simulation and measurement of the CMOS op amp are almost identical, they are presented simultaneously. The only differences found are in the parasitics that the actual measurement introduces in the op amp circuit and the limited bandwidths of the instrumentation.

The categories of op amp measurements and simulations discussed include: open-loop gain, open-loop frequency response (including the phase margin), input-offset voltage, common-mode gain, power-supply rejection ratio, common-mode input- and output-voltage ranges, open-loop output resistance, and transient response including slew rate. Configurations and techniques for each of these measurements will be presented in this section.

Simulating or measuring the op amp in an open loop configuration is one of the most difficult steps to perform successfully. The reason is the high differential gain of the op amp. Fig. 8.5-1 shows how this step might be performed. V_{os} is an external voltage whose value is adjusted to keep the dc value of v_{OUT} between the power supply limits. Without V_{os} the op amp will be driven to the positive or negative power supply for either the measurement or simulation cases. The resolution necessary to find the cor-

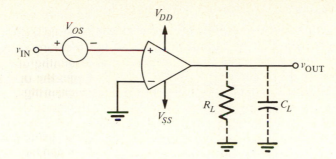

Figure 8.5-1   Open-loop mode with offset compensation.

rect value of V_{OS} usually escapes the novice designer. It is necessary to be able to find V_{OS} to the accuracy of the magnitude of the power supply divided by the low-frequency differential gain (typically in the range of millivolts). Although this method works well for simulation, the practical characteristics of the op amp make the method difficult to use for measurement.

A method more suitable for measuring the open-loop gain is shown in the circuit of Fig. 8.5-2. In this circuit it is necessary to select the reciprocal RC time constant a factor of $A_V(0)$ times less than the anticipated dominant pole of the op amp. Under these conditions, the op amp has total dc feedback which stabilizes the bias. The dc value of v_{OUT} will be exactly the dc value of v_{IN}. The true open-loop frequency characteristics will not be observed until the frequency is approximately $A_V(0)$ times $1/RC$. Above this frequency, the ratio of v_{OUT} to v_{IN} is essentially the open-loop gain of the op amp. This method works well for both simulation and measurement.

Simulation or measurement of the open-loop gain of the op amp will characterize the open-loop transfer curve, the open-loop output-swing lim-

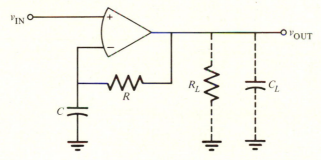

Figure 8.5-2   A method of measuring open-loop characteristics with dc bias stability.

its, the phase margin, the dominant pole, the unity-gain bandwidth, and other open-loop characteristics. The designer should connect the anticipated loading at the output in order to get meaningful results. In some cases, where the open-loop gain is not too large, the open-loop gain can be measured by applying v_{IN} in Fig. 8.5-3 and measuring v_{OUT} and v_I. In this configuration, one must be careful that R is large enough not to cause a dc current load on the output of the op amp.

The dc input-offset voltage can be measured using the circuit of Fig. 8.5-4. If the dc input-offset voltage is too small, it can be amplified by using a resistor divider in the negative-feedback path. One must remember that V_{OS} will vary with time and temperature and is very difficult to precisely measure experimentally. Interestingly enough, V_{OS} cannot be simulated. The reason is that the input-offset voltage is not only due to the bias mismatches as discussed for the two-stage comparator and op amp but is due to device and component mismatches. Presently, most simulators do not have the ability to predict device and component mismatches.

The common-mode gain is most easily simulated or measured using Fig. 8.5-5. It is seen that if V_{OS} fails to keep the op amp in the active region, this configuration will fail, which is often the case in experimental measurements. More often than not, the designer wishes to measure or simu-

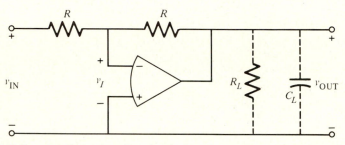

Figure 8.5-3 Configuration for simulating the open-loop frequency response for moderate gain op amps.

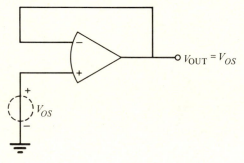

Figure 8.5-4 Configuration for measuring the input offset voltage.

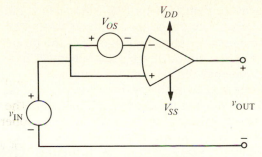

Figure 8.5-5   Configuration for simulating the common-mode gain.

late the CMRR. The common-mode gain could be derived from CMRR and the open-loop gain if necessary. A method of measuring the CMRR of an op amp is given in Fig. 8.5-6. The method involves a sequence of two measurements. The first measurement is to set V_{DD} to V_{DD} + 1 volt, V_{SS} to V_{SS} + 1 volt, and V_{OUT} to 1 volt by applying -1 volt to $-V_{OUT}$. This is equivalent to applying a common-mode input signal of 1 volt to the amplifier with the nominal supply values. The value at V_{OS} is measured and designated as V_{OS1}. Next, V_{DD} is set to V_{DD} − 1 volt, V_{SS} to V_{SS} − 1 volt, and V_{OUT} to −1 volt by applying +1 volt to $-V_{OUT}$. Measure V_{OS} designated as V_{OS2}. The CMRR can be found as

$$\text{CMRR} = \frac{2000}{|V_{OS1} - V_{OS2}|} \tag{1}$$

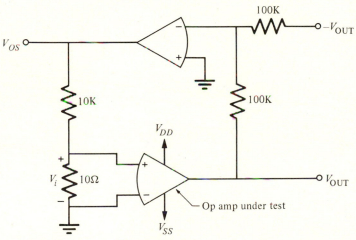

Figure 8.5-6   Circuit used to measure CMRR and PSRR.

This measurement is for a single frequency and must be modified to obtain the frequency response of the CMRR. The modification would involve placing small sinusoidal signals of the proper phase in series with the power supplies. Unfortunately this scheme is not very suitable for measuring the frequency response of the common-mode gain.

The better way to measure the CMRR would be to measure first the differential voltage gain in dB and then the common-mode voltage gain in dB by applying a common-mode signal to the input. The CMRR in dB could be found by subtracting the common-mode voltage gain in dB from the differential-mode voltage gain in dB. If the measurement system is associated with a controller or computer, this could be done automatically.

Neither of the two methods presented above is suitable for simulation. The objective of simulation is to get an output which is equal to CMRR or can be related to CMRR. Fig. 8.5-7(a) shows a method which can accomplish this objective. Two identical voltage sources designated as V_{cm} are placed in series with both op amp inputs where the op amp is connected in the unity-gain configuration. A model of this circuit is shown in Fig. 8.5-7(b). It can be shown that

$$\frac{V_{out}}{V_{cm}} = \frac{\pm A_c}{1 + A_v - (\pm A_c/2)} \cong \frac{|A_c|}{A_v} = \frac{1}{\text{CMRR}} \tag{2}$$

Computer simulation can be used to calculate Eq. (2) directly. If the simulator has a post-processing capability, then it is usually possible to plot the reciprocal of the transfer function so that CMRR can be directly plotted. Figure 8.5-8(a) shows the simulation results of the magnitude of the CMRR for the op amp of Ex. 8.3-1 and Fig. 8.5-8(b) gives the phase response of the CMRR. It is seen that the CMRR is quite large for frequencies up to 100 kHz.

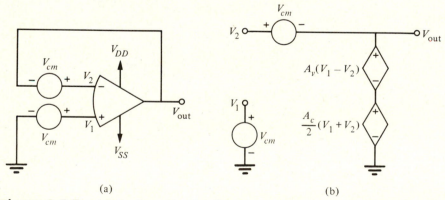

(a)   (b)

Figure 8.5-7   (a) Configuration for direct measurement of CMRR. (b) Model for (a).

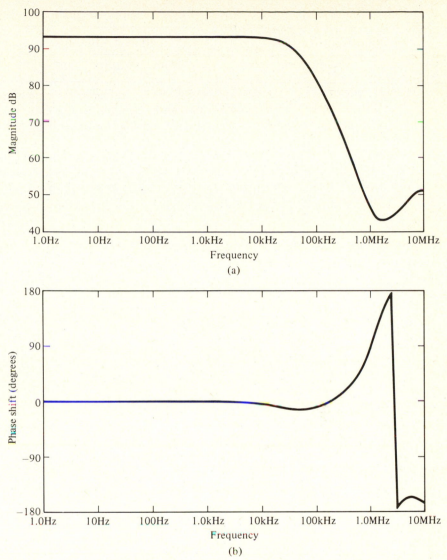

Figure 8.5-8 CMRR frequency response of Example 8.3-1 with $C_L = 20$ pF. (a) Magnitude response. (b) Phase response.

The configuration of Fig. 8.5-6 can also be used to measure the power-supply rejection ratio, PSRR. The procedure is to first set V_{DD} to $V_{DD} + 1$ volt and V_{OUT} to 0 volts by grounding $-V_{OUT}$. In this case, V_i is the input offset voltage for $V_{DD} + 1$ volt. Measure V_{OS} under these conditions and designate it as V_{OS3}. Next set V_{DD} to $V_{DD} - 1$ volt and V_{OUT} to 0 volts and measure V_{OFF}

and designate it as V_{OS4}. The PSRR of the V_{DD} supply is given as

$$\text{PSRR of } V_{DD} = \frac{2000}{|V_{OS3} - V_{OS4}|} \tag{3}$$

Similarly for the V_{SS} rejection ratio, change V_{SS} and keep V_{DD} constant while V_{OUT} is at 0 volts. The above formula can be applied in the same manner. Again, it is seen that this approach is not suitable for measuring PSRR as a function of frequency.

Fig. 8.5-9 shows a configuration similar to Fig. 8.4-9 that is suitable for measuring the PSRR as a function of frequency. A small sinusoidal voltage is inserted in series with V_{DD} (V_{SS}) to measure PSRR$^+$ (PSRR$^-$). From Eq. (12) of Sec. 8.4 it was shown that

$$\frac{V_{out}}{V_{dd}} \cong \frac{1}{\text{PSRR}^+} \quad \text{or} \quad \frac{V_{out}}{V_{ss}} \cong \frac{1}{\text{PSSR}^-} \tag{4}$$

This procedure was the method by which PSRR was calculated for the two-stage op amp in Sec. 8.4.

The input and output common-mode voltage range can be defined for both the open-loop and closed-loop mode of the op amp. For the open-

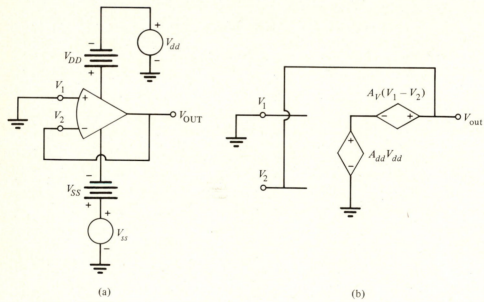

(a)   (b)

Figure 8.5-9 (a) Configuration for the direct measurement of PSRR. (b) Model of (a) with $V_{ss} = 0$.

loop case, only the output CMR makes sense. One of the configurations of Figs. 8.5-1 or 8.5-2 can be used to measure the output CMR. Typically, the open-loop, output CMR is about half the power-supply range. Because the op amp is normally used in a closed-loop mode, it makes more sense to measure or simulate the input and output CMR for this case. The unity-gain configuration is useful for measuring or simulating the input CMR. Fig. 8.5-10 shows the configuration and the anticipated results. The linear part of the transfer curve where the slope is unity corresponds to the input common-mode voltage range. The initial jump in the voltage sweep from negative values of v_{IN} to positive values is due to the turn-on of M5. In the simulation of the input CMR of Fig. 8.5-10, it is also useful to plot the current in M1 because there may be a small range of v_{IN} before M1 begins to conduct after M5 is turned on (see Fig. 8.5-17(a) for example).

In the unity-gain configuration, the linearity of the transfer curve is limited by the CMR. Using a configuration of higher gain, the linear part of the transfer curve corresponds to the output-voltage swing of the amplifier. This is illustrated in Fig. 8.5-11 for an inverting gain of 10 configuration.

The output resistance can be measured by connecting a load resistance R_L to the op amp output in the open-loop configuration. The measuring configuration is shown in Fig. 8.5-12. The voltage drop caused by R_L at a constant value of v_{IN} can be used to calculate the output resistance as

$$R_{out} = R_L \left(\frac{V_{01}}{V_{02}} - 1 \right) \tag{5}$$

An alternate approach is to vary R_L until $V_{02} = V_{01}/2$. Under this condition $R_{out} = R_L$. If the op amp must be operated in the closed-loop mode, then

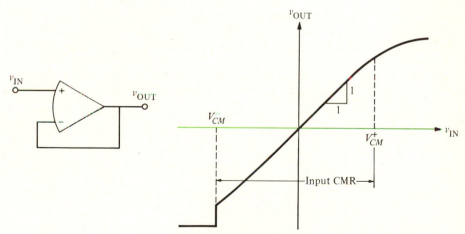

Figure 8.5-10   Measurement of the input common-mode voltage range.

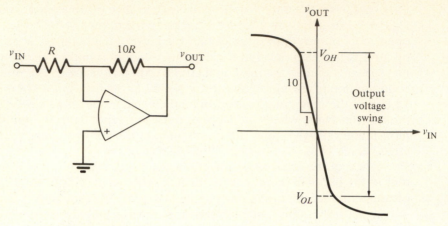

Figure 8.5-11   Measurement of the output voltage swing.

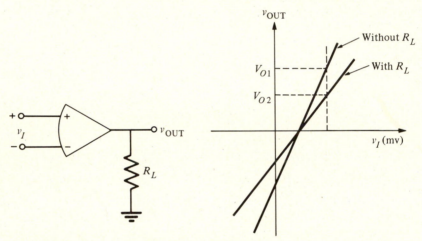

Figure 8.5-12   Measurement of the open-loop low-frequency gain.

the effects of the feedback on the measured output resistance must be considered. The best alternative is to use Fig. 8.5-13, where the value of the
open-loop gain A_v is already known. In this case, the output resistance is

$$R_{\text{out}} = \left(\frac{1}{R_o} + \frac{1}{100R} + \frac{A_v}{100R_o} \right)^{-1} \simeq \frac{100R_o}{A_v} \qquad (6)$$

It is assumed that A_v is in the range of 1000 and that R is greater than R_o.
Measuring R_{out} and knowing A_v allows one to calculate the output resistance of the op amp R_o from Eq. (6). Other schemes for measuring the output impedance of op amps can be found in the literature [8, 9].

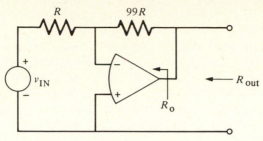

Figure 8.5-13   An alternative method of measuring the open-loop, output resistance R_o.

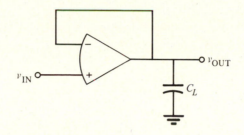

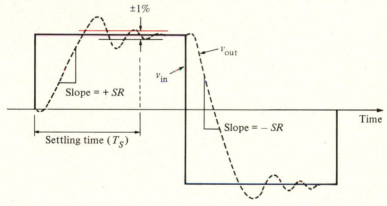

Figure 8.5-14   Measurement of slew rate *(SR)* and 1% settling time.

The configuration of Fig. 8.5-14 is useful for measuring the slew rate and the settling time. Fig. 8.5-14 gives the details of the measurement. For best accuracy, the slew rate and settling time should be measured separately. If the input step is sufficiently small (<0.5 V) the output should not slew and the transient response will be a linear response. The settling time can be easily measured. (Appendix C shows how the unity-gain step response can be related to the phase margin, making this configuration a quick method of measuring the phase margin.) If the input step magnitude is sufficiently large, the op amp will slew by virtue of not having enough current to charge or discharge the compensating and/or load capacitances. The slew rate is determined from the slope of the output waveform during the rise or fall of the output. The output loading of the op amp should be present during the settling-time and slew-rate measurements. The unity-gain configuration places the severest requirements on stability and slew rate because its feedback is the largest, resulting in the largest values of loop gain, and should always be used as a worst-case measurement.

Other simulations (such as noise, tolerances, process-parameter variations, and temperature) can also be performed. At this point, one could breadboard the op amp. However, if the accuracy of the simulation models is sufficient this step is questionable. An example of using SPICE to simulate a CMOS op amp is given in the following.

Example 8.5-1

Simulation of the CMOS Op Amp of Ex. 8.3-1.

The op amp designed in Example 8.3-1 and illustrated in Fig. 8.3-2 is to be analyzed by SPICE to determine if the specifications are met. The device parameters to be used are those of Tables 3.1-2 and 3.2-1 along with C_{GBO} = 200 pF/m, C_{GSO} = C_{GDO} = 350 pF/m, CJ_N = 300 μF/m^2, CJSW$_N$ = 500 pF/m, CJ_P = 150 μF/m^2, CJSW$_P$ = 400 pF/m, MJ = 0.5, MJSW$_N$ = 0.3, MJSW$_P$ = 0.25, LD = 0.8μ, and TOX = 0.08μ. In addition to verifying the specifications of Example 8.3-1, we will simulate PSRR$^+$ and PSRR$^-$.

The op amp will be treated as a subcircuit in order to simplify the repeated analyses. Table 8.5-1 gives the SPICE subcircuit description of Fig. 8.3-1. While the values of AD, AS, PD, and PS could be calculated from Fig. 8.3-4, often the design has not yet been geometrically defined. In this case, one can make an educated estimate of these values by using the following approximations.

$$AS = AD \cong W[L1 + L2 + L3]$$
$$PS = PD \cong 2W + 2[L1 + L2 + L3]$$

where $L1$ is the minimum allowable distance between the polysilicon and a contact in the moat (Rule 10C of Table 2.6-1), $L2$ is the length of a minimum-size square contact to moat (Rule 10A of Table 2.6-1), and $L3$ is the minimum allowable distance between a contact to moat and the edge of the moat (Rule 10D of Table 2.6-1).

Table 8.5-1

SPICE Subcircuit Description of Fig. 8.3-2.

```
.SUBCKT OPAMP 1 2 6 8 9
M1 4 2 3 3 NMOS1 W= 43U L=10U AD=0.3N AS=0.3N PD= 50U PS= 50U
M2 5 1 3 3 NMOS1 W= 43U L=10U AD=0.3N AS=0.3N PD= 50U PS= 50U
M3 4 4 8 8 PMOS1 W= 10U L=10U AD=0.3N AS=0.3N PD= 20U PS= 20U
M4 5 4 8 8 PMOS1 W= 10U L=10U AD=0.3N AS=0.3N PD= 10U PS= 20U
M5 3 7 9 9 NMOS1 W= 38U L=10U AD=0.3N AS=0.3N PD= 40U PS= 40U
M6 6 5 8 8 PMOS1 W=344U L=10U AD=1.3N AS=1.3N PD=350U PS=350U
M7 6 7 9 9 NMOS1 W=652U L=10U AD=2.3N AS=2.3N PD=660U PS=660U
M8 7 7 9 9 NMOS1 W= 38U L=10U AD=0.3N AS=0.3N PD= 40U PS= 40U
CC 5 6 4.4P
.MODEL NMOS1 NMOS VTO=1 KP=17U GAMMA=1.3 LAMBDA=0.01 PHI=0.7
+PB=0.80 MJ=0.5 MJSW=.3 CGBO=200P CGSO=350P CGDO=350P CJ=300U
+CJSW=500P CJSW=500P LD=0.8U TOX=80N
.MODEL PMOS1 PMOS VTO=-1 KP=8U GAMMA=0.6 LAMBDA=0.02 PHI=0.6
+PB=0.50 MJ=0.5 MJSW=.25 CGBO=200P CGSO=350P CGDO=350P CJ=150U
+CJSW=400P LD=0.8U TOX=80N
IBIAS 8 7 8.8U
.ENDS
```

The first analysis to be made involves the open-loop configuration of Fig. 8.5-1. A coarse sweep of v_{IN} is made from -5 V to $+5$ V to find the value of v_{IN} where the output makes the transition from V_{SS} to V_{DD}. Once the transition range is found, v_{IN} is swept over values that include only the transition region. The result is shown in Fig. 8.5-15. From this data, the value of V_{OS} in Fig. 8.5-1 can be determined. While V_{OS} need not make v_{OUT} exactly zero, it should keep the output in the linear range so that when SPICE calculates the bias point for small-signal analysis, reasonable results are obtained. In this example V_{OS} was not included (or left equal to zero) since the output remained in the linear region without it.

At this point, the designer is ready to begin the actual simulation of the op amp. In the open-loop configuration the voltage-transfer curve, the frequency response, and the small-signal gain, input and output resistances can be simulated. The SPICE input deck using the PC version of SPICE (PSPICE) is shown in Table 8.5-2. Fig. 8.5-16 shows the results of this simulation. The open-loop voltage gain is 16,387 (determined from the output file), *GB* is 0.9 MHz, output resistance is 224 KΩ (determined from the output file), power dissipation is 1.7 mW (determined from the output file), phase margin for a

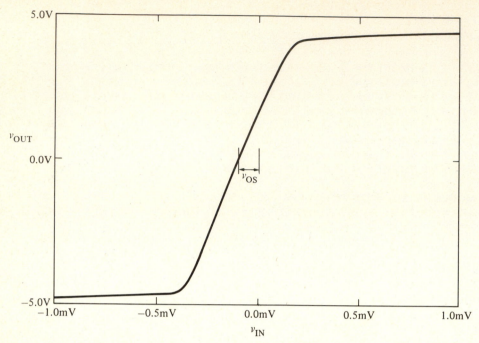

Figure 8.5-15 Open-loop transfer characteristic of Example 8.5-1 illustrating V_{OS}.

Table 8.5-2

PSPICE Input File for the Open-Loop Configuration.
EXAMPLE 8.5-1 OPEN LOOP CONFIGURATION
.OPTION LIMPTS = 1000
VIN+ 1 0 DC 0 AC 1.0
VDD 4 0 DC 5.0
VSS 0 5 DC 5.0
VIN- 2 0 DC 0
CL 3 0 20P
X1 1 2 3 4 5 OPAMP
$\vdots$
(Subcircuit of Table 8.5-1)
$\vdots$
.OP
.TF V(3) VIN+
.DC VIN+ -0.005 0.005 100U
.PRINT DC V(3)
.AC DEC 10 1 10MEG
.PRINT AC VDB(3) VP(3)
.PROBE   (This entry is unique to PSPICE)
.END

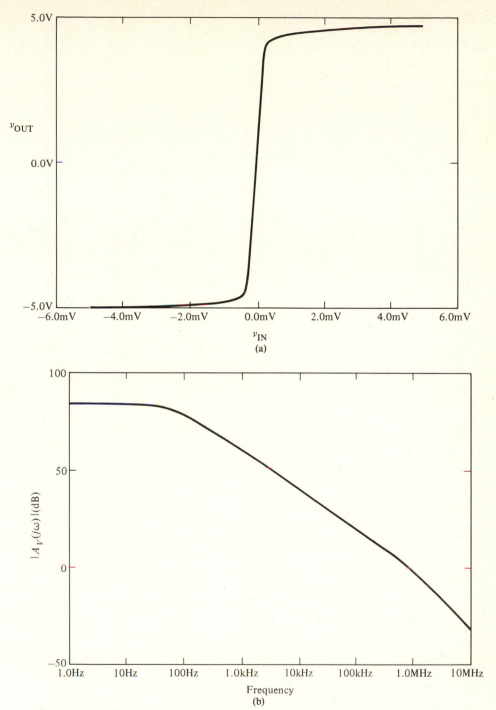

Figure 8.5-16 (a) Open-loop transfer characteristic of Example 8.5-1. (b) Open-loop transfer magnitude response of Example 8.5-1.

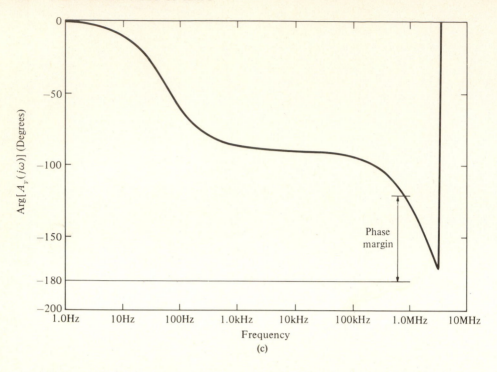

Figure 8.5-16 (c) Open loop transfer phase response of Example 8.5-1.

20 pF load is 58 degrees, and the open-loop output voltage swing is +4 V to −4.2 V.

The next configuration is the unity-gain configuration of Fig. 8.5-10. From this configuration, the input CMR, PSRR$^+$, PSRR$^-$, slew rate, and settling time can be determined. Table 8.5-3 gives the SPICE input file to accomplish this (PSRR$^+$ and PSRR$^-$ must be done on separate runs). The results of this simulation are shown in Fig. 8.5-17. The input CMR range is −3.2 V to +4.2 V as seen on Fig. 8.5-17(a). Note that the lower limit of the input CMR range is determined by when the current in M1 reaches its quiescent value. PSRR$^+$ and PSRR$^-$ are shown in Figs. 8.5-17(b) through 8.5-17(e). The large-signal and small-signal transient response was made by applying a 4 V pulse and a 0.2 V pulse to the unity-gain configuration. The results are illustrated in Fig. 8.5-17(f) and (g). From this data both the positive and negative slew rates are determined to be approximately 1.9 V/μs, which is within specification. The settling time to within ±5% is approximately 0.5 μs as determined from the output file. The rela-

Table 8.5-3

Input File for the Unity-Gain Configuration.

EXAMPLE 8.5-1 UNITY GAIN CONFIGURATION.
.OPTION LIMPTS = 501
VIN+ 1 0 PWL(0 −2 10N −2 20N 2 2U 2 2.01U −2 4U −2 4.01U −.1 6U −.1
+6.01U .1 8U .1 8.01U −.1 10U −.1)
VDD 4 0 DC 5.0 AC 1.0
VSS 0 5 DC 5.0
CL 3 0 20P
X1 1 3 3 4 5 OPAMP
⋮
(Subcircuit of Table 8.5-1)
⋮
.DC VIN+ −5 5 0.1
.PRINT DC V(3)
.TRAN 0.05U 10U 0 10N
.PRINT TRAN V(3) V(1)
.AC DEC 10 1 10MEG
.PRINT AC VDB(3) VP(3)
.PROBE   (This entry is unique to PSPICE)
.END

tively large-value compensation capacitor is preventing the 20 pF load from causing significant ringing in the transient response.

The specifications resulting from this simulation are compared to the design specifications in Table 8.5-4. It is seen that the design is almost satisfactory. Slight adjustments can be made in the W/L ratios or dc currents to bring the amplifier within the specified range. The next step in simulation would be to vary the values of the model parameters, typically K', V_T, γ, and λ to insure that the specifications are met even if the process varies.

The measurement schemes of this section will work reasonably well as long as the open-loop gain is not large. If the gain should be large, techniques applicable to bipolar op amps must be employed. A test circuit for the automatic measurement of integrated-circuit op amps in the frequency domain has been developed and described [10]. This method supplements the approaches given in this section. Data books and technical literature are also a good source of information on this topic.

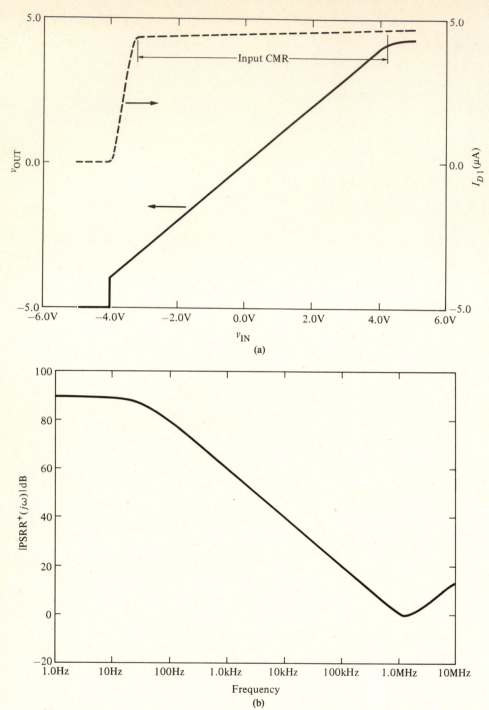

Figure 8.5-17 (a) Input CMR simulation of Example 8.5-1. (b) PSRR⁺ magnitude response of Example 8.5-1. (c) PSRR⁺ phase response of Example 8.5-1. (d)

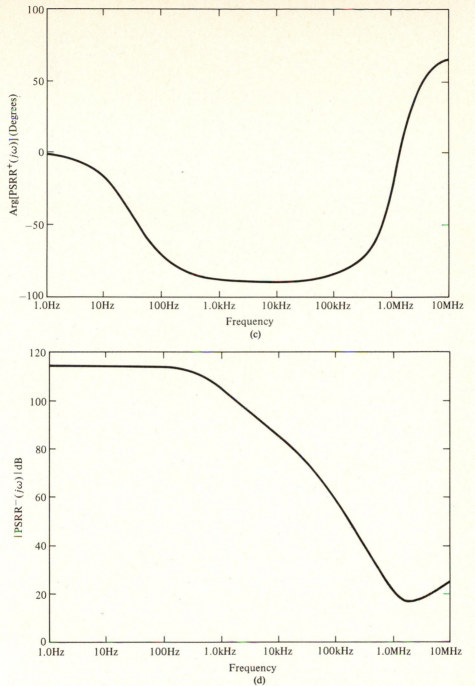

(c)

(d)

PSRR⁻ magnitude response of Example 8.5-1. (e) PSRR⁻ phase response of Example
8.5-1. (f) Unity-gain transient response of Example 8.5-1. (g) Unity-gain small signal transient response of Example 8.5-1.

443

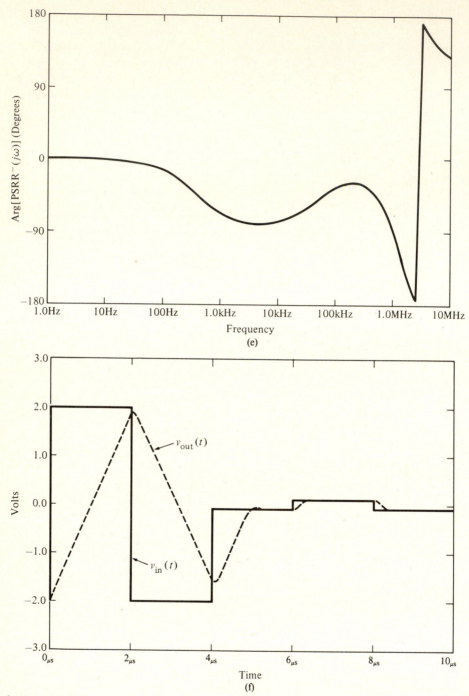

Figure 8.5-17 (e—f)

444

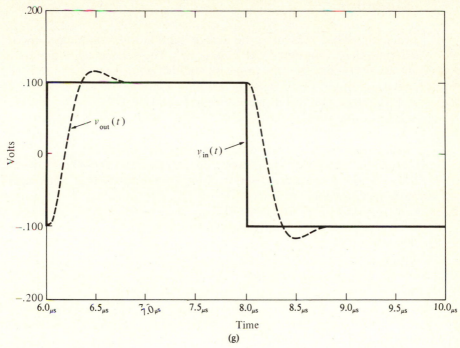

Figure 8.5-17   (g)

Table 8.5-4

Comparison of the Simulation with Specifications of Ex. 8.3-1.

Specification	Design (Ex. 8.3-1)	Simulation (Ex. 8.5-1)
Open Loop Gain	> 4000	16,387
GB (MHz)	1 MHz	0.9 MHz
Input CMR (Volts)	±3	+4.3V, −4V
Slew Rate (V/μsec)	> 2	±1.9
P_{diss} (mW)	< 10	1.70
V_{out} range (V)	± 4V	+4V, −4.2V
$PSRR^{+}(0)$ (dB)	-	89
$PSRR^{-}(0)$ (dB)	-	114
Settling Time (μs)	-	0.5
Output resistance (Kilohms)	-	224

8.6 *Summary*

This chapter has presented the design, simulation, and measurement considerations of unbuffered CMOS op amps. The general approach to designing op amps concentrates first on establishing dc conditions that are process insensitive. This results in defining some of the ratios of the devices and establishing constraints between the device ratios. Next the ac performance is achieved by selecting dc current levels and the remaining device ratios. Constraints are then developed in order to achieve satisfactory frequency response. This procedure for designing simple CMOS op amps was seen to be reasonably straightforward, and a design procedure was developed for the two-stage CMOS op amp that insures a first-cut design for most specifications.

One important aspect of the op amp is its stability characteristics, which are specified by the phase margin. Several compensation procedures that allow the designer to achieve reasonably good phase margins even with large capacitive loads were discussed. The stability of the op amp was also important in the settling time of the pulse response. The Miller compensation method of pole splitting, used together with a nulling resistor to eliminate the effects of the RHP zero, was found to be satisfactory.

The design of the CMOS op amp given in Secs. 8.3 and 8.4 resulted in a first-cut design for a two-stage op amp or a cascode op amp. The two-stage op amp was seen to give satisfactory performance for most typical applications. The cascode configuration of Sec. 6.3 was used to improve the performance of the two-stage op amp in the areas of gain, stability, and PSRR. If all internal nodes of an op amp are low impedance, compensation can be accomplished by a shunt capacitance to ground at the output. This configuration is self-compensating for large capacitive loads. Although the output resistance of the cascode op amp was generally large, this is not a problem if the op amp is to drive capacitive loads.

Secs. 8.3 and 8.4 show how the designer can obtain the approximate values for the performance of the op amp. However, it is necessary to simulate the performance of the CMOS op amp to refine the design and to check to make sure no errors were made in the design. Refining the design also means to vary the process parameters in order to make sure the op amp still meets its specifications under given process variations. Finally, it is necessary to be able to measure the performance of the op amp when it is fabricated. So, techniques of simulating and measurement applicable to the CMOS op amp were presented.

This chapter has presented the principles and procedures by which the reader can design op amps for applications not requiring low-output resistance. This information serves as the basis for improving the performance of op amps, the topic to be considered in the next chapter.

PROBLEMS — *Chapter 8*

1. Develop a macromodel for the op amp of Fig. 8.1-2 which models the low frequency gain $A_v(0)$, the unity-gain bandwidth GB, the output resistance R_{out}, and the output-voltage swing limits V_{OH} and V_{OL}. Your macromodel should be compatible with SPICE and should contain only resistors, capacitors, controlled sources, independent sources, and diodes.

2. Develop a macromodel for the op amp of Fig. 8.1-2 that models the low-frequency gain $A_v(0)$, the unity-gain bandwidth GB, the output resistance R_{out}, and the slew rate SR. Your macromodel should be compatible with SPICE and should contain only resistors, capacitors, controlled sources, independent sources, and diodes.

3. Show that the controlled source of Fig. 8.1-4 designated as $V_1/CMRR$ is in fact a suitable model for the common-mode behavior of the op amp.

4. Show how to incorporate the PSRR effects of the op amp into the model of the nonideal effects of the op amp given in Fig. 8.1-4.

5. Derive the relationship for GB given in Eq. (17) of Sec. 8.2.

6. For an op amp model with two poles and one RHP zero, prove that if the zero is 10 times larger than GB, then in order to achieve a 45° phase margin, the second pole must be placed at least 1.22 times higher than GB.

7. For an op amp model with two poles and one RHP zero, prove that if the zero is 10 times higher than GB, then in order to achieve a 60° phase margin, the second pole must be placed at least 2.2 times higher than GB.

8. For an op amp model with three poles and no zero, prove that if the highest pole is 10 times GB, then in order to achieve 60° phase margin, the second pole must be placed at least 2.2 times GB.

9. Derive the relationships given in Eqs. (33) through (36) in Sec. 8.2.

10. Physically explain why the RHP zero occurs in the Miller compensation scheme illustrated in the op amp of Fig. 8.2-8. Why does the RHP zero have a stronger influence on a CMOS op amp than on a similar type BJT op amp?

11. Derive Eq. (43) of Sec. 8.2.

12. For the two-stage op amp of Fig. 8.2-8, find W_1/L_1, W_6/L_6, and C_c if $GB = 1$ MHz, $|p_2| = 5\ GB$, $z = 3\ GB$ and $C_L = C_2 = 20$ pF. Use the parameter values of Table 3.1-2 and consider only the two-pole model of the op amp. The bias current in M5 is 40 μA and in M7 is 320 μA.

13. In Fig. 8.2-10, assume that $R_I = 150$ KΩ, $R_{II} = 100$ KΩ, $g_{mII} = 500\ \mu$S, $C_I = 1$ pF, $C_{II} = 5$ pF, and $C_c = 30$ pF. Find the value of R_z and the

locations of all roots for (a) the case where the zero is moved to infinity and (b) the case where the zero cancels the highest pole.

14. Express all of the relationships given in Eqs. (1) through (9) of Sec. 8.3 in terms of the large-signal model parameters and the dc values of drain current.

15. Develop the relationship given in step 5 of Table 8.3-2.

16. Show that the relationship between the W/L ratios of Fig. 8.3-1 which guarantees that $V_{GS4} = V_{GS6}$ is given by

$$\frac{S_6}{S_4} = 2\frac{S_7}{S_5}$$

where $S_i = W_i/L_i$.

17. Draw a schematic of the op amp similar to Fig. 8.3-1 but using p-channel input devices. Discuss the anticipated differences in performance and design between this op amp architecture and the one in Fig. 8.3-1.

18. Derive the relationship given in Eq. (18) of Sec. 8.3.

19. For the p-channel input, CMOS op amp of Fig. P8.19, calculate the open-loop, low-frequency differential gain, the output resistance, the

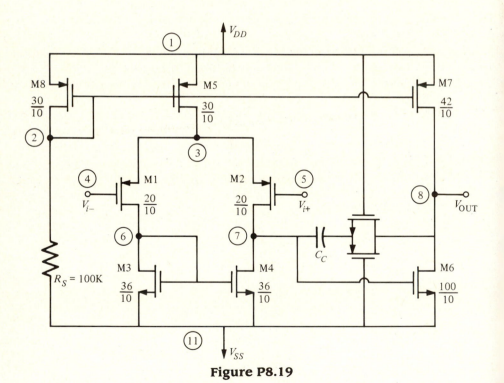

Figure P8.19

power consumption, the power-supply rejection ratio, the input common-mode range, the output-voltage swing, the slew rate, the common-mode rejection ratio, and the unity-gain bandwidth for a load capacitance of 20 pF. Assume the model parameters of Table 3.1-2 except let $K_N' = 24.75\ \mu A/V^2$, $K_P' = 12.4\ \mu A/V^2$, $\gamma_N = \gamma_P = 0.5\ V^{1/2}$, and $\lambda_N = \lambda_P = 0.01\ V^{-1}$. Use the simulation program SPICE to find the phase margin and the 1% settling time for no load and for a 20 pF load.

20. Design the values of W and L for each transistor of the CMOS op amp in Fig. P8.20 to achieve a differential voltage gain of 4000. Assume that $K_N' = 10\ \mu A/V^2$, $K_P' = 5\ \mu A/V^2$, $V_{TN} = -V_{TP} = 1\ V$, and $\lambda_N = \lambda_P = 0.01\ V^{-1}$. Also, assume that the minimum device dimension is 10 μm and choose the smallest devices possible. Design C_c and R_z to give GB = 1 MHz and to eliminate the influence of the RHP zero. How much load capacitance should this op amp be capable of driving without suffering a degradation in the phase margin? What is the slew rate of this op amp? Assume $V_{DD} = -V_{SS} = 5V$ and $R_B = 100$ kilohms.

21. Use the electrical model parameters of the previous problem to design W_3, L_3, W_4, L_4, W_5, L_5, C_c, and R_z of Fig. P8.21 if $W_1 = L_1 = W_2 = L_2 = 10\ \mu m$ to obtain a low-frequency, differential-voltage gain of 5000 and a GB of 1 MHz. All devices should be in saturation under normal operating conditions and the effect of the RHP should be cancelled. How much load capacitance should this op amp be able to drive before

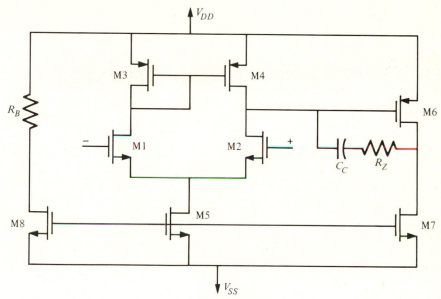

Figure P8.20

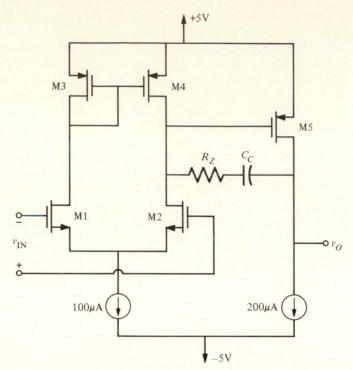

Figure P8.21

suffering a degradation in the phase margin? What is the slew rate of this op amp?

22. On a log-log plot with the vertical axis having a range of 10^{-3} to 10^{+3} and the horizontal axis having a range of 1 μA to 100 μA, plot the low-frequency gain $A_v(0)$, the unity-gain bandwidth GB, the power dissipation P_{diss}, the slew rate SR, the output resistance R_{out}, the magnitude of the dominant pole $|p_1|$, and the magnitude of the RHP zero z, all normalized to their respective values at $I_B = 1$ μA as a function of I_B from 1 μA to 100 μA for the CMOS op amp shown in Fig. P8.22.

23. Repeat Example 8.4-1 to find new values of W_1 and W_2 which will give a voltage gain of 10,000.

24. Devise a scheme to generate the bias current I_{CM} of Fig. 8.4-3 that is proportional to the common-mode input voltage.

25. Find the differential-voltage gain of Fig. 8.4-3 where the output is taken at the drains of MC1 and MC2. Assume that $K_N' = 25.75$ μA/V^2, $K_P' = 10.125$ μA/V^2, and $\lambda_N = \lambda_P = 0.02$ V^{-1}. Ignore the bulk effects.

26. Discuss the influence that the pole at the gate of M6 in Fig. 8.4-5 will have on the Miller compensation of this op amp.

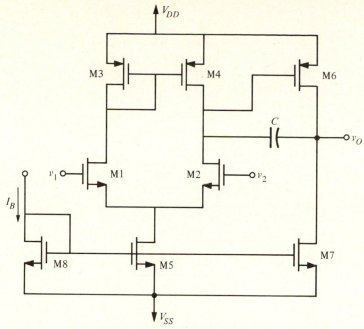

Figure P8.22

27. Verify Eqs. (4) through (8) of Sec. 8.4 for the two-stage op amp of Fig. 8.4-6 having a cascode second stage. If the second stage bias current is 50 μA, what is the output resistance of this amplifier using the parameters of Table 3.1-2?

28. Verify Eqs. (9) and (10) of Sec. 8.4 and give an expression for the overall differential-voltage gain of Fig. 8.4-6.

29. An internally-compensated, cascode op amp is shown in Fig. P8.29. (a) Derive the common-mode input range. (b) Find W_1/L_1, W_2/L_2, W_3/L_3, and W_4/L_4 when I_{BIAS} is 80 μA and the input CMR is -3.5 V to 3.5 V. Use $K_N' = 25\ \mu A/V^2$, $K_p' = 11\ \mu A/V^2$ and $|V_T| = 0.8$ to 1.0 V.

30. Develop an expression for the small-signal differential-voltage gain and output resistance of the cascode op amp of Fig. P8.29.

31. Sketch the frequency response of PSRR$^+$ and PSRR$^-$ of the two-stage op amp designed in Example 8.3-1.

32. Verify the voltage gain of the folded, cascode op amp given in Eq. (29) of Sec. 8.4.

33. Verify the relationships given in Eqs. (1) and (3) of Sec. 8.5.

34. Sketch a circuit configuration suitable for measuring the following op amp characteristics: (a) slew rate, (b) transient response, (c) input CMR, (d) output voltage swing. Also, sketch the anticipated result of each measurement.

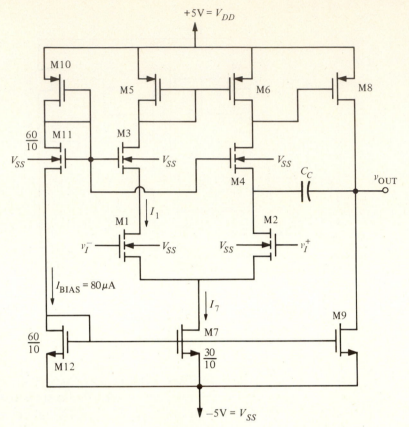

Figure P8.29

35. Using two identical op amps, show how to use SPICE in order to obtain a voltage which is proportional to CMRR rather than the inverse relationship given in Sec. 8.5.

36. Repeat the above problem for PSRR.

37. Use SPICE to simulate the comparator of Example 7.3-4. The open-loop voltage-transfer curve, the power dissipation, the propagation-time delay, the common-mode input range, the output-voltage range, the slew rate, and the settling time are to be simulated with a load capacitance of 2 pF. In addition to the model parameters used in the example, assume that CBD and CBS are 30 fF, PB is 0.87V, MJ is 0.5, CGSO and CGBO are 0.4pf/m, LD is 0.8μm, and TOX is 0.1μm.

38. Use SPICE to simulate the op amp of Fig. P8.38. The differential-frequency response, power dissipation, phase margin, common-mode input range, output-voltage range, slew rate, and settling time are to

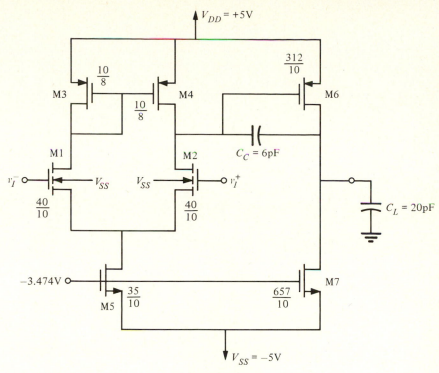

Figure P8.38

be simulated with a load capacitance of 20 pF. Use the following model parameters.

	V_T (V)	K' ($\mu A/V^2$)	λ (V^{-1})	γ ($V^{1/2}$)	$2|\phi_F|$ (V)	PB (V)	CBX (fF)	MJ	CGXO (pF/m)	LD (μm)	TOX (μm)
NMOS	1.0	24.75	0.01	0.5	0.6	0.87	30	.5	0.4	0.8	0.1
PMOS	−1.0	10.125	0.01	0.5	0.6	0.87	30	.5	0.4	0.8	0.1
							X = D or S		X = D or S		

39. Use SPICE to simulate the op amp of Example 8.4-2. The differential-frequency response, power dissipation, phase margin, common-mode input range, output-voltage range, slew rate, and settling time are to be simulated with a load capacitance of 20 pF. Use the model parameters of Table 3.1-2.

40. The CMOS op amp shown in Fig. P8.40 was designed, fabricated, and tested. It worked satisfactorily except that when the op amp was used in a unity-gain configuration, the positive peak of a ±3 V sinusoid

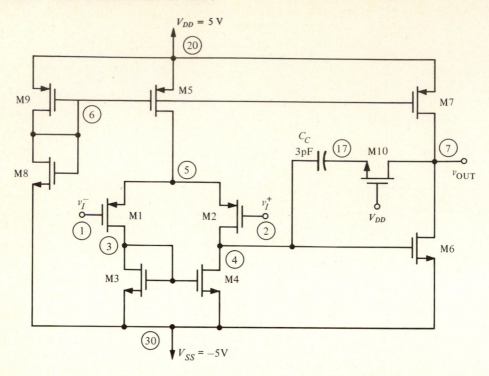

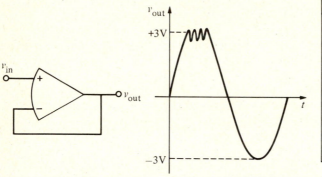

Figure P8.40

Transistor	W/L Ratio (μm/μm)
M1	40/20
M2	40/20
M3	50/10
M4	50/10
M5	77/10
M6	200/10
M7	154/10
M8	5/50
M9	100/10
M10	10/10

was unstable as illustrated. What caused this instability and how can it be fixed? Assume that the electrical parameters of this op amp are $V_{TN} = -V_{TP} = 0.7$ V, $K_N' = 28$ μA/V^2, $K_P' = 8$ μA/V^2, $\lambda_N = \lambda_P = 0.01$ V^{-1}, $\lambda_N = 0.35$ V$^{1/2}$, $\lambda_P = 0.9$ V$^{1/2}$, and $2|\phi_F| = 0.5$ V.

41. Repeat Ex. 8.5-1 for the op amp of Ex. 8.3-2.

42. Fig. P8.42 shows a possible scheme for simulating the CMRR of an op amp. Find the value of V_{out}/V_{cm} and show that it is approximately equal

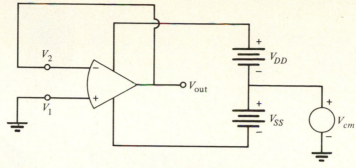

Figure P8.42

to 1/CMRR. What problems might result in an implementation of this circuit?

REFERENCES

1. Y.P. Tsividis and P.R. Gray, "An Integrated NMOS Operational Amplifier with Internal Compensation," *IEEE Journal of Solid-State Circuits,* Vol. SC-11, No. 6 (December 1976) pp. 748–753.
2. W.C. Black, et. al., "A High Performance Low Power CMOS Channel Filter," *IEEE Journal of Solid-State Circuits,* Vol. SC-15, No. 6, (December 1980) pp. 929–938.
3. S. Masuda, Y. Kitamura, S. Ohya, and M. Kikuchi, "CMOS Sampled Differential, Push-Pull Cascode Operational Amplifier," *Proceedings of the 1984 International Conference on Circuits and Systems,* Montreal, Canada, (May 1984) pp. 1211–1214.
4. B.K. Ahuja, "An Improved Frequency Compensation Technique for CMOS Operational Amplifiers," *IEEE Journal of Solid-State Circuits,* Vol. SC-18, No. 6 (December 1983) pp. 629–633.
5. D.B. Ribner and M.A. Copeland, "Design Techniques for Cascode CMOS Op Amps with Improved PSRR and Common-Mode Input Range," *IEEE Journal of Solid-State Circuits,* Vol. SC-19, No. 6 (December 1984) pp. 919–925.
6. P.R. Gray and R.G. Meyer, "MOS Op Amp Design—An Overview," *IEEE Journal of Solid-State Circuits,* Vol. SC-17, No. 6 (December 1982) pp. 969–982.
7. T.C. Choi, R.T. Kaneshiro, R.W. Brodersen, P.R. Gray, W.B. Jett, and M. Wilcox, "High-Frequency CMOS Switched-Capacitor Filters for Communications Applications," *IEEE Journal of Solid-State Circuits,* Vol. SC-18, No. 6 (December 1983) pp. 652–664.
8. W.G. Jung, *IC Op Amp Cookbook,* (Indianapolis, IN: Howard W. Sams and Co., 1974.)
9. J.G. Graeme, G.E. Tobey, and L.P. Huelsman, *Operational Amplifiers—Design and Applications,* (New York: 1974) McGraw-Hill.
10. W.M.C. Sansen, M. Steyaert, and P.J.V. Vandeloo, "Measurement of Operational Amplifier Characteristics in the Frequency Domain," *IEEE Transactions on Instrumentation and Measurement,* Vol. IM-34, No. 1 (March 1985) pp. 59–64.

chapter 9

*H*igh-Performance CMOS Op Amps

In the previous chapter we introduced the analysis and design of general unbuffered CMOS op amps with an eye to developing the principles associated with the design of CMOS op amps. However, in many applications the performance of the unbuffered CMOS op amp is not sufficient. In this chapter, CMOS op amps with improved performance will be considered. These op amps should be capable of meeting the specifications of most designs.

Typically, the areas where increased performance is desired include lower output resistance, larger output-signal swing, increased slew rate, increased gain bandwidth, lower noise, lower power dissipation, and/or lower input-offset voltage. Of course, not all of these characteristics will be obtained at the same time. In many cases, simply including the buffer of

Fig. 8.1-1 will achieve the desired performance. We shall examine several types of buffers that can be used to increase the capabilities of the unbuffered CMOS op amp.

The first topic of this chapter is reducing the output resistance of the op amp in order to drive resistive loads. The first section introduces output stages that use MOS devices only. When a unity-gain output stage is added to the two-stage op amp, the stability of the op amp may be worsened. A push-pull output op amp with crossover control is used to provide a greatly increased the output current without suffering large quiescent dissipation. Another technique uses shunt feedback to reduce the output resistance of the op amp.

The second approach to reducing the output resistance of the op amp uses the substrate BJT. The output resistance of the common collector BJT is equal to $1/g_m$, which will be low because g_m of the BJT is large. The BJT creates several problems, including the need to provide a current drive to the base of the BJT and a symmetrical sink/source output current capability. The ideas of the first two sections are employed in the third section to implement high-speed and/or high-frequency op amps.

The next section deals with low-noise op amps and what design trade-offs must be made in order to minimize noise. This is followed by a presentation of the chopper-stabilized op amp—a technique suitable for reducing the input-offset voltage or input equivalent noise of the op amp. The last two sections deal with op amps oriented to specific applications. The first is low power op amps that achieve quiescent power dissipations of 50 μW or less. The last section describes an op amp circuit that is useful for clocked circuits. This type of op amp is not required to amplify continuous-time signals.

The topics of this chapter illustrate the method of optimizing one or more performance specifications at the expense of others to achieve a high performance in a given area. They serve as a reminder that the design of a complex circuit such as an op amp is by no means unique and that the designer has many degrees of freedom as well as a choice of different circuit architectures that can be used to enhance performance for a given application.

9.1 *Op Amps Using an MOS Output Stage*

The op amps of the previous chapter all had a high output impedance and were classified as unbuffered op amps. While these op amps could drive a moderate load capacitance, they were not able to drive low-resistance output loads. In this section, we shall examine methods of improving both the ability to drive large load capacitances and small load resistances. Our goal will be to accomplish these objectives with reduced power dissipation in the op amp. The op amps of this section could be classified as low power with the capability to drive high-capacitive/low-resistive output loads.

A very simple modification of the two-stage op amp will enable it to drive large currents into a load capacitor without increasing power dissipation in the second stage. Fig. 9.1-1 shows the modification of the p-channel input, unbuffered CMOS op amp that results in the ability to both source and sink current at the output under dynamic conditions without having to establish a large dc current in the output stage. The primary difference is taking the signal available at the drain of M3 and applying it to a common source stage, M8, which results in current through M9 that is mirrored in M7 and is available for sourcing current to the output load. M6 provides the ability to sink current from the output load.

It is of interest to examine the frequency response of Fig. 9.1-1 in more detail. A small-signal equivalent model of Fig. 9.1-1 is shown in Fig. 9.1-2. A_I indicates the current path formed by M7 through M9. This causes the current of M7 to be a function of v_{in} as discussed. The nodal equations for this circuit can be written as follows. For the input node,

$$g_{m1}v_i - v_1[sC_I + (1/R_I)] + (v_o - v_1)sC_c = 0 \qquad \textbf{(1)}$$

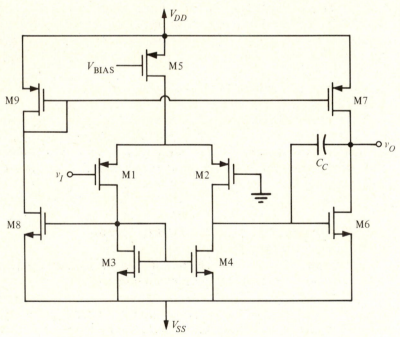

Figure 9.1-1   A current, push-pull, CMOS op amp.

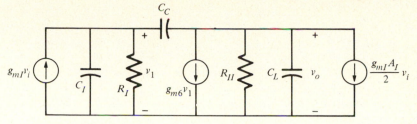

Figure 9.1-2   Small-signal model for Fig. 9.1-1.

For the output node,

$$(v_o - v_1)sC_c + g_{mII}v_1 + v_o[sC_L + (1/R_{II})] + (g_{mI}A_I/2)v_i = 0 \qquad (2)$$

where $g_{m1} = g_{m2} = g_{mI}$, $g_{m6} = g_{mII}$, $R_I = (g_{ds2} + g_{ds4})^{-1}$, and $R_{II} = (g_{ds6} + g_{ds7})^{-1}$. It can be shown that the transfer function v_o/v_i can be written as

$$\frac{v_o}{v_i} = \frac{A_{vo}[(s/z) - 1]}{(s^2/p_1 p_2) - s(1/p_1) - s(1/p_2) + 1} \qquad (3)$$

where

$$A_{vo} \simeq -g_{mI}g_{mII}R_I R_{II} \qquad (4)$$
$$P_1 \simeq -(R_I R_{II}C_c g_{mII})^{-1} \qquad (5)$$
$$P_2 \simeq -(g_{mII}/C_L) \qquad (6)$$

and

$$z \simeq [(A_I/2) + g_{mII}R_1]/\{R_I[C_c - 0.5A_I(C_c + C_I)]\} \qquad (7)$$
$$\simeq -g_{mII}/[(0.5A_I - 1)C_c]$$

The dc gain and poles are unaffected by the extra current path, but the zero now becomes a function of the current gain around the M7–M9 loop. When $A_I = 2$, the zero appears in the left-half plane at infinity, and as A_I increases it starts to move towards the origin. A_I is frequency dependent and thus the zero is difficult to determine analytically. A more detailed computer analysis suggests that the added transistors have an effect on the zero through their parasitic capacitances. At any rate, the extra current path can help stabilize the op amp as well as give more drive flexibility.

The straightforward approach of adding one of the output stages discussed in Sec. 6.4 to the unbuffered op amps will result in an op amp capable of driving not only a large capacitor but also a low resistance. However, we must examine the stability of the new configuration in order to accom-

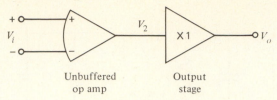

Unbuffered   Output
op amp   stage

Figure 9.1-3   Block diagram representation of a buffered op amp.

plish this objective. Fig. 9.1-3 shows the resulting configuration in block diagram form. The first amplifier is the unbuffered op amp of Chapter 8. The output stage is a unity-gain stage similar to those discussed in Sec. 6.4. The problem at hand is how to compensate the three-stage op amp (assuming that the unbuffered op amp has two stages). With no compensation, the open-loop voltage gain of the op amp is given as

$$\frac{V_o(s)}{V_i(s)} = \frac{-A_{vo}}{(\frac{s}{p_1'} - 1)(\frac{s}{p_2'} - 1)(\frac{s}{p_3'} - 1)} \tag{8}$$

where p_1' and p_2' are the uncompensated poles of the unbuffered op amp and p_3' is the pole due to the output stage. We shall assume that $|p_1'| < |p_2'| < |p_3'|$. It should also be noted that $|p_3'|$ will decrease as C_L is increased and increase as R_L is decreased. If Miller compensation is applied around the second and third stage, the new poles shown by squares in Fig. 9.1-4(a) result. This method has a potential problem in that the locus of the roots of p_1 and p_2 as C_c is increased bend toward the $j\omega$ axis and can lead to poor phase margin.

If the Miller compensation is applied around the second stage, then the closed-loop roots of Fig. 9.1-4(b) occur. The bending locus is no longer present. However, the output pole has not moved to the left on the negative real axis as was the case in Fig. 9.1-4(a). Which approach one chooses depends upon the anticipated output loading and the desired phase margin. The nulling technique can be used to control the zero as was done for the two-stage op amp.

A CMOS op amp capable of delivering 160 mW of power to a 100 ohm load while only dissipating 7 mW of quiescent power is shown in Fig. 9.1-5 [1]. This amplifier consists of three stages. The first stage is the differential amplifier of Sec. 6.2. The output driver consists of a crossover stage and an output stage. The two inverters consisting of M1, M3 and M2, M4 are the crossover stage. The purpose of this stage is to provide gain, compensation, and drive to the two output transistors, M5 and M6. The output stage is a transconductance amplifier designed to have a gain of unity when loaded with a specified load resistance.

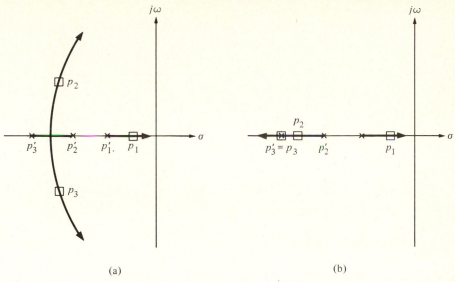

(a)   (b)

Figure 9.1-4   Root locus of the poles of the op amp with Miller compensation applied (a) around the second and third stage, or (b) around the second stage.

The dc transfer function of the two inverters is qualitatively shown in Fig. 9.1-6. Curves a and b result from inverters M1, M3 and M2, M4 and are the drive voltages to the output devices, M5 and M6, respectively. The crossover voltage is defined as

$$V_c = V_b - V_a \qquad (9)$$

where V_b and V_a are the input voltages to the inverters that result in cutting off M5 and M6, respectively. V_c must be greater than zero for low quiescent power dissipation but not too large to prevent unacceptable crossover distortion. In order to prevent a "glitch" in the output waveform during slewing, V_c should be carefully designed. An approach that yields satisfactory crossover distortion and ensures that $V_c \geq 0$ is to ratio the inverters so that the drives to the output stage are matched. Thus, V_c is designed to be small and positive by proper ratioing of M2 to M4 and M1 to M3. Worst-case variations of a typical process result in a maximum of V_c less than 110 mV and greater than zero.

The compensation of this amplifier uses the method depicted in Fig. 9.1-4(b). However, since the output stage is working in class B operation, two compensation capacitors are required, one for the M1, M3 inverter and one for the M2, M4 inverter. The advantage of this configuration is that the

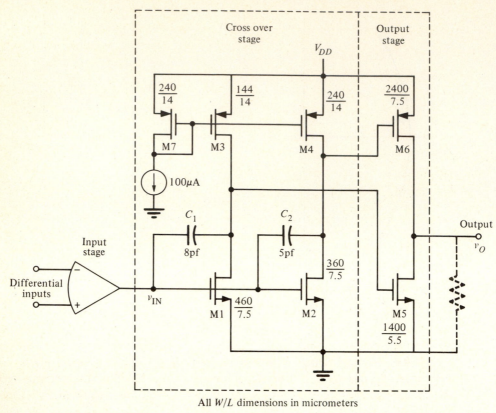

All W/L dimensions in micrometers

Figure 9.1-5   A CMOS op amp with low-impedance drive capability.

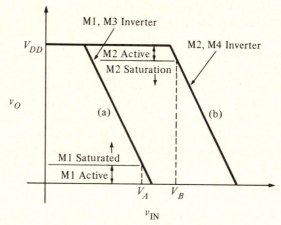

Figure 9.1-6   Idealized dc transfer characteristic of the crossover inverters.

462

output loading does not cause p_2 to move back to the origin. Of course, p_3 must be above p_2 in order for this compensation scheme to work successfully. For low values of R_L this requirement will be satisfied. As R_L is increased, the unity-gain frequency of the output stage becomes larger and p_3 will move on the negative real axis away from the origin. Thus the amplifier is conditionally stable with respect to the output resistive loading.

The CMOS op amp of Fig. 9.1-5 was fabricated in a standard CMOS process using minimum gate lengths of 5.5 μm and 7.5 μm for the NMOS and PMOS devices, respectively. The results are shown in Table 9.1-1. The amplifier dissipated only 7 mW of quiescent power and was capable of providing 160 mW of peak output power.

A third example of a buffered CMOS op amp using a MOS output stage is shown in Fig. 9.1-7 [2]. This example consists of the combination of an unbuffered op amp and a negative-feedback output stage. If M8 is replaced by a short and M8A through M13 are ignored, the output stage is essentially that of Fig. 6.4-7. The output of the unbuffered op amp drives the inverter (M16, M17) which in turn drives the negative terminal of the error amplifiers, which in turn drive the output devices, M6 and M6A. In most cases, the unbuffered op amp is simply a transconductance differential amplifier and M16 and M17 form the second stage inverter with the capacitor C_D serving as the Miller compensation of the first two stages of the op amp. The amplifier A1 and transistor M6 form the unity gain amplifier for the positive half of the output-voltage swing. Similarly, the amplifier A2 and transistor M6A form the unity-gain amplifier for the negative half of the output-voltage swing. Since the output amplifier is operating in a class AB mode, the operation of the negative half-swing circuit is an inverted mirror image of the positive half-swing circuit. Components performing similar functions in each circuit are designated by the additional subscript A for the negative half of the output swing.

Table 9.1-1

Performance Results for the CMOS Op Amp of Fig. 9.1-5.

Specification	Performance
Supply Voltage	$\pm$6V
Quiescent Power	7mW
Output Swing (100Ω Load)	8.1Vpp
Open-Loop Gain (100Ω Load)	78.1dB
Unity Gainbandwidth	260kHz
Voltage Spectral Noise Density at 1kHz	1.7μV/$\sqrt{\text{Hz}}$
PSRR at 1kHz	55dB
CMRR at 1kHz	42dB
Input Offset Voltage (Typical)	10mV

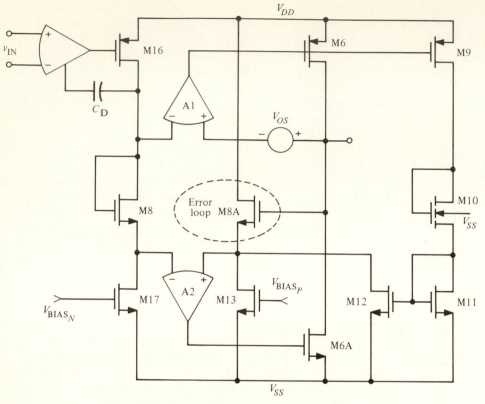

Figure 9.1-7   Block diagram of power amplifier.

Fig. 9.1-8 shows the circuit of the amplifier A1. It is seen that this is simply a two-stage op amp with the current-sink load of the second-stage inverter provided by the output stage of amplifier A2. The current in the output driver, M6, is typically controlled by the current mirror developed in the differential amplifier of the positive unity-gain amplifier and matches the current set in the negative-output driver device M6A by the negative unity-gain amplifier. However, if an offset occurs between amplifiers A1 and A2 (V_{os}), then the current balance between the output drivers M6 and M6A no longer exists and the current flow through these devices is uncontrolled. The feedback loop consisting of transistors M8A, M9, M10, M11, M12, and M13, stabilizes the current in M6 and M6A if an offset voltage V_{os} occurs between A1 and A2. The feedback loop operates as follows. Assume that an offset voltage exists as shown in Fig. 9.1-7. This offset voltage causes the output of A1 to fall, causing the current in M6 and M9 to increase. The increase of current through M9 is mirrored into a current increase through M8A. The increase of current in M8A causes an increase

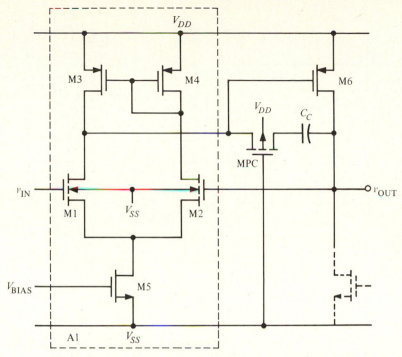

Figure 9.1-8 Positive output stage unity-gain amplifier.

in the gate-source voltage of M8A which balances out the offset voltage in the error loop consisting of M8, V_{OS}, and M8A. In this manner, the currents in M6 and M6A can be balanced.

Because the output-stage current feedback is not unity gain, some current variation in transistors M6 and M6A occurs. Offsets between amplifiers A1 and A2 can produce a 2:1 variation in dc current over temperature and process variations. This change in output current ΔI_o can be predicted by assuming that v_{OUT} is at ground and any offset between amplifiers A1 and A2 can be reflected as a difference between the inputs of A1. The result is given as

$$\Delta I_o = -g_{m6A}A_2 \left(V_{OS} - \left(\frac{2\beta_9\beta_{12}}{\beta_{8A}\beta_6\beta_{11}} \right)^{1/2} \left\{ \left[I_{B1} \left(\frac{\beta_6\beta_{11}}{\beta_9\beta_{12}} + \frac{\beta_5\beta_6}{2\beta_7\beta_3} \right) + \Delta I_o \right]^{1/2} \right.\right.$$
$$\left.\left. - \left[I_{B1} \left(\frac{\beta_6\beta_{11}}{\beta_9\beta_{12}} + \frac{\beta_5\beta_6}{2\beta_7\beta_3} \right) \right]^{1/2} \right\} \right) \tag{10}$$

where $I_{B1} = I_{17}$ and β is defined in Eq. (8) of Sec. 3.1.

Because transistor M6 can supply large amounts of current, care must be taken to ensure that this transistor is off during the negative half-cycle

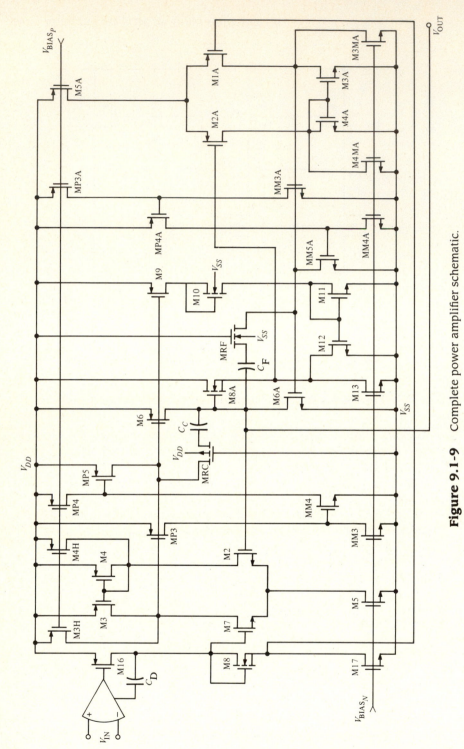

Figure 9.1-9 Complete power amplifier schematic.

of the output-voltage swing. For large negative swings, the drain of transistor M5 pulls to V_{SS}, turning off the current source that biases the error amplifier A1. As the bias is turned off, the gate of transistor M6 floats and tends to pull towards V_{SS}, turning on transistor M6. Fig. 9.1-9 shows the complete schematic of the output amplifier of Fig. 9.1-7. This circuit includes the means to ensure that M6 remains off for large negative voltage swings. As transistor M5 turns off, transistors M3H and M4H pull up the drains of transistors M3 and M4, respectively. As a result, transistor M6 is turned off and any floating nodes in the differential amplifier are eliminated. Positive-swing protection is provided for the negative half-cycle circuit by transistors M3HA and M4HA, which operate in a manner similar to that described above for the negative swing-protection circuit. The swing-protection circuit will degrade the step response of the power amplifier because the unity-gain amplifier not in operation is completely turned off.

Short-circuit protection is also included in the design of the amplifier. From Fig. 9.1-9, we see that transistor MP3 senses the output current through transistor M6, and in the event of excessively large output currents, the biased inverter formed by transistors MP3 and MN3 trips, thus enabling transistor MP5. Once transistor MP5 is enabled, the gate of transistor M6 is pulled up towards the positive supply V_{DD}. Therefore, the current in M6 is limited to approximately 60 mA.

The op amp of Fig. 9.1-9 is compensated by methods studied in Chapter 8. Each amplifier, A1 and A2, is individually compensated by the Miller method (C_C and C_F) including a nulling resistor (MRC and MRF). C_D is used to compensate the second stage as discussed in Sec. 8.2.

The total amplifier circuit is capable of driving 300 Ω and 1000 pF to ground. The unity-gain bandwidth is approximately 0.5 MHz and is limited by the 1000 pF load capacitance. The output stage has a bandwidth of approximately 1 MHz. The performance of this amplifier is summarized in Table 9.1-2. The component sizes of the devices in Fig. 9.1-9 are given in Table 9.1-3.

9.2   Op Amps Using MOS/BJT Output Stages

In the last section we considered op amps that employ MOS devices. The resulting output stages were successful in driving both low-resistance and high-capacitance loads with minimum quiescent power dissipation. However, the circuitry required to accomplish the objective became more complex. One method of avoiding the complexity is to make use of the substrate BJT discussed in Sec. 2.5. Because the collector is on ac ground, the only configuration possible is that of the emitter-follower. Fortunately, this configuration would be the one selected for an output stage. Unless a twin-well CMOS technology is available, only an npn or pnp is available. This

Table 9.1-2

Performance Characteristics of the Op Amp of Fig. 9.1-9.

Specification	Simulated Results	Measured Results
Power Dissipation	7.0mW	5.0mW
Open Loop Voltage Gain	82dB	83dB
Unity Gainbandwidth	500kHz	420kHz
Input Offset Voltage	0.4mV	1mV
$PSRR^+(0)/PSRR^-(0)$	85dB/104dB	86dB/106dB
$PSRR^+(1kHz)/PSRR^-(1kHz)$	81dB/98dB	80dB/98dB
THD ($V_{in} = 3.3V_{pp}$)		
$\quad R_L = 300\Omega$	0.03%	0.13%(1kHz)
$\quad C_L = 1000pF$	0.08%	0.32%(4kHz)
THD ($V_{in} = 4.0V_{pp}$)		
$\quad R_L = 15K\Omega$	0.05%	0.13%(1kHz)
$\quad C_L = 200pF$	0.16%	0.20%(4kHz)
Settling Time (0.1%)	$3\mu s$	$<5\mu s$
Slew Rate	$0.8V/\mu s$	$0.6V/\mu s$
1/f Noise at 1kHz	—	$130nV/\sqrt{Hz}$
Broadband Noise	—	$49nV/\sqrt{Hz}$

Table 9.1-3

Component Sizes for the Op Amp of Fig. 9.1-9.

Transistor/Capacitor	$\mu m/\mu m$ or pF	Transistor/Capacitor	$\mu m/\mu m$ or pF
M16	184/9	M8A	481/6
M17	66/12	M13	66/12
M8	184/6	M9	27/6
M1, M2	36/10	M10	6/22
M3, M4	194/6	M11	14/6
M3H, M4H	16/12	M12	140/6
M5	145/12	MP3	8/6
M6	2647/6	MN3	244/6
MRC	48/10	MP4	43/12
C_C	11.0	MN4	12/6
M1A, M2A	88/12	MP5	6/6
M3A, M4A	196/6	MN3A	6/6
M3HA, M4HA	10/12	MP3A	337/6
M5A	229/12	MN4A	24/12
M6A	2420/6	MP4A	20/12
C_F	10.0	MN5A	6/6

means that the output stage will be a combination of BJT and MOS devices. This circumstance precludes the possibility of class B operation because the circuit for the positive and negative parts of the swing would be hard to match, thus resulting in distortion in the output. As a result, most MOS/ BJT output stages are class A or AB.

Fig. 9.2-1 shows a two-stage unbuffered op amp with an MOS/BJT output stage consisting of emitter-follower Q1 and M9. Except for these devices, Fig. 9.2-1 is a two-stage op amp. It is of interest to note that the Miller compensating capacitor is achieved by using the gate-oxide capacitance of a device (M10) with both drain and source connected together. This approach assumes that the dc potentials of the gate-source/drain will keep the channel enhanced. The output stage is biased in a class AB mode because of the reasons discussed above.

We shall first examine the large-signal performance of this circuit. In Sec. 6.1 it was shown that the negative signal swing can reach the negative power supply. The positive swing limit turns out to be more restricted. If we assume that the load resistance is infinite, then the approximate maximum output voltage is

$$v_{OUT}(max) \cong V_{DD} - V_{BE1} = V_{DD} - V_t \ln(I_{C1}/I_{s1}) \tag{1}$$

As the bias current in Q1 is reduced, the maximum output-voltage swing approaches the positive power supply. However, the load resistance is normally small and therefore a more representative expression for the maxi-

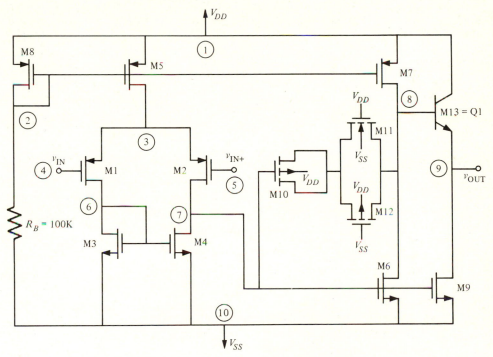

Figure 9.2-1   Low-power, buffered CMOS op amp circuit.

mum output-voltage swing is

$$V_{OUT}(\text{max}) = V_{DD} - V_t \ln \left(\frac{I_{E1}}{I_{s1}} \right) - \left[\frac{2I_7}{K_p' S_7} \right]^{1/2} \qquad (2)$$

To illustrate Eq. (2), assume that Q1 is providing an output current of 1 mA and that the forward current gain β_F is 100 and $I_{s1} = 1$ fA. Also let $K_p' = 13.5$ μA/V^2, $S_7 = 2$, and the bias current in M7 be 105 μA. Putting these values in Eq. (2) gives $v_{OUT}(\text{max}) = 5 - 0.718 - 2.789 = 1.493$.

Most of the loss in signal swing is across the drain-source of M7. If M7 is permitted to enter the active region, then v_{OUT} (max) can be increased until $I_7 \approx I_{E1}/\beta_{F1}$. However, another significant limitation comes from the fact that the BJT requires input current. As M7 pulls the base of Q1 toward V_{DD}, it must continue to source the base current necessary to provide the output current. As the voltage decreases across M7 the current also decreases, causing the current drive for the BJT to be reduced at the peak of the swing. Thus, the positive output swing will be reduced for increased currents. Another characteristic of the BJT that will worsen these results is the fact that β_F reduces for higher currents, causing a demand for more base current in order to continue providing the same output current.

The output slew rate is another aspect of large-signal performance that is of interest. The primary difference between the slew rate and the maximum output swing is that slew rate is typically determined for values of v_{OUT} near the quiescent value. Obviously, the slew rate will worsen as the power supplies are approached because of reduced current drive capability. The positive slew rate of Fig. 9.2-1 can be expressed as

$$SR^+ = \frac{I_{OUT}^+}{C_L} = \frac{(1 + \beta_{F1})I_7}{C_L} \qquad (3)$$

where it has been assumed that the slew rate is due to C_L and not the internal capacitances of the op amp. Assuming that $\beta_{F1} = 100$, $C_L = 1000$ pF, and the bias current in M7 is 105 μA, the positive slew rate becomes 10.6 V/μs. The negative slew rate is determined by assuming that the gate of M9 can be taken to V_{DD}. Therefore, the negative slew rate is

$$SR^- = \frac{\beta_9(V_{DD} + |V_{SS}| - V_{T0})^2}{2C_L} \qquad (4)$$

Assuming a $(W/L)_9$ of 60, $K_N' = 36.72$ μA/V^2, ± 5 V power supplies, $V_{T0} = 0.8$ V, and a load capacitance of 1000 pF gives a negative slew rate of 93.24 V/μs. The reason for the larger slew rate for the negative-going output is that the current is not limited as it is by I_7 for the positive-going output.

The small-signal characteristics of Fig. 9.2-1 are developed using the model shown in Fig. 9.2-2. The controlled source representing the gain of the BJT has been simplified using the techniques in Appendix A. The result is a three-node circuit containing the Miller compensating capacitor C_c and the input capacitance C_π for the BJT. The approximate nodal equations describing this model are given as follows.

$$g_{mI}V_{in} = (G_I + sC_c)V_7 - sC_cV_8 + 0V_9 \tag{5}$$

$$0 = (g_{mII} - sC_c)V_7 + (G_{II} + g_\pi + sC_c + sC_\pi)V_8 \tag{6}$$
$$- (g_\pi + sC_\pi)V_9$$

$$0 \cong g_{m9}V_7 - (g_{m13} + sC_\pi)V_8 + (g_{m13} + sC_\pi)V_9 \tag{7}$$

In the last equation, g_{m13} has been assumed to be larger than g_π and G_3. Using the method illustrated in Sec. 6.3, the approximate voltage-transfer function can be found as

$$\frac{V_9(s)}{V_{in}(s)} = A_{vo}\frac{(\dfrac{s}{z_1}-1)(\dfrac{s}{z_2}-1)}{(\dfrac{s}{p_1}-1)(\dfrac{s}{p_2}-1)} \tag{9}$$

where

$$A_{vo} = \frac{-g_{mI}g_{mII}}{G_IG_{II}} \tag{10}$$

$$z_1 = \frac{1}{\dfrac{C_c}{g_{mII}} - \dfrac{C_\pi}{g_{m13}}\left[1 + \dfrac{g_{m9}}{g_{mII}}\right]} \tag{11}$$

$$z_2 = -\frac{g_{m13}}{C_\pi} + \frac{g_\pi}{C_c}\left[1 + \frac{g_{m9}}{g_{mII}}\right] \tag{12}$$

$$p_1 = \frac{-G_IG_{II}}{g_{mII}C_c}\left[\cfrac{1}{1 + \dfrac{g_{m9}}{\beta_Fg_{mII}} + \dfrac{C_\pi}{C_c}\left(\dfrac{G_IG_{II}}{g_{m13}g_{mII}}\right)}\right] \tag{13}$$

$$p_2 \cong \frac{-g_{m13}g_{mII}}{(g_{mII} + g_{m9})C_\pi} \tag{14}$$

The influence of the BJT output stage can be seen from the above results. The low-frequency, differential gain is unchanged. The presence of the BJT has caused the RHP zero to move further away from the origin, which should help stability. The second zero should normally be an LHP zero and

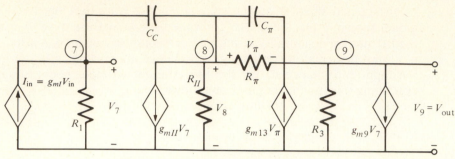

Figure 9.2-2   AC model for Fig. 9.2-1.

can also be used to help stability. The dominant pole p_1 is essentially that of the simple unbuffered two-stage op amp if the quantity in the brackets of Eq. (13) approaches zero, which should normally be the case. The second pole will be above the unity-gain bandwidth. However, the shunt capacitance at the base and collector of Q1 were neglected in the above analysis so that we may expect p_2 to be modified in a more complete analysis.

The primary objective in using the BJT in the output stage was to reduce the small-signal output resistance. This resistance can be calculated from Fig. 9.2-2 after V_{in} is set to zero. The results are

$$r_{out} = \frac{r_{\pi} + R_{II}}{1 + \beta_F} \tag{15}$$

The success of the BJT in the output stage depends on keeping R_{II} small. If the current in M7 and M6 is 105 μA and 104 μA, respectively, the value of R_{II} is approximately 474 kΩ which gives approximately 4.86 Kilohms if β_F is 100. This value is not as low as one might like when driving low resistance loads such as 100 ohms. One method of lowering the output resistance is to use the Darlington configuration which will be illustrated in the next section.

Another consideration the designer must remember to take into account is the need for a dc base current. We noted in the above example that I_{D9} was 105 μA while I_{D6} was 104 μA. The difference is used to provide the base current of Q1 and to define, through β_F, the value of the output biasing current. Unfortunately, the variations in process and the resulting parameters can cause a wide variation in the bias current in the output stage. The current in the output stage I_{C1} can be expressed as

$$I_{C1} \cong \frac{(S_9/S_6)}{1 + (S_9/S_6 \beta_F)} I_{D7} \tag{16}$$

which is 155 μA.

The buffered op amp of Fig. 9.2-1 can be designed by using the design approach for the two-stage op amp to design all devices except M9 and Q1. M9 should be designed to sink the current required for the maximum negative slew rate given by Eq. (4). The only designable parameter of the BJT is the emitter area, which has little effect on the performance of the transistor. Thus one usually chooses a size large enough to dissipate the power but small enough to avoid large device capacitances.

The design of Fig. 9.2-1 resulted in the W and L values of Table 9.2-1. In order to verify the design, the simulation used in Table 9.2-2 was made. The results are shown in Figs. 9.2-3 through 9.2-7. Fig. 9.2-3 shows the open-loop voltage-transfer function. The influence of R_L on performance is very obvious from this figure. On the positive part of the output, the BJT is able to maintain a low output resistance, resulting in high gain. However, on the negative part of the output, the high output resistance of M9 results in a significant loss of gain. Under 1 Kilohm load conditions, the harmonic distortion in this op amp will be significant.

The phase margin of the amplifier with a 100 pF load is seen to be a little over 60° from the open-loop frequency response shown in Fig. 9.2-4. The open-loop common-mode frequency response is shown in Fig. 9.2-5. Figs. 9.2-6 and 9.2-7 give the transient response. We note that positive slew rate is less than the negative slew rate which confirms the anticipated performance. The small-signal transient response of Fig. 9.2-7 indicates a phase margin worse than the 60° of Fig. 9.2-4. Table 9.2-3 summarizes the performance of this op amp. The voltage gain is 81.6 dB and the output resistance is 4.5 Kilohms which agrees with the value calculated previously for similar device currents.

Table 9.2-1

Device Sizes and Currents for the Design of Fig. 9.2-1.

Device	Type	W (μm)	L (μm)	Design Currents
M1	PMOS	60	30	35μA
M2	PMOS	60	30	35μA
M3	NMOS	210	15	35μA
M4	NMOS	210	15	35μA
M5	PMOS	12	6	70μA
M6	NMOS	240	6	104μA
M7	PMOS	18	6	105μA
M8	PMOS	12	6	70μA
M9	NMOS	360	6	155μA
Q1	NPN	Emitter area is 40μm $\times$ 40μm		155μA
M10	NMOS	96	96	0
M11	NMOS	6	6	0
M12	NMOS	6	6	0

Table 9.2-2

Computer Listing for the Simulation of Fig. 9.2-1.

P-CHANNEL INPUT PAIR BUFFERED OPERATIONAL AMPLIFIER
.MODEL NMOS NMOS(LEVEL = 2 KP = 36.72U LAMBDA = 0.01 VTO = 0.8
+ GAMMA = 1.4 CGSO = 4.32E-16 CGDO = 4.32E-16 CGBO = 6.48E-15 RSH = 25
+ CJ = 5.4E-4 CJSW = 5.4E-10 TOX = 650E-10 LD = 0.4U UO = 680 AF = 1.2 KF = 1E-26)
.MODEL PMOS PMOS(LEVEL = 2 VTO = −0.8 KP = 13.5U GAMMA = 0.4 LAMBDA = 0.01
+ CGSO = 4.32E-16 CGDO = 4.32E-16 CGBO = 6.48E-15 RSH = 90 CJ = 1.22E-4
+ CJSW = 1.22E-10 TOX = 650E-10 LD = 0.5U UO = 250 AF = 1.2 KF = 1E-26)
.MODEL NPN NPN IS = 1E-15 BF = 100 NF = 1 VAF = 200 IKF = .01 ISE = 1E-13 NE = 2
+ BR = .1 NR = 1 VAR = 200 IKR = .01 ISC = 1E-13 NC = 1.5 RB = 100 IRB = .1 RBM = 10
+ RE = 1 RC = 10 CJE = 2P VJE = .6 MJE = .33 TF = .1N CJC = 2P VJC = .5 MJC = .5 TR = 10
VDD 1 0 5
VSS 10 0 −5
*VIN 14 0 PULSE(1 −1 0 0 0 5US 10US) (All lines beginning with * are ignored by
SPICE)
*VPOS 5 0 PULSE(1 −1 0 0 0 1US 2US)
VPOS 5 0
*VNEG 4 0 AC 1E-4
*VNEG 4 0 DC 1
*DRAIN AREA = SOURCE AREA = 15U*W , PD = PS = 30U + 2W
M1 6 4 3 1 PMOS W = 60U L = 30U AD = 900P AS = 900P PD = 150U PS = 150U
M2 7 5 3 1 PMOS W = 60U L = 30U AD = 900P AS = 900P PD = 150U PS = 150U
M3 6 6 10 10 NMOS W = 210U L = 15U AS = 787.5P AD = 787.5P
+ PD = 135U PS = 135U
M4 7 6 10 10 NMOS W = 210U L = 15U AS = 787.5P AD = 787.5P
+ PD = 135U PS = 135U
M6 8 7 10 10 NMOS W = 240U L = 6U AD = 900P AS = 900P PD = 150U PS = 150U
M8 2 2 1 1 PMOS W = 12U L = 6U AD = 180P AS = 180P PD = 54U PS = 54U
M5 3 2 1 1 PMOS W = 12U L = 6U AD = 180P AS = 180P PD = 54U PS = 54U
M7 8 2 1 1 PMOS W = 18U L = 6U AD = 270P AS = 270P PD = 66U PS = 66U
M9 9 7 10 10 NMOS W = 360U L = 6U AD = 3600P AS = 3600P PD = 510U PS = 510U
Q1 1 8 9 NPN AREA = 10
M10 11 7 11 11 PMOS W = 96U L = 96U AS = 810P AD = 810P PS = 138U PD = 138U
M11 8 1 11 11 NMOS W = 6U L = 6U AS = 90P AD = 90P PS = 42U PD = 42U
M12 8 10 11 1 PMOS W = 6U L = 6U AS = 90P AD = 90P PS = 42U PD = 42U
CPAR 11 0 4P
CL 9 0 100P
RL 9 0 1K
RBIAS 10 2 100K
*RIN 4 14 1E6
*RF 4 9 1E7
.TF V(9) VPOS
*.SENS V(9)
*.AC DEC 25 10 100MEG
*.PRINT AC VDB(9) VP(9)
*.NOISE V(9) VNEG
*.PRINT NOISE INOISE ONOISE

474

```
*.DC VPOS −5 5 .05
*.PRINT DC V(9)
*.TRAN 1US 100US
.OPTION LIMPTS=405
*.PRINT TRAN V(9)
.END
```

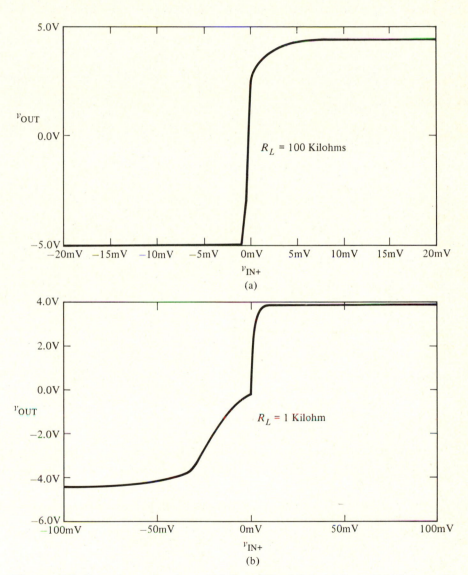

Figure 9.2-3 DC transfer characteristics of Fig. 9.2-1. (a) R_L = 100 kilohms. (b) R_L = 1 kilohm.

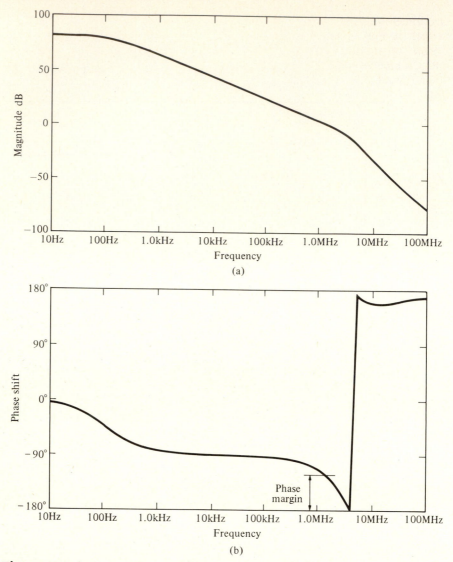

Figure 9.2-4   Open-loop frequency response of Fig. 9.2-1. $C_L = 100$ pF and $R_L = 100$ kilohms. (a) Magnitude response. (b) Phase response.

This section has introduced output stages that use the substrate BJT which is available from the CMOS process. The advantages of a BJT in the output stage depend upon the particular design. The primary advantage is a low resistive output impedance (if R_{ll} is small). The disadvantages are limited output swing and the requirement for a base current causing a slew-rate limitation.

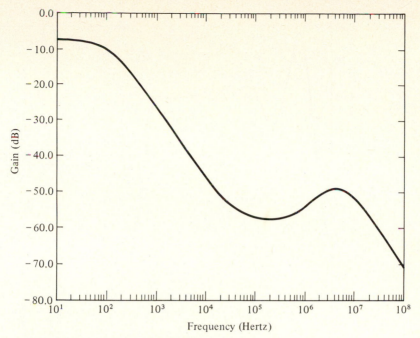

Figure 9.2-5   Open-loop common-mode frequency response (no load).

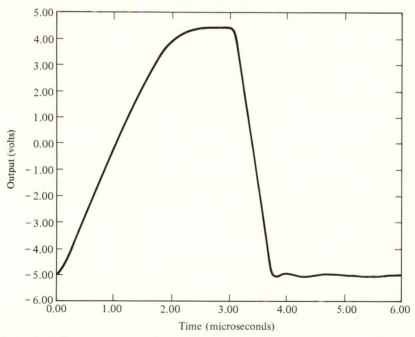

Figure 9.2-6   Transient response of the unity-gain configuration for a −5V to +5V input pulse of 3 μs width (100 pF load).

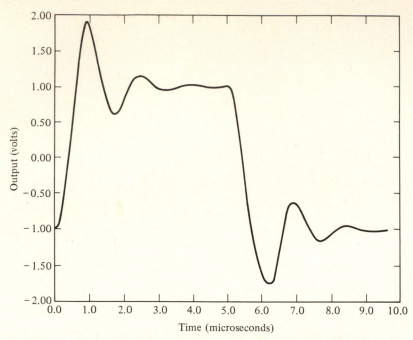

Figure 9.2-7   Transient response of the unity-gain configuration for a −1V to +1V input pulse of 5 μs width (C_L = 100 pF).

Table 9.2-3

Summary of the Simulated Performance of Fig. 9.2-1.

Parameter	Value	Units
Open Loop Low Frequency Gain	12,000 (81.6dB)	V/V
Power Consumption	4	mW
Output Resistance	4500	Ohms
Input Common Mode Voltage Range	−5 to 4.15	V
Output Voltage Swing		
No Load	−5 to 4.4	V
1 Kilohm Load	−3.8 to 4	V
Common Mode Rejection Ratio		
0V input	88.9	dB
1V input	63	dB
Power Supply Rejection Ratio		
Negative	97	dB
Positive	89.5	dB
Unity Gainbandwidth	1.3	MHz
Phase Margin		
No Load	72	Degrees
100pF Load	61	Degrees
Slew Rate	6/−15	V/μs
1% Settling Time with 100pF Load	5	μs

9.3 *High–Speed/Frequency CMOS Op Amps*

The op amps to be discussed in this section will emphasize speed and gain-bandwidth. By speed we mean a minimum time to respond to a step input. This will require a high slew rate for large-signal transitions and a good phase margin to minimize the settling time. Op amps in the category of this chapter should have slew rates in excess of 100 V/μs and unity-gain bandwidths in excess of 20 MHz for large capacitive and low resistive loads (i.e., 100 pF and 1 KΩ). The two approaches presented in this chapter will use the BJT in the Darlington configuration and the push-pull source follower. Obviously, the signal swing will be sacrificed in trying to meet the performance requirements of this section.

Fig. 9.3-1 shows a CMOS op amp designed to work with high-frequency active RC filters [3]. This circuit is a good illustration of the principles applicable to the design of high–speed/frequency op amps. We note the approach uses the two-stage op amp as the basic gain stage. The output

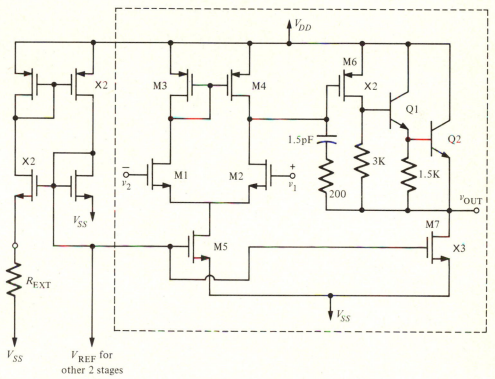

Figure 9.3-1   A high-performance CMOS op amp designed for active filter applications.

stage is a class A stage that uses a Darlington configuration of two substrate npn BJTs and an NMOS current sink. The Darlington configuration is used to increase the β_F of the output BJT so that the source current is not limited by the BJT current gain. The compensation is achieved by using Miller compensation with nulling around *both* the second and third stage.

Using the Darlington configuration will reduce the positive output-voltage swing to at least $2V_{BE}$ below V_{DD}. As was seen from the previous section, the driver transistor M6 will require a nonzero voltage across it to provide the required base current drive for Q1.

The output resistance of this op amp can be found from Fig. 9.3-2 (a). The small-signal model of this circuit is given in Fig. 9.3-2 (b). In this model, $r'_{\pi2}$ is the parallel combination of $r_{\pi2}$ and the 1.5 Kilohm resistor and R is the 3 Kilohm resistor. Writing nodal equations and solving for the out-

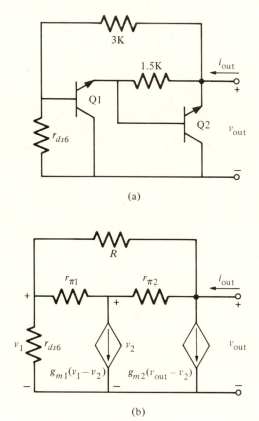

Figure 9.3-2   (a) Circuit for calculating the output resistance of Fig. 9.3-1. (b) Small-signal model of (a).

put resistance gives

$$r_{out} = \frac{(g_{ds6} + G)(g_{\pi 1} + g'_{\pi 2} + g_{m2}) + g_{\pi 1} g'_{\pi 2}}{g_{ds6} G(g_{\pi 1} + g'_{\pi 2} + g_{m1}) + g_{ds6}(g'_{\pi 2} + g_{m2})(g_{\pi 1} + g_{m2})} \cong \frac{r_{ds6}}{g_{m2} R} \quad (1)$$

While the presence of the 3 KΩ resistor is to stabilize the biasing of Q1, it causes the output resistance not to be as small as possible with the Darlington configuration. If r_{ds6} is 42 Kilohms and the dc current in Q2 is 2 mA, then the output resistance at low frequency is about 180 ohms. Fig. 9.3-3 shows a possible modification which will decrease the output resistance to less than 10 ohms. The small-signal output resistance of Fig. 9.3-3 is

$$r_{out} = \frac{R_{ll}}{(1 + \beta_{F1})(1 + \beta_{F2})} \quad (2)$$

where R_{ll} is the resistance to ground seen looking out the base of Q2.

The op amp of Fig. 9.3-1 was fabricated using an ion-implanted CMOS process and a 5–8 ohm-cm, n$^-$ substrate with $\langle 100 \rangle$ orientation. The basic device size was W $= 50$ μm and L $= 6$ μm. Devices indicated with an X2 or X3 consist of 2 or 3 basic devices in parallel. The emitter area of the substrate npn BJTs was 25 μm by 25 μm. The model parameters for the op amp of Fig. 9.3-1 are shown in Table 9.3-1. Fig. 9.3-4 shows a microphotograph of the layout which shows three amplifiers all using the same bias

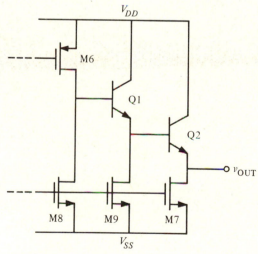

Figure 9.3-3 Modification to the output stage of Fig. 9.3-1 that will decrease the output resistance.

Table 9.3-1

Model Parameters for the Op Amp of Fig. 9.3-1.

MOS	$V_{TO}(V)$	$K'(\mu A/V^2)$	$\gamma(V^{1/2})$	$\lambda(V^{-1})$	$R_D(=R_S)(\Omega/\text{sq.})$	$C_{GB}(fF)$	$C_{BD}(fF)$
NMOS	1.0	48	1.63	0.072	100	175	537
PMOS	−2.0	14.4	.58	0.072	100	175	132

	β_F	$R_B(\Omega)$	$R_C(\Omega)$	$C_{JE}(pF)$	$C_{JC}(pF)$	$V_A(V)$
NPN	100	100	1000	.75	.1	50

Figure 9.3-4 Microphotograph of three op amps of the type shown in Fig. 9.3-1.

circuit. The resistor designated as R_{ext} in Fig. 9.3-1 was external to the circuit in order to allow programmability. The other resistances shown were polysilicon resistors.

Fig. 9.3-5 shows the performance of this op amp. Fig. 9.3-5 (a) shows the open-loop frequency response when R_{ext} is 22 Kilohms. It is seen that the unity-gain bandwidth is 34 MHz. Fig. 9.3-5(a) gives the open-loop phase response and indicates a phase margin of slightly less than 60°. The programmability influence of R_{ext} is given in Fig. 9.3-5 (b), where the magnitude of the op amp is shown for various values of R_{ext}. The common-mode frequency response is given in Fig. 9.3-5(c) and is seen to be near 0 dB. Finally, Fig. 9.3-5(d) shows the equivalent-input-noise voltage spectrum. The performance of the op amp in Fig. 9.3-1 is summarized in Table 9.3-2. The op amp was limited to capacitive loads of less than 200 pF for stability reasons.

Another approach to designing high–speed/frequency op amps is to use a push-pull source-follower output stage. Fig. 9.3-6 shows a low-output-resistance op amp capable of high-frequency and high slew-rate performance [4]. The output buffer consists of transistors M17 through M22. The unbuffered op amp is essentially a cascade of a differential-transconductance input stage and a current-amplifier second stage. The voltage gain is achieved by the high-resistance node at the junction of M7, M12, M18, and M19. The frequency response of the unbuffered amplifier is good because all nodes are low impedance except for the above node. C_L is used to compensate the amplifier by introducing a dominant pole. The output stage is used to buffer the load and provide a low output resistance. The small-signal output resistance is typically

$$r_{out} \cong \frac{1}{g_{m21} + g_{m22}} \tag{3}$$

Depending upon the size of the output devices and their bias current, the output resistance can be less than 1000 ohms.

The transistors M17 and M20 bias M18 and M19, which are complementary to M22 and M21, respectively. Ideally, the gate voltages of M18 and M22 and M19 and M21 compensate each other and therefore the output voltage is zero for a zero voltage input. The measured no-load low-frequency gain was about 65 dB for $I_{B1} = 50 \ \mu A$ and the measured unity-gain bandwidth was about 60 MHz for C_L of 1 pF. The measured output swing at 1 kHz was 1.3 V_{pp} across a 2000 ohm load using ± 3 V power supplies. Although the slew rate with a load capacitance was not measured, it should be large because the output devices can continue to turn on to provide the current necessary to enable the source to "follow" the gate. A large capacitive load will introduce poles into the output stage, eventually deteriorating the stability of the closed-loop configuration.

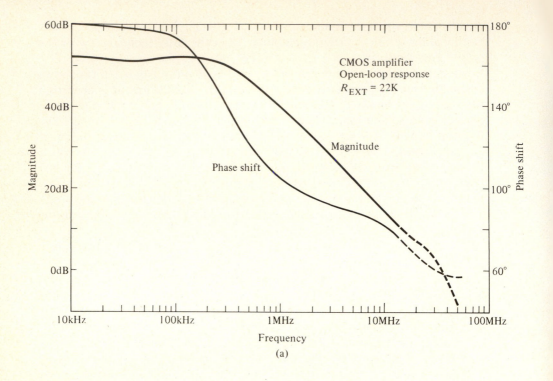

(a)

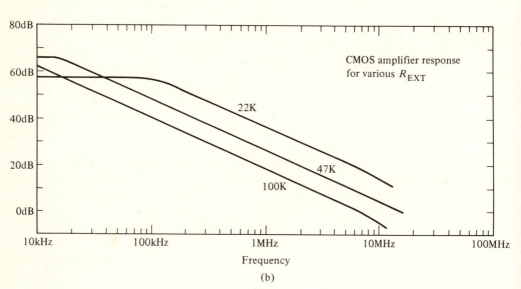

(b)

Figure 9.3-5 (a) Open-loop frequency response of Fig. 9.3-1. (b) Magnitude response of Fig. 9.3-1 for various values of R_{ext}. (c) Common-mode frequency response of Fig. 9.3-1. (d) Equivalent-input-noise voltage of Fig. 9.3-1.

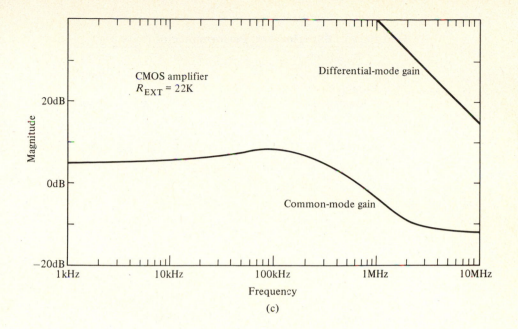

CMOS amplifier
$R_{EXT} = 22K$

Differential-mode gain

Common-mode gain

(c)

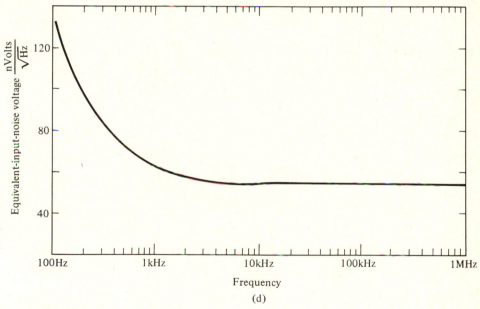

(d)

Figure 9.3-5   (c—d)

Table 9.3-2

Summary of Experimental Performance of Fig. 9.3-1.

Parameter	Value	Units
Power dissipation	32	mW
Open loop gain	450	V/V
Output resistance	100	Ohms
Unity gain bandwidth	34	MHz
Phase margin (C_L = 100pF)	45	Degrees
Settling time (C_L = 100pF)	200	ns
Slew rate (C_L = 100pF)	± 300	V/μs

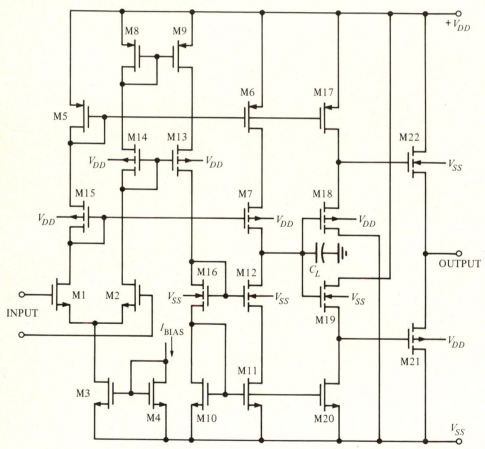

Figure 9.3-6 Low output resistance CMOS operational amplifier circuit suitable for resistive and capacitive loads.

9.4   *Low-Noise Op Amps*

Low-noise op amps are important in several respects. A large percentage of applications for analog CMOS circuits is in the area of telecommunications where the signal-to-noise ratio (S/N) is important. The lower the noise, the better the value of S/N for a given signal level. Another way of looking at this characteristic is from the viewpoint of dynamic range. The dynamic range of a circuit is the ratio of the largest-to-smallest signal that can be processed without distortion. The upper level is typically established by the power supplies and the large-signal swing limits. The lower level is established by the noise or the ripple injected by the power-supply. The noise required for a 100 dB dynamic range assuming a ± 5 V power supply and the ability to swing to within ± 1 V of power supply is found as follows. The maximum signal of 4 V peak is equivalent to 2.83 V rms. Dividing by 10^5 gives a voltage of 28.3 μV rms. Assuming a flat 10 kHz bandwidth gives a noise value of 283 nV/$\sqrt{\text{Hz}}$.

Two approaches are possible for developing an op amp having low-noise characteristics. The first method will be to follow the principles developed earlier (in Chapters 3 and 6), which showed the relationship of the noise to the geometry and process characteristics of the MOS devices. A second method of reducing the noise would be to use techniques such as chopper stabilization, which will be illustrated in the next section of this chapter.

Fig. 9.4-1 shows a CMOS op amp designed to have low noise [5]. This op amp is similar to the two-stage op amp of Sec. 8.3 except for the cascode devices, M8 and M9, which are used to improve the PSRR as discussed in Sec. 8.4. PMOS devices are selected for the input of the differential stage because of their better noise performance (see Sec. 3.2). Fig. 9.4-2 gives a noise model for the op amp of Fig. 9.4-1 ignoring the noise contributed by the dc current sources. Ignoring the current sources is justified because their gates are usually connected to a low impedance. Because of the large resistance to ground seen by the sources of M8 and M9, the contribution of the noise generators at the gates of M8 and M9 can be ignored when compared to the noise contributed by the noise generators at the gates of M1 and M2. The total output-noise voltage spectral density, $\overline{e}_{to}^2$ is found as

$$\overline{e}_{to}^2 = g_{m6}^2 R_{II}^2 [\overline{e}_{n6}^2 + R_I^2 (g_{m1}^2 \overline{e}_{n1}^2 + g_{m2}^2 \overline{e}_{n2}^2 + g_{m3}^2 \overline{e}_{n3}^2 + g_{m4}^2 \overline{e}_{n4}^2)] \qquad (1)$$

The equivalent input-voltage-noise spectral density can be found by dividing Eq. (1) by the differential gain of the op amp $g_{m1} R_I g_{m6} R_{II}$ to get

$$\overline{e}_{eq}^2 = \frac{\overline{e}_{n6}^2}{g_{m1}^2 R_I^2} + 2 e_{n1}^2 \left[1 + \left(\frac{g_{m3}}{g_{m1}} \right)^2 \left(\frac{\overline{e}_{n3}}{\overline{e}_{n1}} \right)^2 \right] \qquad (2)$$

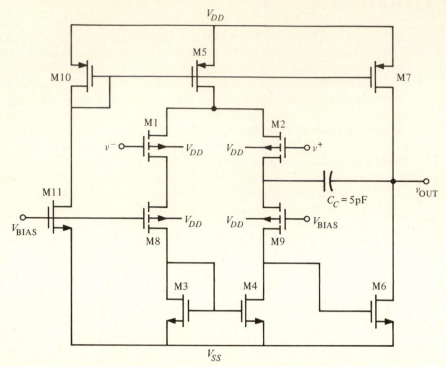

Figure 9.4-1   Low-noise CMOS op amp.

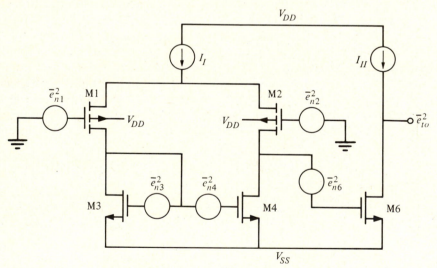

Figure 9.4-2   Noise model for Fig. 9.4-1.

It is seen from Eq. (2) that the noise contribution of the second stage is divided by the gain of the first stage and can therefore be neglected.

To minimize the noise of Fig. 9.4-1, Eq. (2) must be minimized.This is accomplished by making g_{m1} larger than g_{m3} so that the input noise is dominated by the input transistors. PMOS devices were chosen as the input transistors due to their inherent lower noise. The thermal noise contribution can be reduced by increasing the transconductance of the input transistors (by increasing the drain current and/or W/L). The 1/f noise contributed by the input devices can be reduced by increasing their W and L (with W/L constant). The *W/L* values for the op amp of Fig. 9.4-1 are shown in Table 9.4-1. The ratio of L_1 to L_3 is 1/7.5. Table 9.4-2 shows the experimental results when this op amp was fabricated using a CMOS technology. The amplifier has an input referred noise of 130 nV/$\sqrt{\text{Hz}}$ at 100 Hz and a thermal noise of 30 nV/$\sqrt{\text{Hz}}$ above 1 kHz. The dominant pole of this op

Table 9.4-1

Width and Length Ratios of the Op Amp of Fig. 9.4-1.

Device	Width (micrometers)	Length (micrometers)
M1	52	10
M2	52	10
M3	120	75
M4	120	75
M5	350	20
M6	60	20
M7	300	20
M8	120	10
M9	120	10
M10	60	20
M11	10	10

Table 9.4-2

Performance of the Op Amp in Fig. 9.4-1.

Specification	Value	Units
Die Area	$(0.35)^2$	mm
Power Dissipation	1.2	mW
Input device area	1300	μm^2
1/f Noise at 100 Hz	130	nV/$\sqrt{\text{Hz}}$
Broadband (thermal) noise	30	nV/$\sqrt{\text{Hz}}$
Noise corner frequency	1	kHz
Input offset voltage	10	mV
Unity Gain Bandwidth	1.1	MHz
Output Swing	± 4.3	V
Open Loop Gain	80	dB
Settling Time (1%, 4V step)	2	μs

amp is approximately 100 Hz. Assuming a flat noise of 130 nV/$\sqrt{\text{Hz}}$ over a bandwidth of 100 Hz, gives a noise voltage of 13 μV rms. Assuming a peak voltage swing of 4.3 V gives a dynamic range of 107 dB. To accurately convert the noise voltage spectral density to a noise voltage requires integrating the noise spectral density over the bandwidth of the amplifier.

We have seen that the corner frequency (the intersection of the 1/f and thermal noise) is lower for BJTs. Consequently, one would prefer to use BJT devices rather than MOS devices if noise at low frequencies (less than 1 kHz) is important. For example, the corner frequency of a typical BJT may be in the vicinity of 10 Hz compared to 1000 Hz for the typical MOS. Unfortunately, the CMOS process as presented in Chapter 2 does not appear to allow uncommitted BJTs to be fabricated. However, it was recently shown that a good bipolar junction transistor can be achieved which is compatible with any bulk CMOS technology [6]. The results yielded a low-noise differential two-stage op amp with an equivalent input-voltage-noise spectral density of 100 nV/$\sqrt{\text{Hz}}$ for frequencies above 1 Hz. Further details on noise in op amps can be found in the references [7,8].

9.5   *Chopper-Stabilized Op Amps*

One of the serious problems with MOS technology is the large values of offset voltage that occur for a CMOS op amp. Typically the input-offset voltage of a CMOS op amp might vary from $\pm$5 mV to $\pm$20 mV. This offset is due to systematic bias errors and to mismatches between "identical" MOS devices. The auto-zeroing techniques of Sec. 7.5 are probably the best method of reducing the input-offset voltage if the op amp is used in a discrete-time application. Another useful approach in reducing the input-offset voltage uses chopper stabilization. Recent efforts in chopper-stabilized op amps have produced very attractive performance in these areas [9,10].

The concept of chopper stabilization has been used for many years in designing precision dc amplifiers. The principle of chopper stabilization is illustrated in Fig. 9.5-1. Here, a two-stage amplifier and an input signal spectrum are shown. Inserted at the input and the output of the first stage are two multipliers which are controlled by a chopping square wave of amplitude +1 and −1. After the first multiplier, the signal is modulated and translated to the odd harmonic frequencies of the chopping square wave while the undesired signal v_u, which represents sources of noise or distortion, is unaffected. After the second multiplier, the signal is demodulated back to the original one and the undesired signal has been modulated as shown in Fig. 9.5-1(a). This chopping operation results in an equivalent input spectrum of the undesired signal as shown in Fig. 9.5-1(b). Here we see that the spectrum of the undesired signal has been shifted to the odd harmonic frequencies of the chopping square wave. Note that the

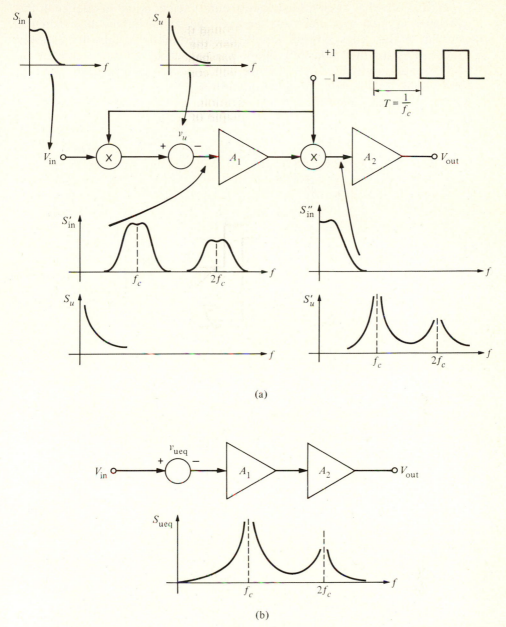

(a)

(b)

Figure 9.5-1 (a) Concept of chopper stabilization and the reduction of the in-band undesired signal v_u. (b) Equivalent undesired-signal input voltage for the circuit in (a).

491

spectrum of v_u has been folded back around the chopping frequency. If the chopper frequency is much higher than the signal bandwidth, then the amount of the undesired signal in the passband of the signal will be greatly reduced. Since the undesired signal will consist of $1/f$ noise and the dc offset of the amplifier, the influence of this source of undesired signal is mixed out of the desired range of operation.

Fig. 9.5-2(a) shows how the principle of chopper stabilization can be

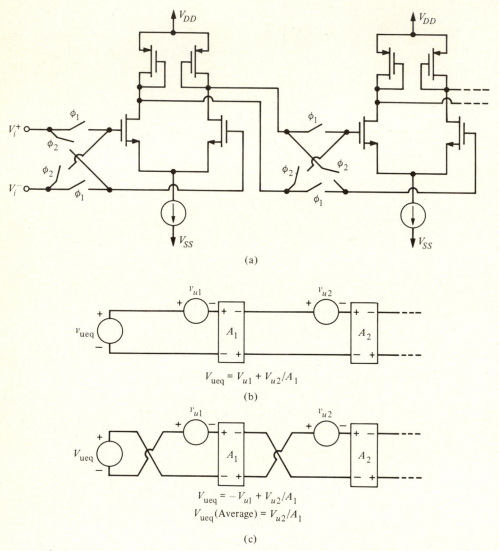

(a)

$$V_{ueq} = V_{u1} + V_{u2}/A_1$$

(b)

$$V_{ueq} = -V_{u1} + V_{u2}/A_1$$
$$V_{ueq}(\text{Average}) = V_{u2}/A_1$$

(c)

Figure 9.5-2   (a) Implementation of the chopper stabilized circuit of Fig. 9.1-1(a). (b) Circuit equivalent in the ϕ_1 phase. (c) Circuit equivalent in the ϕ_2 phase.

applied to a CMOS op amp. The multipliers are implemented by two cross-coupled switches which are controlled by two nonoverlapping clocks. Fig. 9.5-2(b) shows the operation of Fig. 9.5-2 (a). When ϕ_1 is on and ϕ_2 is off, the equivalent undesired-input signal is equal to the equivalent undesired-input signal of the first stage plus that of the second divided by the gain of the first stage. Thus the equivalent noise at the input during this phase is

$$V_{ueq.}(\phi_1) = V_{u1} + \frac{V_{u2}}{A_1} \tag{1}$$

In Fig. 9.5-2(c), ϕ_1 is off and ϕ_2 is on and the equivalent undesired-input signal is equal to the negative of the previous value, assuming that there has been no change in the undesired signal from the previous phase period. Again, the equivalent noise at the input during the ϕ_2 phase is

$$V_{ueq.}(\phi_2) = -V_{u1} + \frac{V_{u2}}{A_1} \tag{2}$$

The average of the equivalent input noise over the entire period can be expressed as

$$V_{ueq.}(aver.) = \frac{V_{ueq.}(\phi_1) + V_{ueq.}(\phi_2)}{2} = \frac{V_{u2}}{A_1} \tag{3}$$

Through the chopping effect, the equivalent undesired-input signal has been removed. If the voltage gain of the first stage is high enough, then the contribution of the undesired signal from the second stage can be neglected. Consequently, the equivalent input noise (particularly the $1/f$ noise) of the chopper-stabilized op amp will be greatly reduced. This method could be used to further reduce the noise of the op amp in the previous section.

Fig. 9.5-3(a) illustrates the experimental input noise of the op amp with the chopper at two different chopping frequencies and without the chopper. These results show that lower noise and offset voltage will be achieved using the chopper technique. It is seen that the higher the chopping frequency, the lower the noise. Fig. 9.5-3(b) shows the experimental and calculated results where the chopper amplifier is used in a filter.

The differential implementation has several advantages that are important to practical applications. First, the maximum signal swing is improved by a factor of two because of the differential configuration. Second, the differential configuration has the additional advantage that because the signal path is balanced, signals injected due to the power supply and the clock are greatly reduced. Also, if MOS devices are used differentially as active resistors, the nonlinearity effects are cancelled. Disadvan-

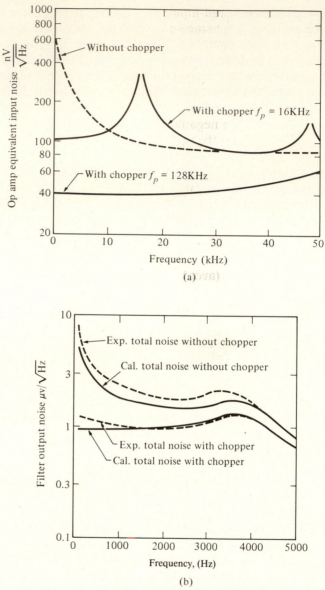

Figure 9.5-3 (a) Experimentally observed operational amplifier equivalent-input noise for various chopping frequencies. (b) Experimentally observed filter output noise with and without chopper, compared to theoretically predicted total noise.

tages of the differential configuration include more complex circuitry and more thermal noise from the increased number of switches.

It is of interest to examine more closely the implementation of the differential-in, differential-out op amp. One possible implementation of the differential-in, differential-out op amp is the differential amplifier stages shown in Fig. 9.5-2(a). There are several disadvantages of this configuration which include insufficient gain and level shifting. An alternate approach to implementing a differential-in, differential-out op amp is shown in Fig. 9.5-4 [11]. A combination of two single-ended output op amps can be used to implement a differential-in, differential-out op amp.

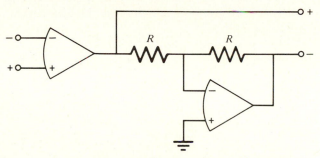

Figure 9.5-4 Implementation of a differential-in, differential-out op amp using two differential-in, single-ended out op amps.

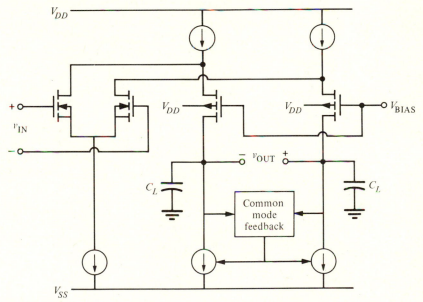

Figure 9.5-5 Conceptual implementation of a differential-in, differential-out op amp.

The concept of a direct implementation of a differential-in, differential-out is shown in Fig. 9.5-5 [12]. This implementation uses a folded cascode configuration to produce the differential output signal, v_{OUT}. The common-mode output signal of op amps with a differential output can be undefined and thus cause the op amp to drift from the region where high gain is achieved. To prevent this, it is usually necessary to provide some form of common-mode feedback in order to stabilize the common-mode output signal. In Fig. 9.5-5, the common-mode output signal is sampled and fed back to the current-sink loads of the folded cascode.

Fig. 9.5-6 shows the implementation of the conceptual realization of the differential-input, differential-output op amp of Fig. 9.5-5. The two PMOS transistors, MP1 and MP1A, provide the bias current to the amplifier.

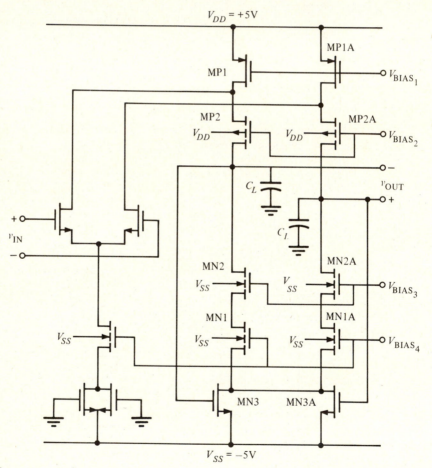

Figure 9.5-6 Complete schematic of the fully differential folded cascode amplifier.

In order to maintain a reasonably high output resistance, the channel length of these two devices is made longer than that of the cascode elements MP2 and MP2A. The high-impedance current-sink loads at the output of the amplifier are realized by the cascoded current sink MN1 through MN2A. The resulting output resistance can be greater than 1 MΩ. MN3 and MN3A are biased in their active region and sample the common-mode output signal and feed back a correcting common-mode signal into the source terminals of MN1 and MN1A. The cascode devices, MN1 through MN2A, amplify this compensating signal to restore the common-mode output voltage to its original level.

In order to obtain the maximum possible output swing, it is necessary to use the techniques presented in Sec. 5.3 for reducing v_{OUT}(sat) of the cascode current mirrors. The performance of the op amp of Fig. 9.5-6 using ±5 V power supplies is shown in Table 9.5-1. This op amp has been used to build filters at 260 kHz [12]. It would also be suitable for implementing the chopper-stabilization concepts discussed in this section.

9.6 *Micropower Op Amps*

In this section, we will turn our attention to amplifiers operating in the weak-inversion region of transistor operation. This class of amplifiers has gained increased attention in recent years [13–17] due to the need for such low-power devices in battery-powered, human-implantible, biomedical devices. The usefulness of amplifiers operating in this region is not only the very low power-supply currents that they draw but also their very low power-supply voltage operation. Our first task in this section is to develop the small-signal equations for transistors operating in weak inversion. These will be applied to understanding a few basic amplifier architectures that work well using micro-power techniques.

First, consider the equations that model the large-signal behavior of transistors operating at very low current densities. Assuming operation in

Table 9.5-1

Performance of the Op Amp of Fig. 9.5-6 with ±5 V Power Supplies.

Specification	Performance
Unity gain frequency (2pF load)	80 MHz
Settling time to 0.1% of final value (2pF load with 2.5V step)	40 ns
Power dissipation	10 mW
Die Area	200 mils²
Open loop gain	1500

the saturation region, subthreshold drain current was given in Eq. (7) of Sec. 3.5 as

$$i_D = \frac{W}{L} I_{DO} \exp\left(\frac{qv_{GS}}{nKT}\right) \tag{1}$$

From this equation, the transconductance can be easily derived as

$$g_m = \frac{I_D}{nkT/q} \tag{2}$$

This result is very interesting in that it shows a linear relationship between transconductance and drain current. Furthermore, the transconductance is independent of device geometry. These two characteristics set the sub-threshold region apart from the strong-inversion region, where the relationship between g_m and I_D is a square-law one and also a function of device geometry. In fact, the transconductance of the MOS device operating in the weak-inversion region looks very much like that of a bipolar transistor.

Equation (1) shows no dependence of drain current on drain-source voltage. If such were the case, then the device output impedance would be infinite (which is obviously not correct). The dependence of i_D upon v_{DS} can be approximated in the same way as it was for the simple strong-inversion model, where the drain current is modulated by the term $1 + \lambda v_{DS}$. Note that the weak-inversion λ may not necessarily be the same as that extracted from strong-inversion measurements. The expression for output resistance in weak inversion is

$$r_o \cong \frac{1}{\lambda I_D} \tag{3}$$

Like the transconductance, the output resistance is also independent of device aspect ratio, W/L (at constant current). Since λ is a function of channel length, it is the only control the designer has on the gain ($g_m r_o$) of a single stage operating in weak inversion.

With these things in mind, consider the simple op amp shown in Fig. 9.6-1. The dc gain of this amplifier is

$$A_{vo} = g_{m2} g_{m6} \left(\frac{r_{o2} r_{o4}}{r_{o2} + r_{o4}}\right) \left(\frac{r_{o6} r_{o7}}{r_{o6} + r_{o7}}\right) \tag{4}$$

In terms of device parameters this gain can be expressed as

$$A_{vo} = \frac{1}{n_2 n_6 (kT/q)^2 (\lambda_2 + \lambda_4)(\lambda_6 + \lambda_7)} \tag{5}$$

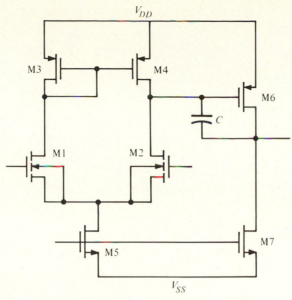

Figure 9.6-1   Op amp operating in weak inversion.

The gainbandwidth g_{m1}/C is

$$GB = \frac{I_{D1}}{(n_1 kT/q)C} \tag{6}$$

It is interesting to note that while the dc gain of the op amp is independent of I_D, the GB is not. The slew rate of this amplifier is

$$SR = \frac{I_{D5}}{C} = 2\frac{I_{D1}}{C} = 2GB\left(n_1 \frac{kT}{q}\right) = 2GBn_1 V_t \tag{7}$$

Example 9.6-1

Gain and GB Calculations for Subthreshold Op Amp.

Calculate the gain, GB, and SR of the op amp shown in Fig. 9.6-1. The current, I_{D5} is 200 nA. The device lengths are 10 μm. Values for n are 1.5 and 2.5 for p-channel and n-channel transistors respectively. The compensation capacitor is 5 pF. Use Table 3.1-2 as required. Assume that the temperature is 27 °C.

Using Eq. (5), the gain is

$$A_v = \frac{1}{(1.5)(2.5)(0.026)(2)(0.01 + 0.02)(0.01 + 0.02)}$$
$$A_v = 5698$$

The gain bandwidth is

$$GB = \frac{50 \times 10^{-9}}{2.5(0.026)(5 \times 10^{-12})} = 153846 \text{ rps} \cong 24.5 \text{ kHz}$$
$$SR = (2)(153846)(2.5)(0.026) = 0.02 \text{ V}/\mu s$$

Consider the circuit shown in Fig. 9.6-2 as an alternative to the amplifier just described. The differential gain of the first stage is

$$A_{vo} = \frac{g_{m2}}{g_{m4}} \tag{8}$$

In terms of device parameters, this gain can be expressed as

$$A_{vo} = \frac{I_{D2}n_4V_t}{I_{D4}n_2V_t} = \frac{I_{D2}n_4}{I_{D4}n_2} \cong 1 \tag{9}$$

Essentially no gain is available from the first stage. The second stage does, however, provide a reasonable amount of gain. The total gain of the circuit is best calculated assuming that the device pairs M3,M8; M4,M6; and M9,M7 act as current mirrors. Therefore

$$A_{vo} = \frac{g_{m1}(S_6/S_4)}{(g_{ds6} + g_{ds7})} = \frac{(S_6/S_4)}{(\lambda_6 + \lambda_7)n_1V_t} \tag{10}$$

At room temperature and for typical device lengths, gains on the order of 60 dB can be obtained. The gain bandwidth can be expressed

$$GB = \frac{g_{m1}}{C}\left(\frac{S_6}{S_4}\right) = \frac{g_{m1}b}{C} \tag{11}$$

where the coefficient b is the ratio of $(W/L)_6$ to $(W/L)_4$. If higher gains are required using this basic circuit, two things can be done. The first is to bleed off some of the current flowing in devices M3 and M4 so that their transconductances will be lower than the input devices thus giving the first stage

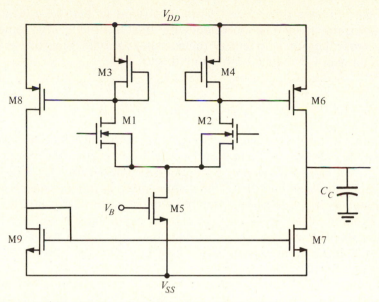

Figure 9.6-2 Push-pull output op amp for operation in weak inversion.

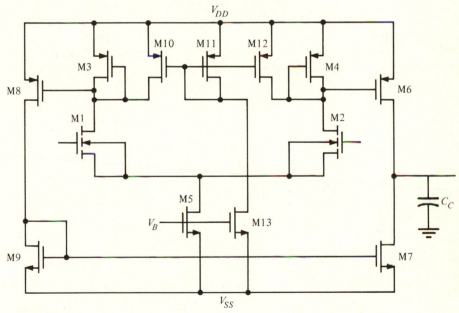

Figure 9.6-3 Op amp with gain improvement technique for operation in weak inversion.

a gain greater than one. This is illustrated in Fig. 9.6-3. Only a small improvement can be obtained with this technique. The other way to increase the gain is to replace the output stage with a cascode as illustrated in Fig. 9.6-4. The gain of this amplifier is

$$A_{vo} = \frac{1/n_n}{V_t^2(\lambda_p^2 n_p + \lambda_n^2 n_n)} \tag{12}$$

A simple calculation shows that this cascode amplifier can achieve gains greater than 100 dB and all of the gain is achieved at the output.

The disadvantage of the amplifiers presented thus far is their inability to provide large output currents while still maintaining micro-power consumption when quiescent. An interesting solution to this problem has been presented in the literature [13]. It is shown in Fig. 9.6-5. The basic idea in this architecture is to provide a boost of tail current (which is ultimately available at the output through mirroring) whenever there is a differential input voltage.

Assume that devices M11 through M14 are equal to M3 and M4 and that devices M15, M16, M17, M20, M21, and M22 are all equal. Further

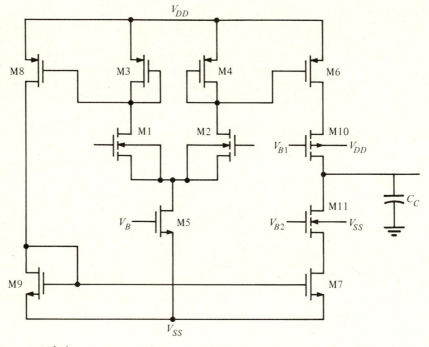

Figure 9.6-4   Op amp using cascode output for gain improvement in weak inversion.

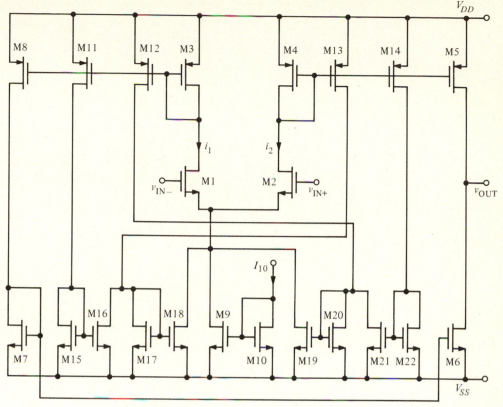

Figure 9.6-5 Dynamically-biased op amp.

assume that M17 through M20 are related in the following way

$$(W/L)_{18} = A(W/L)_{17} \tag{13}$$
$$(W/L)_{19} = A(W/L)_{20} \tag{14}$$

During quiescent conditions, the currents i_1 and i_2 are equal; therefore the current in M16 is equal to the current supplied by M13. As a result no current flows in M17, and thus none in M18. Similarly, no current flows in M19. Therefore no additional current is provided for the differential stage. When a differential input voltage is supplied that unbalances the currents i_1 and i_2, the tail current supplied to the differential stage is increased by the amount $A|i_1 - i_2|$. The current available at the output is

$$i_{OUT} = b \frac{I_{10} \exp(v_{IN}/n_n V_t)}{(A + 1) - (A - 1) \exp(v_{IN}/n_n V_t)} \tag{15}$$

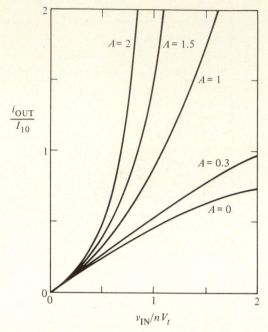

Figure 9.6-6 Parametric curves showing normalized output current versus input voltage for different values of A.

where b is the ratio of M5 to M4 (and also M8 to M3 where M7 equals M6). Figure 9.6-6 shows parametric curves of normalized output current as a function of input voltage for various values of A. This configuration can be very useful when very low quiescent current is required and considerable transient currents may be needed to drive capacitors in sampled-data filter applications.

When operating transistors in the subthreshold region, the gate-source voltage applied for proper circuit operation can easily be below the threshold voltage by 100 mV or more. The v_{DS} saturation voltage is typically below 100 mV also. As a result of these small voltage drops, the op amp operating in weak inversion can easily function with a 1.5 volt supply, provided that signal swings are kept small. Because of this low-voltage operation, circuits operating in weak inversion are very amenable to implantable, biomedical applications, where battery size and capacity are limited.

9.7 Dynamic Op Amp Circuits

A dynamic circuit is one that operates in the desired mode for only a finite period of time. The concept is similar to that used in dynamic memories which must occasionally be refreshed. In many circuit applications, the operation is not continuous in time but rather is clocked (discrete-time). A large number of circuits use a two-phase clock that is not overlapping, as shown in Fig. 9.7-1. During phase 1, designated by ϕ_1, certain switches in the circuit are closed, causing the circuit to work in one mode. During phase 2, designated by ϕ_2, other switches are closed, causing the circuit to work in a different mode. It is important that there is a finite period of time between phase 1 and phase 2 in which all switches are open. The period of the clocks is given as T and is the time required for a sequential operation of all clock phases. Although some circuits have more than two clock phases, we shall consider only the two-phase clock of Fig. 9.7-1. These considerations lead to the widely used switched-capacitor technique [18].

In most cases, the circuit is only required to function during one of the two clock phases and is reset or is idle during the other. As a result, it need not operate in a continuous-time mode. This leads to an interesting class of circuits, which we will consider in this section. Now, if a circuit needs to operate for only a limited amount of time, different design techniques can be used. For example, a capacitor can be charged to a desired voltage and act like a battery for a short period of time. This would simplify biasing, which in turn can lead to better circuit performance. For example, we shall show the design of a dynamic op amp in this chapter that has a gain-bandwidth that exceeds those op amps considered so far.

We shall begin by considering a dynamic resistor replacement called switched resistor [19]. In this technique a capacitor C_G is connected between the gate and source of a MOS transistor as shown in Fig. 9.7-2. If the voltage across the drain-source of the MOS transistor is sufficiently small, the MOS transistor simulates a linear resistance. The switches are closed during the phase indicated on the schematic. During the ϕ_2 phase,

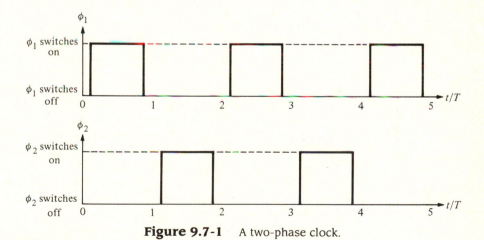

Figure 9.7-1   A two-phase clock.

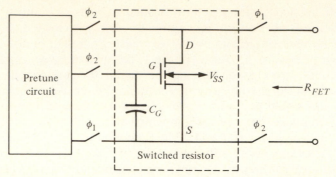

Figure 9.7-2 A switched resistor realization of a resistance R_{FET}.

C_G is charged to the voltage required to obtain the desired resistance by a pretuning circuit. During the ϕ_1 phase, the MOS transistor simulates a resistor R_{FET}. In order to make this voltage range as large as possible, the W/L ratio of the device is made very small, resulting in a long-channel device. Typical voltage ranges of ± 100 mV can give less than 1% harmonic distortion. The resistance of the switched resistor is given by Eq. (2) of Sec. 5.1.

It is necessary to recharge the capacitor after a finite amount of time has elapsed. If we model the gate-source resistance by a resistor R_G then the time constant of the voltage on the gate-source capacitor C_G is given as

$$\tau_G = R_G(C_G + C_{GS}) \tag{1}$$

where C_{GS} is the gate-source capacitance of the MOS transistor. The gate-source voltage $v_{GS}(t)$ can be written as

$$v_{GS}(t) = v_{GS}(0)e^{-(t/\tau_G)} \tag{2}$$

The drain-source resistance of the MOS transistor as a function of time $R_{FET}(t)$ can be expressed as

$$R_{FET}(t) = R_{FEQ}\left(\frac{v_{GS}(0) - V_T}{v_{GS}(0)e^{-(t/\tau_G)} - V_T}\right) \tag{3}$$

where R_{FEQ} is the desired resistance of the MOS transistor between the drain and source terminals when $v_{GS} = 0$ V. The time at which it is necessary to recharge C_G can be found from Eq. (3). If R_G is 10^8 ohms, $C_G = 10$ pF, and $C_{GS} = 0.1$ pF, then the time constant of Eq. (1) is approximately 1 millisecond. If $v_{GS}(0)/V_T = 3$, then a 1% increase in $R_{FET}(t)$ with respect to R_{FEQ} will occur after approximately 6.6 microseconds. This means that to keep the resistor $R_{FET}(t)$ within 1% of the desired value, it must be recharged after

6.6 microseconds. This circuit would be suitable for a 100 KHz clock with $T = 10$ microseconds and C_G would be charged after slightly less than 5 microseconds because the duty cycle of the clock is slightly less than 50%. One might feel that the value 10^8 ohms is too small for the resistance seen looking into the gate and source of the MOS transistor. This is true. However, the source/drain diffusions of the switch connected to the gate cause R_G to be essentially that of a back-biased source/drain diffusion-to-substrate junction.

One problem with the switched resistor of Fig. 9.7-2 is that the voltage across the source-drain terminals is restricted because the channel is nonlinear at larger source-drain voltages. A method of extending this voltage range is shown in Fig. 9.7-3 [20]. (The principle of this circuit was explained in Sec. 5.2.) The method is illustrated by assuming that M1 and M2 are matched and that the voltage v_{DS} is small enough so that M1 and M2 are in the nonsaturated region. The current i can be written as

$$i = \frac{\mu C_{ox} W}{L} \left[(v_S + V_C - V_T)(v_D - v_S) - \frac{(v_D - v_S)^2}{2} \right]$$
$$+ \frac{\mu C_{ox} W}{L} \left[(v_D + V_C - V_T)(v_D - v_S) - \frac{(v_D - v_S)^2}{2} \right] \quad (4)$$

Eq. (4) can be written as

$$i = \frac{2\mu C_{ox} W}{L} (V_C - V_T)(v_D - v_S) \quad (5)$$

Eq. (5) shows that the MOS realization of R_{FET} has been linearized. Fig. 9.7-4 shows a switched-resistor implementation of Fig. 9.7-3. It can be seen that two identical capacitors are charged to a voltage V_C and then con-

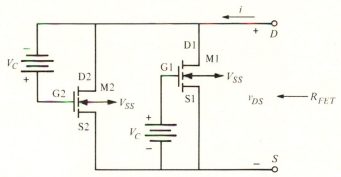

Figure 9.7-3   A continuous-time resistor realization with increased signal swing.

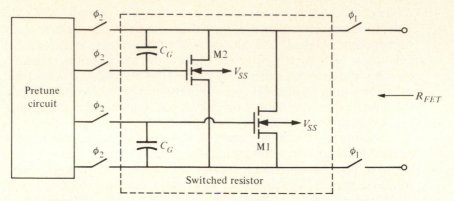

Figure 9.7-4   Switched-resistor implementation of Fig. 9.7-3.

nected appropriately to M1 and M2. Unfortunately, the signal swing is now so large that V_T will change appreciably because V_{SB} changes. This will introduce nonlinearities into the R_{FET} realization as V_{SB} becomes large. Fig. 9.7-5 shows the relative deviation from linearity for the continuous-time resistor realization of Fig. 9.7-3. In this figure $V_{SB} = 7.5$ V, $V_C - V_T = 7.5$ V, and the bulk threshold parameter (see Sec. 3.1) has the values of 0.2, 0.3, 0.4, and 0.5 $V^{1/2}$.

The continuous-time version of Fig. 9.7-4 requires at least four more transistors and will introduce parasitics that prevent the performance

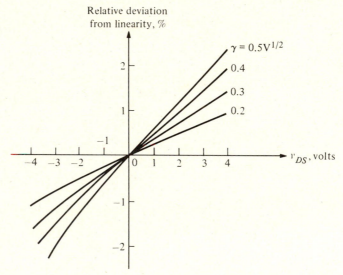

Figure 9.7-5   Relative deviation from linearity in percent for the continuous-time resistor realization of Fig. 9.7-3.

achieved by Fig. 9.7-4. The designer must be aware of some problems with the switched-resistor technique, including switch feedthrough and frequency limitations of long-channel devices. As the capacitor C_G becomes larger, the considerations of Sec. 5.1 indicate that the feedthrough problem will lessen. As the channel length of the MOS transistor becomes long it begins to act like a distributed RC transmission line, creating a frequency limitation. One should try to avoid long-channel devices for high-frequency applications or replace a single long-channel device with several shorter-channel devices. Unfortunately each device will require its own capacitor C_G thus creating a pretuning problem.

The dynamic-circuit concept of this section can be applied to many of the circuits we have previously studied. For example, consider the inverter using a current-source load illustrated in Fig. 6.1-4. A dynamically-biased version of this inverter is shown in Fig. 9.7-6. During the ϕ_2 phase, the circuit is connected dc-wise as a voltage divider. The voltage at the output is defined by the considerations given for Fig. 5.2-2 in Sec. 5.2. C_B is charged to the voltage necessary to cause M2 to provide the dynamic bias current through M1 during the ϕ_1 phase. C_{OS} is a capacitor that biases the input so the operating point is not changed when v_{IN} is connected to the circuit during the ϕ_1 phase. Depending upon the width of the ϕ_2, the circuit will dissipate much less power and not require an input bias compared with the continuous-time version in Fig. 6.1-4.

An excellent example of dynamic or clocked circuits can be shown for the conventional CMOS push-pull, cascode operational amplifier similar to Fig. 8.4-7 in Sec. 8.4. A simplified schematic of a dynamic, differential,

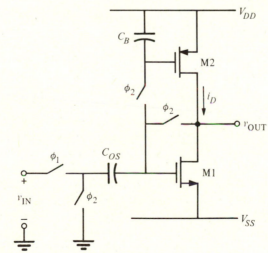

Figure 9.7-6   A dynamically-biased version of Fig. 6.1-4.

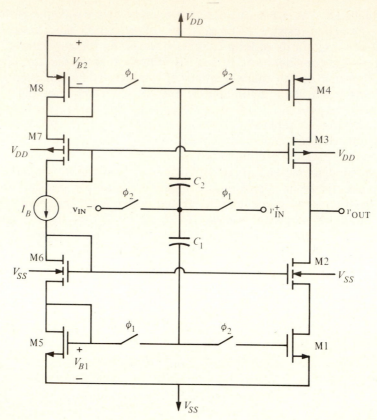

Figure 9.7-7 Simplified schematic of a dynamic, push-pull, cascode op amp.

push-pull, cascode op amp is shown in Fig. 9.7-7 [21]. The NMOS transistors M1 and M2 and the PMOS transistors M3 and M4 constitute the push-pull, cascode output stage. The six switches and the capacitors C_1 and C_2 represent the differential-input stage of the op amp. NMOS transistors M5 and M6 and the PMOS transistors M7 and M8 generate the bias voltages for the op amp. The channel lengths for M1, M2, M3, and M4 are equal to those of M5, M6, M7, and M8, respectively. Therefore, the push-pull, cascode output stage will be properly biased when V_{B1} and V_{B2} are applied to the gates of M1 and M4, respectively.

The basic operation of the dynamic op amp is explained in the following. During ϕ_1, C_1 and C_2 are charged by the bias voltages referenced to v_{IN+}. Typically, v_{IN-} is analog ground in most op amp applications. During ϕ_2, the inverting input voltage, v_{IN-}, is connected to the gates of M1 and M4 with suitable voltage shifts and becomes amplified in the push-pull, cascode output stage.

Advantages of Fig. 9.7-7 include a class-AB push-pull operation which enables fast settling time with little power dissipation. The voltage swing at the input of the push-pull cascode stage is limited by the supply voltage, because large voltage swings will cause the switch-terminal diffusions to become forward biased for bulk CMOS switches. The maximum input-voltage swing is approximately equal to the bias voltages V_{B1} and V_{B2}. If the dynamic op amp must operate on both clock phases the circuit is simply doubled, as shown in Fig. 9.7-8.

When the op amp of Fig. 9.7-8 was fabricated using a 1.5 micron, n-well, double-poly, double-metal, CMOS technology, it was found to have a gain-bandwidth of 127 MHz with a 28 pF load capacitance. The low-frequency gain was 51 dB. The common-mode input range was 1.5 to 3.5 V for a 5 V power supply. The settling time was 10 nanoseconds for a 5 pF

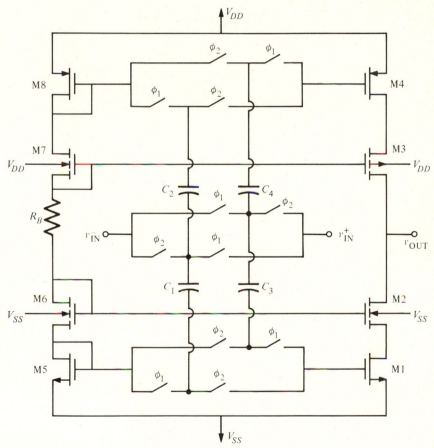

Figure 9.7-8 A version of Fig. 9.7-7 that will operate on both clock phases.

load capacitance. The PSRR for both positive and negative power supplies was 33 dB at 100 kHz. Equivalent input noise was 0.1 microvolts/ (Hertz)$^{1/2}$ at 100 kHz for a Δf of 1 kHz. The power dissipation was 1.6 mW. The key to the high performance of this circuit was the simplification due to dynamic-circuit techniques and the clever use of a high-speed technology.

The dynamic-circuit techniques presented in this section can be used to improve various areas of circuit performance. These areas include power, speed, bandwidth, and signal swing. An example of a dynamic circuit used to reduce the power dissipation in op amps can be found in a class-AB CMOS op amp designed for micropower switched-capacitor circuits [22]. The success of trading continuously-biased MOS devices for switches and capacitor depends upon many factors. With the increasing use of switched-capacitor circuits, dynamic circuits will find more and more applications.

9.8   *Summary*

This chapter has discussed CMOS op amps with performance capabilities that exceed those of the unbuffered CMOS op amps of the previous chapter. The primary difference between the two types is the addition of an output stage to the high-performance op amps. Output stages presented include those that use only MOS devices and those that use both MOS and bipolar devices. Adding the output stage usually introduces more poles in the open-loop gain of the operational amplifier, making it more difficult to compensate. The primary function of the buffered op amps is to drive a low-resistance and a large-capacitance load.

High-speed op amps were developed that tend to result in a buffered op amp that has been optimized for high unity-gain bandwidth and slew rate. Typically, these op amps dissipate more power than op amps with lesser high-speed capability. Op amps having minimum equivalent input-voltage noise were considered. The key to low-noise op amps is to use PMOS devices for the input stage and to maintain the proper ratios of the channel lengths between these devices and their load devices. The $1/f$ noise tends to be the dominant noise in most low-frequency CMOS op amps.

In many applications, it is necessary to reduce the power dissipation of the op amp. This is particularly important when a large number of op amps are to implement an integrated circuit. It was seen that a low-power op amp could be obtained at the expense of frequency response and other desirable characteristics. Most low-power op amps work in the weak-inversion mode in order to reduce dissipation and therefore perform like BJT op amp circuits.

The CMOS op amp designs of this chapter are a good example of the principle of trading decreased performance in one area for increased performance in another. Depending on the application of the op amp, this trade-off leads to increased system performance. The circuits and techniques presented in this chapter should find application in many practical areas of CMOS analog-circuit design.

PROBLEMS — *Chapter 9*

1. Verify the root locations for the transfer function of the op amp shown in Fig. 9.1-1 as given in Eqs. (5) through (7) of Sec. 9.1.
2. Design the current, push-pull CMOS op amp of Fig. 9.1-1 to achieve a low-frequency gain of 10,000, a unity-gain bandwidth of 1 MHz, and a symmetrical slew rate of ± 10 V/μs. What is the maximum and minimum output-voltage swing of your design?
3. Compare the quiescent currents of a class-A output stage (such as the one used in the two-stage op amp of Sec. 8.3) with the push-pull output stage of Fig. 9.1-1 if the output is to swing to within 1.25 volts of the positive and negative rails with a 20 Kilohm load. Assume that the output stage is not slew-rate limited.
4. An MOS output stage is shown in Fig. P9.4. Draw a small-signal model and calculate the ac voltage gain at low frequency. Assume that bulk effects can be neglected.

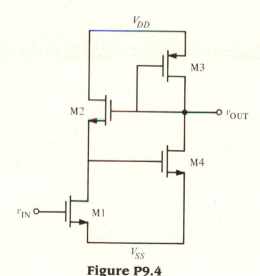

Figure P9.4

5. Find the voltage gain, poles, and zero of Fig. 9.1-1 if $K_N' = 25\ \mu A/V^2$, $K_P' = 10\ \mu A/V^2$, $\lambda = 0.02\ V^{-1}$, $C_{bs} = C_{bd} = 100$ fF, $C_{ox} = 1$ fF/μm^2, $I_{SS} = 20\ \mu A$, and $V_{DD} = -V_{SS} = 5$ V. Assume the W values shown below where all $L = 10\ \mu m$.

Device	M1	M2	M3	M4	M5	M6	M7	M8
Width (μm)	10	10	50	50	100	75	25	50

6. Assume that $K_N' = 47\ \mu A/V^2$, $K_P' = 17\ \mu A/V^2$, $V_{TN} = 0.7$ V, $V_{TP} = -0.9$ V, $\gamma_N = 0.85\ V^{1/2}$, $\gamma_P = 0.25\ V^{1/2}$, $2|\phi_F| = 0.62$ V, $\lambda_N = 0.05\ V^{-1}$, and $\lambda_P = 0.04\ V^{-1}$. Use SPICE to simulate Fig. 9.1-5 and obtain the simulated results of Fig. 9.1-6.

7. Use SPICE to plot the total harmonic distortion (THD) of the output stage of Fig. 9.1-5 as a function of the RMS output voltage at 1 kHz for an input-stage bias current of 20 μA. Use the SPICE model parameters given in the previous problem.

8. What type of BJT is available with a bulk CMOS p-well technology? A bulk CMOS n-well technology?

9. List the advantages and disadvantages of using a substrate BJT in the output stage of a CMOS op amp.

10. Verify the relationships developed for the low-power, buffered CMOS op amp of Fig. 9.2-1 given in Eqs. (10) through (14) of Sec. 9.2.

11. Verify Eq. (15) of Sec. 9.2.

12. Use SPICE to obtain simulations similar to Figs. 9.2-3 through 9.2-7 for the high-performance op amp of Fig. 9.3-1. Assume that the W/L ratios of M1 through M5 are 20 μm/20 μm and that the devices indicated as X2 or X3 are 2 or 3 identical devices in parallel. Use the model parameters of Table 9.3-1.

13. Redesign Fig. 9.3-1 with $R_{ext} = 22$ Kilohms to obtain a low-frequency voltage gain of 1000, keeping the power consumption constant and minimizing the reduction in the unity-gain bandwidth.

14. Given the op amp in Fig. P9.14, find the quiescent currents flowing in the op amp and the small-signal voltage gain, ignoring any loading produced by the output stage. Assume $K_N' = 25\ \mu A/V^2$ and $K_P' = 10\ \mu A/V^2$. Find the small-signal output resistance assuming that $\lambda = 0.04$ V^{-1}.

15. List the important principles in minimizing the equivalent input-noise voltage of an op amp.

16. For the transistor amplifier in Fig. P9.16, what is the equivalent input-noise voltage due to thermal noise? Assume the transistor has a dc drain current of 20 μA, $W/L = 150\ \mu m$/10 μm, $K_N' = 25\ \mu A/V^2$, and R_D is 100 Kilohms.

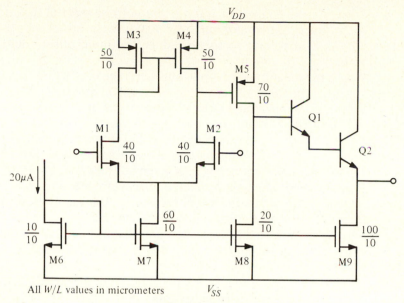

All *W/L* values in micrometers V_{SS}

Figure P9.14

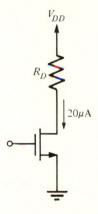

Figure P9.16

17. Assume that *KF* of Eq. (12) of Sec. 3.2 is 10^{-28} Farad·Amperes and C_{ox} is 0.45×10^{-7} F/cm^2. Use the parameters of Table 3.1-2 and assume that $I_{10} = 5\ \mu A$, $I_5 = 30\ \mu A$, and $I_7 = 85\ \mu A$ and calculate the equivalent input-noise-voltage spectral density at 100 Hz of Fig. 9.4-1. Assume that $1/f$ noise is predominant at 100 Hz. Use the W and L values of Table 9.4-1.

18. Calculate the equivalent input-noise-voltage spectral density at 100 Hz for the op amp designed in Ex. 8.3-1.

19. Why is it important to try to reduce the input-offset voltage of op amps?

20. Why is the magnitude of input-offset voltage of a BJT op amp typically less than the input-offset voltage of a CMOS op amp?

21. What are the advantages and disadvantages of using chopper stabilization to cancel input-offset voltage in CMOS op amps?

22. Fig. P9.22 shows the concept of using double sampling for noise reduction. Plot the equivalent input-noise spectrum $\bar{v}^2_{\mu eq.}$ of this circuit if the input noise spectrum $\bar{v}^2_n$ is given as shown.

23. Propose and implement a scheme (give a circuit) to reduce the input-offset voltage of the op amp in Fig. P9.23. This circuit implements an integrator when a two-phase, nonoverlapping clock is applied.

24. Fig. P9.24 shows a differential-in, differential-out op amp. Develop an expression for the small-signal, differential-in, differential-out voltage gain.

25. Find an expression for the small-signal output resistance of the differential-in, differential-out op amp of Fig. P9.24.

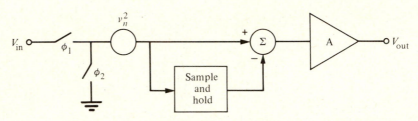

(a) Concept of double sampling noise reduction

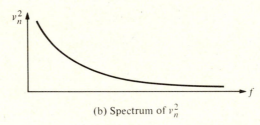

(b) Spectrum of v^2_n

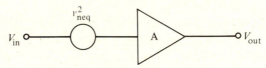

(c) Equivalent circuit of (a)

Figure P9.22

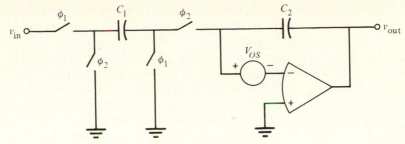

Figure P9.23

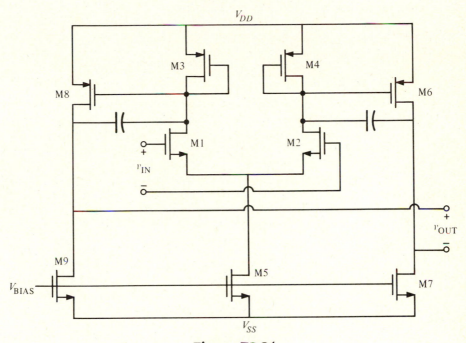

Figure P9.24

26. Develop an expression for the small-signal, differential-in, differential-out voltage gain of the op amp given in Fig. 9.5-6.
27. Verify Eqs. (4), (5), (6), and (7) of Sec. 9.6.
28. Repeat Ex. 9.6-1 if $I_{D5} = 400$ nA. What is quiescent power dissipation if $V_{DD} = -V_{SS} = 3$ V?
29. If $I_{10} = 200$ nA, what is the quiescent power dissipation of Fig. 9.6-5 if $V_{DD} = 1.5$ V and $V_{SS} = 0$?
30. Verify Eq. (3) of Sec. 9.7.
31. Develop an expression for the time T when the switched resistor of Fig. 9.7-2 should be recharged in terms of $R_{FEQ}/R_{FET}(T)$ and $V_{GS}(0)/V_T$. Plot the curve of T/τ_g for values of $R_{FET}(T)/R_{FEQ}$ of 1.01 and 1.05.

32.   What effect will V_{SB} have on the switched resistor of Fig. 9.7-2?

33.   Show how the batteries of Fig. 9.7-2 can be implemented using continuous-time techniques.

34.   Show how the circuit of Fig. 9.7-6 can be modified to obtain a dynamic amplifier having a gain of (a) $-C_1/C_2$ and (b) $+C_1/C_2$. If either C_{OS} or C_B can be replaced by the additional switches and capacitances C_1 or C_2, then do so in your realization.

35.   For the switched-resistor implementation of Fig. 9.7-4, find the slowest clock speed that can be used for a duty factor of 45% assuming that $R_{FET}(T)$ stays within 1% of R_{FEQ}, $C_G = 20$ pF, $C_{GS} = 0.1$ pF, $V_{GS}(0) = 2$ V, and $V_T = 1$ V.

REFERENCES

1.   D.G. Maeding, "A CMOS Operational Amplifier with Low Impedance Drive Capability," *IEEE Journal of Solid-State Circuits*, Vol. SC-18, No. 2 (April 1983) pp. 227–229.

2.   K.E. Brehmer and J.B. Wieser, "Large Swing CMOS Power Amplifier," *IEEE Journal of Solid-State Circuits*, Vol. SC-18, no. 6 (Dec. 1983), pp. 624–629.

3.   W.J. Parrish, "An Ion Implanted CMOS Amplifier for High Performance Active Filters," (Ph.D. Dissertation, Dept. of EECS, University of California, Santa Barbara, CA, June 1976).

4.   M. Milkovic, "Current Gain High-Frequency CMOS Operational Amplifiers," *IEEE Journal of Solid-State Circuits*, Vol. SC-20, No. 4 (August 1985) pp. 845–851.

5.   R.D. Jolly and R.H. McCharles, "A Low-Noise Amplifier for Switched Capacitor Filters," *IEEE Journal of Solid-State Circuits*, Vol. SC-17, No. 6 (December 1982) pp. 1192–1194.

6.   E.A. Vittoz, "MOS Transistors Operated in the Lateral Bipolar Mode and Their Application in CMOS Technology," *IEEE Journal of Solid-State Circuits*, Vol. SC-18, No. 3 (June 1983) pp. 273–279.

7.   P.R. Gray and R.G. Meyer, *Analysis and Design of Analog Integrated Circuits*, Second Edition, (New York: John Wiley & Sons, 1984).

8.   Jiri Dostal, *Operational Amplifiers*, (New York: Elsevier, 1981).

9.   K.C. Hsieh and P.R. Gray, "A Low Noise Chopper-Stabilized Differential Switched Capacitor Filtering Technique," *ISSCC Digest of Technical Papers*, (February 1981) pp. 128–129.

10.   K.C. Hsieh, P.R. Gray, D. Senderowicz, and D.G. Messerschmitt, "A Low-Noise Chopper-Stabilized Switched-Capacitor Filtering Technique," *IEEE Journal of Solid-State Circuits*, Vol. SC-16, No. 6 (Dec. 1981) pp. 708–715.

11.   M. Banu and Y. Tsividis, "Fully Integrated Active RC Filters in MOS Technology," *IEEE Journal of Solid-State Circuits*, Vol. SC-18, No. 6 (December 1983) pp. 644–651.

12.   T.C. Choi, R.T. Kaneshiro, R.W. Brodersen, P.R. Gray, W.B. Jett, and M. Wilcox, "High-Frequency CMOS Switched-Capacitor Filters for Communications Application," *IEEE Journal of Solid-State Circuits*, Vol. SC-18, No. 6 (December 1983) pp. 652–664.

13. M.G. Degrauwe, J. Rijmenants, E.A. Vittoz, and H.J. De Man, "Adaptive Biasing CMOS Amplifiers," *IEEE Journal of Solid-State Circuits*, Vol. SC-17, No. 3 (June 1982) pp. 522–528.
14. M. Degrauwe, E. Vittoz, and I. Verbauwhede, "A Micropower CMOS-Instrumentation Amplifier," *IEEE Journal of Solid-State Circuits*, Vol. SC-20, No. 3 (June 1985) pp. 805–807.
15. P. Van Peteghem, I. Verbauwhede, and W. Sansen, "Micropower High-Performance SC Building Block for Integrated Low-Level Signal Processing," *IEEE Journal of Solid-State Circuits*, Vol. SC-2, No. 4 (August 1985) pp. 837–844.
16. D.C. Stone, J.E. Schroeder, R.H. Kaplan, and A.R. Smith, "Analog CMOS Building Blocks for Custom and Semicustom Applications," *IEEE Journal of Solid-State Circuits*, Vol. SC-19, No. 1 (February 1984) pp. 55–61.
17. F. Krummenacher, "Micropower Switched Capacitor Biquadratic Cell," *IEEE Journal of Solid-State Circuits*, Vol. SC-17, No. 3 (June 1982) pp. 507–512.
18. P.E. Allen and E. Sanchez-Sinencio, *Switched Capacitor Circuits*, (New York: Van Nostrand Reinhold, 1984).
19. R.L. Geiger, P.E. Allen, and D.T. Ngo, "Switched-Resistor Filters—A Continuous Time Approach to Monolithic MOS Filter Design," *IEEE Transactions on Circuits and Systems*, Vol. CAS-29, No. 5 (May 1982) pp. 306–315.
20. M. Banu and Y. Tsividis, "Floating Voltage-Controlled Resistors in CMOS Technology," *Electronics Letters*, Vol. 18, No. 15 (July 22, 1982) pp. 678–679.
21. S. Masuda, Y. Kitamura, S. Ohya, and M. Kikuchi, "CMOS Sampled Differential, Push-Pull Cascode Operational Amplifier," *Proceedings of the 1984 International Conference on Circuits and Systems*, (Montreal, Canada: May 1984) pp. 1211–1214.
22. F. Krummenacher and E. Vittoz, "Class AB CMOS Amplifier for Micropower SC Filters," *Electronics Letters*, Vol. 17, No. 13 (June 25, 1981) pp. 433–435.

chapter 10

CMOS Digital-Analog and Analog-Digital Converters

The subject of this chapter is the means of converting between analog and digital signals. In Sec. 1.1, the definition and distinction between analog and digital signals was presented. Fig. 1.3-1 illustrated the typical block diagram of a signal processing system. It was noted that there are areas where signal processing is done on analog signals and areas where the signal processing is done on digital signals. Consequently, it is necessary to be able to convert back and forth between the two types of signals. Therefore analog to digital and digital to analog converters are an important part of any signal processing system.

From the viewpoint of Table 1.1-2, analog-digital (A/D) and digital-analog (D/A) converters are at the systems level. They typically contain one or more comparators, digital circuitry, switches, integrators, a sample-and-hold, and/or passive components. Another important component of the A/

D or D/A converter is a precise voltage reference. In this chapter, the voltage reference will be assumed to be external. However, in the next chapter, methods of designing a precise voltage reference will be considered. The D/A and A/D converters presented here do not represent all possible approaches, but have been selected to be compatible with CMOS technology.

The D/A converter is presented first because it is generally part of an A/D converter. This presentation begins with a general introduction and characterization of converters and then subdivides D/A converters into scaling converters and serial converters. A/D converters are categorized as low-speed (serial), medium-speed, high-speed, and high-performance. The last category of A/D converters includes several types that are compatible with digital audio or other high-bit, fast converters.

10.1   *Introduction and Characterization of Digital-Analog Converters*

This section examines the digital-to-analog conversion aspect of this important interface. Fig. 10.1-1 illustrates how analog-to-digital (A/D) and digital-to-analog (D/A) converters are used in data systems [1]. In general, an A/D conversion process will convert a sampled-and-held analog signal to a digital word that represents the sampled analog signal. Often, many analog inputs are multiplexed to the A/D converter. The D/A conversion process is essentially the reciprocal of the A/D process. Digital words are applied to the input of the D/A to create from a reference voltage an analog output signal that represents the digital word.

The principles of D/A converters are presented first, followed by their performance characterization. To understand the application and design of D/A converters we must first understand how they are specified. Their specifications define the accuracy of the converter, its static performance, and its dynamic performance. Because the A/D converter is the inverse of the D/A converter, the characterization of the D/A converter will be applicable to the A/D converter.

Fig. 10.1-2a shows a conceptual block diagram of a D/A converter. The inputs are a digital word of N bits (b_1, b_2, b_3, . . . , b_N) and a reference voltage V_{REF}. The voltage output v_{OUT} can be expressed as

$$v_{OUT} = KV_{REF}D \qquad (1)$$

where K is a scaling factor and the digital word D is given as

$$D = \frac{b_1}{2^1} + \frac{b_2}{2^2} + \frac{b_3}{2^3} + \cdots + \frac{b_N}{2^N} \qquad (2)$$

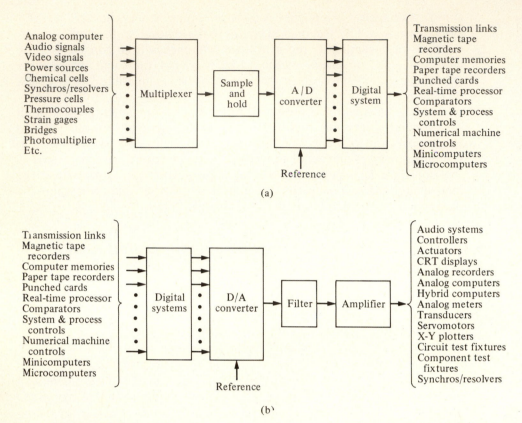

(a)

(b)

Figure 10.1-1 (a) A/D converters and (b) D/A converters in signal processing systems.

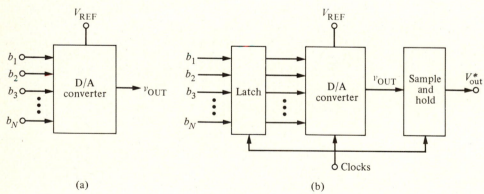

(a)

(b)

Figure 10.1-2 (a) Conceptual block diagram of a D/A converter. (b) Clocked D/A converter.

N is the total number of bits of the digital word and b_i is the ith bit coefficient and is either 0 or 1. Therefore, the output of a D/A converter can be expressed by combining Eqs. (1) and (2) to get

$$V_{OUT} = KV_{REF} \left[\frac{b_1}{2^1} + \frac{b_2}{2^2} + \frac{b_3}{2^3} + \cdots + \frac{b_N}{2^N} \right] \tag{3}$$

or

$$V_{OUT} = KV_{REF} [b_1 2^{-1} + b_2 2^{-2} + b_3 2^{-3} + \cdots + b_N 2^{-N}] \tag{4}$$

In many cases the digital word is synchronously clocked. In this case latches must be used to hold the word for conversion and a sample-and-hold circuit is needed at the output, as shown in Fig. 10.1-2(b). The sample-and-hold circuit consists of a circuit such as that shown in Fig. 10.1-3. The sample mode occurs when the switch is closed and the analog signal is sampled on a capacitor C_H. During the switch-open cycle or the hold mode,

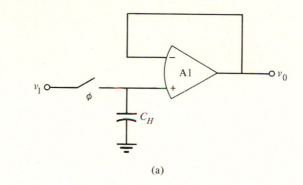

(a)

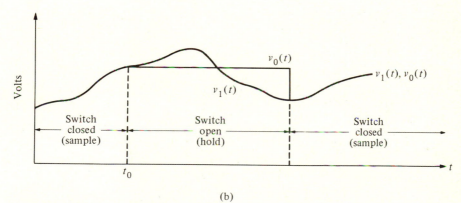

(b)

Figure 10.1-3   (a) Simple sample-and-hold circuit. (b) Waveforms illustrating the operation of the sample and hold.

the voltage at the time t_0 is available at the output. An alternate version of the sample-and-hold circuit with higher performance is shown in Fig. 10.1-4. The sample-and-hold circuit must be able to rapidly track changes in the input voltage in the sample mode and not discharge the capacitor when in the hold mode.

The basic form of the D/A converter is shown in Fig. 10.1-5. The various blocks include binary switches, a scaling network, and an output amplifier. The scaling network and binary switches operate on the externally-supplied reference voltage to create the digital word as either a voltage or charge signal and the output amplifier converts this signal to a voltage signal that can be sampled without affecting the value of the conversion.

The characterization of the D/A converter is very important in understanding its design. The characteristics of the D/A converter can be divided into static and dynamic properties. The ideal static behavior of a D/A converter is shown in Fig. 10.1-6. The digital input word is on the horizontal axis, which consists of all possible combinations of that word. Fig. 10.1-6 is for a 3-bit digital word. The vertical scale is the analog output of the ideal D/A converter. The maximum analog-output signal is designated as the full

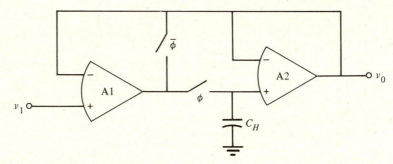

Figure 10.1-4   An improved sample-and-hold circuit.

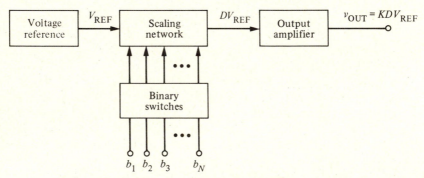

Figure 10.1-5   Block diagram of a D/A converter.

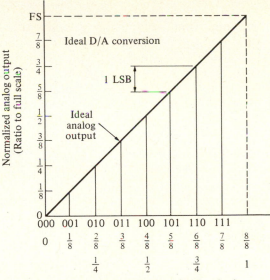

Figure 10.1-6   Ideal input-output characteristics for a three-bit D/A converter. (By permission from D.H. Sheingold, *Analog-Digital Conversion Handbook*, Analog Devices, Inc., 1972.)

scale (FS) value. For each digital word, there should be a unique analog-output signal. Any deviations from Fig. 10.1-6 fall into the category of static-conversion errors.

Static-conversion errors can affect the following characterstics: integral linearity, differential linearity, absolute linearity, resolution, zero and full-scale error, and monotonicity. *Integral linearity* is the maximum deviation of the output of a D/A (for any given input code) from a straight line drawn from its ideal minimum to its ideal maximum output. Integral linearity can be expressed as a percentage of the full-scale range or in terms of the least significant bit (LSB). Integral linearity has several subcategories, which include absolute, best-straight-line, and end-point linearity [2].

Absolute linearity is measured by assuming that the output of a D/A will begin at zero and end at full scale. The actual outputs are compared with a line drawn through these two points as illustrated by Fig. 10.1-7(a). Absolute linearity emphasizes the zero and full-scale errors. The *zero error* is the difference between the actual output and zero when the digital word for a zero output is applied; the *full-scale error* is the difference between the actual and the ideal voltage when the digital word for a full-scale output is applied. Another error closely related to full-scale error is gain error. *Gain error* is the difference between the gains of the actual static and ideal

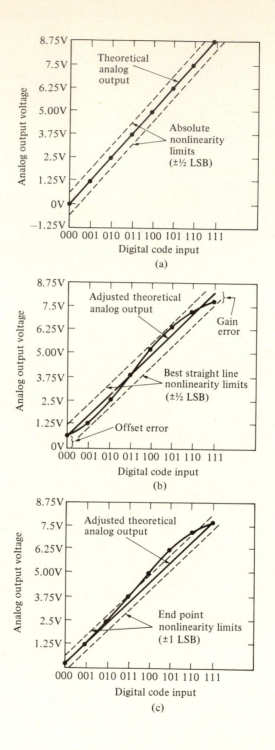

(a)

(b)

(c)

input-output characteristics. Gain errors exist when the slope of the actual characteristic is not parallel to the slope of the line indicative of the ideal characteristic. Thus, a disadvantage of the absolute linearity measurement is that zero and full-scale errors can contribute to the nonlinearity of an otherwise linear D/A.

Another subcategory of integral linearity is *best-straight-line linearity*. This depicts the accuracy of a D/A in terms of the deviation from the ideal output range without regard to zero or full-scale errors. Fig. 10.1-7(b) shows the characteristics of a D/A converter that has a large full-scale error but good linearity. The last subcategory of integral linearity is *end-point linearity* and is illustrated in Fig. 10.1-7(c). End-point linearity uses a straight line through the actual end points instead of the ideal end points.

Differential linearity is a measure of the separation between adjacent levels. Differential linearity measures bit-to-bit deviations from ideal output steps, rather than along the entire output range. If V_{cx} is the actual voltage change on a bit-to-bit basis and V_s is the ideal change, then the differential linearity can be expressed as

$$\text{Differential linearity} = \left[\frac{V_{cx} - V_s}{V_s}\right] \times 100\% \tag{5}$$

For an N-bit D/A converter and a full-scale voltage range of V_{FSR},

$$V_s = \frac{V_{FSR}}{2^{N-1}} \tag{6}$$

It is evident that the differential linearity can also be specified as a fraction of the LSB. Fig. 10.1-8 shows how differential linearity differs from integral linearity. Fig. 10.1-8(a) is for a D/A converter having a ± 2 LSB integral linearity and a ± 0.5 LSB differential linearity. Fig. 10.1-8(b) is for a D/A converter having a ± 0.5 LSB integral linearity and a ± 1 LSB differential linearity.

Monotonicity in a D/A converter means that as the digital input to the converter increases over its full-scale range, the analog output never exhibits a decrease between one conversion step and the next. In other words, the slope of the transfer characteristic is never negative in a monotonic

←——————————————————

Figure 10.1-7   Examples of various types of linearity for a three-bit D/A converter. (a) Absolute linearity. (b) Best-straight-line linearity. (c) End-point linearity. (By permission from D. Gilbert, "Understanding D/A Accuracy Specs.," *Electronic Products,* Vol. 24, No. 3, July 1981, pp. 61–63, © Hearst Business Communications, Inc., 1981.)

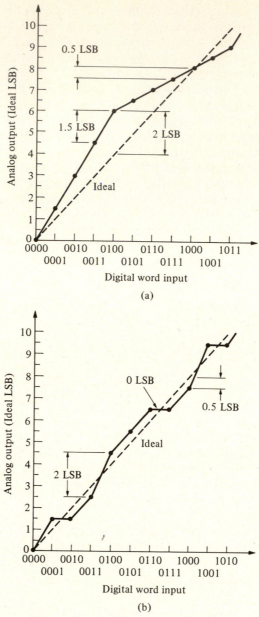

Figure 10.1-8 Integral and differential linearity for a D/A converter. (a) D/A converter with ±2 LSB integral linearity and ±½ LSB differential linearity. (b) D/A converter with ±½ LSB integral linearity and ±1 LSB differential linearity. ([10], © Van Nostrand Reinhold, 1984)

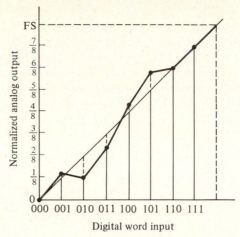

Figure 10.1-9   Example of a three-bit D/A converter that is not monotonic. (By permission from D. H. Sheingold, *Analog-Digital Conversion Handbook*, Analog Devices, Inc., 1972.)

converter. Fig. 10.1-9 gives an example of a D/A converter that is not monotonic. Another term used to characterize a D/A converter is resolution. *Resolution* is defined as the smallest input digital code for which an analog output level is produced. Theoretically, the resolution of an N-bit D/A converter is 2^N discrete analog-output levels. However, the actual resolution will be less if internal noise excursions exceed the quantization level or if component drift causes switching errors such as those observed when the most significant bit (MSB) changes. These sources of error will reduce the maximum theoretical resolution of the converter.

The dynamic characteristics of the D/A converter are observed when the input digital word is changed. The time required for the output of the converter to respond to a bit change is called *settling time* and is similar to the settling time for op amps, as defined in Sec. 8.1. Settling time for D/A converters depends on the type of converter and can range from as much as 100 microseconds to less than 100 nanoseconds.

10.2   *Voltage- and Charge-Scaling D/A Converters*

A very important component of a D/A converter is the accurate scaling of the reference voltage V_{REF}. The analog output is equal to this scaled value which is determined by the digital word. Scaling is usually accomplished by the use of passive components, e.g., resistors or capacitors. Resistors are used to accomplish voltage scaling whereas capacitors are used to accomplish charge scaling. These two categories of D/A converters will be examined in this section.

Voltage scaling uses series resistors connected between V_{REF} and ground to selectively obtain voltages between these limits. For an N-bit converter, the resistor string would have at least 2^N segments. These segments can all be equal or the end segments may be partial values, depending upon the requirements. Fig. 10.2-1(a) shows a 3-bit voltage scaling D/A converter. An op amp can be used to buffer the resistor string from loading. Each tap is connected to a switching tree whose switches are controlled by the bits of the digital word. If the ith bit is 1, then the switches

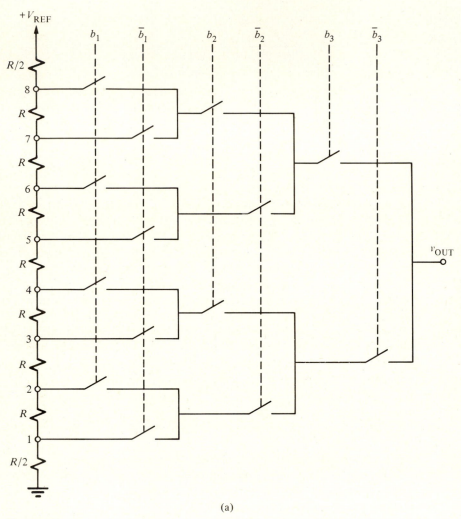

(a)

Figure 10.2-1   (a) Implementation of a 3-bit voltage-scaling D/A converter. (b) Input-output characteristics of Fig. 10.2-1(a).

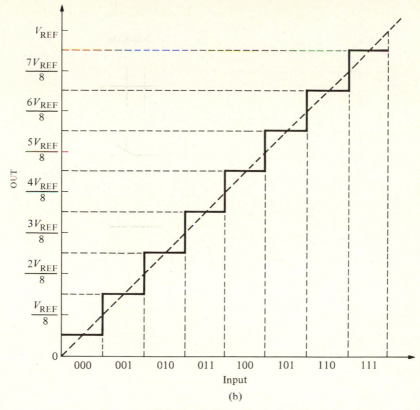

Figure 10.2-1 (*continued*)

controlled by b_i are closed. If the ith bit is 0, then the switches controlled by $\overline{b}_i$ are closed.

The voltage-scaling D/A converter of Fig. 10.2-1(a) works as follows. Suppose that the digital word to be converted is $b_1 = 1$, $b_2 = 0$, and $b_3 = 1$. Following the sequence of switches, we see that v_{OUT} is equal to $\frac{11}{16}$ of V_{REF}. In general, the voltage at any tap n of Fig. 10.2-1(a) can be expressed as

$$V_n = \frac{V_{\text{REF}}}{8} (n - 0.5) = \frac{V_{\text{REF}}}{16} (2n - 1) \tag{1}$$

Fig. 10.2-1(b) shows the input-output characteristics of the D/A converter of Fig. 10.2-1(a). It may be desirable to connect the bottom tap to ground, so that a well-defined output is available when the digital word is zero.

If the number of bits is large, then one can use the configuration shown in Fig. 10.2-2. Here a single switch is connected between each node of the resistor string and the output. Which switch is closed depends upon the

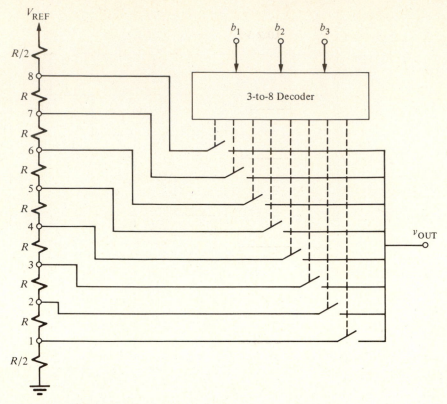

Figure 10.2-2 Alternate realization of Fig. 10.2-1(a).

logic circuit, which will consist of an N-to-2^N decoder or similar circuitry. This configuration reduces the series resistance of the switches and the effect of the parasitic capacitances at each switch node to ground. An area-performance tradeoff may be made, resulting in some bits being determined directly and the rest indirectly by the logic decoder.

Example 10.2-1

Accuracy Requirements of a Voltage-Scaling D/A Converter

Find the accuracy requirements for the resistor string of Fig. 10.2-1(a) as a function of the number of bits N if the resistor string is a 5 μm wide polysilicon strip. If the relative accuracy is 2%, what is the largest number of bits than can be resolved to within ± 0.5 LSB?

The ideal voltage to ground across k resistors can be expressed as

$$V_k = \frac{kR}{2^N R} V_{REF}$$

The worst case variation in this voltage can be found by assuming that all resistors above this point in the string are maximum and below this point are minimum. Therefore, the worst case lowest voltage to ground across k resistors is

$$V_k' = \frac{kR_{min} V_{REF}}{(2^N - k)R_{max} + kR_{min}}$$

The difference between the ideal and worst case voltages can be expressed as

$$\left| \frac{V_k}{V_{REF}} - \frac{V_k'}{V_{REF}} \right| = \left| \frac{kR}{2^N R} - \frac{kR_{min}}{(2^N - k)R_{max} + kR_{min}} \right|$$

If we assume that this difference should be less than 0.5 LSB, then the desired relationship can be obtained as

$$\left| \frac{k}{2^N} - \frac{kR_{min}}{(2^N - k)R_{max} + kR_{min}} \right| < \frac{0.5}{2^N}$$

The relative accuracy of the resistor R can be expressed as $\Delta R/R$ which gives $R_{max} = R + 0.5\Delta R$ and $R_{min} = R - 0.5\Delta R$. Normally, the worst case occurs when k is midway in the resistor string or $k = 0.5(2^N)$. Assuming a relative accuracy of 2% and substituting the above values gives

$$|0.01| < 2^{-N}$$

Solving this equation for N gives $N = 6$ as the largest integer.

It is seen that the voltage-scaling D/A structure is very regular and thus well-suited for MOS technology. An advantage of this architecture is that it guarantees monotonicity, for the voltage at each tap cannot be less than the tap below. The area required for the voltage-scaling D/A converter is large if the number of bits is eight or more. Also, the conversion speed of the converter will be sensitive to parasitic capacitances at each of its internal nodes.

Charge-scaling D/A converters operate by binarily dividing the total charge applied to a capacitor array. Fig. 10.2-3(a) shows an illustration of a charge-scaling D/A converter. A two-phase, nonoverlapping clock is used for this converter. During ϕ_1 the top and bottom plates of all capacitors in the array are grounded. Next, during ϕ_2, the capacitors associated with bits that are 1 are connected to V_{REF} and those with bits that are 0 are connected to ground. The output of the D/A converter is valid during ϕ_2. The resulting situation can be described by equating the charge in the capacitors to V_{REF} (C_{eq}) to the charge in the total capacitors (C_{tot}). This is expressed as

$$V_{REF}C_{eq} = V_{REF}\left(b_1C + \frac{b_2C}{2} + \frac{b_3C}{2^2} + \cdots + \frac{b_NC}{2^{N-1}} \right)$$

$$= C_{tot} \, v_{OUT} = 2C \, v_{OUT} \qquad (2)$$

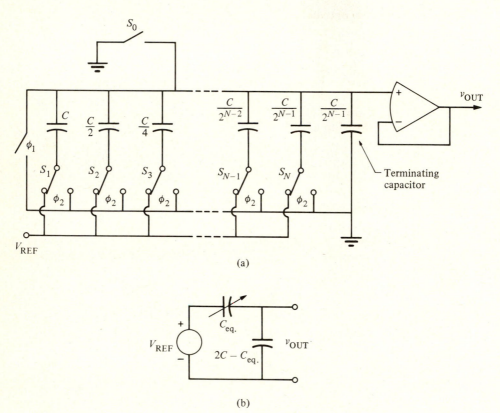

(a)

(b)

Figure 10.2-3   (a) Charge-scaling D/A converter. Switches S_i close to V_{ref} if $b_i = 1$ or ground if $b_i = 0$ during ϕ_2. All switches are connected to ground during ϕ_1. (b) Equivalent circuit of (a).

From Eq. (2) we may solve for V_{out} as

$$V_{OUT} = [b_1 2^{-1} + b_2 2^{-2} + b_3 2^{-3} + \cdots + b_N 2^{-N}] V_{REF} \qquad (3)$$

Another approach to understanding Fig. 10.2-3(a) is to consider the capacitor array as a capacitive attenuator (illustrated in Fig. 10.2-3(b)). As before $C_{eq.}$ consists of the sum of all capacitances connected to V_{REF} and $C_{tot.}$ ($= 2C$) is the sum of all the capacitors in the array.

The D/A of Fig. 10.2-3(a) can have bipolar operation if the bottom plates of all array capacitors are connected to V_{REF} during ϕ_1. During ϕ_2, the bottom plate of a capacitor is connected to ground for $b_i = 1$ or connected to V_{REF} for $b_i = 0$. The resulting output voltage is

$$V_{OUT} = [b_1 2^{-1} + b_2 2^{-2} + b_3 2^{-3} + \cdots + b_N 2^{-N}] (-V_{REF}) \qquad (4)$$

The decision to connect all capacitors to ground or to V_{REF} will require an additional bit called a *sign bit*. By including a sign bit bipolar behavior for the digital word D will be obtained. If V_{REF} is also bipolar then a four-quadrant D/A converter results.

The accuracy of the capacitor and the area required are both factors that limit the number of bits used. The accuracy is seen to depend upon the capacitor ratios. The ratio error for capacitor in MOS technology can be as low as 0.1%. If the capacitor ratios were able to have this accuracy, then the D/A converter of Fig. 10.2-3(a) should be capable of a 10-bit resolution. However, this implies that the ratio between the MSB and LSB capacitors will be 1024:1 in the extreme case which is undesirable from an area viewpoint. Also, the 0.1% capacitor ratio accuracy is applicable only for ratios in the neighborhood of unity. As the ratio increases, the capacitor ratio accuracy decreases. These effects are considered in the following example.

Example 10.2-2

Influence of Capacitor Ratio Accuracy on Number of Bits

Use the data of Fig. 2.4-2 to estimate the number of bits possible for a charge scaling D/A converter assuming a worst case approach and that the worst conditions occur at the midscale (1 MSB).

From Fig. 10.2-3(b) the ideal output voltage of the charge-scaling D/A converter can be expressed as

$$\frac{V_{OUT}}{V_{REF}} = \frac{C_{eq}}{2C}$$

Assume the worst-case output voltage is given as

$$\frac{V'_{OUT}}{V_{REF}} = \frac{C_{eq}(min)}{[2C - C_{eq}](max) + C_{eq}(min)}$$

The difference between the ideal output and the worst case output can be written as

$$\left| \frac{V_{OUT}}{V_{REF}} - \frac{V'_{OUT}}{V_{REF}} \right| = \left| \frac{C_{eq}}{2C} - \frac{C_{eq}(min)}{[2C - C_{eq}](max) + C_{eq}(min)} \right|$$

If we assume that the worst-case condition occurs at midscale, then C_{eq} is equal to C. Therefore the difference between the ideal output and the worst-case output is

$$\left| \frac{V_{OUT}}{V_{REF}} - \frac{V'_{OUT}}{V_{REF}} \right| = \left| \frac{1}{2} - \frac{C(min)}{C(max) + C(min)} \right|$$

Replacing $C(max)$ by $C + 0.5\Delta C$ and $C(min)$ by $C - 0.5\Delta C$ and setting the difference between the ideal and worst-case output voltage equal to ± 0.5 LSB results in the following equation.

$$\left| \frac{\Delta C}{2C} \right| = \frac{1}{2^N}$$

From the data presented in Fig. 2.4-2, it is reasonable to assume that the relative accuracy of the capacitor ratios will increase with the number of bits. Let us assume a unit capacitor of 50 μm by 50 μm and a relative accuracy of approximately 0.1%. Solving for N in the above equation gives approximately 11 bits. However, the 0.1% figure corresponds to ratios of 16:1 or 4 bits. In order to get a solution, we estimate the relative accuracy of capacitor ratios as

$$\frac{\Delta C}{C} \cong 0.001 + 0.0001N$$

Using this approximate relationship, a 9-bit D/A converter should be realizable.

The total amount of capacitance required for a given-size converter can be reduced by using the capacitor divider approach as shown in Fig. 10.2-4(a)[3]. In this configuration, the portion of the array that lies to the

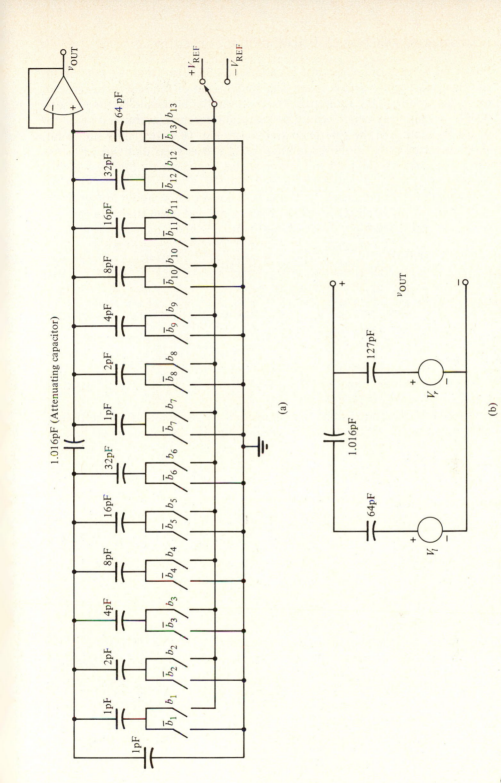

Figure 10.2-4 (a) A 13-bit two-stage MDAC. Note that whether or not the ϕ_2 switches close depends upon the state of the binary variable, b_i. (b) An equivalent circuit of (a).

537

left of the attenuating capacitor is called the LSB array, and the portion to the right is called the MSB array. The smallest bit of the MSB array would give a $\frac{1}{128}$ scaling of the reference. The next smaller bit would be the highest bit of the LSB array which would be $\frac{1}{2}$ of the reference attenuated by $\frac{1}{128}$ to give a $\frac{1}{256}$ scaling of the reference and so on. Fig. 10.2-4(a) shows a 13-bit D/A converter with bipolar capability for V_{REF}. It is seen that the 1.016 pF capacitor acts as a 64:1 divider which scales the lower six bits by a factor of 64.

An equivalent circuit of the 13-bit D/A converter is shown in Fig. 10.2-4(b). Two voltage sources are shown which depend upon the state of each of the switches. The right-hand voltage source is given as

$$V_r = \sum_{i=1}^{7} \frac{\pm b_i V_{REF}}{127} \tag{5}$$

where the polarity of V_{REF} depends upon the polarity of the digital word. The left-hand voltage source is given as

$$V_1 = \sum_{i=8}^{13} \frac{\pm b_i V_{REF}}{64} \tag{6}$$

where in Eqs. (5) and (6) any unused inputs are assumed to be grounded. The overall output of the D/A converter of Fig. 10.2-4(a) can be written as

$$V_{OUT} = \frac{\pm V_{REF}}{128} \left[\sum_{i=1}^{7} b_i C_i + \sum_{i=8}^{13} b_i \frac{C_i}{64} \right] \tag{7}$$

The charge-scaling D/A converters are sensitive to capacitance loading at the output. If this capacitor is designated as C_L, then Eq. (7) is modified as

$$V_{OUT} = \left[1 - \frac{C_L}{128} \right] \left[\frac{V_{REF}}{128} \right] \left[\sum_{i=1}^{7} b_i C_i + \sum_{i=8}^{13} b_i \frac{C_1}{64} \right] \tag{8}$$

where we see that C_L has caused an error of $[1 - (C_L/128)]$. If C_L is 1% of the total ladder capacitance, then a 1% error is introduced by the capacitor C_L.

The accuracy of the capacitor attenuator must also be small enough for the divider approach to work. A deviation of the 1.016 pF attenuating capacitance from the desired ratio of 1.016:1 introduces both gain and linearity errors. Assuming a variation of $\pm \Delta$ in the 1.016 pF capacitor modifies the output given in Eq. (7) to

$$V_{OUT} = \left[\frac{V_{REF}}{128}\right]\left[1 - \frac{\Delta C}{128}\right]\left[\sum_{i=1}^{7} b_i C_i + (1 + \Delta C)\sum_{i=8}^{13} b_i \frac{C_i}{64}\right] \quad (9)$$

If we assume that $\Delta C = \pm 0.016$ (i.e. 1.6% error), then the gain term has an error of

$$\text{Gain error term} = [1 - (\Delta C/128)] = 1 - \left[\frac{1}{(64)(128)}\right] \quad (10)$$

which is negligible. The linearity error term is given from Eq. (9) by

$$\text{Linearity error term} = \Delta C\sum_{i=8}^{13}\frac{b_i C_i}{64} \quad (11)$$

The worst-case error occurs for all $b_i = 1$ and is essentially ΔC. It is also important to keep $+V_{REF}$ and $-V_{REF}$ stable and equal in amplitudes. This influences the stability and gain tracking of the D/A converter.

A different two-stage, 8-bit D/A converter is shown in Fig. 10.2-5. An op amp is connected in its inverting configuration with the $2C$ capacitor fed back from the output to the inverting input of the op amp. Because the input node is a virtual ground during operation, effects due to parasitics associated with it are eliminated.

The voltage-scaling and charge-scaling approaches to implementing D/A converters can be combined to result in converters having a resolution that exceeds the number of bits of the separate approaches [4]. An M-bit resistor string and a K-bit binary-weighted capacitor array can be used to achieve an $N = (M + K)$-bit conversion. Fig. 10.2-6 gives an example of such a converter where $M = 4$ and $K = 8$. The resistor string R_1 through $R_{(2M)}$ divides an inherently monotonic V_{REF} into 2^M nominally identical voltage segments. The binary-weighted capacitor array C_1 through C_{K+1} is used to subdivide any one of these voltage segments into 2^K levels. This is accomplished by the following sequence of events. First, the switches S_F, S_B, and S_{1B} through $S_{K,B}$ are closed, connecting the top and bottom plates of the capacitors C_1 through C_K to ground. If the output of the D/A converter is applied to any circuit having a nonzero threshold, switch S_B could be connected to this circuit rather than ground to cancel this threshold effect. After switch S_F is opened, the buses A and B are connected across the resistor whose lower and upper voltage is V'_{REF} and $V'_{REF} + 2^{-M}V_{REF}$, respectively, where

$$V'_{REF} = V_{REF}[b_1 2^{-1} + b_2 2^{-2} + \cdots + b_{(M-1)}2^{(M-1)} + b_M 2^{-M}] \quad (12)$$

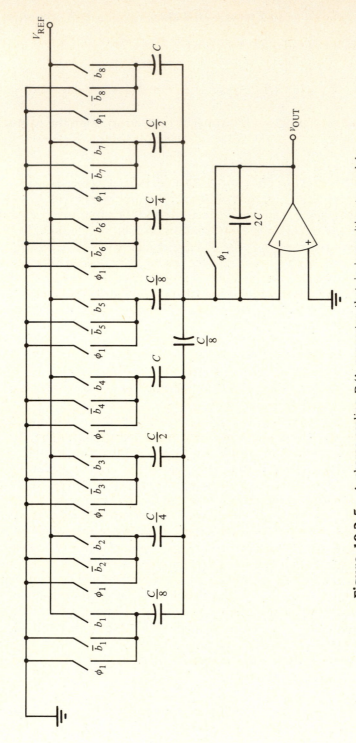

Figure 10.2-5 A charge-scaling D/A converter that is insensitive to nodal-capacitive parasitics.

Although the resistor string could take the form of Fig. 10.2-1(a), the configuration used in Fig. 10.2-6 has the advantage of switching both buses A and B. This causes any switch nonidealities, such as clock feedthrough, to be canceled. After the first M bits are determined, we have the equivalent circuit of Fig. 10.2-7(a). The final step is to decide whether or not to connect the bottom plates of the capacitors to bus A or bus B. This is determined by the M + 1 through M + K bits of the digital word being converted. The equivalent circuit for the analog output voltage of the D/A converter of Fig. 10.2-6 is shown in Fig. 10.2-7(b).

The D/A converter of Fig. 10.2-6 has the advantage that, because the resistor string is inherently monotonic, the first M bits will be monotonic regardless of any resistor mismatch. This implies that the capacitor array has to be ratio-accurate to only K bits and still be able to provide $(M + K)$-bit monotonic conversion. The conversion speed for the resistors is faster than the capacitors because no precharging is necessary. Therefore some interesting tradeoffs can be made between area and speed of conversion. Techniques such as trimming the resistor string using polysilicon fuses can help to improve the integral linearity of this approach.

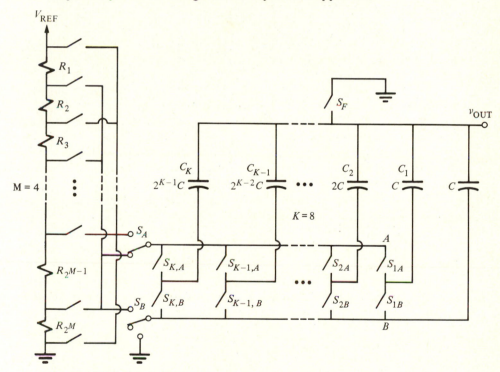

Figure 10.2-6   A D/A converter using a combination of voltage-scaling and charge-scaling techniques. The 4 MSBs are accomplished by voltage-scaling and the 8 LSBs are accomplished by charge-scaling.

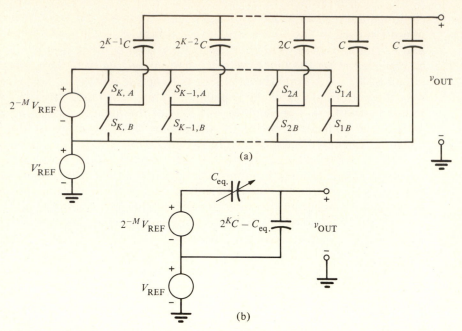

Figure 10.2-7   (a) Equivalent circuit of Fig. 10.2-6 for the voltage-scaling part. (b) Equivalent circuit of the entire converter of Fig. 10.2-6.

Other combinations of voltage-scaling and charge-scaling techniques are also possible. Instead of using resistive string techniques for the MSBs and binary-weighted capacitors for the LSBs, one can use binary-weighted capacitors for the MSBs and the resistor string for the LSBs. For this case the resistors must be trimmed to maintain absolute linearity. Fig. 10.2-8 illustrates how such a D/A converter could be implemented. The three to four MSBs will probably have to be trimmed, using a technique such as polysilicon fuses. The trimmed components are indicated by dashed arrows through the appropriate capacitors in Fig. 10.2-8.

Combining the voltage-scaling and charge-scaling approaches allows for some interesting performance tradeoffs. Let us assume that K is the number of bits in the capacitor array of the charge-scaling approach and that M is the number of bits in the resistor array of the voltage-scaling approach. The total number of bits will be designated as N. Let us consider two cases. The first case assumes that the resistors are the MSBs and the capacitors are the LSBs. If we let DL be the measure of differential linearity, then we may write

$$DL = 2^K \left[\frac{\Delta C}{C} + \frac{\Delta R}{R} \right]$$ (13)

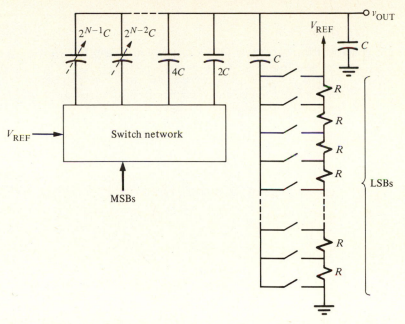

Figure 10.2-8 An illustration of a charge-scaling–voltage-scaling D/A converter.

where $\Delta C/C$ and $\Delta R/R$ are the accuracies of the capacitors and resistors. If we designate IL as the measure of integral linearity, then we may write

$$IL = 0.5 \left[2^K \frac{\Delta C}{C} + 2^N \frac{\Delta R}{R} \right] \qquad (14)$$

The second case assumes that the resistors are the LSBs and the capacitors are the MSBs. In that case we have,

$$DL = 2^N \left[\frac{\Delta C}{C} \right] + \frac{\Delta R}{R} \qquad (15)$$

and

$$IL = 0.5 \left[2^N \frac{\Delta C}{C} + 2^M \frac{\Delta R}{R} \right] \qquad (16)$$

These relationships illustrate the choices that face the designer of D/A converters.

Example 10.2-3

Area Minimization of D/A Converters Using both Voltage and Current Scaling

If A_R is the area of the resistors and A_C is the area of the capacitors, show that for minimum area

$$K - M = \frac{\ln(A_R/A_C)}{\ln(2)}$$

The total area of the resistors plus capacitors can be expressed as

$$A_T = A_R 2^M + A_C 2^K = A_R 2^M + A_C 2^{N-M}$$

Differentiating A_T with respect to M and equating to zero gives

$$\frac{A_R}{A_C} = 2^{K-M}$$

Taking the natural log of this expression verifies the above relationship for minimum area. It is of interest to plot this relationship as a function of A_R/A_C. Fig. 10.2-9 shows the results. When the area of the unit resistor is equal to the unit capacitor, the number of bits implemented by resistors and capacitors should be the same. As the area of the unit resistor increases, the number of bits implemented by resistors decreases with respect to the number of bits implemented by capacitors. Similarly, the number of resistor bits increases as the area of the unit resistor decreases.

The voltage- and charge-scaling converters are seen to be compatible with CMOS technology and to provide D/A converters with a resolution up to 9 bits or more. Some of the techniques (such as self-calibration, which will be described in the last section) will allow the scaling converters to be applicable to 12 bits or more.

10.3   *Serial D/A Converters*

The category of D/A converters considered in this section is the serial D/A converter. A serial D/A converter is one in which the conversion is done sequentially. Typically, one clock pulse is required to convert one bit. Thus, N clock pulses would be required for the typical serial, N-bit D/A converter.

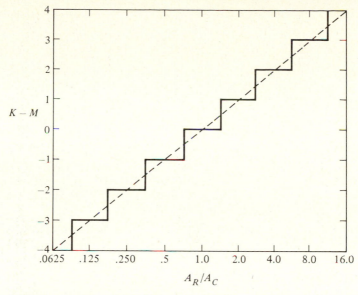

Figure 10.2-9 Relationship of the scaling decomposition to the ratio of a unit resistor area to a unit capacitor area.

The two types of converters that will be examined here are the serial, charge-redistribution and the algorithmic D/A converters.

Fig. 10.3-1 shows the simplified schematic of a serial, charge-redistribution D/A converter. We see that this converter consists of four switches, two equal-valued capacitors and a reference voltage. The function of the switches is as follows: S_1 is called the redistribution switch and places C_1 in parallel with C_2, causing their voltages to become identical through charge redistribution; switch S_2 is used to precharge C_1 to V_{REF}, if the ith bit b_i is a 1, and switch S_3 is used to precharge C_1 to zero volts if the ith bit is 0; switch S_4 is used at the beginning of the conversion process to initially discharge C_2. The following example will illustrate the operation of this converter.

Example 10.3-1

Operation of the Serial D/A Converter

Assume that $C_1 = C_2$ and that the digital word to be converted is given as $b_1 = 1$, $b_2 = 1$, $b_3 = 0$, and $b_4 = 1$. The conversion starts with the closure of switch S_4, so that $v_{C2} = 0$. Since $b_4 = 1$, then switch S_2 is closed causing $v_{C1} = V_{REF}$. Next, switch S_1 is closed causing $v_{C1} = v_{C2} = 0.5\ V_{REF}$. This completes the conversion of the LSB. Fig. 10.3-2

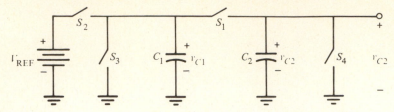

Figure 10.3-1 Simplified schematic of a serial, charge redistribution D/A converter.

illustrates the waveforms across C_1 and C_2 during this example. Going to the next most LSB, b_3, switch S_3 is closed, discharging C_1 to ground. When switch S_1 closes, the voltage across both C_1 and C_2 is 0.25 V_{REF}. Because the remaining two bits are both 1, C_1 will be connected to V_{REF} and then connected to C_2 two times in succession. The final voltage across C_1 and C_2 will be $(^{13}\!/_{16})\,V_{REF}$. This sequence of events will require nine sequential switch closures to complete the conversion.

From the above example it can be seen that the serial D/A converter requires considerable supporting external circuitry necessary to make the decision on which switch to close during the conversion process. Although the circuit for the conversion is extremely simple, several sources of error will limit the performance of this type of D/A converter. These sources of error include the capacitor parasitic capacitances, the switch parasitic capacitances, and the clock feedthrough errors. The capacitors C_1 and C_2 must be matched to within the LSB accuracy. This converter has the advantage of monotonicity and requires very little area for the portion shown in Fig. 10.3-1. An eight-bit converter using this technique has been fabricated and has demonstrated a conversion time of 13.5 μsec [5].

A second approach to serial D/A conversion is called algorithmic [6]. Fig. 10.3-3 illustrates the pipeline approach to implementing an algorithmic D/A converter. Fig. 10.3-3 consists of unit delays and weighted summers. It can be shown that the output of this circuit is

$$V_{out}(z) = [b_1 z^{-1} + 2^{-1} b_2 z^{-2} \qquad\qquad (1)$$
$$+ \cdots + 2^{-(N-1)} b_N z^{-N} + 2^{-N} b_{N+1} z^{-(N+1)}] V_{REF}$$

where b_i is either ± 1. Fig. 10.3-3 shows that it takes $n + 1$ clock pulses for the digital word to be converted to an analog signal, even though a new digital word can be converted on every clock pulse.

The complexity of Fig. 10.3-3 can be reduced using the techniques of

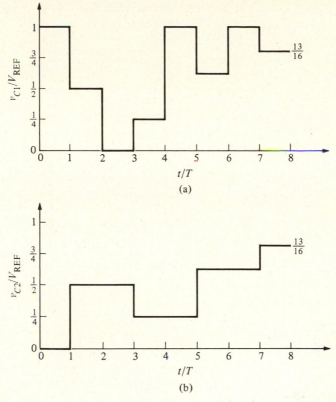

Figure 10.3-2 Waveforms of Fig. 10.3-1 for the conversion of the digital word 1101. (a) Voltage across C_1. (b) Voltage across C_2.

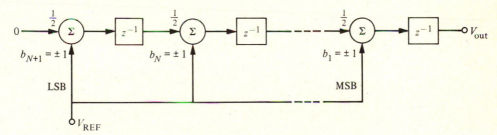

Figure 10.3-3 Pipeline approach to implementing an algorithmic D/A converter.

replication and iteration. Here we shall consider only the iteration approach. Eq. (1) can be rewritten as

$$V_{out}(z) = \frac{b_i z^{-1} V_{REF}}{1 - 0.5z^{-1}} \qquad (2)$$

where all b_i have been assumed to be identical. The fact that each b_i is either ± 1 will be determined in the following realization. Fig. 10.3-4 shows a block diagram realization of Eq. (2). It consists of two switches, A and B. Switch A is closed when the ith bit is 1 and switch B is closed when the ith bit is 0. Then $b_i V_{REF}$ is summed with one-half of the previous output and applied to the sample-and-hold circuit that outputs the result for the ith-bit conversion. The following example illustrates the conversion process.

Example 10.3-2
D/A Conversion Using the Algorithm Method

Assume that the digital word to be converted is 11001 in the order of MSB to LSB. The conversion starts by zeroing the output (not shown on Fig. 10.3-4). Fig. 10.3-5 is a plot of the output of this example. T is the period for one conversion. The process starts with the LSB, which in this case is 1. Switch A is closed and V_{REF} is summed with zero to give an output of $+V_{REF}$. On the second conversion, the bit is zero so that switch B is closed. Thus $-V_{REF}$ is summed with $(1/2)V_{REF}$ giving $-(1/2)V_{REF}$ as the output. On the third conversion, the bit is also 0 so that $-V_{REF}$ is summed with $-(1/4)V_{REF}$ to give an output of $-(5/4)V_{REF}$. On the fourth conversion, the bit is 1 which causes V_{REF} to be summed with $-(5/8)V_{REF}$ giving $+(3/8)V_{REF}$ at the output. Finally, the MSB is unity which causes V_{REF} to be summed with $(3/16)V_{REF}$ giving the final analog output of $+(19/16)V_{REF}$. Because the actual V_{REF} of this example is $\pm V_{REF}$ or $2V_{REF}$, the analog value of the digital word 11001 is 19/32 times $2V_{REF}$ or $(19/16)V_{REF}$.

The algorithmic converter has the primary advantage of being independent of capacitor ratios; it is often called a ratio-independent algorithmic D/A converter. It is necessary for the gain of the 0.5 amplifier of Fig. 10.3-4 to be equal to 0.5 $\pm$ 0.5 LSB in order to be able to resolve to the LSB. Because the gain of the 0.5 amplifier is usually determined by capacitor ratios, the algorithmic converter is not truly independent of capacitor ratios. The algorithmic converter will be presented again under the subject of serial A/D converters. Two serial D/A converters have been presented. The serial D/A converter is seen to be very simple but to require

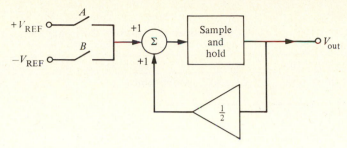

Figure 10.3-4 Equivalent realization of Fig. 10.3-3 using iterative techniques.

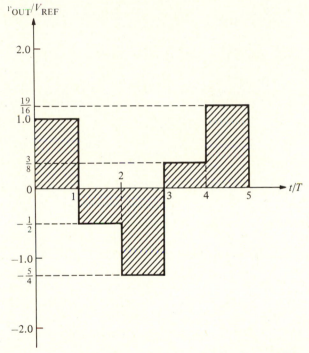

Figure 10.3-5 Output waveform for Fig. 10.3-4 for the conditions of Example 10.3-2.

a longer time for conversion. In some applications, these characteristics are advantageous.

This section and the previous one have presented D/A converter techniques compatible with MOS technology. Table 10.3-1 gives a summary of these D/A converters. In the following sections, we shall examine the complementary subject of A/D converters. Many of the converters that will be discussed use the D/A converters developed and illustrated in this section.

Table 10.3-1

Comparision of D/A Conversion Techniques Compatible with MOS Technology.

D/A Converters	Figure	Advantage	Disadvantage
Voltage Scaling	10.2-1(a)	Monotonic	Large area, sensitive to parasitic capacitances
Charge Scaling	10.2-2(a)	Fast	Large element spread, nonmonotonic
Charge Scaling	10.2-4(a)	Minimum area	Nonmonotonic, divider must be accurate
Voltage Scaling, Charge Scaling	10.2-6	Monotonic, simple	Must trim for absolute accuracy
Charge Scaling Voltage Scaling	10.2-8	Monotonic	Must trim for absolute accuracy
Serial, Charge Redistribution	10.3-1	Simple, minimum area	Slow, requires complex external circuitry, precise capacitor ratios
Serial, algorithmic	10.3-3	Simple, minimum area	Slow, requires complex external circuitry, precise capacitor ratios

10.4 *Introduction and Characterization of Analog-to-Digital Converters*

This section examines the principles and characteristics of analog-to-digital converters. The objective of an A/D converter is to determine the output digital word corresponding to an analog input signal. The A/D converter will require a sample-and-hold circuit at the input because it is not possible to convert a changing analog input signal. We shall see that A/D converters make use of D/A converters presented in the previous sections. The types of A/D converters considered include the low-speed (serial), medium-speed, high-speed, and high-performance A/D converters.

Fig. 10.4-1 shows the block diagram of a basic A/D converter. The input to the A/D comes from a sample-and-hold circuit and is designated as V_{in}^*. This input, along with a reference voltage V_{REF} is used to determine the digital word that best represents the sampled analog input signal. The means by which the conversion is accomplished may be different from that suggested in Fig. 10.4-1, where a set of fixed reference levels V_i are applied to comparators whose outputs X_i are then decoded into an output digital word. Regardless of the means of conversion, the A/D converter is a device that converts a continuous range of input amplitude levels into a discrete, finite set of digital words.

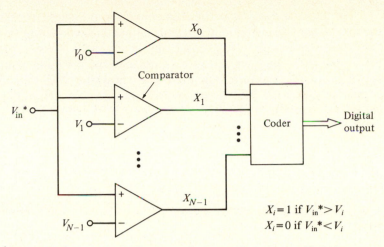

$$X_i = 1 \text{ if } V_{in}{}^* > V_i$$
$$X_i = 0 \text{ if } V_{in}{}^* < V_i$$

Figure 10.4-1 Block diagram of a general analog-to-digital converter.

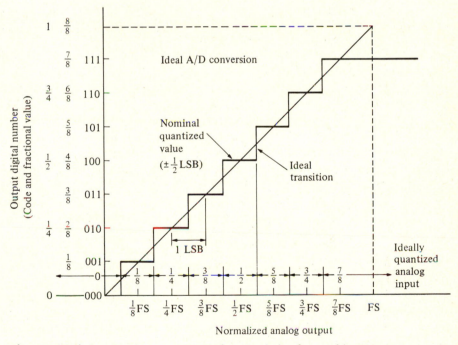

Figure 10.4-2 Ideal input-output characteristics for a 3-bit A/D converter. (By permission from D.H. Sheingold, *Analog-Digital Conversion Handbook*, Analog Devices, Inc., 1972.)

The characterization of the D/A converter in Section 10.1 is also valid for the A/D converter if the input and output definitions are interchanged. Fig. 10.4-2 shows the ideal input-output characteristics of an A/D converter. Fig. 10.4-3 shows various examples of the previous definitions now applied to A/D converters. Fig. 10.4-3(a) shows the offset error in an A/D converter. Fig. 10.4-3(b) gives an example of gain error. Fig. 10.4-3(c) shows an example of absolute linearity error. Finally, Fig. 10.4-3(d) indicates differential linearity for an A/D converter.

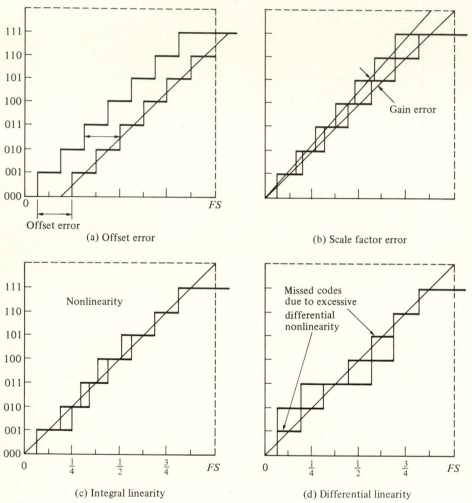

(a) Offset error

(b) Scale factor error

(c) Integral linearity

(d) Differential linearity

Figure 10.4-3 Characterization of a 3-bit A/D converter. (a) Offset error. (b) Scale factor (gain) error. (c) Integral linearity. (d) Differential linearity. (By permission from D.H. Sheingold, *Analog-Digital Conversion Handbook,* Analog Devices, Inc., 1972.)

The above characteristics of an A/D converter are classified as static characteristics. Dynamic characteristics are primarily associated with the speed of operation. Because either the input or the output of a converter is a digital word, sampling or discrete-time signals are inherent. Therefore, the speed of the converter is of interest. The *conversion time* is the time from the application of the start convert signal to the availability of the completed output signal (digital or analog). Typically, a D/A converter can provide an analog output signal as soon as the digital word is applied (except for the serial converters). On the other hand, A/D converters may require several clock cycles after the application of the analog input signal before the output digital word is available.

Because a sample-and-hold circuit is a key aspect of the A/D converter, it is worthwhile to examine it in greater detail than we did in Sec. 10.1. Fig. 10.4-4 shows the waveforms of a practical sample-and-hold circuit. The *acquisition time,* indicated by t_a, is the time during which the sample-and-hold circuit must remain in the sample mode to ensure that the subsequent hold-mode output will be within a specified error band of the input level that existed at the instant of the sample-and-hold conversion. The acquisition time assumes that the gain and offset effects have been removed. The *settling time,* indicated by t_s, is the time interval between the sample-and-hold transition command and the time when the output transient and subsequent ringing have settled to within a specified error band. Thus the minimum sample-and-hold time would be

$$T_{\text{sample}} = t_s + t_a \qquad\qquad (1)$$

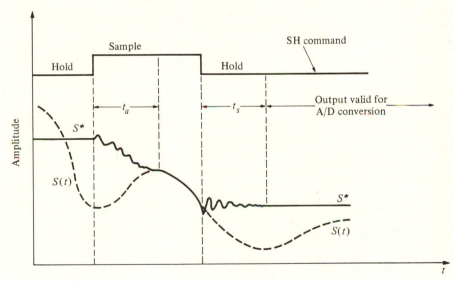

Figure 10.4-4   Waveforms for a sample-and-hold circuit.

The minimum conversion time for an A/D would be equal to T_{sample} and the maximum sample rate is

$$f_{sample} = \frac{1}{T_{sample}} \qquad (2)$$

The dynamic performance of the converter will depend largely on the dynamic characteristics of the op amps and comparators. Therefore the slew rate, settling time, and overload recovery time of these circuits will be of importance.

In addition to the dynamic characteristics of converters, there are characteristics having to do with stability of operation that define the immunity of the converter to time, temperature, power supplies, and component aging. These characteristics are typically expressed in the change of the converter performance parameter per unit change in the affecting influence. Examples include the *temperature coefficient of linearity, temperature coefficient of gain,* and *temperature coefficient of differential nonlinearity.* Also of importance to the stability of a converter is the voltage reference that will be presented in Sec. 11.1. This voltage reference can be supplied externally or internally to the integrated-circuit converter. It is important

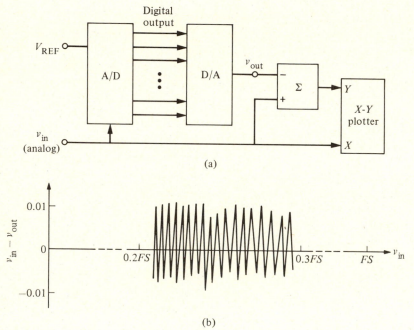

Figure 10.4-5 (a) A method for testing A/D and D/A converters. (b) Typical output for the error voltage of (a).

that the reference provide the stability necessary for the proper operation of the converter.

It is of interest to consider how one can test A/D and D/A converters. Testing is divided into static and dynamic tests for D/A and A/D converters. Most test configurations demand the ability to resolve the analog signal to within ±0.5 LSB, which can be very demanding if the number of bits is large. Techniques for testing both types of converters can be found in more detail in the literature [7–9]. Often, a microprocessor is used to perform the tests on the converter. One simple means of testing both the D/A and A/D converter is shown in Fig. 10.4-5(a). In this configuration, the A/D converter is cascaded with the D/A converter. The output of the D/A converter is subtracted from the input of the A/D converter, resulting in an error voltage that can be plotted on an x-y recorder as a function of the amplitude of the analog input signal. A typical portion of the error voltage for a 12-bit converter is shown in Fig. 10.4-5(b). The order of the A/D and D/A converters in Fig. 10.4-5(a) can be interchanged so that the input and output are digital words which can be compared, and several methods of presenting the performance are possible (such as a parity check algorithm or a scatter plot of the input word plotted against the output word).

10.5   *Serial A/D Converters*

The serial A/D converter is similar to the serial D/A converter in that it performs serial operations until the conversion is complete. We shall examine two architectures called the single slope and the dual slope. Fig. 10.5-1 gives the block diagram of a single-slope, serial A/D converter. This

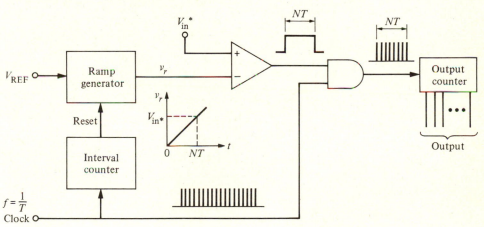

Figure 10.5-1   Block diagram of a single-slope, serial A/D converter.

type of converter consists of a ramp generator, an interval counter, a comparator, an AND gate, and a counter that generates the output digital word. At the beginning of a conversion cycle, the analog input is sampled and held and applied to the positive terminal of the comparator. The counters are reset and a clock is applied to both the interval counter and the AND gate. On the first clock pulse, the ramp generator begins to integrate the reference voltage V_{REF}. If V_{in}^* is greater than the initial output of the ramp generator, then the output of the ramp generator, which is applied to the negative terminal of the comparator, begins to rise. Because V_{in}^* is greater than the output of the ramp generator, the output of the comparator is high and each clock pulse applied to the AND gate causes the counter at the output to count. Finally, when the output of the ramp generator is equal to V_{in}^*, the output of the comparator goes low and the output counter is now inhibited. The binary number representing the state of the output counter can now be converted to the desired digital word format.

The single-slope A/D converter can have many different implementations. For example, the interval counter can be replaced by logic to detect the state of the comparator output and reset the ramp generator when its output has exceeded V_{in}^*. The serial A/D converter has the advantage of simplicity of operation. A disadvantage of the single-slope A/D converter is that it is subject to error in the ramp generator and is unipolar. Another disadvantage of the single-slope A/D converter is that a long conversion time is required if the input voltage is near the value of V_{REF}.

The second type of serial A/D converter is called the dual-slope converter. A block diagram of a dual-slope A/D converter is shown in Fig. 10.5-2. The basic advantage of this architecture is that it eliminates the dependence of the conversion process on the linearity and accuracy of the slope. Initially, v_{int} is zero and the input is sampled and held. (In this scheme, it is necessary for V_{in}^* to be positive.) The conversion process begins by resetting the positive integrator by integrating a positive voltage (not shown)

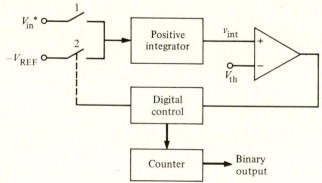

Figure 10.5-2 Block diagram of a dual-slope A/D converter.

until the output of the integrator is equal to the threshold V_{th} of the comparator. Next, switch 1 is closed and V_{in}^* is integrated for N_{ref} number of clock cycles. Fig. 10.5-3 illustrates the conversion process. It is seen that the slope of the voltage at V_{int} is proportional to the amplitude of V_{in}^*. The voltage $v_{int}(t)$ during this time is given as

$$v_{int}(t) = K \int_0^{N_{ref}T} V_{in}^* \, dt + v_{int}(0) = K N_{ref} T V_{in}^* + V_{th} \tag{1}$$

where T is the clock period. At the end of N_{ref} counts, the carry output of the counter is applied to switch 2 and causes $-V_{REF}$ to be applied to the integrator. Now the integrator integrates negatively with a constant slope, because V_{REF} is constant. When $v_{int}(t)$ becomes less than the value of V_{th}, the counter is stopped and binary count can be converted into the digital word. This is demonstrated by considering $v_{int}(t)$ during the time designated as t_2 on Fig. 10.5-3. This voltage is given as

$$v_{int}(t) = v_{int}(0) + K \int_0^{N_{out}T} (-V_{REF}) \, dt \tag{2}$$

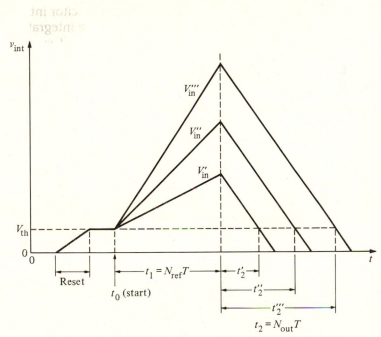

Figure 10.5-3 Waveforms of the dual slope A/D converter of Fig. 10.5-2. $V_{in}'' > V_{in}' > V_{in}'$.

However, when $t = N_{out}T$, then Eq. (2) becomes

$$V_{int}(N_{out}T) = [KN_{ref}TV^*_{in} + V_{th}] - KV_{REF}N_{out}T \tag{3}$$

Because $V_{int}(N_{out}T) = V_{th}$, then Eq. (3) can be solved for N_{out} giving

$$N_{out} = N_{ref}\frac{V^*_{in}}{V_{REF}} \tag{4}$$

It is seen that N_{out} will be some fraction of N_{ref} where that fraction corresponds to the ratio of V^*_{in} to V_{REF}.

The output of the serial, dual-slope D/A converter (N_{out}), is not a function of the threshold of the comparator, the slope of the integrator, or the clock rate. Therefore, it is a very accurate method of conversion. The only disadvantage is that it takes a worst-case time of $2(2^N)T$ for a conversion where N is the number of bits of the A/D converter. The positive integrator of this scheme can be replaced by a switched-capacitor integrator [10]. Fig. 10.5-4(a) shows the implementation of a positive integrator. ϕ_1 and ϕ_2 are nonoverlapping clocks which cause the designated switch to be closed when they are in the high state. During the ϕ_1 phase, C_1 is charged to the value of $v_1(t)$. Normally, $v_1(t)$ is an analog signal which has been sampled

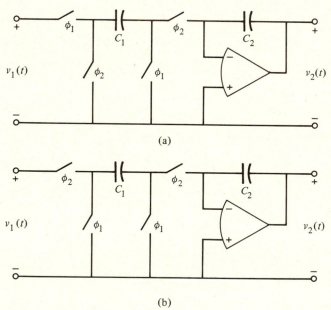

(a)

(b)

Figure 10.5-4   (a) Noninverting switched-capacitor integrator. (b) Inverting switched-capacitor integrator.

and held. During the ϕ_2 phase, C_1 is inverted and discharged by the virtual ground of the op amp onto the capacitor C_2. The charge across C_2 at time nT, $q_2[nT]$, can be written as

$$q_2[nT] = q_2[(n-1)T] + q_1[(n-1)T] \tag{5}$$

where $q_2[(n-1)T]$ is the charge across C_2 at the previous clock period (when C_1 was charged) and $q_1[(n-1)T]$ is the charge on C_1 at the previous clock period. Eq. (5) can be written as

$$C_2 v_2[nT] = C_2 v_2[(n-1)T] + C_1 v_1[(n-1)T] \tag{6}$$

Using z-domain notation, we can write Eq. (6) as

$$C_2 V_2(z) = C_2 z^{-1} V_2(z) + C_1 z^{-1} V_1(z) \tag{7}$$

Solving for $V_2(z)/V_1(z)$ gives

$$H(z) = \frac{V_2(z)}{V_1(z)} = \frac{C_1}{C_2}\left[\frac{z^{-1}}{1-z^{-1}}\right] \tag{8}$$

Replacing z by $e^{j\omega T}$ gives

$$H(e^{j\omega T}) = \frac{\omega_o}{j\omega}\left[\frac{(\omega T/2)}{\sin(\omega T/2)}\right]\exp(j\omega T/2) \tag{9}$$

where ω_o is defined as $C_1/(TC_2)$. It can be shown that if $\omega T/2$ is much less than one ($f \ll f_c$ where f_c is the clock frequency) that Eq. (9) reduces to

$$H(e^{j\omega T}) \cong \frac{\omega_o}{j\omega} \tag{10}$$

which is the frequency-domain response of a noninverting integrator. Consequently, the circuit of Fig. 10.5-4(a) can be used to replace the positive integrator of Fig. 10.5-3 providing the clock frequency of the switched-capacitor integrator of Fig. 10.5-4(a) is greater (e.g., 10-50 times) than the highest frequency of the signal being integrated. Fig. 10.5-4(b) gives a switched-capacitor realization of an inverting integrator which will have the approximate transfer function of Eq. (10) times a minus sign if the clock frequency is greater than the highest signal frequency.

The above two examples of serial A/D converters are representative of the architecture and resulting performance. Other forms of serial conversion exist in the literature [11, 12]. The serial A/D converter is expected to be slow but to provide a high resolution. Typical values for serial A/D converters are conversion frequencies of less than 100 Hz and greater than 12 bits.

10.6   *Medium-Speed A/D Converters*

A second category of A/D converters is classified as medium-speed A/D converters. This class of A/D converters converts an analog input into an N-bit digital word in approximately N clock cycles. Consequently, the conversion time is less than that of the serial converters, without a significant increase in the circuit complexity. The medium-speed A/D converters examined here will include the successive-approximation converters (which use a combination of voltage-scaling and charge-scaling D/A converters), serial D/A converters, and algorithmic D/A converters.

Fig. 10.6-1 illustrates the architecture of a successive-approximation A/D converter. This converter consists of a comparator, a D/A converter, and digital control logic. The function of the digital control logic is to determine the value of each bit in a sequential manner based on the output of the comparator. To illustrate the conversion process, assume that the converter is unipolar (only analog signals of one polarity can be applied). The conversion cycle begins by sampling the analog input signal to be converted. Next, the digital control circuit assumes that the MSB is 1 and all

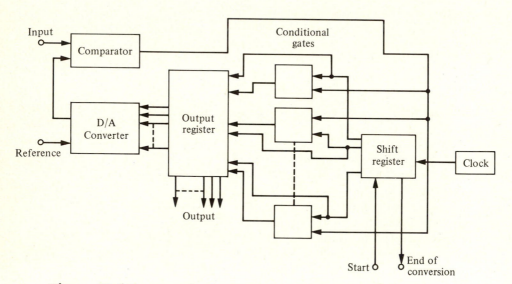

Figure 10.6-1   Example of a successive-approximation A/D converter architecture. (R. Hnatek, *A User's Handbook of D/A and A/D Converters*, John Wiley and Sons, Inc., New York, NY, © 1976, John Wiley and Sons, Inc.).

other bits are zero. This digital word is applied to the D/A converter, which generates an analog signal of 0.5 V_{REF}. This is then compared to the sampled analog input V_{in}^*. If the comparator output is high, then the digital control logic makes the MSB 1. If the comparator output is low, the digital control logic makes the MSB 0. This completes the first step in the approximation sequence. At this point the value of the MSB is known. The approximation process continues by once more applying a digital word to the D/A converter, with the MSB having its proven value, the next lower bit a "guess" of 1, and all the other remaining bits having a value of 0. Again, the sampled input is compared to the output of the D/A converter with this digital word applied. If the comparator is high, the second bit is proven to be 1. If the comparator is low, the second bit is 0. The process continues in this manner until all bits of the digital word have been decided by successive approximation.

Fig. 10.6-2 shows how the successive-approximation sequence works in converging to the analog output of the D/A converter closest to the sampled analog input. It is seen that the number of cycles for conversion to an N-bit word is N. It is also observed that as N becomes large, the ability of

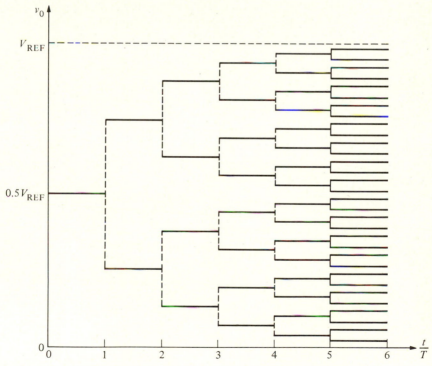

Figure 10.6-2   The successive-approximation process.

the comparator to distinguish between almost identical signals must increase. Bipolar A/D conversion can be achieved by using a sign bit to choose either $+V_{REF}$ or $-V_{REF}$.

The digital control logic is often called a successive-approximation register (SAR). An example of a 5-bit SAR is shown in Fig. 10.6-3. This SAR has the advantage of compatibility with a bit-slice approach which makes it attractive for integrated-circuit implementation. The bit-slice consists of a shift register (SR), and AND gate (G)., registers or flip-flops (FF), and an analog switch (AS).

Fig. 10.6-4 shows an example of a successive-approximation A/D converter that uses the voltage-scaling and charge-scaling D/A converter of Fig. 10.2-6. The extra components in addition to the D/A converter include a comparator and a SAR. From the concepts of Chapter 7, we know that the comparator should have a gain greater than $(V_L 2^{M+K}/V_{REF})$ where V_L is the minimum output swing of the comparator required by the logic circuit it drives. For example, if $M + K = 12$ and $V_L = V_{REF}$, then the comparator must have a voltage gain of at least 4096, where M equals the number of bits scaled by voltage and K equals the number of bits scaled by charge.

The conversion operation is described as follows. With S_F closed, the bottom plates of the capacitors are connected through switch S_B to V_{in}^*. The voltage stored on the capacitor array at the end of the sampling period is actually V_{in}^* minus the threshold voltage of the comparator, which removes the threshold as a source of offset error. After switch S_F is opened, a successive-approximation search among the resistor string taps is performed to find the segment in which the stored sample lies. Next, buses A and B are switched to the ends of the resistor defining this segment. Finally, the capacitor bottom plates are switched in a successive-approximation sequence until the comparator input voltage converges back to the threshold voltage. The sequence of comparator outputs is a digital code corresponding to the unknown analog input signal. By driving the capacitor array directly through the MOS switches, there are no offset errors if enough time is allowed for the switching transients to settle. Also the parasitic capacitors of all switches except S_F do not cause errors because every node is driven to a final voltage which is independent of the capacitor parasitics after the switch transients have settled. The A/D converter in Fig. 10.6-4 is capable of 12-bit, monotonic conversion with a differential linearity of less than $\pm\frac{1}{2}$ LSB and a conversion time of 50 microseconds [4].

A successive-approximation A/D converter using the serial D/A converter of Fig. 10.3-1 is shown in Fig. 10.6-5. This converter works by converting the MSB a_N first. (The ith bit is denoted as d_i for D/A conversion and a_i for A/D conversion.) The control logic takes a very simple form because the D/A input string at any given point in the conversion is just the previously encoded word taken LSB first. For example, consider the point during the A/D conversion where the first K MSBs have been decided. To decide

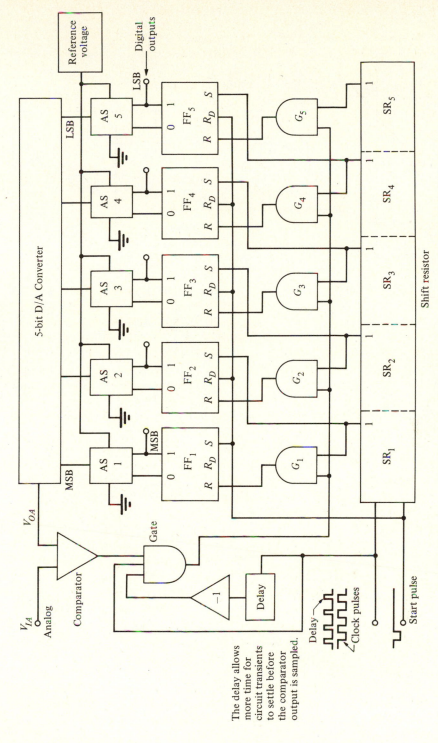

Figure 10.6-3 A 5-bit successive-approximation A/D converter with shift register control. (E. R. Hnatek, *A User's Handbook of D/A and A/D Converters*, John Wiley and Sons, Inc., New York, NY, © 1976, John Wiley and Sons, Inc.).

The delay allows more time for circuit transients to settle before the comparator output is sampled.

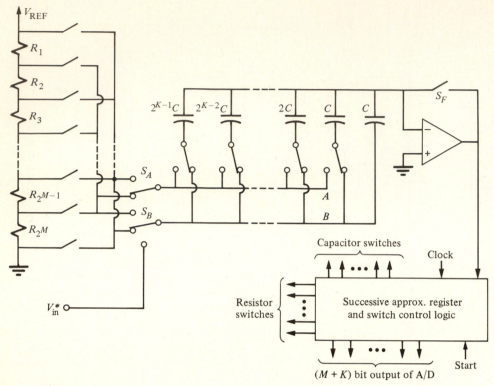

Figure 10.6-4 A voltage-scaling, charge-scaling, successive-approximation A/D converter.

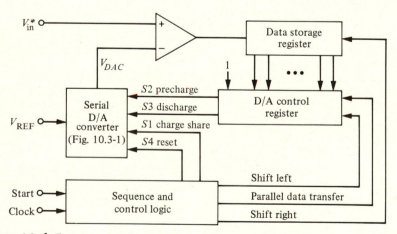

Figure 10.6-5 A successive-approximation A/D converter using the serial D/A converter of Fig. 10.3-1.

Table 10.6-1

Conversion Sequence for the Serial D/A Converter of Fig. 10.6-5.

D/A Conversion Number	D/A Input Word						Comparator Output	Number of Charging Steps
	d_1	d_2	d_3	$\cdots$	d_{N-1}	d_N		
1	1	—	—		—	—	a_N	2
2	1	a_N	—		—	—	a_{N-1}	4
3	1	a_{N-1}	a_N		—	—	a_{N-2}	6
.	.	.	.		.	.	.	.
.	.	.	.		.	.	.	.
.	.	.	.		.	.	.	.
N	1	a_2	a_3	$\cdots$	a_{N-1}	a_N	a_1	2N
	Total number of charging steps = $N(N+1)$							

the $(K+1)$ MSB, a $(K+1)$-bit word is formed in the D/A control register by adding a 1 as the LSB to the K-bit word already encoded in the data storage register. A $(K+1)$-bit D/A conversion then establishes the value of a_{N-K} by comparison with the unknown voltage V_{in}^*. The bit is then stored in the data-storage register and the next serial D/A conversion is initiated. The conversion sequence is shown in detail in Table 10.6-1. Fig. 10.6-6 illustrates a four-bit A/D conversion for $V_{in}^* = (^{13}\!/_{16})V_{REF}$. Altogether, $N(N+1)$ clock cycles are required for a N-bit A/D converter using the configuration of Fig. 10.6-5.

An algorithmic A/D converter patterned after the algorithmic D/A converter of the previous section is shown in Fig. 10.6-7. This N bit, A/D converter consists of N stages and N comparators for determining the signs of the N outputs. Each stage takes its input, multiplies it by 2 and adds or subtracts the reference voltage depending upon the sign of the previous output. The comparator outputs form an N-bit digital representation of the bipolar analog input to the first stage. The operation of the algorithmic A/D converter can be demonstrated by the following example:

Example 10.6-1

Illustration of the Operation of the Algorithmic A/D Converter

Assume that the sampled analog input to a 4-bit algorithmic A/D converter is 1.50 V. If V_{REF} is equal to 5 V, then the conversion proceeds as follows. Since V_{in}^* (= 1.50 V) is positive, the output of the comparator of stage 1 is high, which corresponds to a digital 1. Stage 1 then multiplies this value by two to get 3 V and subtracts V_{REF} to obtain an output of -2.00 V. Stage 2 input sees a negative value

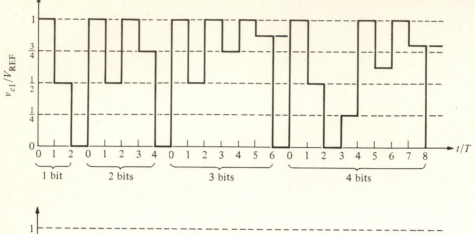

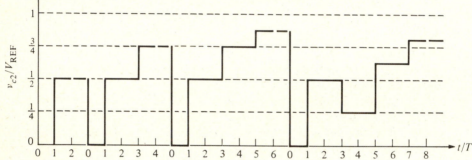

Figure 10.6-6   Illustration of the operation of the successive-approximation A/D converter of Fig. 10.6-5 for the conversion of the sampled analog input voltage of $(^{13}/_{16})$ V_{ref}. The digital word out is $b_0 = 1$, $b_1 = 1$, $b_2 = 0$, and $b_3 = 1$.

which causes the comparator of this stage to be low which is equivalent to a digital 0. Stage 2 then multiplies -2.00 V by 2 and adds the 5.0 V reference to output a value of 1.00 volts. Because the output of stage 2 is positive, the comparator of stage 3 is high, which causes the 1.00 volt to be multiplied by 2 and subtracted by 5 giving a stage 3 output of -3.00 V. The conversion ends when the comparator of the fourth stage goes low because of the negative input voltage from stage 3.

The digital output word is 1010 for this example. To determine whether this is correct, we use the following formula.

$$V_{analog} = V_{REF}[b_1 2^{-1} + b_2 2^{-2} + b_3 2^{-3} + \cdots + b_N 2^{-N}]$$

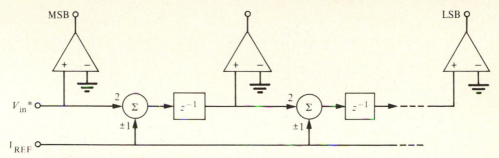

Figure 10.6-7   Pipeline implementation of the algorithmic A/D converter.

where b_i is $+1$ if the ith-bit is 1 and -1 if the ith-bit is 0. In this example, we see that

$$V_{\text{analog}} = 5\left[\frac{1}{2} - \frac{1}{4} + \frac{1}{8} - \frac{1}{16}\right] = 5(0.3125) = 1.5625$$

It is seen that the value of V_{analog} would eventually converge on the value of 1.50.

The algorithmic A/D converter of Fig. 10.6-7 has the disadvantage that the time needed to convert a sample is N clock cycles, although one complete conversion can be obtained at each clock cycle. The algorithmic A/D converter is considered to be ratio-independent because the performance does not depend upon the ratio accuracy of a capacitor or resistor array. The accuracy of the multiplication by two of the pipeline configuration of the algorithmic A/D converter in Fig. 10.6-7 must be accurate to within 1 LSB.

The iterative reduction of Fig. 10.6-7 can be applied to the A/D converter in a manner similar to that done for the algorithmic, pipeline D/A converter. The analog output of the ith stage can be expressed as

$$V_{oi} = [2V_{o,i-1} - b_iV_{\text{REF}}]z^{-1} \tag{1}$$

where b_i is $+1$ if the ith-bit is 1 and -1 if the ith-bit is 0. This equation can be implemented with the circuit in Fig. 10.6-8(a). The next step is to incorporate the ability to sample the analog input voltage at the start of the conversion. This step is shown in Fig. 10.6-8(b) [13]. In this implementation, $-V_{\text{REF}}$ has been replaced with ground for simplicity. The iterative version of the algorithmic A/D converter consists of a sample-and-hold circuit, a gain-of-two amplifier, a comparator, and a reference-subtraction circuit.

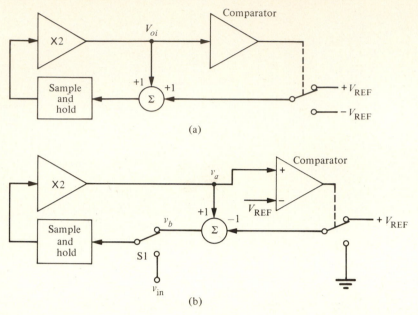

Figure 10.6-8 (a) Realization of Eq. (1). (b) Implementation of the iterative, algorithmic A/D converter.

The operation of the converter consists of first sampling the input signal by connecting switch S1 to v_{in}. V^*_{in} is then applied to the gain of two amplifier. To extract the digital information from the input signal, the resultant signal, denoted as v_a, is compared to the reference voltage. If v_a is larger than V_{REF}, the corresponding bit is set to 1 and the reference voltage is then subtracted from v_a. If v_a is less than V_{REF}, the corresponding bit is set to 0 and v_a is unchanged. The resultant signal, denoted by v_b, is then transferred by means of switch S1 back into the analog loop for another iteration. This process continues until the desired number of bits have been obtained, whereupon a new sampled value of the input signal will be processed. The digital word is processed in a serial manner with the MSB first. An example illustrates the process.

Example 10.6-2

Conversion Process of an Iterative, Algorithmic A/D Converter

The iterative, algorithmic A/D converter of Fig. 10.6-8(b) is to be used to convert an analog signal of $0.8V_{REF}$. Fig. 10.6-9 shows the

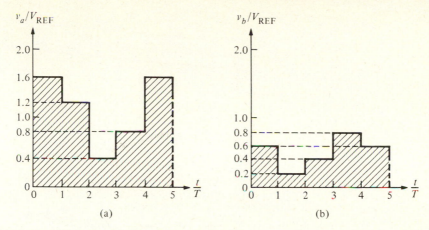

Figure 10.6-9 Waveforms for Ex. 10.6-2 at (a) v_a and (b) v_b of Fig. 10.6-8 (b).

waveforms for v_a and v_b during the process. T is the time for one iteration cycle. In the first iteration, the analog input of $0.8V_{REF}$ is applied by switch S1 and results in a value of v_a of $1.6V_{REF}$ which corresponds to a value of v_b of $0.6V_{REF}$ and the MSB as 1. During the next iteration, v_b is multiplied by two to give v_a of $1.2V_{REF}$. Thus the next bit is also 1 and v_b is $0.2V_{REF}$. v_a during the third iteration is $0.4V_{REF}$, making the next bit 0 and a value of $0.4V_{REF}$ for v_b. The fourth iteration gives v_a as $0.8V_{REF}$, which gives $v_b = 0.8V_{REF}$ and the fourth bit as 0. The fifth iteration gives $v_a = 1.6V_{REF}$, $v_b = 0.6V_{REF}$ and the fifth bit as 1. This procedure continues as long as desired. The digital word after the fifth iteration is 11001 and is equivalent to an analog voltage of $0.78125V_{REF}$.

The iterative, algorithmic A/D converter requires very little precision hardware. Its implementation in a monolithic technology can therefore be area efficient. A distinct advantage over the pipeline configuration is that the amplifiers having a gain of two are identical because only one amplifier is used in an iterative manner. Thus only one accurate gain-of-two amplifier is required. Sources of error of this A/D converter include low operational amplifier gain, finite input-offset voltage in the operational amplifier, charge injection from the MOS switches, and capacitance voltage dependence.

A 12-bit A/D converter using the approach of Fig. 10.6-8(b) has been integrated and is shown in Fig. 10.6-10. Experimental performance resulted

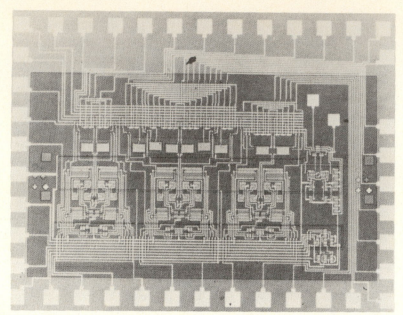

Figure 10.6-10 Microphotograph of an iterative implementation of the algorithmic A/D converter ([13], © 1984 IEEE).

in a differential linearity and integral linearity of 0.019% (0.8 LSB) and 0.034% (1.5 LSB), respectively, for a sample rate of 4 kHz. These values increased to 0.022% (0.9 LSB) and 0.081% (3.2 LSB) for a sample rate of 8 kHz.

The successive-approximation and algorithmic A/D converters have been presented as examples of medium-speed A/D converters. The successive-approximation A/D converter is a very general realization for the medium-speed A/D converter. It can make use of any of the previous D/A converters, as we have illustrated. If serial D/A converters are used, the conversion time of the successive-approximation converter is increased and the area required is decreased. In general, medium-speed A/D converters can have conversion rates that fall within the 10^4 to 10^5 conversions/sec. range. They are also capable of 8 to 12 bits of untrimmed accuracy. The number of bits can be increased if trimming is permitted.

10.7 *High Speed and High Performance A/D Converters*

In many applications, it is necessary to have a smaller conversion time than is possible with the previous A/D converter architectures. This has led to the development of high-speed A/D converters that use parallel techniques to achieve short conversion times. The ultimate conversion speed is one clock cycle, which would typically consist of a set-up and convert phase. Some of the high-speed architectures compromise speed with area and require more than one clock cycle but less than the N clock cycles required for the medium-speed A/D architectures. Another method of improving the speed of the converter is to increase the speed of the individual components. Typically the comparator sample time T_{sample} [see Eq. (1) of Sec. 10.4] is the limiting factor for the speed. In this presentation, we shall consider the parallel, time-interweaved, 2 step, and ripple approaches to implementing a high-speed, A/D converter.

Fig. 10.4-1 represents a general block diagram of a high-speed A/D converter known as the parallel or flash A/D converter. An example of this converter is illustrated in Fig. 10.7-1. Fig. 10.7-1 is a 3-bit, parallel A/D converter. V_{REF} is divided into eight values as indicated on the figure. Each of these values is applied to the positive terminal of a comparator. The output of the comparators are taken to a digital decoding network that

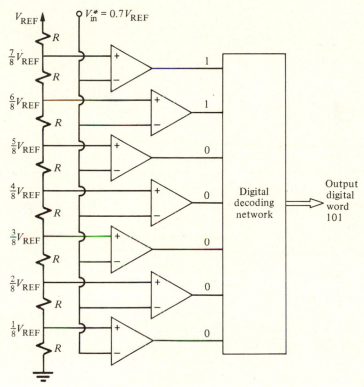

Figure 10.7-1   A 3-bit, parallel A/D converter.

determines the digital output word from the comparator outputs. For example, if V_{in}^* is 0.7 V_{REF}, then the top two comparator's outputs are 1, and the bottom five are 0. The digital decoding network would identify 101 as the corresponding digital word. Many versions of this basic concept exist. For example, one may wish to have the voltage at the taps be in multiples of $V_{REF}/16$ with $V_{REF}/8$ voltage differences between the taps. Also, the resistor string can be connected between $+V_{REF}$ and $-V_{REF}$ to achieve bipolar conversion.

The parallel A/D converter of Fig. 10.7-1 converts the analog signal to a digital word in one clock cycle which has two phase periods. During the first phase period, the analog input voltage is sampled and applied to the comparator inputs. During the second phase period the digital decoding network determines the correct output digital word and stores it in a register/buffer. Thus, the conversion time is limited by how fast this sequence of events can occur. Typical clock frequencies can be as high as 10 MHz for CMOS technology. This gives a theoretical conversion time of 100 nanoseconds. Unfortunately, the sample-and-hold time is larger than these values and prevents these conversion times from being realized. Another problem is that as N increases, the number of comparators required is $2^N - 1$. For N greater than 6, too much area is required. Other methods we shall discuss give almost the same conversion times with much more efficient utilization of chip area.

One method of achieving small system-conversion times is to use the slower A/D converters in parallel. This is called time-interweaving (shown in Fig. 10.7-2). Here, M successive-approximation A/D converters are used in parallel to complete the N-bit conversion of one analog signal per clock cycle. The sample-and-hold circuits consecutively sample and apply the input analog signal to their respective A/D converters. N clock cycles later, the A/D converter provides a digital word out. If $M = N$, then a digital word is converted at every cycle. If one examines the chip area used for the parallel A/D converter architecture compared with the time-interweaved method for $M = N$, the minimum area is likely to be somewhere between these extremes.

Combining the parallel approach with a series approach results in an A/D converter architecture with high speed and reasonable area. Fig. 10.7-3 shows a $2M$-bit A/D converter using two M-bit parallel A/D converters. The method first converts the M MSBs followed by the conversion of the M LSBs. Consequently, only $2^{M+1} - 2$ comparators are required to convert a $2M$-bit digital word. For the configuration of Fig. 10.7-3, the analog input is applied to the left-hand string of $2^M - 1$ comparators during the first clock phase and the M MSBs are decoded during the second clock phase. During the third clock phase, the M MSBs are converted into an analog equivalent, subtracted from V_{REF}, multiplied by a gain of 2^M, and applied to the right-hand string of $2^M - 1$ comparators. Finally, during the fourth

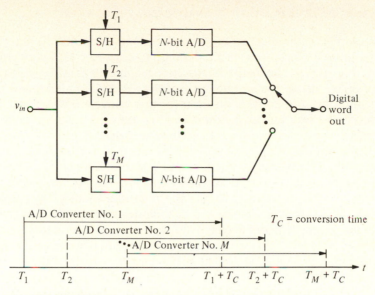

Figure 10.7-2 A time-interweaved A/D converter array.

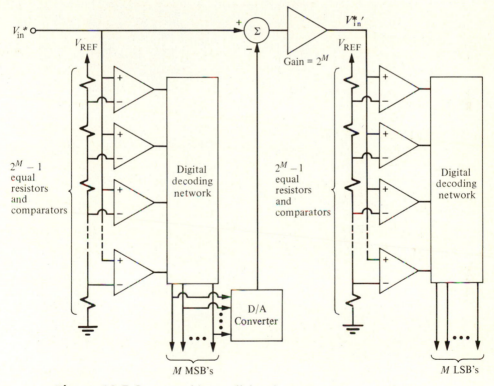

Figure 10.7-3 A $2M$-bit, parallel-series A/D converter configuration.

clock phase, the M LSBs are decoded. Thus, if the clock has two phases, then in two clock cycles a $2M$-bit digital word has been converted.

Fig. 10.7-4 shows the microphotograph of a parallel-series, 8-bit, A/D converter using the architecture of Fig. 10.7-3 implemented using MOS technology. The conversion time for this converter was in the range of 2 μsecs. and was limited by the sampling time of the comparators and the settling times for the logic decoding network. It is possible to make the conversion time of Fig. 10.7-3 in 3 phases if necessary because the second and third phases can be combined into a single phase.

Another approach to achieving high-speed, A/D conversion is shown in Fig. 10.7-5. This method has the potential of very-high-speed conversion. The approach for an 8-bit converter uses a resistor string of 256 equal-valued resistors between V_{REF} and ground. The 256-resistor string is divided into 16 equal parts, each consisting of 16 series resistors. The voltage across each segment of 16 series resistors is applied to a string of 15

Figure 10.7-4 Microphotograph of the implementation of Fig. 10.7-3 for $M = 4$. ([10], © 1984 Van Nostrand Reinhold)

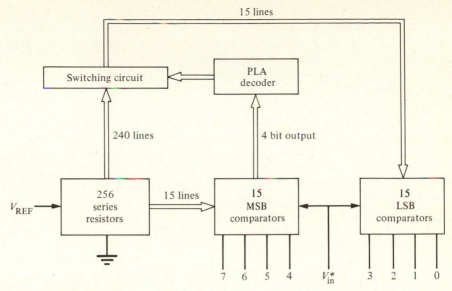

Figure 10.7-5   Block diagram of an 8-bit, ripple A/D converter.

comparators that will provide the information necessary to decode the 4 MSBs. The remaining 240 taps in the resistor string are taken to a switching network controlled by the 4 MSBs. The switching network selects which of the 16 series-resistor segments the analog signal is contained within. The taps of the selected segment are taken to another string of 15 comparators that supplies the information necessary to decode the 4 LSBs. Although it is possible to let the circuit "ripple" at its own speed, it is better to clock the above sequence in order to prevent the possible influence of the LSBs on the MSBs.

Many of the previous A/D converters use methods which were available before the advent of CMOS technology. These converters happened to be compatible with CMOS technology and therefore have been presented. The performance of such converters was not necessarily optimized for CMOS technology. Many of these converters require trimming in order to realize a larger number of bits. Recently, several methods have been introduced which optimize the performance of A/D converters implemented in CMOS technology. These converters can convert at least 12 bits at a frequency of 10 kHz or greater with no trimming needed. Such converters offer numerous applications in the area of digital audio.

Two methods of achieving higher performance A/D conversion will be presented in this section. The first uses self-calibrating methods to extend the accuracy of the successive-approximation architecture [14,15]. The second uses an oversampling method which converts the analog signal without requiring precision analog components [16,17]. Some of the methods already discussed, such as the iterative algorithmic A/D converter, could also be included in this section.

A block diagram of a self-calibrating A/D converter using the basic architecture of Fig. 10.6-4 is shown in Fig. 10.7-6. This circuit consists of an N-bit charge-scaling array called the main DAC, an M-bit voltage-scaling array called the sub DAC, and a voltage-scaling array called a calibration DAC. The calibration DAC must have several more bits of resolution than the sub DAC. Digital control circuits govern capacitor switching during the calibration cycle and store the nonlinearity correction terms in data registers. The ratio errors of the sub DAC and overall quantization errors accumulate during digital computation of error voltages. To overcome these errors at least one bit of additional resolution is needed during the calibration cycle. This extra bit is used to achieve final linearity with 1 LSB of an ideal straight line or within 0.5 LSB of an ideal staircase converter

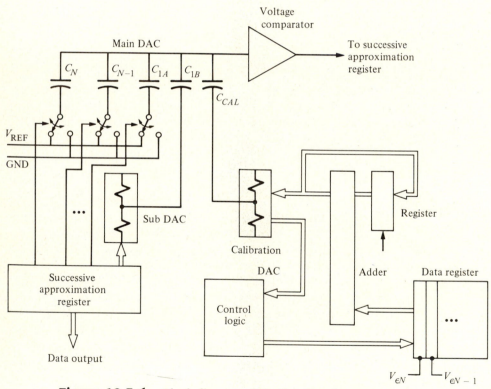

Figure 10.7-6   Block diagram of a self-calibrating A/D converter.

response. In practice, two extra bits are used in order to have a margin of safety.

Fig. 10.7-7 shows an N-bit weighted capacitor DAC. Each weighted capacitor C_n is assumed to be off by a factor of $(1 + \epsilon_n)$ from the ideal value due to process variations. Therefore the actual capacitance C_n can be expressed as

$$C_n = 2^{n-1}C(1 + \epsilon_n), \qquad n = 1A,\ 1B,\ 2,\ \ldots,\ N \qquad (1)$$

where C is defined as the total capacitance divided by 2^N or

$$C = \frac{C_{total}}{2^N} = \frac{C}{2^N} \sum_{i=1A}^{N} 2^{i-1}(1 + \epsilon_i) = C + C \sum_{i=1A}^{N} 2^{i-1}\epsilon_i \qquad (2)$$

Eq. (2) results in the relationship

$$\sum_{i=1A}^{N} 2^{i-1}\epsilon_i = 0 \qquad (3)$$

The output voltage of a charge-scaling DAC such as shown in Fig. 10.7-7(a) can be expressed in terms of capacitor values and corresponding digital input codes as

$$V_o = \frac{V_{REF}}{C_{total}} \sum_{i=1B}^{N} C_i D_i = \frac{V_{REF}}{2^N \left[1 + \sum_{i=1A}^{N} 2^{i-1}\epsilon_i \right]} \sum_{i=1B}^{N} 2^{i-1}(1 + \epsilon_i) D_i \qquad (4)$$

Using Eq. (3) the output voltage expressed in Eq. (4) becomes

$$V_o = \frac{V_{REF}}{2^N} \sum_{i=1B}^{N} 2^{i-1}(1 + \epsilon_i) D_i \qquad (5)$$

The ideal output voltage $V_{o,ideal}$ can be obtained by taking $\epsilon_i = 0$ for all i in Eq. 5 to get

$$V_{o,ideal} = \frac{V_{REF}}{2^N} \sum_{i=1B}^{N} 2^{i-1} D_i \qquad (6)$$

error voltage V_{error} is the difference between the ideal and actual output voltages and is given as

$$V_{error} = V_o - V_{o,ideal} = \frac{V_{REF}}{2^N} \sum_{i=1B}^{N} 2^{i-1}\epsilon_i D_i \qquad (7)$$

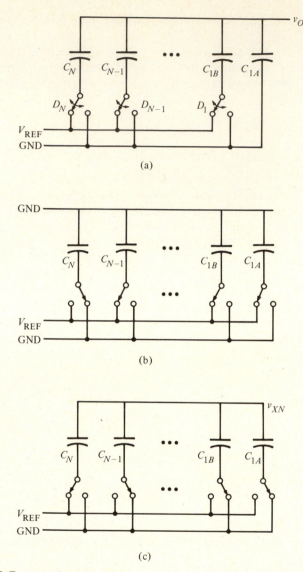

Figure 10.7-7 (a) Charge-scaling D/A. (b) Precharge cycle. (c) Charge redistribution and error acquisition.

Defining the error voltage V_{en} due to the nth capacitor mismatch as

$$V_{en} = \frac{V_{REF}}{2^N} 2^{n-1} \epsilon_n, \qquad n = 1B, 2, \ldots, N \tag{8}$$

causes the error voltage of Eq. (7) to be expressed as

$$V_{error} = \sum_{i=1B}^{N} V_{ei} D_i \tag{9}$$

During the calibration cycle, the individual error voltages V_{en} are measured and digitized by the calibration DAC and then stored in the RAM. During the normal conversion cycle, the total error voltage V_{error} is computed by Eq. (9) in digital form and converted to analog voltage by the same calibration DAC. This total error voltage is subtracted from the main DAC through the coupling capacitor C_{cal} to correct the initial linearity error.

The calibration cycle begins by measuring the error voltage due to the MSB capacitor C_N. This is done by sampling the reference voltage V_{REF} on all the capacitors except the MSB capacitor, as shown in Fig. 10.7-7(b). At this time, the sampled charge in the capacitor array is

$$Q = -V_{REF} \sum_{i=1A}^{N-1} C_i = -CV_{REF} \sum_{i=1A}^{N-1} 2^{i-1}(1 + \epsilon_i) \tag{10}$$

Next, the charge is redistributed by reversing the switching configuration, as shown in Fig. 10.7-7(c). The residual charge at the top plate of the capacitor array is

$$Q_{XN} = V_{REF} \left[C_N - \sum_{i=1A}^{N-1} C_i \right] = CV_{REF} \left[2^{N-1} \epsilon_N - \sum_{i=1A}^{N-1} 2^{i-1} \epsilon_i \right] \tag{11}$$

Rewriting Eq. (3) gives

$$\sum_{i=1A}^{N} 2^{i-1} \epsilon_i = 2^{N-1} \epsilon_N + \sum_{i=1A}^{N-1} 2^{i-1} \epsilon_i = 0 \tag{12}$$

Substituting Eq. (12) into Eq. (11) results in

$$Q_{XN} = CV_{REF} 2^N \epsilon_N \tag{13}$$

Therefore the residual voltage V_{XN} is given as

$$V_{XN} = \frac{Q_{XN}}{C_{total}} = 2V_{\epsilon N} \tag{14}$$

Errors due to smaller capacitors are measured in the same manner. In each case, a successive-approximation search using the sub DAC is employed.

It can be shown that the general relation between the residual voltages V_{XN} and the error voltages $V_{\epsilon n}$ is

$$V_{\epsilon} = 0.5\left(V_{xn} - \sum_{i=n+1}^{N} V_{\epsilon}\right), \qquad n = 1B, 2, \ldots, N-1 \tag{15}$$

In the digital domain we can express Eq. (15) as

$$DV_{\epsilon N} = 0.5DV_{XN} \tag{16}$$

and

$$DV_{\epsilon n} = 0.5\left(DV_{XN} - \sum_{i=n+1}^{N} DV_{\epsilon i}\right) \tag{17}$$

where $DV_{\epsilon n}$ and DV_{XN} are the digitized error voltages and the digitized residual voltages, respectively. Therefore, by digitizing the residual voltages by using the sub DAC, the correction terms $DV_{\epsilon N}, DV_{\epsilon N-1}, \ldots, DV_{\epsilon 1B}$ can be computed subsequently by Eqs. (16) and (17) using a two's-complement adder and a shift register. All these correction terms are stored in the digital memory.

During subsequent normal conversion cycles, the calibration logic is disengaged. The converter works the same way as an ordinary successive-approximation converter except that error-correction voltages are added or subtracted by proper adjustment of the calibration DAC digital input code. When the nth bit is being tested, the corresponding correction term $DV_{\epsilon n}$ is added to the correction terms accumulated from the first bit through the $(n-1)$th bit. If the bit decision is 1, then the added result is stored in the accumulator. Otherwise, $DV_{\epsilon n}$ is dropped, leaving the accumulator with the previous result. The content of the accumulator is converted to an analog voltage by the calibration DAC. This voltage is then subtracted from the main DAC output voltage through the capacitor C_{cal}. The overall operation precisely cancels the nonlinearity due to capacitor mismatches by subtracting V_{error} in Eq. (9) from the main DAC. The only extra operation involved in a normal conversion cycle is one two's-complement addition.

Fig. 10.7-8 shows a microphotograph of a self-calibrating, 15 bit, CMOS A/D converter. The chip dimensions are 2.8 mm $\times$ 4.2 mm using a 6 micrometer process. It is anticipated that a 16-bit converter using 5 micrometer rules would require 15 mm^2. The results of the converter measured at room temperature are shown in Table 10.7-1. The comparator used a folded cascode topology similar to that discussed in Sec. 8.4. The self-correcting technique is an important step in the development of high-performance A/D converters with high resolution.

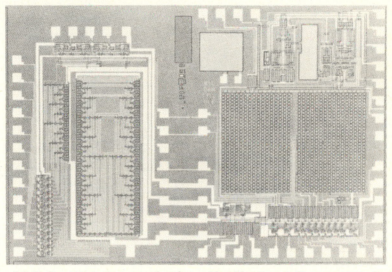

Figure 10.7-8 Microphotograph of a self-calibrating A/D converter. Overall chip dimensions are 2.8 mm x 4.2 mm for a 6 micrometer CMOS process. ([14], © 1984 IEEE)

Table 10.7-1

Performance Characteristics of Fig. 10.7-8 at Room Temperature.

Performance Characteristic	Value	Units
Supply Voltages	± 5	V
Resolution	15	Bits
Linearity	15	Bits
Offset	$< \pm 0.25$	LSB
Conversion Time for ± 0.5 LSB Linearity		
12 Bit	12	μs
15 Bit	80	μs
RMS Noise	40	μV
Power Dissipation (excluding logic)	20	mW
Die Area (excludes logic)	7.5	mm^2

A second approach to achieving a high-performance A/D converter is based on the concept of oversampling. An analog signal is sampled at a rate significantly above the bandwidth of the signal. Each time the analog signal is sampled, a single bit is produced which is related to some property of the analog signal (such as amplitude). This single-bit sequence is applied to a digital, low-pass filter resulting in a digital word equivalent of the analog signal with the bit resolution of the digital filter. This approach has the advantage that critical analog signal-processing functions that are difficult to accomplish with CMOS technology can be implemented as digital circuits with any accuracy desired. The results are A/D converters with at least 12-bit resolution and a conversion rate of 10 kHz or greater [17].

Fig. 10.7-9(a) shows an implementation of the oversampled A/D conversion technique using a delta-sigma modulator [16,18]. A sample of the input is summed with $\pm V_{REF}$ to an integrator. The output of the integrator is compared with a voltage V_{th} to determine the state of the comparator.

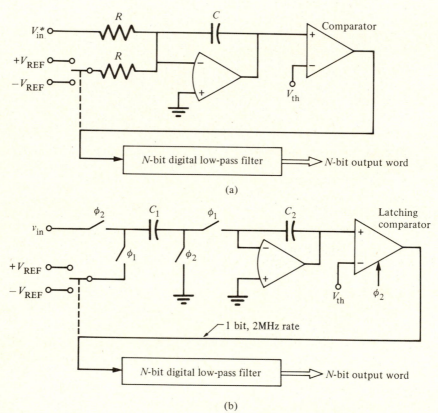

(a)

(b)

Figure 10.7-9 Oversampled A/D converter. (a) Continuous time version. (b) Switched-capacitor (discrete-time) version.

The state of the comparator determines whether $+V_{REF}$ or $-V_{REF}$ is compared with V_{in}^* during the next sample. If the output of the comparator is high, then $-V_{REF}$ is summed with the next input sample; if the output of the comparator is low, then $+V_{REF}$ is summed with the next input sample. The actual implementation of the delta-sigma modulator is given in Fig. 10.7-9(b) using the switched capacitor integrator of Fig. 10.5-4(a). Instead of grounding the input ϕ_1 switch, this switch terminal is taken to $\pm V_{REF}$ as determined by the comparator. Capacitor C_1 is charged differentially by v_{in} during ϕ_1 and $\pm V_{REF}$ during ϕ_2. The comparator is a latching comparator that runs at the oversampling rate.

A special-purpose digital low-pass filter operates on the 1-bit signal to remove frequencies and quantization noise above the desired signal bandwidth. The output of the filter is a multibit digital representation at a lower (decimated) sampling rate. Basic delta-sigma modulator systems generate severe noise components for certain input values. In order to remove this noise, a squarewave dither frequency within the stopband of the decimating filter randomizes the delta-sigma noise. The squarewave dither frequency does not appear at the output of the filter or prevent arbitrary dc inputs from being passed through the filter.

The output of the delta-sigma modulator is downsampled by a finite-impulse-response digital low-pass filter. One possible architecture of the digital filter is shown in Fig. 10.7-10. This architecture uses a 256:1 decimation factor and the one-bit digital input. A 1024-point impulse response of which a symmetric half is stored in a read-only memory, is distributed to four accumulators. No explicit multiplications are necessary because of the one-bit input signal. The impulse-response coefficients were designed to satisfy the dual objectives of quantization-noise removal and anti-alias filtering. Because the finite-impulse-response low-pass filters are relatively insensitive to coefficient roundoff, six-bit coefficients are adequate to define the impulse response.

A prototype integrated circuit [16] displayed approximately 12 bits of linearity at a 8 kHz output rate with a single 5 V power supply and a total die area of 4.8 mm². The oversampling rate for this converter was 2 MHz. A similar method using a pulse-density modulator running at an oversample rate of 12 MHz resulted in 13 bit A/D accuracy at 20 kHz, which corresponds to a signal-to-noise ratio of 80 dB.

Analog-to-digital conversion techniques compatible with CMOS technology have been presented in Secs. 10.5 through 10.7. The major categories included the serial, medium-speed, high-speed, and high-performance A/D converters. An interesting trend in A/D converters is the development of self-calibrating, high-bit, A/D converters. It was shown that CMOS A/D converters are on the threshold of implementing audio signals, which should open up new areas of signal processing applications. Table 10.7-2 compares the performance of the various A/D converters considered.

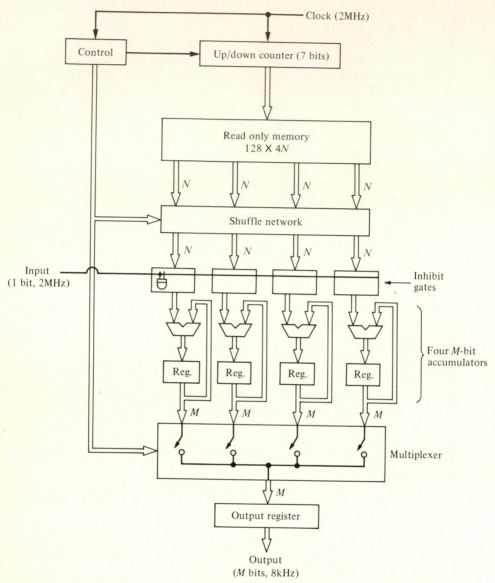

Figure 10.7-10 Architecture of a 1024-point finite-impulse-response, low-pass, digital filter. Typical N = 6 and M = 15. ([16], © 1985 IEEE)

Table 10.7-2

Comparison of the Performance of the Various Types of A/D Converters.

A/D Converter Type	Performance Characteristics
Serial	1–100 conversions/second, 12–14 bit accuracy, requires no element-matching, a stable voltage reference is necessary.
Medium Speed	10,000–100,000 conversions/second, 8–10 bits of untrimmed accuracy, 12–14 bits monotonocity, 12–14 bits trimmed accuracy.
High Speed	$10^6 - 2 \times 10^7$ conversions/second, 7–8 bits of accuracy, requires large area.
High Performance	8,000–20,000 conversions/second, 12–15 bit accuracy, requires no precision analog components, minimizes noise and offset.

10.8 *Summary*

Analog-to-digital and digital-to-analog converters are used to convert between analog and digital signal formats. All converters use some form of scaling a voltage reference to achieve the conversion. The accuracy of the scaling generally determines the performance of the converter. The one exception to this was found in the over-sampled A/D converter discussed in the previous section.

Digital-to-analog converters are divided into parallel and serial. The parallel type D/A converters were represented by the scaling converters. Resistors were used to achieve voltage scaling with resolutions of up to 6 bits. Capacitors were used for charge scaling and could achieve resolutions of up to 9 bits or more. Trimming techniques could increase the resolution but are not practical for high-volume, monolithic integrated circuits.

Analog-to-digital converters were categorized by speed into low-, medium-, and high-speed converters. Lumped in with high-speed converters were converters that have been optimized for CMOS technology. These converters are represented by the algorithmic, self-calibration, and over-sampled types.

While the accuracy of the converter is limited by its scaling capability, there are other sources of errors. One significant source is the comparator and its performance. Other sources include the voltage reference, sample-and-hold, and integrator imperfections such as offset and clock feedthrough.

PROBLEMS — *Chapter 10*

1. Plot the analog output versus the digital word input for a three-bit D/A converter that has ± 1 LSB differential linearity and ± 1 LSB integral linearity. Assume an arbitrary analog full-scale value.

2. Repeat the above problem for ± 1.5 LSB differential linearity and ± 0.5 LSB integral linearity.

3. Repeat Prob. 1, for ± 0.5 LSB differential linearity and ± 1.5 LSB integral linearity.

4. What is the necessary relative accuracy of resistor ratios in order for a voltage-scaling D/A converter to have a 8-bit resolution?

5. What is the necessary relative accuracy of capacitor ratios in order for a charge-scaling D/A converter to have 11-bit resolution?

6. Develop the equivalent circuit of Fig. 10.2-3(b) from Fig. 10.2-3(a).

7. Develop Eqs. (5) through (7) for Fig. 10.2-4(a).

8. Verify the value of the 1.016 pF capacitor in the D/A converter of Fig. 10.2-4(a).

9. Design an eight-bit, two-stage charge-scaling D/A converter similar to Fig. 10.2-4(a) using two four-bit sections with a capacitive attenuator between the stages. Give all capacitances in terms of C, which is the smallest capacitor of the design.

10. What is v_{C1} in Fig. 10.3-1 after the following sequence of switch closures? S_4, S_3, S_1, S_2, S_1, S_3, S_1, S_2, and S_1.

11. Repeat the above problem if $C_1 = 1.05C_2$.

12. In Sec. 10.3, show how Eq. (2) can be derived from Eq. (1). Also show in the block diagram of Fig. 10.3-4 how the initial zeroing of the output can be accomplished.

13. Assume that the amplifier with a gain of 0.5 in Fig. 10.3-4 has a gain error of ΔA. What is the maximum value ΔA can be in Example 10.3-2 without causing the conversion to be in error?

14. Repeat Example 10.3-2 for the digital word 10101.

15. Develop the z-domain transfer function, H(z), for the inverting, switched-capacitor integrator of Fig. 10.5-4(b). If the clock frequency is greater than the highest signal frequency, show that this frequency-domain transfer function reduces to $-\omega_o/j\omega$ where $\omega_o = \dfrac{C_1}{TC_2}$.

16. If the sampled, analog input applied to an 8-bit successive-approximation converter is $0.7V_{REF}$, find the output digital word.

17. Assume that the input of Example 10.6-1 is $0.8V_{REF}$ and find the digital output word to 6 bits.

18. Assume that the input of Example 10.6-1 is $0.3215V_{REF}$ and find the digital output word to 8 bits.

19. Repeat Example 10.6-1 if the gain of two amplifiers actually have a gain of 2.1.

20. Continue Example 10.6-2 out to the 10th bit and find the equivalent analog voltage.

21. Repeat Example 10.6-2 if the gain of two amplifier actually has a gain of 2.1.

22. Why are only 2^N-1 comparators required for a N-bit flash A/D con-

verter? Give a logic diagram for the digital decoding network of Fig. 10.7-1 which will provide the correct digital output word.

23. What are the comparator outputs in order of the higher to lower if V_{in}^* is $0.7V_{REF}$ for the A/D converter of Fig. 10.7-1?

REFERENCES

1. B.M. Gordon, "Linear Electronic Analog/Digital Conversion Architecture, Their Origins, Parameters, Limitations, and Applications," *IEEE Transactions on Circuits and Systems*, Vol. CAS-25, No. 7 (July 1978) pp. 391–418.
2. B. Gilbert, *Electronic Products*, Vol. 24, No. 3 (July 1983) pp. 61–63.
3. K.B. Ohri and M.J. Callahan, Jr., "Integrated PCM Codec," *IEEE Journal of Solid-State Circuits*, Vol. SC-14, no. 1 (Feb. 1979) pp. 38–46.
4. B. Fotouhi and D.A. Hodges, "High-Resolution A/D Conversion in MOS/LSI," *IEEE Journal of Solid-State Circuits*, Vol. SC-14, No. 6 (December 1979) pp. 920–926.
5. R.E. Suarez, P.R. Gray, and D.A. Hodges, "All-MOS Charge Redistribution Analog-to-Digital Conversion Techniques—Part II," *IEEE Journal of Solid-State Circuits*, Vol. SC-10, No. 6 (December 1975) pp. 379–385.
6. R.H. Charles and D.A. Hodges, "Charge Circuits for Analog LSI," *IEEE Transactions on Circuits and Systems*, Vol. CAS-25, No. 7 (July 1978) pp. 490–497.
7. D.H. Sheingold, *Analog-Digital Conversion Handbook*, (Norwood, MA: Analog Devices, 1972).
8. J.R. Naylor, "Testing Digital/Analog and Analog/Digital Converters," *IEEE Transactions on Circuits and Systems*, Vol. CAS-25, No. 7 (July 1978) pp. 527–538.
9. J. Doernberg, H.S. Lee, and D.A. Hodges, "Full-Speed Testing of A/D Converters," *IEEE Journal of Solid-State Circuits*, Vol. SC-19, No. 6 (December 1984) pp. 820–827.
10. P.E. Allen and E. Sanchez-Sinencio, *Switched Capacitor Circuits*, (New York: Van Nostrand Reinhold 1984).
11. G.F. Landsburg, "A Charge-Balancing Monolithic A/D Converter," *IEEE Journal of Solid-State Circuits*, Vol. SC-12, No. 6 (December 1977) pp. 662–673.
12. E.R. Hnatek, *A User's Handbook of D/A and A/D Converters*, (New York: John Wiley & Sons, 1976).
13. P.W. Li, M.J. Chin, P.R. Gray, and R. Castello, "A Ratio-Independent Algorithmic Analog-to-Digital Conversion Technique," *IEEE Journal of Solid-State Circuits*, Vol. SC-19, No. 6 (December 1984) pp. 828–836.
14. H.S. Lee, D.A. Hodges, and P.R. Gray, "A Self-Calibrating 15-bit CMOS A/D Converter," *IEEE Journal of Solid-State Circuits*, Vol. SC-19, No. 6 (December 1984) pp. 813–819.
14. H.S. Lee and D.A. Hodges, "Self-Calibration Technique for A/D Converters," *IEEE Transactions on Circuits and Systems*, Vol. CAS-30, No. 3 (March 1983) pp. 188–190.
16. M.W. Hauser, P.J. Hurst, and R.W. Brodersen, "MOS ADC-Filter Combination that does not Require Precision Analog Components," *Proceedings of 1985 IEEE International Solid-State Circuits Conference*, (February 13, 1985) pp. 80–81.

17.   H.L. Fiedler and B. Hoefflinger, "A CMOS Pulse Density Modulator for High-Resolution A/D Converters," *IEEE Journal of Solid-State Circuits,* Vol. SC-19, No. 6 (December 1984) pp. 995–996.

18.   J.A. Betts, *Signal Processing, Modulation, and Noise,* (New York: Elsevier, 1971) pp. 159–163.

chapter 11

*C*MOS *Analog Circuits and Systems*

The subject of this chapter is the design of various CMOS analog circuits and systems. This material—a continuation of the previous chapter, which was devoted to analog-digital converters—represents the highest level of complexity in the hierarchical viewpoint of analog integrated-circuit design presented in Table 1.1-2. The actual boundary between circuits and systems in Table 1.1-2 is fuzzy and depends on individual interpretation. Typically, the complexity of the design determines its classification. It will be observed in this chapter that op amps and comparators can be treated as ideal blocks, thus allowing us to concentrate on the subject at hand without having to be distracted by details that have already been covered (in Chapters 7–9).

Another measure of the level of design is found in simulation. Most of the circuits and systems presented in this chapter place severe demands on a device simulator such as SPICE. It will be seen that one would like to use a higher-level simulator for the circuits and systems. The combination of analog and digital circuits seen in the last chapter will also be present here. From a simulation viewpoint, a simulator is needed that can simultaneously simulate both analog and digital circuits.

The important bandgap-voltage reference is the first topic of this chapter. This reference uses the material in Sec. 5.5 along with the op amp to implement a very stable voltage reference. Next, some nonlinear analog circuits are considered, including four-quadrant multipliers, modulators, and waveshaping circuits. The ideal diode can be realized using an op amp, comparator, and switches. The last subject is oscillators and timers. A bistable realization is shown that allows voltage-controlled amplitude and frequency modulation.

This chapter concludes the presentation of analog circuit design with emphasis on CMOS technology. Through the hierarchy in Table 1.1-2 and the organization of the material, the principles of design have been illustrated, and many specific techniques useful for design have been presented. The actual process of design often starts from the top of Table 1.1-2 and works down (a top-down approach). It is felt that the bottom-up approach presented in this text is reversible and is in fact the best way to teach the complex subject of circuit design.

11.1 *Reference Sources*

In this section we present a technique that results in references which have very little dependence upon temperature and power supply. The *bandgap reference* can generate references having a temperature coefficient on the order of 10 ppm/°C over the temperature range of 0 °C to 70 °C [1,2]. The principle behind the bandgap reference is illustrated in Fig. 11.1-1. A voltage V_{BE} is generated from a pn-junction diode having a temperature coefficient of approximately -2.2 mV/°C at room temperature. Also generated is a thermal voltage V_t ($V_t = kT/q$) that has a temperature coefficient of $+0.085$ mV/°C at room temperature. If the V_t voltage is multiplied by a constant K and summed with the V_{BE} voltage, then the output voltage is given as

$$V_{REF} = V_{BE} + KV_t \qquad (1)$$

Differentiating Eq. (1) with respect to temperature and using the temperature coefficients for V_{BE} and V_t leads to a value of K that should theoretically give zero temperature dependence. In order to achieve the desired performance, it is necessary to develop the temperature dependence of V_{BE} in

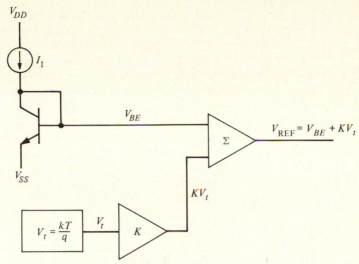

Figure 11.1-1   General principle of the bandgap reference.

more detail. One can see that since V_{BE} can have little dependence upon the power supply (i.e. the bootstrapped references of Sec. 5.5), the power-supply dependence of the bandgap reference will be quite small.

To understand thoroughly how the bandgap reference works, we must first develop the temperature dependence of V_{BE} [3]. Consider the relationship for the collector-current density in a bipolar transistor

$$J_C = \frac{qD_n n_{po}}{W_B} \exp\left(\frac{V_{BE}}{V_t}\right) \tag{2}$$

where

$$J_C = \text{collector current density (A/m}^2)$$

$$n_{po} = \text{equilibrium concentration of electrons in the base}$$

$$D_n = \text{average diffusion constant for electrons}$$

$$W_B = \text{base width}$$

The equilibrium concentration can be expressed as

$$n_{po} = \frac{n_i^2}{N_A} \tag{3}$$

where

$$n_i^2 = DT^3 \exp(-V_{GO}/V_t) \tag{4}$$

The term D is a temperature independent constant and V_{GO} is the bandgap voltage (1.205 volts). Combining Eqs. (2) through (4) result in the following equation for collector current density

$$J_c = \frac{qD_n}{N_A W_B} DT^3 \exp\left(\frac{V_{BE} - V_{GO}}{V_t}\right) \tag{5}$$

$$= AT^\gamma \exp\left(\frac{V_{BE} - V_{GO}}{V_t}\right) \tag{6}$$

In Eq. (6), the temperature independent constants of Eq. (5) are combined into a single constant A. The coefficient of temperature γ is slightly different from 3 due to the temperature dependence of D_n.

A relation for V_{BE} can be developed from Eq. (6) and is given as

$$V_{BE} = \frac{kT}{q} \ln\left(\frac{J_c}{AT^\gamma}\right) + V_{GO} \tag{7}$$

Now consider J_c at a temperature T_0.

$$J_{co} = AT_0^\gamma \exp\left[\frac{q}{kT_0}(V_{BEO} - V_{GO})\right] \tag{8}$$

The ratio of J_c to J_{co} is

$$\frac{J_c}{J_{co}} = \left(\frac{T}{T_0}\right)^\gamma \exp\left[\frac{q}{k}\left(\frac{V_{BE} - V_{GO}}{T} - \frac{V_{BEO} - V_{GO}}{T_0}\right)\right] \tag{9}$$

Eq. (9) can be rearranged to get V_{BE}

$$V_{BE} = V_{GO}\left(1 - \frac{T}{T_0}\right) + V_{BEO}\left(\frac{T}{T_0}\right) + \frac{\gamma kT}{q} \ln\left(\frac{T_0}{T}\right) + \frac{kT}{q} \ln\left(\frac{J_c}{J_{co}}\right) \tag{10}$$

By taking the derivative of Eq. (10) at T_0 with respect to temperature, (assuming that J_c has a temperature dependence of T^α), the dependence of V_{BE} on temperature is clearly seen to be

$$\left.\frac{\partial V_{BE}}{\partial T}\right|_{T=T_0} = \frac{V_{BE} - V_{GO}}{T_0} + (\alpha - \gamma)\left(\frac{k}{q}\right) \tag{11}$$

At 300 °K the change of V_{BE} with respect to temperature is approximately -2.2 mV/°C. We have thus derived a suitable relationship for the V_{BE} term shown in Fig. 11.1-1. Now, it is also necessary to develop the relationship for ΔV_{BE} for two bipolar transistors having different current densities. Using the relationship given in Eq. (7), a relationship for ΔV_{BE} can be given as

$$\Delta V_{BE} = \frac{kT}{q} \ln\left(\frac{J_{C1}}{J_{C2}}\right) \tag{12}$$

Therefore

$$\frac{\partial \Delta V_{BE}}{\partial T} = \frac{V_t}{T} \ln\left(\frac{J_{C1}}{J_{C2}}\right) \tag{13}$$

In order to achieve zero temperature coefficient at T_0, the variations of V_{BE} and ΔV_{BE} as given in Eqs. (11) and (13) must add up to zero. This is expressed mathematically as

$$0 = K'' \left(\frac{V_{t0}}{T_0}\right) \ln\left(\frac{J_{C1}}{J_{C2}}\right) + \frac{V_{BE0} - V_{GO}}{T_0} + \frac{(\alpha - \gamma)V_{t0}}{T_0} \tag{14}$$

where K'' is a circuit constant adjusted to make Eq. (14) true.

$$0 = K \left(\frac{V_{t0}}{T_0}\right) + \frac{V_{BE0} - V_{GO}}{T_0} + \frac{(\alpha - \gamma)V_{t0}}{T_0} \tag{15}$$

Solving for K yields

$$K = \frac{[V_{GO} - V_{BE0} + (\gamma - \alpha)V_{t0}}{V_{t0}} \tag{16}$$

The term K ($K = K'' \ln [J_{C1}/J_{C2}]$) is under the designer's control, so that it can be designed to achieve zero temperature coefficient. Rearranging Eq. (16) yields

$$KV_{t0} = V_{GO} - V_{BE0} + V_{t0}(\gamma - \alpha) \tag{17}$$

Noting that K in Eq. (17) is the same as that in Eq. (1), as both are constants required to achieve a zero temperature coefficient, then substituting of Eq. (17) into Eq. (1) gives

$$V_{REF}\Big|_{T=T_0} = V_{GO} + V_{t0}(\gamma - \alpha) \tag{18}$$

For typical values of $\gamma = 3.2$ and $\alpha = 1$, $V_{REF} = 1.262$ at 300 °K. A typical family of reference-voltage variations as a function of T for various values of T_o is shown in Fig. 11.1-2.

A conventional CMOS bandgap reference for a p-well process is illustrated in Fig. 11.1-3. The input-offset voltage of the otherwise ideal op amp (V_{OS}) has been included in the circuit. Transistors Q1 and Q2 are assumed to have emitter-base areas of A_{E1} and A_{E2}, respectively. If we assume for the present that V_{OS} is zero, then the voltage across R_1 is given as

$$V_{R1} = V_{BE2} - V_{BE1} = V_t \ln \left(\frac{J_2}{J_{S2}}\right) - V_t \ln \left(\frac{J_1}{J_{S1}}\right) = V_t \ln \left(\frac{I_2 A_{E1}}{I_1 A_{E2}}\right) \qquad \textbf{(19)}$$

However, the op amp also forces the relationship

$$I_1 R_2 = I_2 R_3 \qquad \textbf{(20)}$$

The reference voltage of Fig. 11.1-3 can be written as

$$V_{REF} = V_{BE2} + I_1 R_2 = V_{BE2} + V_{R1} \left(\frac{R_2}{R_1}\right) \qquad \textbf{(21)}$$

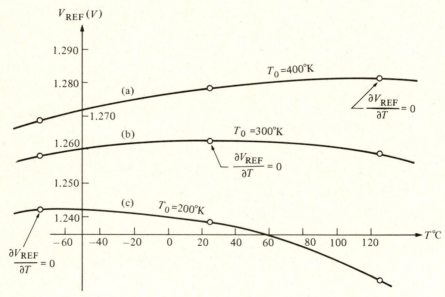

Figure 11.1-2   Variation of band-gap reference output voltage with temperature. ([3], © 1984 John Wiley and Sons, Inc.)

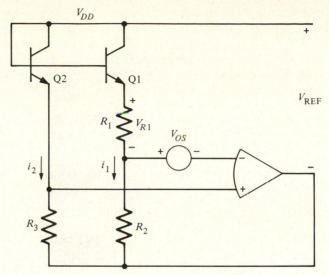

Figure 11.1-3   A conventional, p-well, CMOS, bandgap reference.

Substituting Eq. (20) into Eq. (19) and the result into Eq. (21) gives

$$V_{REF} = V_{BE2} + \left(\frac{R_2}{R_1}\right) V_t \ln \left(\frac{R_2 A_{E1}}{R_3 A_{E2}}\right) \tag{22}$$

Comparing Eq. (22) with Eq. (1) defines the constant K as

$$K = \left(\frac{R_2}{R_1}\right) \ln \left(\frac{R_2 A_{E1}}{R_3 A_{E2}}\right) \tag{23}$$

Thus, the constant K is defined in terms of resistor and emitter-base area ratios. It can be shown that if the input-offset voltage is not zero, that Eq. (22) becomes

$$V_{REF} = V_{BE2} - \left(1 + \frac{R_2}{R_1}\right) V_{OS} + \frac{R_2}{R_1} V_t \ln \left[\frac{R_2 A_{E1}}{R_3 A_{E2}} \left(1 - \frac{V_{OS}}{I_1 R_2}\right)\right] \tag{24}$$

It is clear that the input-offset voltage of the op amp should be small and independent of temperature in order not to deteriorate the performance of V_{REF}.

The dependence of V_{REF} upon power supply can now be investigated. In Eq. (24), the only possible parameters which may depend upon power supply are V_{BE2}, V_{OS} and I_1. Since V_{BE2} and I_1 are derived from V_{REF}, the only way in which V_{REf} can depend upon the power supply is through a finite power-supply rejection ratio of the op amp (manifesting itself as a variation in V_{OS}). If the PSRR of the op amp is large, then Fig. 11.1-3 is for all practical purposes a power supply independent reference as well as a temperature independent reference.

Example 11.1-1

The Design of a Bandgap-Voltage Reference

Assume that $A_{E1} = 10 A_{E2}$, $V_{BE2} = 0.7$ V, $R_2 = R_3$, and $V_t = 0.026$ V at room temperature. Find R_2/R_1 to give a zero temperature coefficient at room temperature. If $V_{OS} = 10$ mV, find the change in V_{REF}. Note that $I_1R_2 = V_{REF} - V_{BE2} - V_{OS}$.

Using the values of V_{BE2} and V_t in Eq. (1) and assuming that $V_{REF} = 1.262$ V gives a value of K equal to 21.62. Eq. (23) gives $R_2/R_1 = 9.39$. In order to use Eq. (24), we must know the approximate value of V_{REF} and iterate if necessary. Assuming V_{REF} to be 1.262, we obtain from Eq. (24) a new value $V_{REF} = 1.153$ V. The second iteration makes little difference on the result because V_{REF} is in the argument of the logarithm.

The temperature dependence of the conventional bandgap reference of Fig. 11.1-3 is capable of realizing temperature coefficients in the vicinity of 50 ppm/°C. Unfortunately, there are several important second-order effects that must be considered in order to approach the 10 ppm/°C behavior [4]. One of these effects, as we have already seen, is the input-offset voltage V_{OS} of the op amp. We have seen in Eq. (24) how the magnitude of V_{OS} can contribute a significant error in the output of the reference circuit. Furthermore, V_{OS} is itself a function of temperature and will introduce further deviations from ideal behavior. A further source of error is the temperature coefficient of the resistors. Other effects include the mismatch in the betas of Q1 and Q2 and the mismatch in the finite base resistances of Q1 and Q2. Yet another source of complication is that the silicon bandgap voltage varies as a function of temperature over wide temperature ranges. A scheme for compensating the V_{GO} curvature and cancelling V_{OS}, the mismatches in β, and the mismatches in base resistance, has permitted temperature coefficients of the reference circuit to be as small as 13 ppm/°C over the range of 0 °C to 70 °C [4].

The bandgap principle can be used to implement a reference circuit where the MOS devices are working in the weak-inversion mode. The resulting reference circuit dissipates very little power. Because the current in the MOS transistor in the weak-inversion (subthreshold) mode is proportional to the exponential of qv_{GS}/nkT [(or v_{GS}/nV_t) as developed in Sec. 3.5], the thermal-voltage generator of Fig. 11.1-1 can be implemented by the difference in two gate-source voltages superimposed across a resistor. A pn junction is still needed to achieve the V_{BE} portion of Eq. (1). A voltage reference using this concept was proposed by Tsividis and Ulmer [5]. Unfortunately, the circuit is dependent upon the parameter n in the argument of the exponential, which is dependent upon temperature itself.

A low-power, CMOS bandgap-voltage reference that eliminates the dependence of V_{REF} upon n is shown in Fig. 11.1-4 [6]. This reference is suitable for a p-well, CMOS technology. The current mirrors, M1 through M4, form a closed loop with an initial loop gain greater than unity. Therefore, the current in both branches increases until equilibrium is achieved at which time the loop gain is reduced to unity by the voltage V_{RI} across R_1. If we assume that M1 through M4 operate in the weak inversion region and that V_{DD} is high enough to ensure drain current saturation, then V_{RI} can be expressed as

$$V_{RI} = V_t \ln \left(\frac{S_1 S_4}{S_2 S_3} \right) \tag{25}$$

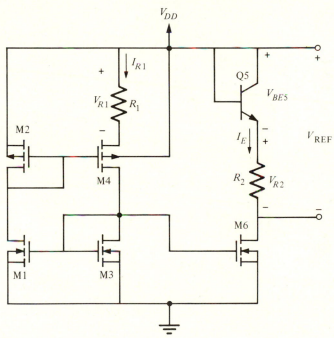

Figure 11.1-4 A low-power, voltage reference suitable for a p-well, CMOS technology.

where $S_i = W_i/L_i$. Note that V_{R1} depends upon only the thermal voltage and the ratio of the geometry of the devices and is independent of n. I_{R1} and I_E are related by

$$\frac{I_{R1}}{I_E} = \frac{S_3}{S_6} \qquad (26)$$

Solving for I_E results in

$$I_E = \frac{S_6 V_t}{S_3 R_1} \ln\left(\frac{S_1 S_4}{S_2 S_3}\right) \qquad (27)$$

V_{REF} can be written as

$$V_{REF} = V_{BE5} + I_E R_2 = V_{BE5} + \left(\frac{R_2 S_6 V_t}{R_1 S_3}\right) \ln\left(\frac{S_1 S_4}{S_2 S_3}\right) \qquad (28)$$

An expression which emphasizes the temperature dependence of V_{BE} at a constant emitter current is given from [7] or from Eq. (10) as

$$V_{BE}(T) = V_{GO}\left(1 - \frac{T}{T_0}\right) + V_{BE0}\left(\frac{T}{T_0}\right) + \frac{\gamma kT}{q}\ln\left(\frac{T_0}{T}\right) \qquad (29)$$

where V_{GO} is the extrapolated bandgap voltage of silicon, V_{BE0} is the V_{BE} of a diode-connected transistor at $T = T_0$, and γ is a constant dependent on the diode fabrication and temperature characteristics. It can be shown that the condition for dV_{REF}/dT to be zero is given by

$$\frac{R_2 S_6}{R_1 S_3}\ln\left(\frac{S_1 S_4}{S_2 S_3}\right) = \frac{q(V_{GO} - V_{BE0})}{kT_0} + \gamma \qquad (30)$$

This results in a reference voltage of

$$V_{REF} = V_{GO} + \frac{\gamma kT}{q}\left[1 + \ln\left(\frac{T_0}{T}\right)\right] \qquad (31)$$

In order to ensure proper operation of the voltage reference of Fig. 11.1-4, the following precautions should be observed. First, the devices must be in weak inversion even at the highest temperature of operation. Secondly, leakage currents, particularly in the n-channel devices, must be minimized to prevent those currents from becoming a major source of error at higher temperature. Thirdly, this circuit has two stable states. One is with zero current in all devices. This state must be avoided by providing circuitry to start current flowing and thereby allowing the reference to reach a stable state. If the start-up circuit is properly designed, it will turn off when the stable state is reached [see Fig. 5.5-8(a)]. Lastly, the output resistance of the devices must be large enough to ensure their proper operation as current mirrors. This can be accomplished by using long devices, or the cascode or Wilson current mirrors presented in Chapter 5. Fig. 11.1-4 is capable of a TC_F of 70 ppm/°C over the temperature range of 0 °C to 65 °C with only 10 microwatts of total power dissipation.

Although other techniques have been used to develop power-supply and temperature-independent references, the bandgap circuit has proven the best to date. In this section we have used the bandgap concept to develop precision references. As the requirement for higher precision increases, the designer will find it necessary to begin including second-order and sometimes third-order effects that might normally be neglected. These higher-order effects require the designer to be familiar with the physics and operation of the MOS devices.

11.2 *Analog Multipliers*

Often, when performing analog signal processing, one requires a circuit that gives as its output, the product of two variables (voltage or current) [8]. Such circuits have long been used in technologies other than CMOS [9,10]. In this section, an analog multiplier suitable for implementation in CMOS will be developed. First consider what function a multiplier accomplishes.

In Fig. 11.2-1(a) a set of parametric curves are shown that illustrate the multiplication of two variables, v_α and v_β, in one quadrant (both variables positive). The product of the two variables (with v_α as the horizontal axis and v_β as the parametric parameter as labeled on each curve) is given as the vertical coordinate v_o. Generally it is desirable to achieve multiplication in at least two quadrants, as shown in Fig. 11.2-1(b), and more typically in four quadrants, as shown in Fig. 11.2-1(c). The multiplier ultimately developed in this section will operate in all four quadrants and thus is called a *four-quadrant multiplier*.

The development that follows will assume: there is no channel-length modulation effect and all transistors (n-channel only) have their substrates tied to their sources. Furthermore, the simple square-law device characteristic will be used.

Consider the differential pair shown in Fig. 11.2-2. This circuit was analyzed previously in Sec. 6.2. It is straightforward to take the results derived in that section and derive the following relationship (Prob. 9),

$$i_1 - i_2 = \frac{\beta_1}{2} v_I \left[\frac{4i_{SS}}{\beta_1} - v_I^2 \right]^{1/2} \tag{1}$$

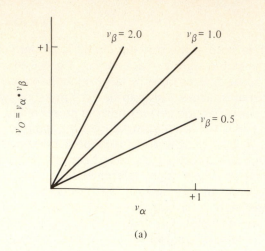

(a)

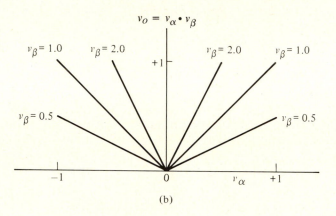

(b)

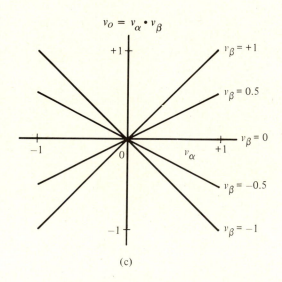

(c)

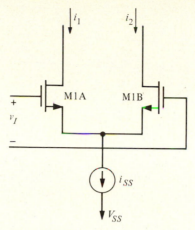

Figure 11.2-2   Differential pair.

If v_I is small such that

$$v_I^2 \ll \left[\frac{4i_{SS}}{\beta_1} \right] \tag{2}$$

then the relationship given in Eq. (1) is approximately

$$\Delta i_A = i_1 - i_2 \cong \frac{\beta_1}{2} v_I \left[\frac{4i_{SS}}{\beta_1} \right]^{1/2} \tag{3}$$

Based upon the above approximation, Δi_A is a linear function of v_I and a square-law function of i_{SS}. If two stages like that in Fig. 11.2-2 are cascaded, then Δi can be made a linear function of both i_{SS} and v_I. Such is the desired function of a multiplier.

Consider the circuit shown in Fig. 11.2-3. The relationship defining Δi_B is

$$\Delta i_B = i_3 - i_4 = \frac{\beta_2}{2} v_X \left[\frac{4i_{SS}}{\beta_2} - v_X^2 \right]^{1/2} \tag{4}$$

$$v_X = -R_1(i_1 - i_2) = -R_1(\Delta i_A) \tag{5}$$

Figure 11.2-1   (a) One-quadrant multiplication. (b) Two-quadrant multiplication. (c) Four-quadrant multiplication.

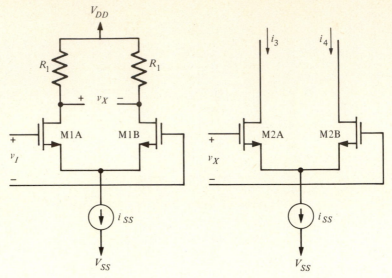

Figure 11.2-3 Cascaded differential stages.

Substituting Eq. (5) into Eq. (4) gives

$$\Delta i_B = \frac{\beta_2}{2}(-R_1 \Delta i_A)\left[\frac{4 i_{SS}}{\beta_2} - R_1^2 \Delta i_A^2\right]^{1/2} \tag{6}$$

$$\Delta i_B = -R_1 i_{SS} v_I (\beta_1 \beta_2)^{1/2}\left[1 - \frac{\beta_1 \beta_2 R_1^2 v_I^2}{4}\right]^{1/2} \tag{7}$$

This relationship illustrates that Δi_B is now a linear function of i_{SS}, and a square-law function of v_I. This square-law relationship can be compensated using the circuit shown in Fig. 11.2-4. It can be shown (see Problem 10) that

$$\Delta i_C = i_{C1} - i_{C2} = v_{CI}(\beta_3 I_{DX})^{1/2}\left[1 - \frac{\beta_3 v_{CI}^2}{4 I_{DX}}\right]^{1/2} \tag{8}$$

where

$$I_{DX} = i_{C1} + i_{C2} \tag{9}$$

Equation (8) is in the same form as Eq. (7). If v_{CI} is taken as the input of the circuit shown in Fig. 11.2-3 and the second term under the radical of Eqs. (7) and (8) is equated as follows

$$\frac{\beta_1 \beta_2 R_1^2}{4} = \frac{\beta_3}{4 I_{DX}} \rightarrow \frac{\beta_1 \beta_2 R_1^2 I_{DX}}{\beta_3} = 1 \tag{10}$$

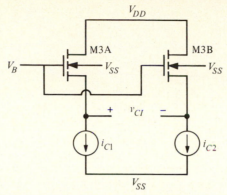

Figure 11.2-4   Precompensation circuit.

then the ratio of Eq. (7) to (8) is

$$\frac{i_3 - i_4}{i_{C1} - i_{C2}} = \frac{v_I[-R_1 i_{SS}(\beta_1 \beta_2)^{1/2}]}{v_{CI}(I_{DX}\beta_3)} \tag{11}$$

Noting again that v_{CI} is the new input v_I, then we have

$$i_3 - i_4 = -R_1 i_{SS} \left[\frac{\beta_1 \beta_2}{I_{DX}\beta_3}\right]^{1/2} (i_{C1} - i_{C2}) \tag{12}$$

$$\Delta i_B = -R_1 \left[\frac{\beta_1 \beta_2}{I_{DX}\beta_3}\right]^{1/2} i_{SS} \Delta i_C \tag{13}$$

Equation (13) shows that the output current Δi_B is the linear multiple of i_{SS} and Δi_C. In order to implement a voltage multiplier, i_{SS} and Δi_C must be made linear functions of input voltages. This can be accomplished for small ranges of input voltages using a differential transistor pair as shown in Fig. 11.2-2.

Consider the multiplier circuit shown in Fig. 11.2-5. The output i_O of this multiplier is

$$i_O = i_3 - i_4 = K v_\alpha v_\beta \tag{14}$$

where K is merely a constant. With the given restriction that v_β be greater than zero, the multiplier only multiplies in two quadrants.

A four-quadrant multiplier can be implemented using two two-quadrant multipliers. To see how a four-quadrant multiplier works, consider the circuit shown in Fig. 11.2-6. This circuit is made from two two-quadrant

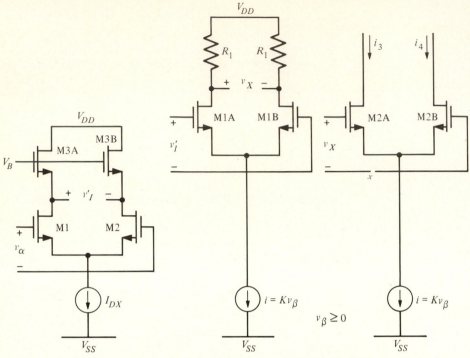

Figure 11.2-5 Two-quadrant multiplier.

multipliers cascoded with the output of another two-quadrant multiplier. As shown, the output currents are easily shown to be

$$i_1 - i_2 = KV_\alpha i_A \tag{15}$$
$$i_3 - i_4 = K(-V_\alpha)i_B \tag{16}$$

Taking the difference between Eqs. (15) and (16) gives

$$i_1 + i_3 - (i_2 + i_4) = KV_\alpha(i_A - i_B) \tag{17}$$

Noting that $i_A - i_B$ is the product of V_β and I leads to the result

$$i_1 + i_3 - (i_2 + i_4) = KV_\alpha V_\beta I \tag{18}$$

By making the summation of currents as given in Eq. (18) and combining the constant K and the constant tail current I into one constant gives

$$i_{o1} - i_{o2} = KV_\alpha V_\beta \tag{19}$$

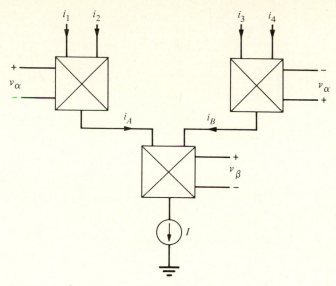

Figure 11.2-6   Cascode connection of two-quadrant multipliers.

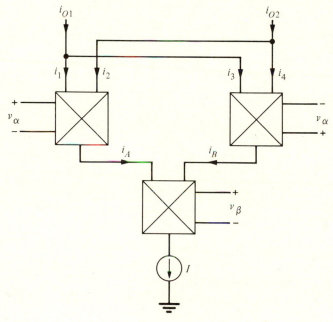

Figure 11.2-7   Four-quadrant multiplication using two-quadrant multipliers.

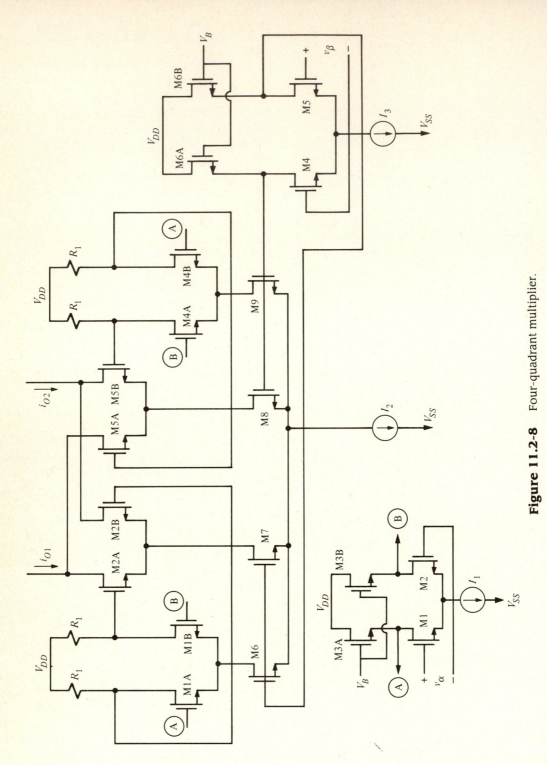

Figure 11.2-8 Four-quadrant multiplier.

as illustrated in Fig. 11.2-7. The inputs V_α and V_β can be either polarity. A four-quadrant multiplier using the two-quadrant multipliers developed in this section is shown in Fig. 11.2-8.

Analog multipliers can be used as analog signal-processing elements in a number of ways. One of the simplest applications is as a frequency doubler. Consider the circuit shown in Fig. 11.2-9. In this circuit an analog multiplier is connected as a squaring circuit in which the output is the square of the input. With both inputs connected to a sinusoidal source, the output is represented mathematically as

$$V_o = \sin^2(\omega t) = \tfrac{1}{2}[1 - \cos(2\omega t)] \qquad (20)$$

Notice that the frequency of the output is twice that of the input. This is illustrated in Fig. 11.2-10.

An analog multiplier can be used to create an analog divider. This is accomplished simply by placing the multiplier in the feedback loop of a

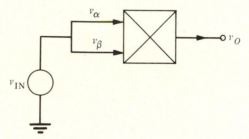

Figure 11.2-9   Multiplier used as a frequency doubler.

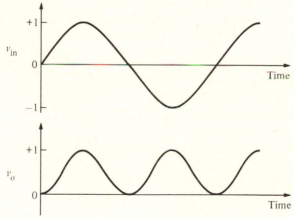

Figure 11.2-10   Frequency doubler input and output.

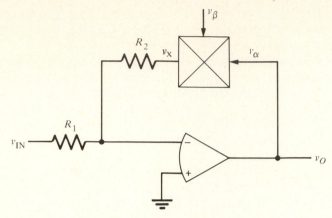

Figure 11.2-11   Analog divider.

high-gain op amp, a technique illustrated in Fig. 11.2-11. Assuming that the op amp has ideal characteristics, the voltage at the output of the multiplier is

$$v_X = -v_{IN}(R_2/R_1) \tag{21}$$

and

$$v_X = v_\alpha v_\beta = v_O v_\beta \tag{22}$$

therefore

$$v_O = -(v_{IN}/v_\beta)(R_2/R_1), \; v_\beta > O \tag{23}$$

By connecting both α and β inputs together to the output of the op amp and inverting the output of the multiplier before applying it to R_2, the square root of v_{IN} is achieved for $v_{IN} > O$.

Multipliers can be used in many other ways in signal-processing applications, making them a useful addition to the category of circuits that are available to the CMOS designer.

11.3   *Waveshaping Circuits*

In many applications of analog signal processing, it is necessary to be able to realize nonlinear voltage-transfer functions. Such realizations typically require diodes. Unfortunately, the diodes available from a CMOS technology are required to have one terminal common to the highest or lowest power supply and do not provide the necessary flexibility. In this section we shall develop methods compatible with CMOS technology for realizing nonlinear voltage-transfer functions. With the resulting circuits, one can accomplish numerous types of waveshaping including half-wave and full-wave rectification, limiting, and other functions such as triangle-to-sine wave conversion.

Fig. 11.3-1 shows the current-voltage characteristics of an ideal diode. When v_D is greater than or equal to zero, the diode is a short circuit. When v_D is less than zero, the diode is an open circuit. Unfortunately, the ideal diode characteristic can only be approximated. Fig. 11.3-2 shows the current-voltage characteristics of a semiconductor diode. The equation of this diode was given in Eq. (24) of Sec. 2.2. However, since the diode of the CMOS process is constrained to have one terminal on V_{DD} or V_{SS}, it is not of general use.

If the gate and drain of an enhancement MOS device are connected together, the current-voltage characteristics of Fig. 11.3-3 result. The equa-

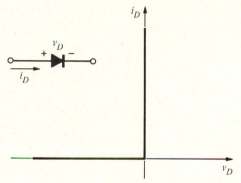

Figure 11.3-1   Voltage-current characteristics of an ideal diode.

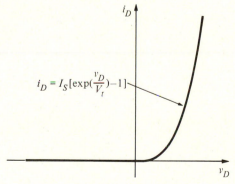

Figure 11.3-2   Voltage-current characteristics of semiconductor diode.

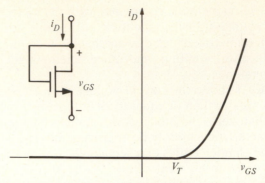

Figure 11.3-3   Voltage-current characteristics of a drain-gate–connected, enhancement MOS transistor.

tion describing this curve is given in Eq. (12) of Sec. 3.1. The interesting feature of this configuration occurs when v_{GS} is negative. The drain and source terminals interchange and we see that new gate-"source" terminals are shorted. The terminal behavior is given as

$$i_D = \begin{cases} (\beta/2)(v_{GS} - V_T)^2, & v_{GS} \geq V_T \\ 0, & v_{GS} < V_T \end{cases} \tag{1}$$

While Eq. (1) is not as good an approximation to Fig. 11.3-1 as the semiconductor diode, in application this difference will turn out to be unimportant. We shall consider several categories of nonlinear circuits and develop implementations compatible with CMOS technology. These categories include half-wave rectifiers, full-wave rectifiers, and breakpoint shifting circuits.

Half-Wave Rectifiers

The half-wave rectifier is an important nonlinear circuit in signal processing. It is used primarily for detection, or converting a dc signal from an ac signal. Fig. 11.3-4(a) shows a half-wave rectifier circuit using a diode. If the diode is ideal, the voltage characteristic of Fig. 11.3-4(b) results. However, if the diode is a semiconductor diode, the dotted characteristic of Fig. 11.3-4(b) is obtained. The difference is due to the voltage drop across the diode necessary to provide the required current. The problem represented by the dotted line in Fig. 11.3-4(b) can be solved through the use of an op amp and negative feedback.

A general principle usable for analog circuit design can be illustrated in Fig. 11.3-5 where an unwanted signal v_N is injected at the op amp output. This unwanted signal can be diminished by applying negative feedback of

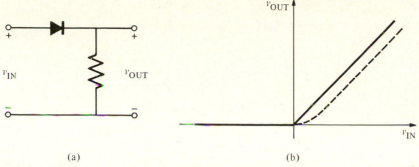

(a)   (b)

Figure 11.3-4 (a) Half-wave rectifier realization. (b) Voltage transfer characteristics of a half-wave rectifier using an ideal diode and practical diode (dotted
line).

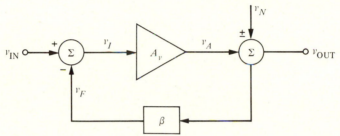

Figure 11.3-5 Block diagram illustrating the use of negative feedback to reduce
unwanted signal v_N.

β around both the amplifier and the unwanted signal. It is easy to show that
the output voltage of Fig. 11.3-5 can be written as

$$V_{OUT} = \left[\frac{A_v}{1 + A_v\beta} \right] V_{IN} \pm \left[\frac{V_N}{1 + A_v\beta} \right] \tag{2}$$

If $A_v\beta \gg 1$, then Eq. (2) becomes

$$V_{OUT} \cong \left(\frac{1}{\beta} \right) V_{IN} \tag{3}$$

Thus, the negative feedback has eliminated the unwanted signal from the
desired signal.

Using this principle on the diode of Fig. 11.3-5 results in the circuit of
Fig. 11.3-6(a) which is an inverting half-wave rectifier. (A noninverting
half-wave rectifier can be achieved by cascading this realization with an
inverter.) Diode D1 is used for the rectification while diode D2 is used to

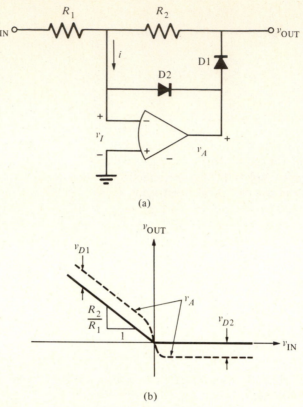

(a)

(b)

Figure 11.3-6 (a) Inverting half-wave rectifier using the principle of Fig. 11.3-5. (b) Voltage transfer characteristics of (a).

keep the op amp output voltage defined when D1 is off. Since diode D1 is within the feedback loop, it should act like an ideal diode. The point at which the diodes change states is determined by the current i. If i is positive, D2 is on and D1 is off. Similarly, if i is negative, D2 is off and D1 is on. The point at which they switch states is at $i = 0$. If $v_{IN} < 0$, then i is negative, and D1 is on and D2 is off. The output voltage can be written as

$$v_{OUT} = v_A - v_{D1} = -A_v v_I - v_{D1}$$

$$= -A_v \left[\frac{R_2 v_{IN}}{R_1 + R_2} + \frac{R_1 v_{OUT}}{R_1 + R_2} \right] - v_{D1} \qquad (4)$$

Solving for v_{OUT} gives

$$v_{OUT} = \frac{-A_v R_2/(R_1 + R_2)}{1 + [A_v R_1/(R_1 + R_2)]} v_{IN} - \frac{v_{D1}}{1 + [A_v R_1/(R_1 + R_2)]} \qquad (5)$$

If $A_v R_1/(R_1 + R_2)$ is much greater than one, then Eq. (5) simplifies to

$$v_{OUT} \cong -\frac{R_2}{R_1} v_{IN} \tag{6}$$

If $v_{IN} > 0$, then i is positive and D1 is off and D2 is on. The output voltage v_{OUT} is zero since any current produced by v_{IN} goes through D2. The solid line in Fig. 11.3-6(b) gives the resulting half-wave rectifier transfer function. The dashed line shows the actual output voltage of the op amp.

The transition from the state where D1 is on and D2 is off to the state where D1 is off and D2 is on is called the *breakpoint*. Ideally, the breakpoint should be defined as a given value of v_{IN} (or v_{OUT}). The breakpoint can be developed by finding an expression for the current i that flows in the diode D2. The direction of i is not important because we set i equal to zero to find the breakpoint. For example, consider Fig. 11.3-6(a). The current i can be written as

$$i = \frac{v_{IN}}{R_1} + \frac{v_{OUT}}{R_2} \tag{7}$$

We know that at the breakpoint, $i = 0$ and $v_{OUT} = 0$. Substituting these values into Eq. (7) shows that the breakpoint is given by $v_{IN} = 0$. The input-offset voltage of the op amp V_{os} will cause an uncertainty in the breakpoint that will limit the lower value of signals that can be successfully rectified.

Unfortunately, the circuit of Fig. 11.3-6(a) is not compatible with CMOS technology. However, if the diodes are each replaced by a drain-gate–connected enhancement device ("MOS diode") of Fig. 11.3-3, similar performance is achieved. Even though the MOS diode is a less perfect realization of Fig. 11.3-1, the feedback principle of Fig. 11.3-5 overcomes this imperfection to give a practical realization. The arrow of the MOS diode should be in the same direction as the arrow of the semiconductor diode. Fig. 11.3-7(a) gives a CMOS compatible realization of Fig. 11.3-6(a). Fig. 11.3-7(b) illustrates the voltage-transfer characteristics. The dashed characteristics represent the actual output voltage of the op amp. The difference between these curves is greater because the drop across the MOS diode is greater than the semiconductor diode. One might be concerned about the body effect on V_T, but the negative-feedback loop keeps this effect from becoming important.

If negative feedback is used, then the MOS diode can be substituted almost on a 1:1 basis for semiconductor diode circuits. One concern that may become important is the accuracy with which the slope of the half-wave rectifier can be defined. In both Figs. 11.3-6(a) and 11.3-7(a), this slope was given as $-R_2/R_1$. However, the ratio of resistors for CMOS tech-

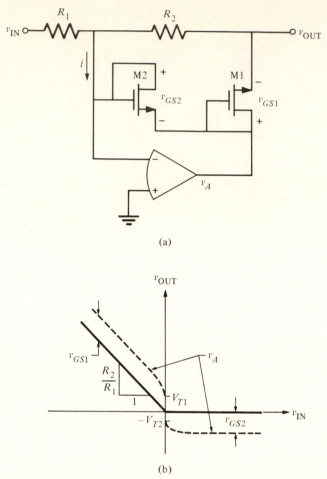

(a)

(b)

Figure 11.3-7 (a) Implementation of Fig. 11.3-6(a) compatible with CMOS technology. (b) Voltage transfer characteristics of (a).

nology falls in the several percent range as given by Table 2.4-1. If more accuracy is desired, one must try to use capacitor ratios that are accurate to one percent or less. Fig. 11.3-8 shows noninverting and inverting charge amplifiers that use capacitors, switches, and op amps [11,12]. ϕ_1 and ϕ_2 are nonoverlapping clocks that close the indicated switch during their high value. For Fig. 11.3-8(a), C_1 is charged to v_{IN} and C_2 is discharged during ϕ_1. During ϕ_2, C_1 is inverted and discharged into C_2. The resulting charge equations at the end of the ϕ_2 phase period are

$$C_2 v_{OUT} \lfloor (n - 0.5)T \rfloor = C_1 v_{IN} [(n - 1)T] \tag{8}$$

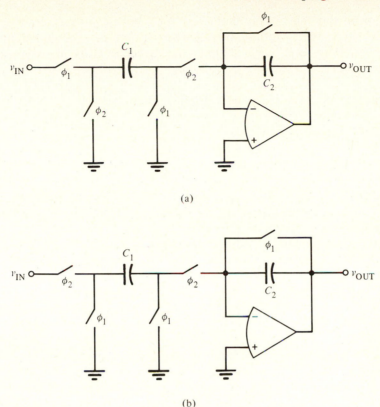

(a)

(b)

Figure 11.3-8   Charge amplifiers. (a) Noninverting with a voltage gain of C_1/C_2. (b) Inverting with a voltage gain of $-C_1/C_2$.

Using discrete-time notation gives

$$C_2 V_{OUT}(z) = z^{-1/2} C_1 V_{IN}(z) \tag{9}$$

Solving for the voltage transfer function results in

$$\frac{V_{OUT}(z)}{V_{IN}(z)} = \frac{C_1}{C_2} z^{-1/2} \tag{10}$$

It is seen that the circuit of Fig. 11.3-8(a) is simply an amplifier with a gain of C_1/C_2 and a phase shift of $-0.5\omega T$ radians when z is replaced by $e^{j\omega T}$. It is necessary to sample and hold the output or input to get a full-clock-period delay. If the output of Fig. 11.3-8(a) was sampled and held, then the output of the sample and hold would be the new amplified input sample during the entire clock phase. Fig. 11.3-8(b) gives an inverting configuration of the charge amplifier. During ϕ_1, both C_1 and C_2 are discharged. In the subsequent ϕ_2 phase, the circuit functions as a charge amplifier with no delay. If the amplifier is sampled during the ϕ_2 clock phase, the gain of the amplifier is

$$\frac{V_{OUT}(z)}{V_{IN}(z)} = -\frac{C_1}{C_2} \tag{11}$$

The two charge amplifiers (switched-capacitor amplifiers) of Fig. 11.3-8 are the basis of obtaining a realization of the half-wave rectifier. Consider the circuit of Fig. 11.3-9(a). Except for the comparator, this circuit is similar to Fig. 11.3-8(a) if $\phi_A = \phi_1$ and $\phi_B = \phi_2$ or to Fig. 11.3-8(b) if $\phi_A = \phi_2$ and $\phi_B = \phi_1$. A comparator is used to determine when $v_{IN} = 0$ and to close the switch connected between terminals A and B. The objective of the switch between terminals A and B is to cause the gain of the amplifier to be zero. This can be accomplished by connecting A and B to C and D. It can also be accomplished by replacing the E-C connection by the comparator-controlled switch. Fig. 11.3-9(b) shows the various possible functions of this circuit. If $\phi_A = \phi_1$, $\phi_B = \phi_2$, and for the comparator connected as shown, a negative-output, noninverting, half-wave rectifier is obtained. If $\phi_A = \phi_2$ and $\phi_B = \phi_1$, then a positive-output, inverting, half-wave rectifier results. If the comparator input terminals are reversed, then the first case gives a positive-output, noninverting, half-wave rectifier and the second gives a negative-output, inverting, half-wave rectifier. Consequently, the circuit of Fig. 11.3-9(a) has a great deal of generality.

The half-wave rectifiers presented above have been used in CMOS circuits and have given good results. If a sample-and-hold circuit is used with the switched capacitor, the overall transfer function will be multiplied by $\sin(\omega T)/\omega T$ where T is the period of the sample-and-hold circuit. One advantage of the switched-capacitor circuits is that the autozeroing schemes of Sec. 7.5 can be used.

Full-Wave Rectifiers

The full-wave rectifier is a simple extension of the half-wave rectifier. One possible realization is shown in Fig. 11.3-10(a). The resuting voltage-transfer function is shown in Fig. 11.3-10(b). The operation of this circuit is easy to see. The voltage-transfer characteristic from v_{IN} to v_A is simply an inverting, negative-output, half-wave rectifier. If v_A is multiplied by a gain of -2, its contribution to v_{OUT} is a noninverting, positive-output, half-wave rectifier with a gain of $+2$. When this contribution is summed with $-v_{IN}$, the output of Fig. 11.3-10(b) results. A negative output, full-wave rectifier can be implemented by simply reversing the direction of the MOS diodes (see Problem 15).

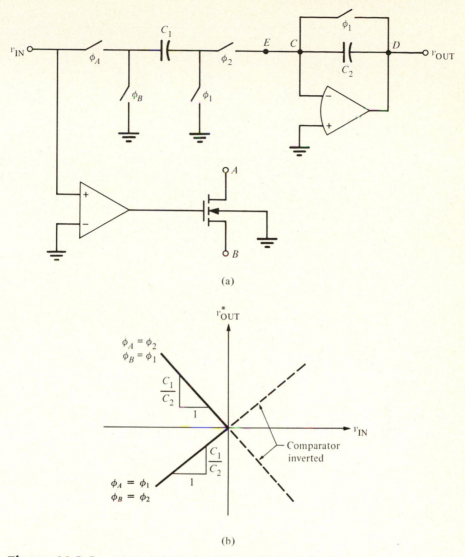

(a)

(b)

Figure 11.3-9 (a) Switched-capacitor realization of a half-wave rectifier. (b) Voltage transfer characteristics of (a).

Another continuous-time approach to implementing the full-wave rectifier is shown in Fig. 11.3-11(a) [13]. When v_{IN} is positive, M1 is on and M2 is off. Thus v_{IN} is coupled to the output through the follower M1. If v_{IN} is negative, M1 is off and M2 is on. $-v_{IN}$ is coupled to the output through the follower M2 to obtain the full-wave rectifier curve of Fig. 11.3-11(b). This circuit has several disadvantages that must be overcome before it

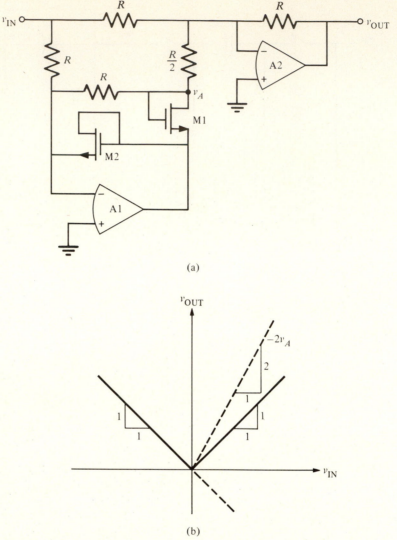

(a)

(b)

Figure 11.3-10   (a) CMOS realization of a noninverting full-wave rectifier. (b) Voltage transfer characteristics.

becomes useful. A finite amount of $\pm v_{IN}$ is necessary before M1 and M2 switch completely on or off. The op amp A2 and M3 can be used to compensate this gate-source offset voltage.

The full-wave rectifier can also be realized by the switched-capacitor amplifier method. Fig. 11.3-12(a) shows a switched-capacitor realization of Fig. 11.3-10(a). The gain through the upper switched capacitor and the amplifier is -1 for all values of v_{IN}. The gain through the lower switched

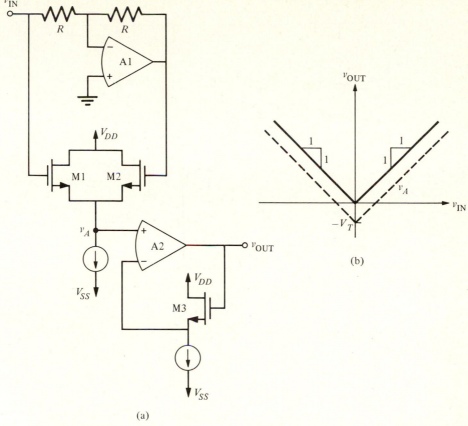

(a)

(b)

Figure 11.3-11 (a) A realization of a full-wave rectifier not using diodes. (b) Voltage transfer characteristics of (a).

capacitor to the output is $+2$ when switch S1 is on or $v_{IN} > 0$. The resulting positive-output, full-wave rectifier characteristic is shown in Fig. 11.3-12(b). A negative-output, full-wave rectifier can be obtained by reversing the input terminals of the comparator A2. It would be more efficient to remove the switches enclosed within the dashed lines and connect switch S1 between points A and B. Again, the discrete-time characteristics of this realization would permit autozeroing of the comparator offset voltage.

Breakpoint Shifting

In many cases, the breakpoint is to be shifted away from zero. This can be accomplished as shown in Fig. 11.3-13(a) for negative-output, inverting,

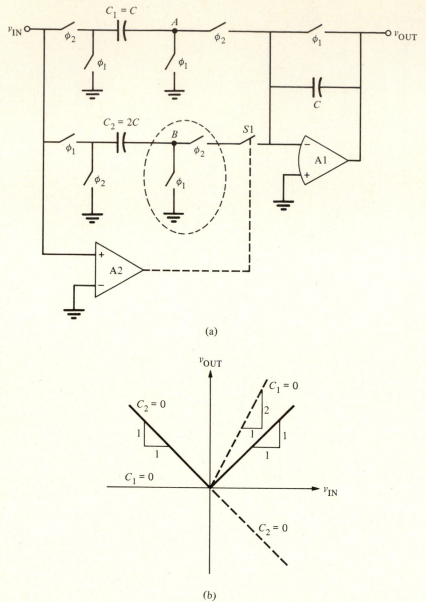

(a)

(b)

Figure 11.3-12 (a) Switched-capacitor realization of a noninverting full-wave rectifier. (b) Voltage transfer characteristics of (a).

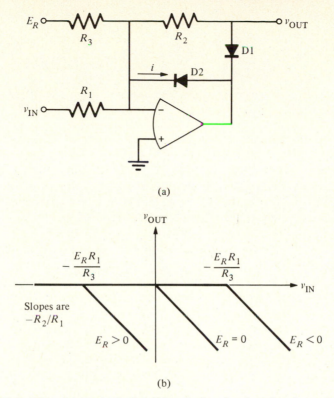

(a)

(b)

Figure 11.3-13   (a) Method of shifting the breakpoint of an inverting half-wave rectifier. (b) Voltage transfer characteristics of (a).

half-wave rectifier. A dc voltage E_R has been connected to the inverting input through a resistor R_3. Using the principles explained above, we can solve for i as

$$i = \frac{v_{IN}}{R_1} + \frac{E_R}{R_3} + \frac{v_{OUT}}{R_2} \tag{12}$$

Because the breakpoint occurs when $i = 0$ and $v_{OUT} = 0$, Eq. (12) gives the breakpoint in terms of v_{IN} as

$$v_{IN}(BP) = -\frac{E_R R_1}{R_3} \tag{13}$$

Fig. 11.3-13(b) shows the voltage-transfer characteristics of Fig. 11.3-13(a). Mathematically, we may express the output voltage of Fig. 11.3-

13(a) as

$$V_{OUT} = \begin{cases} -\dfrac{R_2 V_{IN}}{R_1} + \dfrac{R_2 E_R}{R_1}, & V_{IN} > -\dfrac{R_1 E_R}{R_3} \\[3ex] 0, & V_{IN} < -\dfrac{R_1 E_R}{R_3} \end{cases} \qquad (14)$$

We note that E_R can be positive or negative and shift the breakpoint to the left or right, respectively. The same principle can be applied to any of the previous circuits that use diodes for waveshaping to develop the general capability of establishing a breakpoint at any value of v_{IN}. In any diode circuit, the diode can be either a semiconductor diode or an MOS diode (which extends this technique to CMOS technology).

These concepts can be extended to synthesize a voltage-transfer function by using piecewise linear segments. Fig. 11.3-14(a) shows a number of circuits similar to Fig. 11.3-6(a) applied to a summing amplifier. The result is shown in Fig. 11.3-14(b). Each of the individual outputs implements one line segment. We note that all E_is can be identical and R_{Ei} can be used to design the breakpoints. On the other hand, all R_{Ei}s can be identical and E_i used to design the breakpoints. The slopes of the line segments become increasingly larger and are indicated on the curve. If the diodes are all reversed and the polarities of E_i changed, the piecewise linear curve moves to the fourth quadrant. The remaining two quadrants could be realized by cascading Fig. 11.3-14(a) with an inverter.

The curve of Fig. 11.3-14(b) is monotonically increasing in slope. A monotonically decreasing slope can be obtained by subtracting shifted, half-wave rectifier characteristics from a linear amplifier. Fig. 11.3-15 shows an inverting, monotonically decreasing, piecewise linear approximation. R_0 of Fig. 11.3-15(a) will allow a line segment with nonzero slope to go through the origin of Fig. 11.3-15(b). Using these ideas, all possible monotonically increasing and decreasing slopes are realizable in any of the four quadrants.

Sometimes, it is not necessary to have a precise breakpoint. If the diode is not enclosed with a negative-feedback loop, then the breakpoint will not be well defined. In many cases, this results in a smoother transition between multiple segments. As a consequence, the voltage-transfer function will have less undesired harmonic distortion. The primary disadvantage is that the breakpoints are not easy to design to a given value.

A method which provides a decreasing monotonic approximation is shown in Fig. 11.3-16(a) and (b) [14]. This method does not enclose the diodes within a feedback loop. If we assume that the diodes are ideal, it is easy to see that diode D1 turns on when $v_{IN} \leq -2E_1$ and diode D2 turns on

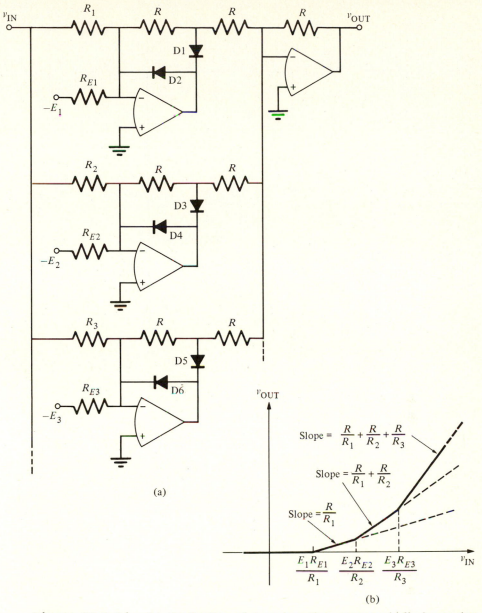

(a)

(b)

Figure 11.3-14 (a) Realization of a noninverting, monotonically increasing, piecewise linear approximation. (b) Voltage transfer characteristics of (a).

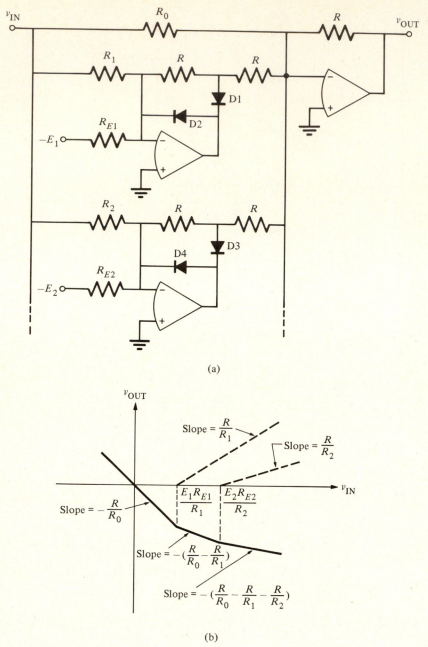

(a)

(b)

Figure 11.3-15 (a) Realization of an inverting, monotonically decreasing piecewise linear approximation. (b) Voltage transfer characteristics.

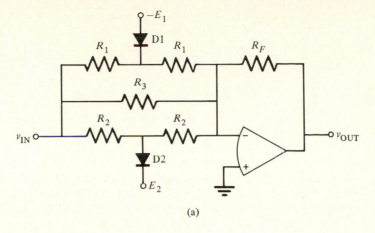

(a)

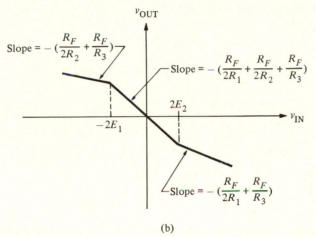

(b)

Figure 11.3-16 (a) Monotonic decreasing realization. (b) Voltage transfer characteristic of (a).

when $v_{IN} \geqq 2E_2$. When v_{IN} is greater than $-2E_1$ and less than $2E_2$, the output voltage of this limiter is given as

$$v_{OUT} = -\left(\frac{R_F}{R_3} + \frac{R_F}{2R_1} + \frac{R_F}{2R_2}\right) v_{IN}, \qquad -2E_1 \leqq v_{IN} \leqq 2E_2 \qquad (15)$$

When $v_{IN} < -2E_1$, diode D1 turns on and the output is given as

$$v_{OUT} = -\left(\frac{R_F}{R_3} + \frac{R_F}{2R_2}\right) v_{IN} + \left(\frac{R_F}{R_1}\right) E_1, \qquad -2E_1 < v_{IN} \qquad (16)$$

When $v_{IN} > 2E_2$, diode D2 turns on and the output is given as

$$V_{OUT} = -\left(\frac{R_F}{R_3} + \frac{R_F}{2R_1}\right) v_{IN} + \left(\frac{R_F}{R_2}\right) E_2, \qquad v_{IN} > 2E_{21} \qquad (17)$$

The transfer function of Fig. 11.3-16(a) is shown in Fig. 11.3-16(b), assuming the diodes are ideal. Nonideal diodes will modify this characteristic slightly. To a first-order consideration, if the ideal diode is assumed to turn on at a voltage of V_{ON} rather than zero, the magnitude of the reference voltages E_1 and E_2 are increased by the value V_{ON}. If the diode forward resistance is added to this model, then R_1 and R_2 in the above equations must be increased by an amount equal to this resistance. This method works well for several breakpoints. If the number of breakpoints increases, it becomes difficult to attribute a portion of the transfer characteristic to a particular diode and reference voltage.

Fig. 11.3-17(a) shows a method similar to that of Fig. 11.3-16(a) for increasing monotonic slopes. Assuming ideal diodes, it is seen that when the current i_1 (i_2) is zero, diode D1 (D2) is at its breakpoint. Solving for the value of v_{IN} which gives this breakpoint results in

$$v_{IN}(BP) = -\left(\frac{R_2}{R_4}\right) E_1 \qquad (18)$$

for diode D1. The breakpoint for diode D2 is

$$v_{IN}(BP) = \left(\frac{R_1}{R_3}\right) E_2 \qquad (19)$$

Between these breakpoints both diodes are off. Therefore the output voltage is

$$V_{OUT} = -\left(\frac{R_F}{R_5}\right) v_{IN}, \qquad -(R_2/R_4)E_1 \leq v_{IN} = (R_1/R_3)E_2 \qquad (20)$$

When $v_{IN} < -(R_2/R_4)E_1$, the output voltage is

$$V_{OUT} = -\left(\frac{R_F}{R_2} + \frac{R_F}{R_5}\right) v_{IN} - \left(\frac{R_F}{R_4}\right) E_1, \qquad -(R_2/R_4)E_1 < v_{IN} \qquad (21)$$

When $v_{IN} > (R_1/R_3)E_2$, the output voltage is

$$V_{OUT} = -\left(\frac{R_F}{R_1} + \frac{R_F}{R_5}\right) v_{IN} - \left(\frac{R_F}{R_3}\right) E_2, \qquad (R_1/R_3)E_2 < v_{IN} \qquad (22)$$

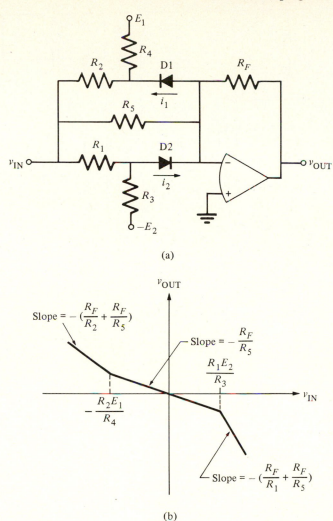

(a)

(b)

Figure 11.3-17 (a) Monotonic increasing realization. (b) Voltage transfer characteristic of (a).

Fig. 11.3-17(b) gives the voltage-transfer function described by the above equations for Fig. 11.3-17(a). The comments about the nonideal diodes made for Fig. 11.3-16(a) are also valid for Fig. 11.3-17(a). Other forms of diode-shaping networks similar to Figs. 11.3-16(a) and 11.3-17(a) are found in the problems and references. These networks are compatible with CMOS technology if the pn diodes are replaced by MOS diodes.

The transfer characteristics of the breakpoint-shifting realizations given above all depend upon resistor ratios. If more accuracy is required,

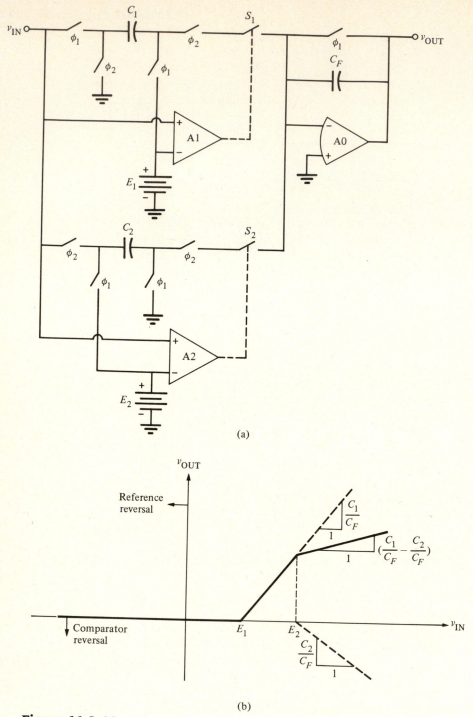

Figure 11.3-18 (a) Switched-capacitor realization of multiple, shifted line segments. (b) Voltage transfer characteristics of (a).

discrete-time circuits that provide transfer characteristics dependent upon capacitor ratios must be used. These circuits use the concepts illustrated in Fig. 11.3-9. Unfortunately, if the other input of the comparator is taken to a reference voltage E_1 the result will not be what is desired. The problem is that when the comparator enables the transfer of the charge on C_1 to C_2, the charge on C_1 is C_1E_1 rather than 0. This results in a step of E_1 in the transfer characteristics. This problem can be solved by the circuit of Fig. 11.3-18(a). This circuit consists of two realizations. C_1, its switches and comparator A1 form a monotonically-increasing realization in the first quadrant while C_2, its switches and the comparator A2 form a monotonically-increasing realization in the fourth quadrant. Both of these implementations require C_F and the op amp A0 to complete their realization. Of course, the output must be sampled and held during the ϕ_2 phase.

The voltage-transfer function of Fig. 11.3-18(a) is shown in Fig. 11.3-18(b). Assuming that E_1 is less than E_2, we see that no charge is transferred to the capacitor C_F until v_{IN} reaches E_1. At this point, comparator A1 turns on switch S1 which will allow charge on C_1 to be transferred to C_F. Because C_1 has been charged with respect to E_1, there is zero charge on C_1 at $v_{IN} = E_1$. As v_{IN} increases further, the output increases with a slope of C_1/C_F. When v_{IN} reaches E_2, comparator A2 turns on switch S2. During ϕ_1, C_2 is charged to E_2. At $v_{IN} = E_2$, C_2 is charged by opposing equal voltages and transfers no charge to C_F. As v_{IN} increases, C_2 begins to transfer charge to C_F, which subtracts from the charge being transferred by C_1. The resulting slope of the segment is $(C_1 - C_2)/C_F$.

By using combinations of these two realizations, a monotonically-increasing or -decreasing transfer function can be realized in any of the four quadrants. For example, if the voltage references are reversed, it shifts the segment to the left. A breakpoint in the first (fourth) quadrant is now in the second (third). If the comparator inputs are reversed, the segment is shifted from the first (second) quadrant to the fourth (third) quadrant. Many other realizations of shifted breakpoints using discrete-time circuits are possible. Most are based on the ideas illustrated in Fig. 11.3-18(b). Several examples will be given to illustrate the capability of the waveshaping circuits which have been presented in this section.

Example 11.3-1

Triangle to Sine Wave Converter

A 3 V peak-to-peak triangle wave is to be converted into a 3 V peak-to-peak sinusoid using a circuit such as Fig. 11.3-18(a). We shall select a 7-segment approximation which is designed as follows. From $0 < t < T/4$, the input voltage can be expressed as

$$v_{in}(t) = \left(\frac{3}{T/4}\right)t = \left(\frac{12}{T}\right)t = (6/\pi)\omega t$$

The desired output voltage is expressed as

$$v_{out}(t) = 3 \sin[\omega t]$$

Substituting the first equation above into the second results in

$$v_{out}(t) = 3 \sin\left[\frac{\pi}{6}v_{in}(t)\right]$$

The breakpoints are arbitrarily selected to get a best match between the 7 line segments and the above equation. The breakpoints and segment information is

Segment	Breakpoint v_{in}	Breakpoint v_{out}	Slope
0	–	–	1.57
1	0.86	1.35	1.16
2	1.91	2.57	0.59
3	2.54	2.94	0.15

The 0 segment goes through the origin and therefore does not have a breakpoint. The other three segments have the same slope as above, but the breakpoints are negative. The seven-segment approximation is plotted in Fig. 11.3-19(a). A realization of this seven-segment approximation of a triangle–to–sine-wave converter is shown in Fig. 11.3-19(b). When $v_{in} = 0$, all the comparators are off and the slope is $+1.57$ and goes through the origin. As v_{in} increases, comparators A1, A2, and A3 successively trip causing the slope to decrease at the specified breakpoints. As v_{in} decreases, comparators A4, A5, and A6 successively trip causing the slope of the realization to decrease in the third quadrant similar to the first quadrant. At 1 kHz, the total harmonic distortion of Fig. 11.3-19(b) was less than 5%.

Example 11.3-2

Quarter-Square Multiplier

Fig. 11.3-20 shows an alternate method of obtaining a four-quadrant multiplier. The blocks in this multiplier consist of summers with various gains, a noninverting and inverting full-wave rectifier and first and third quadrant squaring circuits. All of these blocks can be implemented using circuits considered in previous sections except for the

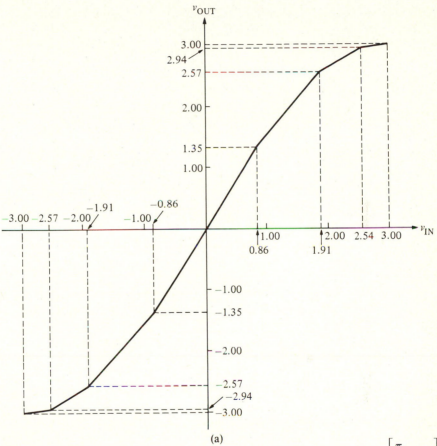

(a)

Figure 11.3-19 (a) Seven segment approximation of $v_{OUT} = 3 \sin\left[\dfrac{\pi}{6} v_{IN}(t)\right]$.

(continued on next page)

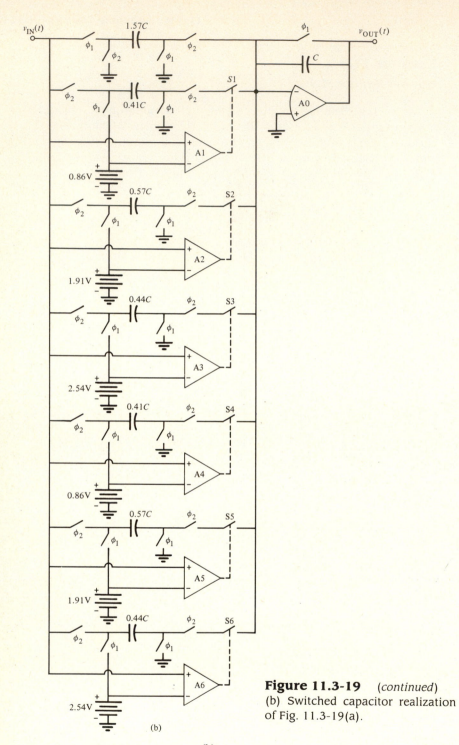

(b)

(b)

Figure 11.3-19 (*continued*)
(b) Switched capacitor realization
of Fig. 11.3-19(a).

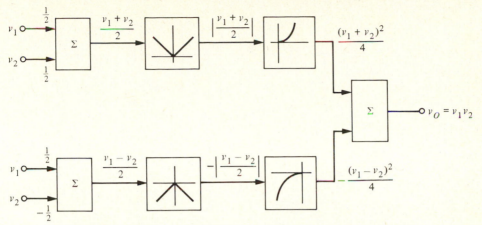

Figure 11.3-20   Block diagram of a quarter-square multiplier.

squaring circuits. A squaring circuit can be easily developed using the ideas of Fig. 11.3-18.

Assume that the output to the squaring circuit is limited at 4 V. A four-segment approximation to the squaring circuit is given in Fig. 11.3-21(a). This monotonically increasing transfer function can be realized by the circuit of Fig. 11.3-21(b) where the breakpoints are at 0.5 V, 1 V, and 1.5 V. Because the previous stage does not permit negative values, we do not need a comparator to stop the segment that goes through the origin. The negative squaring circuit can be realized by reversing both the comparator inputs and the reference voltages.

This section has introduced techniques for shaping analog signals. The methods are compatible with CMOS technology and can be continuous time or discrete time. Discrete-time methods have the advantage of improved accuracy but the disadvantage of more complexity. These techniques can be extended to numerous other applications beyond that illustrated here.

11.4   Oscillators and Waveform Generators

Many signal-processing circuits are designed to operate in some manner on signals. However, there is also a class of circuits which generate the

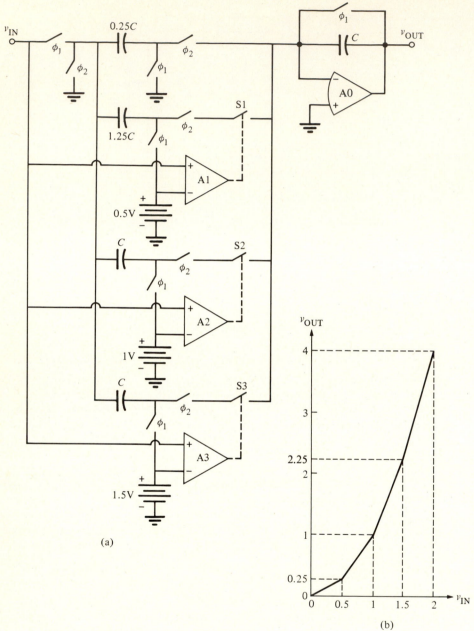

(a)

(b)

Figure 11.3-21 (a) Four segment approximation of $v_{OUT} = v_{IN}^2$. (b) Realization of Fig. 11.3-21(a).

634

Table 11.4-1

Classification of Oscillators.

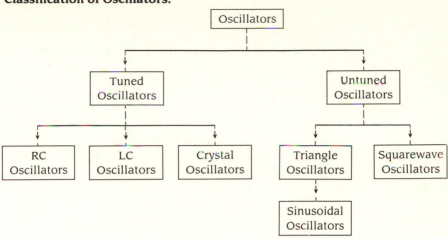

signals in the first place. Such circuits are generally called oscillators and are circuits which convert dc power into a periodic waveform or signal. Oscillators can be classified as shown in Table 11.4-1. The two general classes are tuned and untuned. Tuned oscillators produce nearly sinusoidal outputs whereas untuned oscillators typically have only two stable states. Tuned oscillators can be further divided into RC, LC, and crystal oscillators. We shall only consider RC oscillators since these are most suitable for integrated-circuit technology, although crystal oscillators are often used with the crystal external to the integrated circuit. The outputs of the untuned oscillator are typically square waves and triangle waves. The untuned oscillator can create a sinusoid by applying the triangle wave to a sine-shaping circuit such as the one considered in Ex. 11.3-1. The untuned oscillators are very compatible with integrated-circuit technology, which in part accounts for their widespread use. They are also capable of implementing a voltage-controlled oscillator (VCO) which has many uses in signal processing circuits. Due to lack of space, the consideration of oscillators will be limited. Much more information can be found in the references [15–19].

The following will consider both tuned and untuned oscillators. The principles of operation and an example will be given. We shall restrict our considerations to oscillators that use op amps, capacitors, and resistors. Fig. 11.4-1 shows a block diagram of a single-loop feedback system similar to Fig. 8.2-1. This diagram consists of an amplifier (A), a feedback network (β), and a summing junction. The variables shown are v_S, v_F, v_I, and v_O, which are the source, feedback, input, and output voltages, respectively. The $\pm$ sign next to the feedback input to the summing junction determines whether or not the feedback is positive or negative. The $+$ sign corresponds

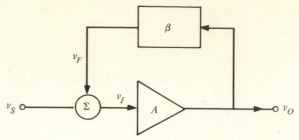

Figure 11.4-1   Block diagram of a single-loop feedback system.

to positive feedback and the $-$ sign to negative feedback. The loop gain L is the voltage gain around the loop when $v_S = 0$ and is expressed as

$$L\,\Big|_{v_S = 0} = \pm A\beta \tag{1}$$

The closed loop gain A_f is equal to v_O/v_S and is given as

$$A_f = \frac{v_O}{v_S} = \frac{A}{1 + A\beta} = \frac{A}{1 - L} \tag{2}$$

where negative feedback has been assumed. The operating principle of a tuned oscillator can be seen from Eq. (2). If there is no input ($v_S = 0$) then for a finite output v_O Eq. (2) must be equal to infinity. This can only happen if $L = 1$. Thus, the criterion for oscillation in a tuned oscillator is

$$\text{Loop gain} = L(j\omega_o) = A(j\omega_o)\beta(j\omega_o) = 1\angle 0° \tag{3}$$

where ω_o is called the radian frequency of oscillation. An alternate form of Eq. (3) is

$$\text{Re}[L(j\omega_o)] + j\text{Im}[L(j\omega_o)] = 1 + j0 \tag{4}$$

or

$$L(j\omega_o) = 1\angle 0° \tag{5}$$

In order to satisfy the above criteria, the oscillator must be able to achieve positive feedback at some frequency ω_o where the magnitude of the loop gain is exactly unity. The analysis of oscillators is simple and consists of calculating the loop gain and using Eq. (3), (4), or (5) to find the frequency of oscillation ω_o and the magnitude of the amplifier gain A necessary to oscillate.

Fig. 11.4-2(a) shows an RC oscillator called the Wien bridge oscillator. This oscillator consists of an amplifier whose gain is K, and a feedback network consisting of R_1, C_1, R_2, and C_2. (Fig. 11.4-2(b) shows how to open the feedback loop in order to calculate the loop gain L. The key principle in opening the feedback loop is to do so at a point in the loop where the resistance looking forward in the loop is much greater (less) than the resistance looking backward in the loop. In this case, the op amp in the noninverting configuration offers infinite resistance looking into the noninverting terminal (point A). Note that the loop could also be broken at point B because resistance looking back into the op amp with negative feedback approaches

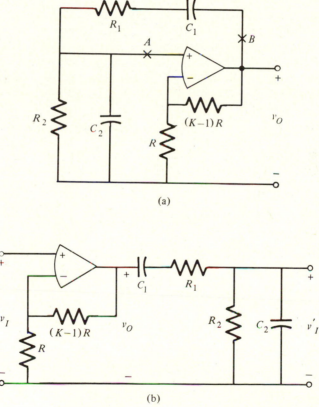

(a)

(b)

Figure 11.4-2   (a) Wien bridge RC oscillator circuit. (b) Open loop version of (a).

zero. Assuming $R_1 = R_2 = R$ and $C_1 = C_2 = C$, the loop gain can be written as

$$G(s) = \frac{K\dfrac{s}{RC}}{s^2 + \dfrac{3}{RC}s + \left[\dfrac{1}{RC}\right]^2} \tag{6}$$

Substituting for s by $j\omega$ and equating to $1 + j0$ results in

$$G(j\omega_o) = \frac{\dfrac{jK\omega_o}{RC}}{\left[\dfrac{1}{RC}\right]^2 - \omega_o^2 + \dfrac{j3\omega_o}{RC}} = 1 + j0 \tag{7}$$

Eq. (7) can be satisfied if $\omega_o = 1/RC$ and $K = 3$, which are precisely the conditions for oscillation of the oscillator of Fig. 11.4-2(a). To obtain a 1 kHz oscillator, we could choose $C = 0.01 \ \mu F$ which gives $R = 15.9 \ K\Omega$.

The above consideration has been ideal because if $K = 3$, the oscillator could oscillate at any amplitude. In practice, the amplitude of the oscillator is determined by nonlinear negative feedback. Fig. 11.4-3 shows a possible voltage-transfer function for the amplifier with a gain K in Fig. 11.4-2(a). For small values of V_0, the gain of the amplifier is designed to be greater

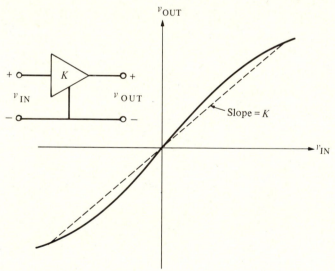

Figure 11.4-3   Amplifier transfer function suitable for amplitude stabilization of an oscillator.

than K ($K = 3$ in the above example). When the gain is greater than K, the amplitude of the sinusoidal oscillation grows. As the amplitude grows, the effective gain becomes smaller. The oscillator amplitude stabilizes at the amplitude at which the effective gain is exactly equal to K. If the amplitude should increase above this level for some reason, the oscillation will begin to decay because the effective gain is less than K. Thus the amplitude stabilization is an important part of the oscillator. If the harmonic content of the sinusoid can be large, the limiting effects of the power supplies can be used (although the waveform will no longer be sinusoidal). Many different schemes have been successfully used to achieve a stable oscillator amplitude. These include piecewise limiters similar to the circuits of the previous section, thermistors, the large-signal characteristics of differential amplifiers, and so on.

Because the RC products determine the oscillator frequency, the accuracy of an untrimmed integrated-circuit RC oscillator may not be sufficient. Switched-capacitor techniques can be used to replace the resistors of the RC oscillators with capacitor-switch equivalents [20,21]. However, the analysis of these circuits requires more background than has been presented in this text. If we restrict ourselves to RC oscillators that use only amplifiers or integrators, we can illustrate the principle with circuits previously considered. Fig. 11.4-4 shows an oscillator that uses a noninverting integrator cascaded with an inverting integrator. This oscillator is called a quadrature oscillator. The loop gain of Fig. 11.4-4 is found as

$$L(s) = -[s^2 R_1 C_2 R_3 C_4]^{-1} \tag{8}$$

If s is replaced by $j\omega$ and Eq. (8) is equated to unity, the frequency of oscillation is

$$\omega_o = \left[\frac{1}{R_1 C_2 R_3 C_4} \right]^{1/2} \tag{9}$$

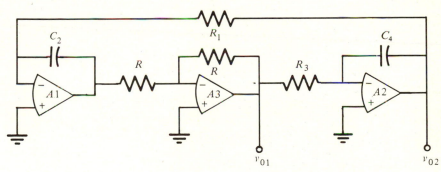

Figure 11.4-4   Schematic of a quadrature oscillator.

The quadrature oscillator provides two sinusoidal voltages that differ by 90 degrees. The sinusoid at v_{o2} leads the sinusoid at v_{o1} by 90 degrees.

A switched-capacitor realization of Fig. 11.4-4 is shown in Fig. 11.4-5. This realization uses the noninverting and inverting integrators of Fig. 10.5-4. By direct substitution, the oscillation frequency of Fig. 11.4-5 is given as

$$\omega_o = \left[\frac{C_1 C_3}{T^2 C_2 C_4} \right]^{1/2} = \left[\frac{C_1 C_3}{C_2 C_4} \right]^{1/2} f_{clock} \tag{10}$$

where T is the period of the clock frequency applied to the switches. (It is seen that a switched-capacitor oscillator is really a frequency divider.)

In general, the tuned oscillator is not widely used in integrated circuits for several reasons. The requirement for accurate RC products or an accurate clock is difficult to accomplish without using external components. The untuned oscillator or relaxation oscillator is much more compatible with integrated-circuit technology. Its operating principle can be seen from the block diagram in Fig. 11.4-6, which consists of an integrator cascaded with a bistable circuit. Although we have not yet discussed the realization of the bistable, let us first consider the operation of Fig. 11.4-6. During the time interval from 0 to T_i, the integrator integrates the voltage L provided by the bistable. If K is the constant of integration, we find that the value of v_T at $t = T_1$ is given as

$$v_T(T_1) = S_- + K \int_0^{T_1} (L_+) \, dt = S_- + KL_+ T_1 = S_+ \tag{11}$$

From Eq. (11) we may solve for the time T_1 to get

$$T_1 = \frac{S_+ - S_-}{KL_+} \tag{12}$$

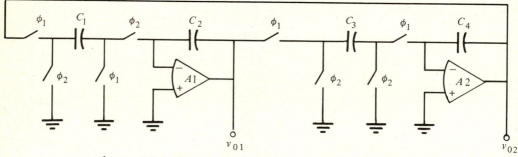

Figure 11.4-5   Switched capacitor realization of Fig. 11.4-4.

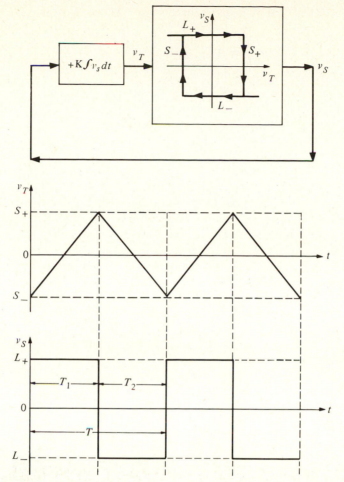

Figure 11.4-6 Block diagram of an untuned oscillator and the resulting waveforms.

We may solve for $v_T(T_2)$ using the same methods to get

$$v_T(T_2) = v_T(T_1) + K \int_{T_1}^{T_2} (L_-) \, dt = S_+ + KT_2L_- = S_-$$ (13)

Solving for T_2 gives

$$T_2 = \frac{S_- - S_+}{KL_-}$$ (14)

The sum of T_1 and T_2 gives the period T and is written as

$$T = \frac{1}{f_o} = T_1 + T_2 = \frac{S_+ - S_-}{K}\left[\frac{1}{L_+} - \frac{1}{L_-}\right] = \frac{4S}{KL} \qquad (15)$$

if $L_+ = -L_- = L$ and $S_+ = -S_- = S$.

The above equations show that the bistable is a key element in the untuned oscillator. Bistables may be clockwise or counterclockwise and can be shifted from the origin. Fig. 11.4-7 shows the two possible generalized bistable characteristics. In this case, $L_+ = V_Y + V_L$, $L_- = V_Y - V_L$, $S_+ = V_X + V_H$, and $S_- = V_X - V_H$. Generally $L_+ = L_-$ and $S_+ = S_-$. Fig. 11.4-8(a) shows a realization of a clockwise (CW) bistable. If the output v_S is equal to V_{DD} (the positive power supply for the op amp), then the voltage at the noninverting input of the op amp is $R_3 V_{DD}/(R_2 + R_3)$. If the input v_T is less than this value then the output is in fact equal to V_{DD}. However, if v_T increases above this value, then the output voltage is at V_{SS} (the negative power supply of the op amp). From this discussion we see that $L_+ = V_{DD}$, $L_- = V_{SS}$, $S_+ = V_{DD}R_3/(R_2 + R_3)$, and $S_- = V_{SS}R_3/(R_2 + R_3)$. If possible, it is desirable to limit the output of the op amp to voltages less than the power supplies. The inverting input of Fig. 11.4-8(a) could be grounded and the voltage v_T applied to the grounded end of R_3 to realize a counterclockwise (CCW) bistable. However, consider Fig. 11.4-8(b), which shows a CCW bistable that limits the op amp swings to less than the power supplies. It can be shown (Prob. 25 with $R_2 = \infty$) that $L_+ = -R_6 V_{SS}/R_7$, $L_- = -R_5 V_{DD}/R_4$, $S_+ = -R_2(L_-)/R_3$, and $S_- = -R_2(L_+)/R_3$. Diodes D1 and D2 serve to

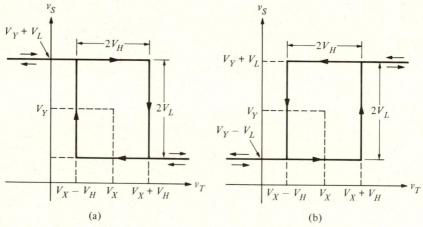

(a) (b)

Figure 11.4-7 (a) CW bistable characteristic. (b) CCW bistable characteristic. V_Y and V_X define the center of the hysteresis characteristic.

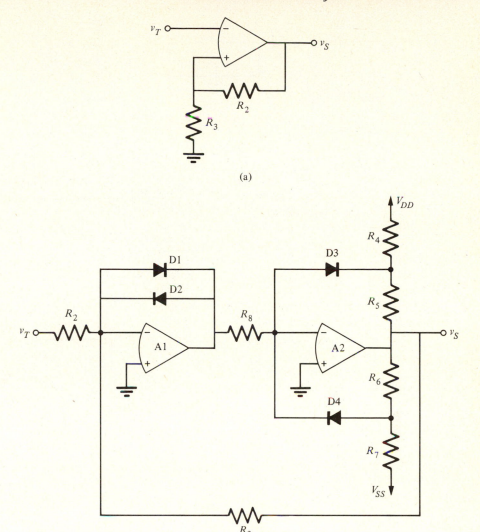

(a)

(b)

Figure 11.4-8   (a) Simple CW bistable. (b) CCW bistable with internal amplitude limiting.

keep the output of op amp A1 from swinging from V_{DD} to V_{SS}, thus permitting quick transition of states.

A bistable circuit can also be implemented using switched-capacitor techniques. Fig. 11.4-9 shows a realization of the bistable characteristics of Fig. 11.4-7. The upper circuit, consisting of op amps 1 and 2, implements

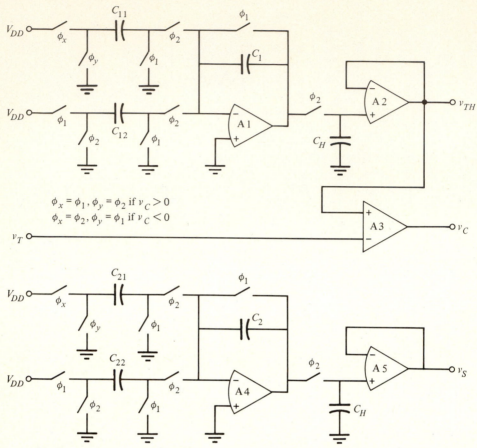

Figure 11.4-9   A switched capacitor implementation of the bistable characteristics of Fig. 11.4-7. ([22], © 1985 IEEE)

a switched-capacitor amplifier with a sample and hold at the output and the ability to change the sign of the gain of the upper input, depending upon the output of comparator 3. The inputs to the comparator are the bistable input and the output of the above amplifier v_{TH}. The lower circuit is similar to the upper amplifier. It can be shown [22] that the discrete time voltages v_{TH}, v_c, and v_{OUT} are given as

$$V_{TH}(z) = \left[\left[\frac{C_{12}}{C_1} \right] V_{DD} - [\text{sgn } V_c(z)] \left[\frac{C_{11}}{C_1} \right] V_{DD} \right] z^{-1} \quad \textbf{(16)}$$

$$= \{ V_x - [\text{sgn } V_c(z)] V_H \} z^{-1}$$

$$V_c(z) = \{ \text{sgn}[V_{TH}(z) - V_T(z)] \} V_{DD} \quad \textbf{(17)}$$

and

$$V_T(z) = \left[\left[\frac{C_{22}}{C_2}\right] V_{DD} - [\text{sgn } V_c(z)]\left[\frac{C_{21}}{C_2}\right] V_{DD}\right] z^{-1}$$

$$= \{V_Y - [\text{sgn } V_c(z)] V_L\} z^{-1}$$
(18)

A CW bistable characteristic is obtained when the ϕ_x and ϕ_y of C_{21} are as shown in Fig. 11.4-9. If ϕ_x and ϕ_y are reversed a CCW bistable is obtained. It is observed that the bistable characteristics are completely general and can be shifted as desired by varying the appropriate V_{DD}, the capacitor ratios, and/or by changing ϕ_x and ϕ_y of C_{21}.

The solid portion of Fig. 11.4-10 shows a simplified implementation of

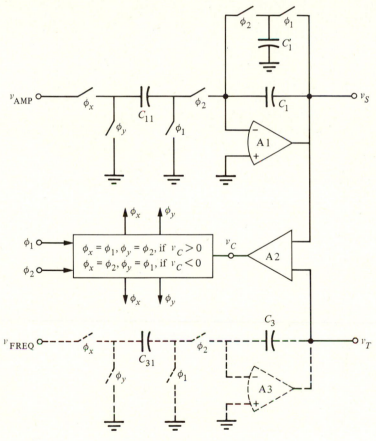

Figure 11.4-10   Simplified CW bistable circuit. C_1' is used to prevent large changes in the op amp output. ([22], © 1985 IEEE)

a CW bistable circuit based on the concepts of Fig. 11.4-9. It can be shown that the second amplifier is not needed for a CW bistable characteristic with $V_X = V_Y = 0$. The output is taken at v_S and the input is applied at v_T. The notation ϕ_X and ϕ_Y implies these clocks are reversed depending upon the sign of v_C. For the case of Fig. 11.4-10, $\phi_X = \phi_1$ and $\phi_Y = \phi_2$ will give a noninverting gain. The phase reversal circuit can be implemented using the ideas shown in Fig. 11.4-11.

The bistable can now be combined with the proper type of integrator (noninverting or inverting) to achieve an implementation of the untuned oscillator block diagram of Fig. 11.4-6. An untuned oscillator using the CCW bistable of Fig. 11.4-8(b) and an inverting integrator is shown in Fig. 11.4-12. The integrating constant K is equal to $1/R_1C_1$ and the various limits of the bistable have already been shown to be $L_+ = -R_6V_{SS}/R_7$, $L_- = -R_5V_{DD}/R_4$, $S_+ = -R_2L_-/R_3$, and $S_- = -R_2L_+/R_3$. For example, if $V_{DD} = -V_{SS} = 15$ volts, $R_1 = 100$ KΩ, $C_1 = 1$ nF, $R_2 = R_3 = 100$ KΩ, $R_4 = R_7 = 30$ KΩ, and $R_5 = R_6 = 20$ KΩ, then $L_+ = -L_- = S_+ = S_- = 10$ volts. From Eq. (15) we find that the frequency of oscillation f_o is 2500 Hz.

Many of the components of Fig. 11.4-12 can be removed at the expense of some of the performance. Fig. 11.4-13(a) shows an example of an untuned oscillator using only a single op amp. The analysis of this oscillator starts by assuming that the output is at L_+. Consequently, the voltage at the positive input terminal of the op amp is $R_3L_+/(R_2 + R_3) = \alpha L_+$. If v_1 is less than this voltage, C_1 will begin to charge to L_+ through R_1. However, when v_1 is equal to αL_+, the output of the op amp switches to L_-. Now the

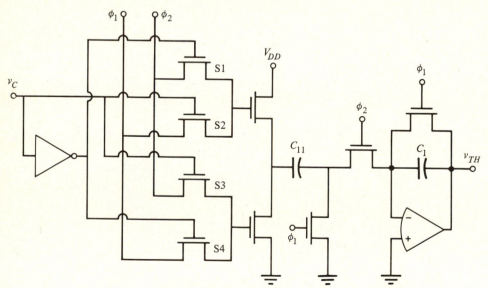

Figure 11.4-11 Implementation of ϕX, ϕY reversal.

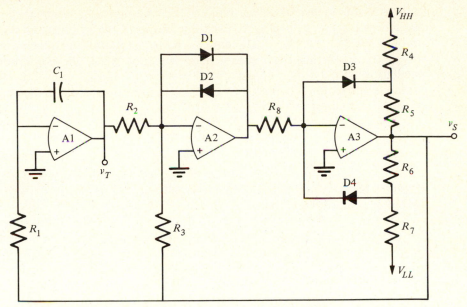

Figure 11.4-12   Untuned oscillator using the bistable of Fig. 11.4-8(b).

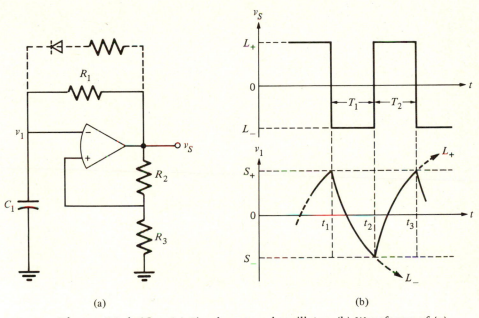

(a)   (b)

Figure 11.4-13   (a) Simple untuned oscillator. (b) Waveforms of (a).

647

voltage at the positive input terminal of the op amp is αL_-. Since this voltage is less than v_1, C_1 begins to discharge toward L_- through R_1. From this analysis we see that $S_+ = \alpha L_+$ and $S_- = \alpha L_-$. The waveform at v_1 is no longer triangular but exponential. It can be shown that the frequency of oscillation is

$$f_o = \frac{1}{2R_1C_1 \ln\left[\dfrac{1+\alpha}{1-\alpha}\right]} \qquad (19)$$

where $\alpha = R_3/(R_2 + R_3)$. The square wave can be made unsymmetrical through the use of diodes and resistors as indicated in Fig. 11.4-13 in the dotted portion of the figure.

One of the advantages of an untuned oscillator is that the frequency can be easily controlled by a voltage. Fig. 11.4-14 shows an example of how this might be accomplished. The output voltage v_S is used to connect the inverting integrator to a positive or negative value of a voltage called v_M. The limits of v_S effectively become $L_+ = -L_- = v_M$. Thus the frequency of the untuned oscillator as developed in Eq. (15) becomes

$$f_o = \frac{v_M}{2R_1C_1(S_+ - S_-)} = \frac{v_M}{4R_1C_1S} \qquad (20)$$

where $S_+ = -S_- = S$. The range of the control or modulating voltage, v_M, will be from the power supplies to the point at where op amp offsets become significant. Consequently, this voltage controlled oscillator (VCO) should have 2 to 3 decades of frequency variation. An alternate VCO realization is shown in Fig. 11.4-15. It can be shown (see Prob. 30) that the frequency of this oscillator is

$$f_o = \frac{i_C}{2C[S_+ - S_-]} \qquad (21)$$

where i_C is controlled by the voltage v_C. Other methods of controlling the oscillator frequency include a diode bridge [23] and current-controlled multivibrators [16]. A low-frequency triangle waveform generator compatible with CMOS technology using the above concepts and having good triangle waveform properties has been described in the literature [24].

The switched-capacitor bistable can also be used to implement an untuned oscillator that can be controlled by a voltage. This untuned oscillator is shown in Fig. 11.4-10 and includes the dotted portion. The amplifier of this realization uses a scheme to avoid requiring the op amp output to be zeroed during each clock cycle. The use of a parallel switched-capacitor

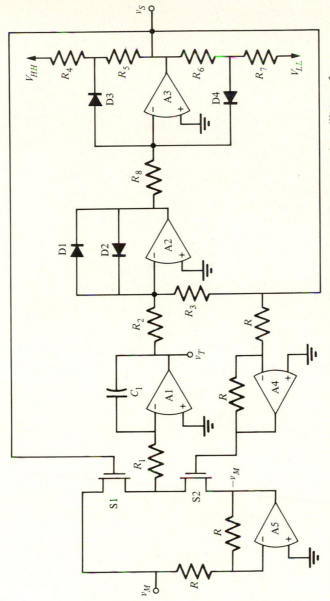

Figure 11.4-14 Voltage controlled oscillator using the untuned oscillator of Fig. 11.4-12.

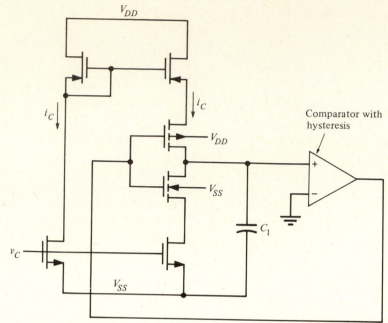

Figure 11.4-15   A VCO compatible with CMOS technology.

resistor realization in shunt with C_1 gives the following discrete-time transfer function from v_{AMP} to v_S.

$$\frac{V_S(z)}{V_{\text{AMP}}(z)} = \frac{-[\text{sgn } V_c(z)]\left[\dfrac{C_{11}}{C_1}\right]z^{-1}}{1 - z^{-1}\left[1 - \dfrac{C_1'}{C_1}\right]} = -[\text{sgn } V_c(z)]\left[\frac{C_{11}}{C_1}\right]z^{-1} \qquad \textbf{(22)}$$

if $C_1 = C_1'$. C_1' opposes the charge being transferred from C_{11} to C_1. Only the charge difference on C_{11} between consecutive samples is transferred. As a result, the op amp output voltage changes only by the difference in the input voltage multiplied by the gain factor of the amplifier. Fig. 11.4-10 functions as a VCO where the amplitude and frequency of the square and triangle waveforms are given as

$$\text{Amplitude} = \left[\frac{C_{11}}{C_1}\right]v_{\text{AMP}}(t) \qquad \textbf{(23)}$$

and

$$f_o = \left[\frac{C_1 C_{31} f_c}{4 C_3 C_{11}} \right] \frac{v_{FREQ}(t)}{v_{AMP}(t)} \tag{24}$$

It is seen that the switched capacitor VCO of Fig. 11.4-10 is capable of both amplitude and frequency modulation. Amplitude modulation is accomplished by connecting both $v_{FREQ}(t)$ and $v_{AMP}(t)$ to the source of modulation. Frequency modulation is accomplished by connecting only $v_{FREQ}(t)$ to the source of modulation and keeping $v_{AMP}(t)$ constant. Fig. 11.4-16 shows the photographs of Fig. 11.4-10 used as an unmodulated, untuned oscillator, an amplitude-modulated oscillator, and a frequency-modulated oscillator (VCO) [22]. As the oscillator frequency approaches the clock frequency, the period of the waveforms of Fig. 11.4-10 experience jitter. This can be solved by using additional circuitry [25].

The block diagram of a CMOS commercially available timer using the above concepts is shown within the dashed box in Fig. 11.4-17(a) [26]. This timer is essentially an identical implementation of the familiar 555 bipolar timer which can be used as monostable or astable multivibrator. The con-

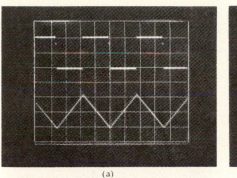

(a)

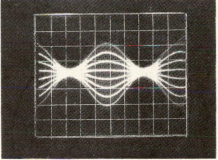

(b)

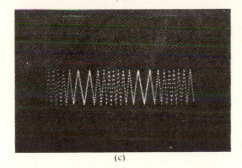

(c)

Figure 11.4-16 (a) Photograph of the square and triangle waveforms of Fig. 11.4-10. (b) Use of Fig. 11.4-10 for amplitude modulation. (c) Use of Fig. 11.4-10 for frequency modulation. ([22], © 1985 IEEE)

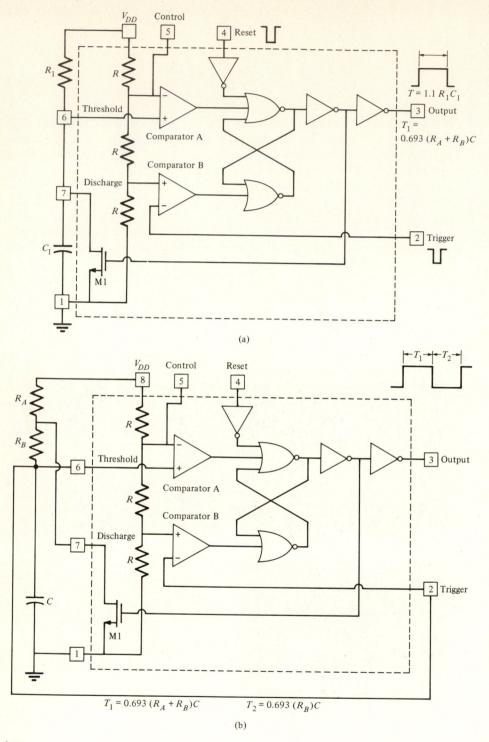

(a)

(b)

nections shown in Fig. 11.4-17(a) generate a single output pulse upon the application of a negative trigger. It is seen that the pulse width depends upon R_1 and C_1 (which are normally external components). The circuit is in its quiescent state when the output is low and M1 is on, which discharges C_1. When a negative-going trigger causes the negative input of comparator B to go below $V_{DD}/3$, the output of this comparator goes low. This causes the flip-flop to shift states, resulting in a high at the output and turning M1 off. Capacitor C_1 begins to charge to V_{DD} through R_1. When the positive input of comparator A reaches $2V_{DD}/3$, the output of this comparator goes high causing the flip-flop to return to its starting state. This causes the output to go low and M1 to discharge C_1 and completes the single output pulse. It can be shown that the width of the output pulse is $1.1R_1C_1$ (see Prob. 31). Fig. 11.4-17(b) shows how the timer can be connected for astable operation.

Fig. 11.4-18 shows the equivalent circuit of the part of Fig. 11.4-17 within the dashed box. The internal bias string consist of three p-well resistors of approximately 100 Kilohms each. The comparators are simple differential-amplifier stages as presented in Section 6.2. Comparator A uses an n-channel input pair while comparator B uses a p-channel input pair. The discharge transistor M16 and the complementary output transistors M14 and M15 are large-area devices designed to handle higher current levels. The substrate BJT, Q12, is included to provide active pull-up to the flip-flop input terminal (node P) in order to assure the quick resetting of the circuit once M11 is turned off. The CMOS version of the 555 timer dissipates about one-fifteenth power as the bipolar version but cannot provide the same output-current drive.

Many other oscillators compatible with integrated circuit technology have not been presented here, including: ring oscillators; constant-current oscillators; multivibrators (both collector- (drain) or emitter- (source) coupled); and so on. The performance of oscillators, including their temperature and power supply stability, has not been covered. These subjects will have to be investigated further by the interested reader using the references and literature available.

11.5 *Summary and Conclusion*

This chapter has presented various types of analog circuits and systems using the concepts of the previous chapters. Most of the topics presented in this chapter have been fabricated or are representative of what can be implemented using CMOS technology. The subjects chosen were selected

←————————————————————

Figure 11.4-17   Block diagram of a CMOS timer used as (a) a monostable, and (b) an astable multivibrator.

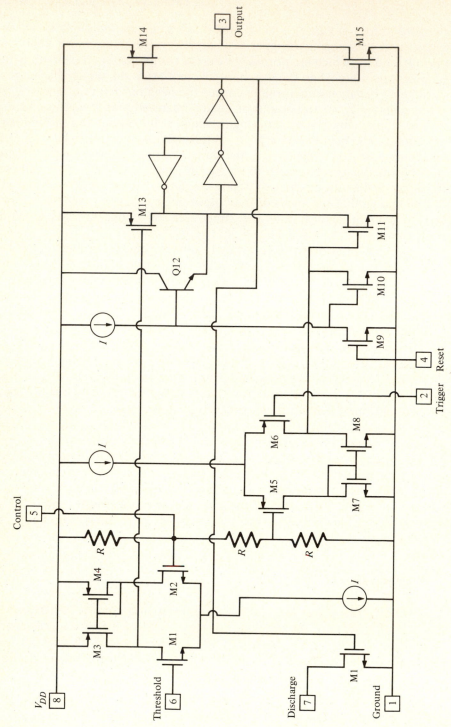

Figure 11.4-18 Equivalent circuit of the 555 CMOS timer.

654

both for their importance to the designer and as illustrative applications of the previous circuit-design principles. Stable voltage references are very important in circuits where precision signal processing must be done. A stable voltage reference was seen to be independent of power supplies and to have a temperature coefficient of less than ± 100 ppm/°C. One application of the voltage reference was found in the subject of digital-to-analog and analog-to-digital conversion. If a stable voltage reference cannot be integrated on the chip, it must be applied externally—which diminishes the advantages of system integration.

Also included in this chapter were the subjects of nonlinear analog-signal operations such as multiplication, division, waveshaping, and oscillation, topics that suggest how the concepts developed earlier might be applied to various signal operations. This subject matter could be extended to include many other areas, such as telecommunication and speech processing. One of the objectives of this book has been to provide the background for the design of large-scale analog systems using CMOS technology.

This chapter also concludes the text on designing CMOS analog integrated circuits. It has been seen that, from a topological viewpoint, CMOS analog circuits are very similar to bipolar analog circuits. However, there is one major difference which has played an important role throughout this entire development. The small signal parameters of the MOS device are functions of *both* the dc variables and geometry of the device. In contrast, the small signal parameters of the bipolar device are *only* dependent upon the dc variables. This difference is fortunate in that it has allowed the designer to make up for some of the inherent deficiencies of MOS devices compared with bipolar devices.

This book was written around the relationships expressed in Table 1.1-2. Hopefully, this Table has been useful to the reader. The objective was to try to illustrate the importance of hierarchy in the analog circuit design process. Unfortunately, the science of analog circuit design is not as well understood as that of digital making it necessary to emphasize the relational aspects of design. Typically, the designer is given a specification at the top of Table 1.1-2. It is the function of the design to break the larger system or circuit into smaller pieces which can then be individually designed and combined to form the whole. Unfortunately, few computer aided design tools exist for analog circuits today. However, this will be one of the areas in which the designer will see major changes in the future. There is no doubt that whatever CAD tools evolve will be based in part on the hierarchy and relationships shown in Table 1.1-2.

PROBLEMS — *Chapter 11*

1. Draw the schematic for a conventional bandgap reference of Fig. 11.1-3 that is compatible with an n-well, CMOS technology.

2. What is the value of dV_{BE}/dt at 27°C if $\alpha = 1$ and $\gamma = 3.2$?
3. Derive Eq. (24) of Sec. 11.1.
4. Repeat Ex. 11.1-1 if zero temperature coefficient is to be attained at 100°C.
5. Repeat Ex. 11.1-1 if $A_{E1} = 5A_{E2}$ and $V_{BE2} = 0.6V$.
6. Show how to use the autozero techniques of Sec. 7.5 to obtain a dynamic bandgap reference voltage which is not dependent upon V_{OS}.
7. Fig. P11.7 shows a simple voltage reference using an enhancement and a depletion NMOS transistor. Ignore the effects of the bulk-source voltage on the depletion transistor. Solve for the value of V_{REF} and find the conditions for zero temperature dependence. What is the value of V_{REF} under these conditions?
8. For the low-power, CMOS reference of Fig. 11.1-4, assume that $V_{BE} = 0.7$ volts at room temperature, $V_{REF} = 1.20$ at room temperature, $S_1/S_3 = S_4/S_2 = 2$, and $S_6/S_3 = 4$. If $R_2 = 353$ Kilohms, find R_1. If $V_{DD} = 3.5$ volts, what is the no-load power dissipation of the reference?
9. Use the results derived in Sec. 6.2 and derive Eq. (1) of Sec. 11.2.
10. Develop the relationship given in Eq. (8) of Sec. 11.2.
11. Show how the cascade of an A/D converter with a D/A converter can realize a multiplier/divider if the reference voltages to both converters are themselves analog signals.
12. Use a four-quadrant, analog multiplier to develop a circuit which creates $-K(v_1)^{0.5}$ when v_1 is greater than zero.
13. The sine of an input voltage v_1 can be approximated by the formula, $\sin(v_1) \cong 1.155v_1 - 0.33v_1^2$, to within $\pm 2\%$ over the range of $0 < v_1 < \pi/2$. Use a minimum number of multipliers and op amps to develop a realization of $\sin(v_1)$ using the above formula.
14. Assume the op amp of Fig. 11.3-6(a) has an input-offset voltage of V_{OS}. Find the value of v_{IN} at which the diodes will switch states for the inverting rectifier of Fig. 11.3-6(a).

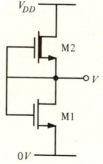

Figure P11.7

15. Reverse the directions of the MOS diodes of Fig. 11.3-10(a) and develop a plot of v_{OUT} versus v_{IN}.

16. Show how autozeroing techniques from Sec. 7.5 can be used for Fig. 11.3-12(a) to eliminate the influence of the input-offset voltage of the comparator, A2.

17. Assume that the diodes of Fig. 11.3-16(a) are ideal and design a limiter with a gain of -1 for -2 V $< v_{IN} < 2$ V and a gain of -0.5 for input voltages outside of this range. Assume that you have power supplies of ± 5 V and express all resistors in terms of R_F.

18. Repeat the above problem if the diodes have a turn-on voltage of 1 V.

19. Assume that the diodes of Fig. 11.3-17(a) are ideal and design a limiter with a gain of -1 for -2 V $< v_{IN} < 2$ V and a gain of -2 for input voltages outside of this range. Assume that you have power supplies of ± 5 V and express all resistors in terms of R_F.

20. Repeat the above problem if the diodes have a turn on voltage of 1 V.

21. Develop a realization of a voltage transfer curve using three line segments defined by the following v_{OUT}, v_{IN} coordinates: (0,0), (0.9,3), (2.5,5), and (10,10). Use resistors, ideal diodes, and op amps.

22. Repeat the above problem using switches, capacitors, comparators, and op amps.

23. Develop a realization of the voltage-transfer curve using four line segments as defined by the following pairs of v_{OUT}, v_{IN} coordinates: $(0,-\infty)$, $(0,-5)$, $(10,0)$, $(0,5)$, and $(0,+\infty)$. Use resistors, ideal diodes, and op amps.

24. Repeat the above problem using switches, capacitors, comparators, and op amps.

25. Fig. P11.25 shows an inverting limiter circuit. Develop a transfer curve giving v_{OUT} as a function of v_{IN} assuming that the diodes are ideal. Be sure to label each breakpoint and slope in terms of the resistors. Identify the regions where each of the diodes is on and off.

26. Repeat the above problem for Fig. P11.26.

27. For the circuit shown in Fig. P11.27, plot v_{OUT} as a function of v_{IN}. What function does this circuit accomplish?

28. Design the values of R_3, R_4, R_5, and R_6 so that the circuit of Fig. P11.27 has the transfer function of Fig. P11.28.

29. Assume that the diodes of the circuit shown in Fig. P11.29 are ideal. Sketch v_{OUT} as a function of v_{IN} as v_{IN} varies from -15 V to $+15$ V. Be sure to carefully label all breakpoints and slopes. Plot the output of this circuit if a ± 15 triangle wave of frequency f_T is applied.

30. Show that the frequency of oscillation of Fig. 11.4-15 is given by Eq. (21).

31. Verify that the period of the CMOS timer of Fig. 11.4-17(a) is given as $T = 1.1 R_1 C_1$.

32. Verify that the width of the output of Fig. 11.4-17(b) is $0.693(R_A +$

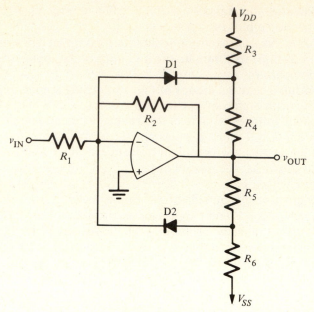

Figure P11.25

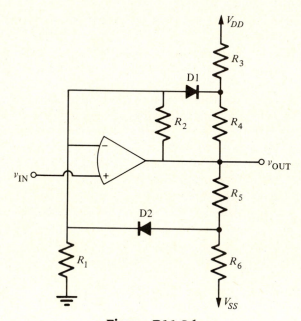

Figure P11.26

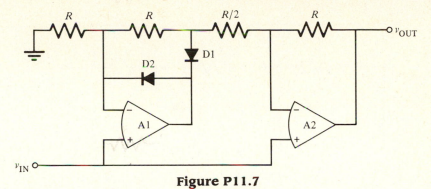

Figure P11.7

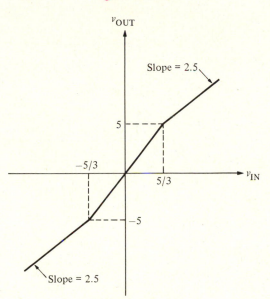

Figure P11.28

$R_B)C$ when the output is high and $0.693\ R_BC$ when the output is low. What is the ratio of the time the output is high to the period of the waveform?

REFERENCES

1. R.J. Widlar, "New Developments in IC Voltage Regulators," *IEEE Journal of Solid-State Circuits*, Vol. SC-6, No.1 (February 1971) pp. 2–7.
2. K.E. Kujik, "A Precision Reference Voltage Source," *IEEE Journal of Solid-State Circuits*, Vol. SC-8, No. 3 (June 1973) pp. 222–226.
3. P.R. Gray and R.G. Meyer, *Analysis and Design of Analog Integrated Circuits*, 2nd Ed. (New York: John Wiley & Sons 1984), Chapter 4.

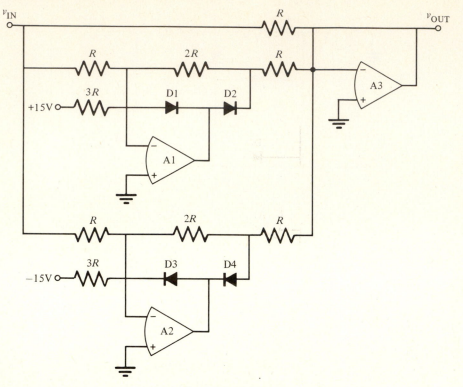

Figure P11.29

4. B.S. Song and P.R. Gray, "A Precision Curvature-Corrected CMOS Bandgap Reference," *IEEE Journal of Solid-State Circuits*, Vol. SC-18, No. 6 (December 1983) pp. 634–643.

5. Y.P. Tsividis and R.W. Ulmer, "A CMOS Voltage Reference," *IEEE Journal of Solid-State Circuits*, Vol. SC-13, No. 6 (December 1982) pp. 774–778.

6. G. Tzanateas, C.A.T. Salama, and Y.P. Tsividis, "A CMOS Bandgap Voltage Reference," *IEEE Journal of Solid-State Circuits*, Vol. SC-14, No. 3 (June 1979) pp. 655–657.

7. E. Vittoz and J. Fellrath, "CMOS Analog Integrated Circuits based on Weak Inversion Operation," *IEEE Journal of Solid-State Circuits*, Vol. SC-12, No. 3 (June 1977) pp. 224–231.

8. R.G. Sparkes and A.S. Sedra, "Programmable Active Filters," *IEEE Journal of Solid-State Circuits*, Vol. SC-8, No. 1, (February 1973) pp. 93–95.

9. B. Gilbert, "A High-Performance Monolithic Multiplier using Active Feedback," *IEEE Journal of Solid-State Circuits*, Vol. SC-9, No. 6, (December 1974) pp. 364–373.

10. J.H. Huijsing, P. Lucas, and B. De Bruin, "Monolithic Analog Multiplier-Divider," *IEEE Journal of Solid-State Circuits*, Vol. SC-17, No. 1 (February 1982) pp. 9–15.

11. K. Martin, "Improved Circuits for the Realization of Switched-Capacitor Filters," *IEEE Transactions on CAS*, Vol. CAS-27, No. 4 (April 1980) pp. 237–244.

12. P.E. Allen and E. Sanchez-Sinencio, *Switched Capacitor Circuits*, (New York: Van Nostrand Reinhold, 1984).

13. K. Yamamoto, S. Fuji, and K. Matsuoka, "A Single Chip FSK Modem," *IEEE Journal of Solid-State Circuits*, Vol. SC-19, No. 6 (December 1984) pp. 855–861.

14. J.G. Graeme, G.E. Tobey, and L.P. Huelsman, *Operational Amplifiers—Design and Application*, (New York: McGraw-Hill 1971), Chapter 7.

15. L. Strauss, *Wave Generation and Shaping*, 2nd. Ed. (New York: McGraw-Hill, 1970).

16. A.B. Grebene, *Bipolar and MOS Analog Integrated Circuit Design*, (New York: John Wiley and Sons, 1984).

17. W.G. Jung, *IC Timer Cookbook*, (Indianapolis, IN: Howard Sams and Co., 1977).

18. J. Millman, *Microelectronics: Digital and Analog Circuits and Systems*, (New York: McGraw-Hill, 1979).

19. A.S. Sedra and K.C. Smith, *Microelectronic Circuits*, (New York: Holt, Rinehart and Winston, 1982).

20. T.R. Viswanathan, K. Singhal, and G. Metzket, "Applications of Switched-Capacitor Resistors in RC Oscillators," *Electronic Letters*, Vol. 14, No. 20 (September 28, 1978) pp. 659–660.

21. E.A. Vittoz, "Micropower Switched-Capacitor Oscillator," *IEEE Journal of Solid-State Circuits*, Vol. SC-14, No. 3 (June 1979) pp. 622–624.

22. P.E. Allen, H.A. Rafat, and S.F. Bily, "A Switched-Capacitor Waveform Generator," *IEEE Transactions on Circuits and Systems*, Vol. CAS-32, No. 1 (January 1985) pp. 103–105.

23. J.V. Wait, L.P. Huelsman, and G.A. Korn, *Introduction to Operational Amplifier Theory and Applications*, (New York: McGraw-Hill, 1975).

24. W.W. Cheng and L.E. Larson, "A Low-Frequency CMOS Triangle Wave Generator," *IEEE Journal of Solid-State Circuits*, Vol. SC-20, No. 2 (April 1985) pp. 649–652.

25. J. Silva-Martínez and Edgar Sánchez-Sinencio, "SC Relaxation Oscillators without Excess Phase Jitter," *Proc. of 1986 IEEE Sym. on Circuits and Systems*, San Jose, CA, May 1986, pp. 813–816.

26. "ICM-7555 CMOS Timer," *Intersil Data Book*, (Cupertino, CA: Intersil Corp. 1980) pp. 6.140–144.

Appendix A

Circuit Analysis for Analog Circuit Design

The objective of this appendix is to provide a systematic approach for analyzing analog circuits. Since much design is done by analysis, this approach will be very useful in the study of analog integrated-circuit design. We will begin with a brief introduction to modeling devices from a general viewpoint, followed by several network-analysis techniques useful in analyzing analog circuits. These techniques include mesh and nodal analysis, superposition, substitution of sources, network reduction, and Miller simplification. Although there are other techniques, these are the ones used most often in analog-circuit analysis.

Modeling is an important part of both analog-circuit analysis and design. *Modeling* is defined as the process by which an electronic component/device is characterized in a manner that lets it be analyzed by either mathematical or graphical methods. Most electronic devices have at least three terminals and the voltage-current relationships among terminals are nonlinear. As a consequence, models are generally categorized as large-signal and small-signal models.

Large-signal models represent the nonlinear behavior of the electronic device. Small-signal models are characterized by having linear relationships between the terminal voltages and currents. Typically a small-signal model is only valid for limited values of amplitude. In fact, the signal amplitude is decreased until the nonlinear relationships can be adequately approximated by linear relationships. However, the advantage of the small-signal model is that analysis is greatly simplified by the inherent linearity between the terminal voltages and currents. In either case, models are only representative of the actual device and may fail to accurately predict the device performance in a given range of terminal voltages and currents.

The most important principle in analyzing analog circuits is to keep the approach as simple as possible. This is particularly important when the analysis is used to assist the design. Complicated expressions do not offer any insight into the relationships between performance and the parameters which determine the performance, so the simplest possible model should always be used. The analysis problem should be broken into parts if the resulting parts are simpler to analyze than the whole. The designer can always turn to computer simulation for a more detailed analysis when necessary, but in using the computer, the designer needs to know whether or not the results make sense, yet another reason for learning to make simple, hand calculations.

In the following analytic techniques, only linear circuits will be considered. This means that the techniques apply only to small-signal models and circuits. This does not represent a serious limitation since much of the performance of analog circuits can be characterized by small-signal analysis.

The first analysis technique to be discussed is the systematic method of writing a set of linear equations which describe the circuit. In many cases, the analog small-signal circuit can be quickly analyzed by a simple chain-type calculation. This is always the case when there is only one signal path from the input to the output. As an example consider Fig. A-1. If we want to find v_{out}/v_{in} it is a simple matter to express this transfer function in a chain form as

$$\frac{V_{out}}{V_{in}} = \left(\frac{V_{out}}{V_1}\right)\left(\frac{V_1}{V_{in}}\right) = (-G_m R_3)\left(\frac{R_2}{R_1 + R_2}\right) = \frac{-G_m R_2 R_3}{R_1 + R_2} \tag{1}$$

In more complex circuits, there is generally more than one signal path from the input to the output. In that case, a more systematic method of writing a set of linear equations which describe the circuit is required. Two well-

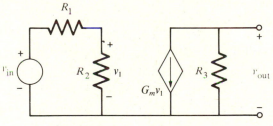

Figure A-1   Chain-type analysis.

known methods use called nodal and mesh equations. They are based on Kirchhoff's laws, which state the sum of the currents flowing into a node must be equal to zero and the sum of the voltage drops around a mesh must be equal to zero. Two examples will follow which illustrates the application of these two familiar approaches.

Example A-1

Nodal Analysis of an Analog Circuit

Consider the circuit of Fig. A-2. The objective is to find v_{out}/i_{in}. Using the techniques of nodal analysis, we may write two equations which represent the sum of the currents at nodes A and B. These equations are

$$i_{in} = (G_1 + G_2)v_1 - G_2 v_{out} \tag{2}$$

and

$$0 = (G_m - G_2)v_1 + (G_2 + G_3)v_{out} \tag{3}$$

Note that $G_i = 1/R_i$ in order to simplify the equations. We may find v_{out} by using Cramer's approach to solving matrix equations, giving

$$V_{out} = \frac{\begin{vmatrix} G_1 + G_2 & i_{in} \\ G_m - G_2 & 0 \end{vmatrix}}{\begin{vmatrix} G_1 + G_2 & -G_2 \\ G_m - G_2 & G_2 + G_3 \end{vmatrix}} = \frac{(G_2 - G_m)i_{in}}{G_1 G_2 + G_1 G_3 + G_2 G_3 + G_m G_2} \tag{4}$$

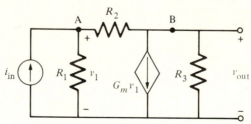

Figure A-2 Nodal analysis.

The desired transfer function is given as

$$\frac{v_{out}}{i_{in}} = \frac{(G_2 - G_m)}{G_1G_2 + G_1G_3 + G_2G_3 + G_mG_2} \tag{5}$$

Example A-2
Mesh Analysis of an Analog Circuit

Consider the circuit of Fig. A-3. We wish to solve for the transfer function, v_{out}/v_{in}. Two mesh currents, i_a and i_b have been identified on the circuit. Note that the choice of meshes can be selected to avoid solving for both i_a and i_b. Summing the voltage drops around these two meshes gives the following two mesh equations.

$$v_{in} = (R_1 + R_2)i_a + R_1i_b \tag{6}$$
$$v_{in} = (R_1 - R_m)i_a + (R_1 + R_3)i_b \tag{7}$$

Using Cramer's approach to solve for i_a gives,

$$i_a = \frac{\begin{vmatrix} v_{in} & R_1 \\ v_{in} & R_1 + R_3 \end{vmatrix}}{\begin{vmatrix} R_1 + R_2 & R_1 \\ R_1 - R_m & R_1 + R_3 \end{vmatrix}} = \frac{R_3v_{in}}{R_1R_2 + R_1R_3 + R_2R_3 + R_mR_1} \tag{8}$$

Since $v_{out} = -R_mi_a$, we can solve for v_{out}/v_{in} as

$$\frac{v_{out}}{v_{in}} = \frac{-R_mR_3}{R_1R_2 + R_1R_3 + R_2R_3 + R_mR_1} \tag{9}$$

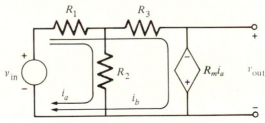

Figure A-3   Mesh analysis.

These two examples illustrate the use of the nodal and mesh equations to analyze circuits. In some cases, a combination of these two approaches may be used to analyze a circuit. This is illustrated in the following example.

Example A-3
Combined Approach to Analyzing an Analog Circuit

The circuit shown in Fig. A-4 is used to model a current source. The objective of this example is to find the small-signal output resistance r_{out} defined as v_o/i_o. Three equations have been written to describe this circuit and to solve for r_{out}. These equations are

$$i_o = A_i i + \frac{v_o - v_1}{R_4} \tag{10}$$

$$i = \frac{-R_3 i_o}{R_1 + R_2 + R_3} \tag{11}$$

and

$$v_1 = i_o \left(\frac{R_3 R_1 + R_3 R_2}{R_1 + R_2 + R_3} \right) \tag{12}$$

The motivation for these equations has been to find an expression for i_o in terms of only i_o or v_o. Thus, Eqs. (11) and (12) replace i and v_1, respectively in Eq. (10) in terms of i_o. Making these replacements results in

$$r_{out} = \frac{v_o}{i_o} = \frac{R_4(R_1 + R_2 + R_3) + R_3 R_4 A_i + R_3(R_1 + R_2)}{R_1 + R_2 + R_3} \tag{13}$$

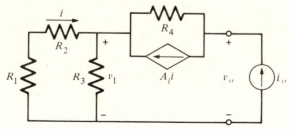

Figure A-4 A combined method of analysis.

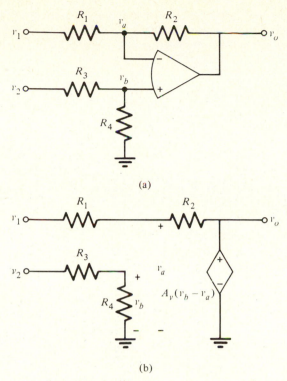

(a)

(b)

Figure A-5   (a) Differential amplifier using an op amp. (b) Small-signal model of (a).

The next circuit technique of interest is called superposition. Superposition can be used in linear circuits where there are one or more excitations contributing to a response. The superposition concept states that the value of a response may be found as the sum of the values of that response produced by each of the excitation sources acting separately. Although superposition is a simple concept, it is often misused, particularly in op amp circuits. The following example will illustrate the use of superposition in an op amp circuit.

Example A-4

Analysis of a Differential Amplifier

A differential amplifier implemented by an op amp is shown in Fig. A-5 (a). The op amp has a finite voltage gain of A_v, an infinite input resisatance and a zero output resistance. Find v_o as a function of v_1 and v_2.

Fig. A-5 (b) shows a small-signal model of Fig. A-5 (a). The output voltage, v_o, can be written as

$$v_o = A_v(v_b - v_a) \tag{14}$$

v_b can be expressed as

$$v_b = \frac{R_4}{R_3 + R_4} v_2 \tag{15}$$

Superposition is used to find v_a, which is a function of both v_1 and v_o. The result is,

$$v_a = \left(\frac{R_2}{R_1 + R_2}\right) v_1 + \left(\frac{R_1}{R_1 + R_2}\right) v_o \tag{16}$$

Substituting Eqs. (15) and (16) into Eq. (14) and simplifying gives

$$v_o = \frac{A_v}{1 + \left(\dfrac{A_v R_1}{R_1 + R_2}\right)} \left[\left(\frac{R_4}{R_3 + R_4}\right) v_2 - \left(\frac{R_2}{R_1 + R_2}\right) v_1 \right] \tag{17}$$

If R_3/R_4 is equal to R_1/R_2, then the amplifier is differential and the influence of the finite op amp voltage gain A_v is given by Eq. (17).

An important principle in the analysis of linear, active circuits is to make any manipulations in the model which will simplify the calculations before starting the analysis. Two techniques which are useful in implementing this principle are called *source rearrangement* and *source substitution*. The concepts are easier demonstrated than defined. Consider the circuit of Fig. A-6 (a). A current source controlled by a voltage v_1 is connected beween two networks, N_1 and N_2, which are unimportant to the application of the concept. The circuit can be simplified by noting that the dependent current source has the effect of taking a current of $G_m v_1$ from node 1 and applying it to node 2. Source rearrangement allows the current source $G_m v_1$ to consist of two sources as shown in Fig. A-6 (b). In this circuit, a current $G_m v_1$ is taken from node 1 and applied to node 3. The same current is then taken from node 3 and applied to node 2. It is seen that the two circuits are identical in performance.

Next we note that the left-hand current source is controlled by the voltage across it. This allows the application of the source-substitution concept

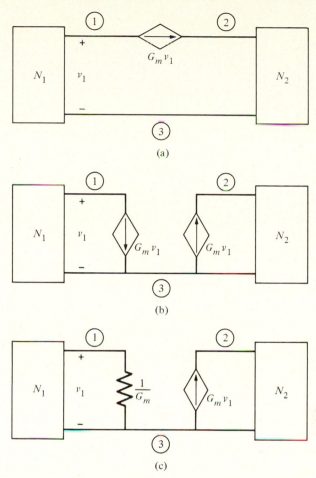

Figure A-6   Current source rearrangement and substitution. (a) Original circuit.
(b) Rearrangement of current surce. (c) Substitution of current source by a resistor
$1/G_m$.

where the left-hand source is replaced by a resistance of value $1/G_m$ ohms.
The final result is shown in Fig. A-6 (c). It can be proven that Figs. A-6 (a)
and (b) are identical. The analysis of Fig. A-6 (c) will generally be much
easier than that for Fig. A-6 (a).

Fig. A-7 (a) through (c) shows the equivalent steps for simplifying a
circuit which has a voltage source connected in parallel with two networks
which is controlled by the current of one of the networks. In this case the
source-rearrangement concept is accomplished by sliding the voltage
source through a node in series with the other branches connected to the
node. An example will demonstrate some of the concepts used in source
rearrangement and source substitution.

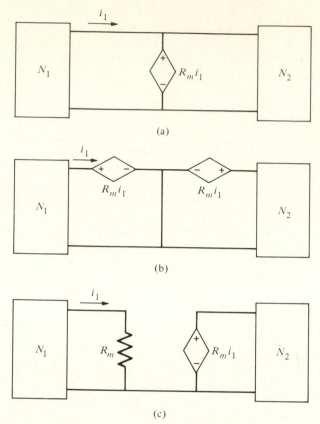

Figure A-7 Voltage source rearrangement and substitution. (a) Original circuit. (b) Rearrangement of voltage source. (c) Substitution of voltage source by a resistor R_m.

Example A-5

Example of Source Rearrangement and Substitution

Fig. A-8 (a) shows a linear active circuit which can be simplified by the source-rearrangement and substitution concepts. The dependent current source $G_m v_1$ is rearranged as illustrated in Fig. A-8 (b). Next, the left-hand dependent current source $G_m v_1$ is substituted by a resistor of value $1/G_m$. The circuit has now been simplified to the point where chain-type calculations can be made.

Another circuit simplification that is possible is called *source reduction.* It is applicable to dependent sources connected in the manner shown in Fig. A-9 (a) and (b). We note that these controlled sources are of the VCVS or CCCS type whereas in the substitution simplification they were VCCS or CCVS types. The circuit-reduction technique for Fig. A-9 (a) states that all currents in N_1 and N_2 remain unchanged if the controlled source, $A_v v_1$, is replaced by a short and if:

1. Each resistance, inductance, reciprocal capacitance, and voltage source in N_1 is multiplied by the factor $1 + A_v$, or
2. Each resistance, inductance, reciprocal capacitance, and voltage source in N_2 is divided by the factor, $1 + A_v$.

The circuit reduction technique for Fig. A-9 (b) states that all voltages in N_1 and N_2 remain unchanged if the controlled source $A_i i_1$ is replaced by a open circuit and if:

1. Each conductance, reciprocal inductance, capacitance, and current source in N_1 is multiplied by the factor, $1 + A_i$, or

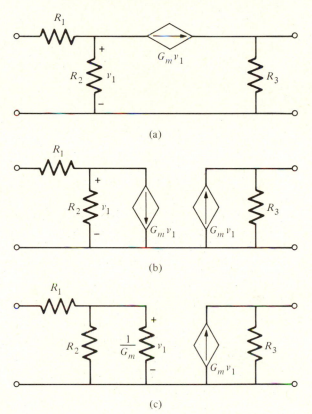

(a)

(b)

(c)

Figure A-8 Current source rearrangement and substitution. (a) Original circuit. (b) Rearrangement. (c) Substitution.

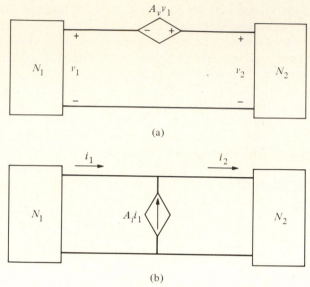

(a)

(b)

Figure A-9 (a) Circuit reduction for a VCVS. (b) Circuit reduction for a CCCS.

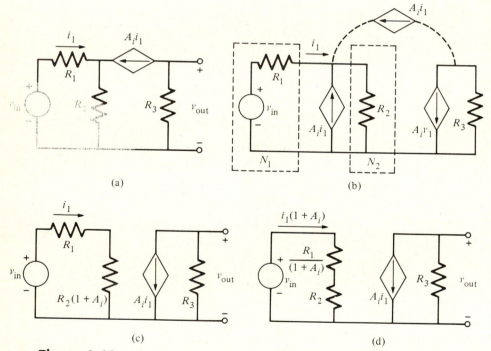

(a)

(b)

(c)

(d)

Figure A-10 Circuit reduction. (a) Original circuit. (b) Identification of N_1 and N_2. (c) Modification of N_2. (d) Modification of N_1.

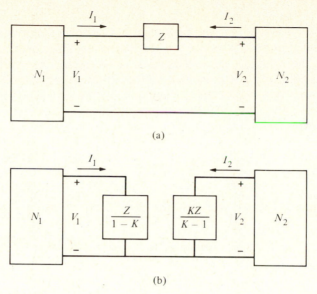

(a)

(b)

Figure A-11 Miller simplification. (a) Original circuit. (b) Equivalent circuit of (a).

2. Each conductance, reciprocal inductance, capacitance, and current source in N_2 is divided by the factor, $1 + A_i$.

An example will illustrate the application of the source-reduction simplification.

Example A-6

Application of the Source Reduction Technique

The circuit of Fig. A-10 (a) is to be simplified using the source-reduction technique. In order to apply the technique, the circuit has been redrawn as shown in Fig. A-10 (b). Note that the presence of a resistor (R_3) in series with the current source does not affect the technique. Fig. A-10 (c) shows the modification of N_1 and Fig. A-10 (d) shows the modification of N_2 by the source-reduction technique. The resulting simplified circuits are suitable for chain-type calculations.

The last simplification technique presented here is called the *Miller simplificaiton*. This technique is good for removing bridging elements in a circuit and accounting for their effect on the forward gain of a circuit. Fig.

A-11 (a) shows the circumstance in which the Miller simplification can be used. We assume that an impedance Z is connected between two networks. We use an impedance since the Miller simplification technique is often used when the bridging element is reactive. We shall also switch to complex-signal notation. The key aspect of the Miller simplification is that V_2 can be expressed in terms of V_1. This expression should be real for best results. If $V_2 = KV_1$, then it can be shown that the impedance seen from N_1, Z_1 can be written as

$$Z_1 = \frac{V_1}{I_1} = \frac{V_i}{(V_1 - V_2)/Z} = \frac{ZV_1}{V_1 - KV_1} = \frac{Z}{1 - K} \qquad (18)$$

and the impedance seen from N_2, Z_2 is

$$Z_2 = \frac{V_2}{I_2} = \frac{V_2}{(V_2 - V_1)/Z} = \frac{ZKV_1}{KV_1 - V_1} = \frac{ZK}{K - 1} \qquad (19)$$

Typically K is negative and greater than unity. An example will illustrate the application of the Miller simplification technique.

Example A-7

Application of the Miller Simplification Technique

Consider the circuit of Fig. A-2. Assuming that R_2 is much greater than R_3, use the Miller simplification technique to remove R_2.

If R_2 is much greater than R_3, then we may assume that most of the current $G_m v_1$ flows through R_3. Therefore, the voltage v_{out} may be expressed as

$$v_{\text{out}} = -G_m R_3 v_1 = K v_1 \qquad (20)$$

Therefore, Fig. A-2 can be redrawn as Fig. A-12 where the resistor R_2' is given as

$$R_2' = \frac{R_2}{1 - K} = \frac{R_2}{1 + G_m R_3} \qquad (21)$$

We have not placed a resistor of value $R_2 K/(K - 1)$ in parallel with R_3 for the following reason. If $G_m R_3$ is greater than unity, then $R_2 K/(K - 1)$ is approximately R_2. However, we used the assumption earlier in the problem that R_2 is greater than R_3. Consequently, it would not make sense to place the resistor R_2 in parallel with R_3.

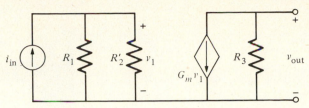

Figure A-12   The result of applying the Miller simplification concept

If the controlled source in Fig. A-2 had been a voltage source rather than a current source, then it would not have been necessary to make the assumption used in Example A-7. In this case, R_2 would have been reflected into both N_1 and N_2. Fig. A-3 shows a case where R_3 could be removed by the Miller simplification and reflected into both N_1 and N_2 according to Eqs. (18) and (19).

Although other simplification techniques could have been included, the above are the ones most useful in analyzing linear active circuits. The material of the book presents many further illustrations of the use of the concepts presented in this appendix. Problems that illustrate the principles developed here can be found at the end of the problems in Chapter 1.

Appendix B

CALCULATOR PROGRAM FOR ANALYZING CMOS CIRCUITS

This program was developed by Randy Conkling and Jim Davis, graduate students at the Georgia Institute of Technology. It calculates the dc drain current and the small-signal conductances and transconductances for a MOSFET given the level 1 or level 2 SPICE model parameters and all terminal voltages. The program allows the storage of three level 1 models and two level 2 models. The minimum system requirements to use this program are an HP-41CV or HP-41C with 4 memory modules.

The operating procedure for the program is as follows.

1. This program uses almost all the available memory in the calculator for program and data storage. Save all current programs in a machine readable form before proceeding further.
2. Key in "XEQ SIZE 100".
3. Delete all programs from memory.
4. Put the calculator in "USER" mode.
5. Load the program (847 lines).
6. Assign "SPICE" to COS key.
7. Key in "XEQ SPICE" to run program (or press COS key in USER mode).
8. Program will respond with the following: (Exact sequence will correspond to the user responses to prompts.)

676

Display	User Response
CREATE FILE?	"Y" OR "N" AND "R/S"
LEVEL 1/2?	"1" OR "2" AND "R/S"
USE FILE?	"Y" OR "N" AND "R/S"
FILE 1/2/3	"1", "2", OR "3" AND "R/S"*
NMOS?	"Y" AND "R/S" FOR NMOS OR "N" AND "R/S" FOR PMOS

9. Program then prompts user for data input.
10. Program displays output, press "R/S" for more output.
11. Voltages and current output for PMOS devices is displayed as absolute values.

The program assumes that the channel width factor, DELTA = 1. The program also assumes that the polysilicon is doped opposite to that of the substrate (TPG = 1) as summarized below.

Gate Type	Substrate Doping	Poly Doping
NMOS	p	n^+
PMOS	n	p^+

The factor NSS is the surface state density and is used in calculating the flat-band voltage. The equations used in this program are from Sections 3.1 and 3.4 and are summarized in Table 3.4-2.

The level 1 model parameters and level 2 model parameters are listed below along with their register location. Examples using these model parameters to solve for the dc current and small signal conductances and transconductances are shown in Table B-1.

LEVEL I MODELS   SUGGESTED MODELS (USER INPUT)

Parameter	Working Regs	Units	Model 1 NMOS (ENH)	Reg	Model 2 PMOS (ENH)	Reg	Model 3 NMOS (DEP)	Reg
NMOS	R88		1	R88	−1	R88	1	R88
W	R28	μm						
L	R29	μm						
VD	R11	volts						
VG	R10	volts						
VS	R12	volts						
VB	R13	volts						
Kprime	R15	a/v^2	17E-6	R46	8E-6	R52	20E-6	R58
Lambda	R16	v^{-1}	0.01	R47	0.02	R53	0.01	R59
VTO	R17	volts	1	R48	−1	R54	−4	R60
Gamma	R	$v^{0.5}$	1.3	R49	0.6	R55	0.37	R61
Phi	R	volts	0.7	R50	0.6	R56	0.64	R62

*Note: Only use file 3 if in level 1.

LEVEL 2 MODELS   SUGGESTED MODELS (USER INPUT)

Parameter	Working Regs	Units	Model 1 NMOS (ENH)	Reg	Model 2 PMOS (ENH)	Reg
NMOS	R88	. . . .	1	R88	-1	R88
W	R28	μm				
L	R29	μm				
VD	R11	volts				
VG	R10	volts				
VS	R12	volts				
VB	R13	volts				
MOB	R14	cm^2/vs	580	R63	230	R75
Cox	R15	F/cm^2	4.3E-8	R64	4.3E-8	R76
Psi	R16	volts	0.35	R65	0.3	R77
UCRIT	R17	v/cm	80E3	R66	80E3	R78
UTRA	R18	. . . .	0.1	R67	0.1	R79
UEXP	R19	. . . .	0.1	R68	0.1	R80
NSS	R21	cm^{-2}	1E10	R69	1E10	R81
Npoly	R22	cm^{-3}	1E20	R70	1E20	R82
Gamma	R23	V$^{0.5}$	1.3	R71	0.6	R83
XJ	R24	μm	1.0	R72	1.0	R84
Nsub	R25	cm^{-3}	1E16	R73	2E15	R85
LD	R26	μm	0.8	R74	0.8	R86

Table B-1

Examples Using the Calculator Program.

SPICE EXAMPLES

$\dfrac{W}{L} = \dfrac{10\mu}{20\mu}$ S 5V ↓ I? G M3 D -0.5927V	**LEVEL 1 (MODEL 2)** $I_D = 46.90\ \mu$A $G_m = 19.371\mu$S $G_{mbs} = 7.502\ \mu$S $G_{ds} = 843.7$ nS	**LEVEL 2 (MODEL 2)** $I_d = 26.209\ \mu$A $G_m = 16.353\ \mu$S $G_{mbs} = 6.059\ \mu$S $G_{ds} = 21.013$ pS

$\dfrac{W}{L} = \dfrac{20\mu}{10\mu}$ -0.9026V   D   ↓   I? -2.42V   G   NMOS   S -5V

LEVEL 1 (MODEL 1) $I_D = 44.178\ \mu$A $G_m = 54.81\ \mu$S $G_{mbs} = 42.58\ \mu$S $G_{ds} = 424.4$ nS	**LEVEL 2 (MODEL 1)** $I_D = 13.022\ \mu$A $G_m = 39.326\ \mu$S $G_{mbs} = 29.222\ \mu$S $G_{ds} = 6.161$ pS

$\dfrac{W}{L} = \dfrac{7\mu}{10\mu}$ 5V   D   NMOS(DEP)

LEVEL 1 (MODEL 3)

$I_D = 114.23\ \mu$a
$G_m = 56.554\ \mu$S
$G_{mbs} = 11.090\ \mu$S
$G_{ds} = 1.0905\ \mu$S

—

PMOS

```
01 LBL^TSPICE
   ΣREG 01
   ENG 4
   CF 01
   CF 02
   CF 03
   ^TCREATE FILE?
   AON
   PROMPT
10 ASTO X
   ^TY
   ASTO Y
   X = Y?
   SF 02
   AOFF
   ^TLEVEL 1/2?
   PROMPT
   1
   X = Y?
20 GTO 01
   SF 03
   LBL 01
   FS?C 02
   GTO 08
   ^TUSE FILE?
   AON
   PROMPT
   ASTO X
   ^TX
30 ASTO Y
   X = Y?
   GTO 03
   AOFF
   CLΣ
   0
   STO 39
   STO 40
   STO 41
   STO 42
40 STO 43
   STO 44
   STO 37
   STO 38
   FS? 03
   XEQ 42
   FS? 03
   GTO 06
   XEQ 15
   GTO 06

50 LBL 15
   1
   STO 14
   STO IND 01
   ^TKPRIME ⇔ ?
   PROMPT
   STO 15
   STO IND 02
   ^TLAMBDA = ?
   PROMPT
60 STO 16
   STO IND 03
   ^TVTO = ?
   PROMPT
   STO 17
   STO IND 04
   ^TGAMMA = ?
   PROMPT
   STO 18
   STO IND 05
70 ^TPHI = ?
   PROMPT
   STO 19
   STO IND 06
   RTN
   LBL 70
   1
   STO 88
   RTN
   LBL 03
80 AOFF
   SF 01
   GTO 08
   LBL 06
   ^TNMOS?
   AON
   PROMPT
   ASTO X
   ^TY
   ASTO Y
90 X = Y?
   SF 04
   AOFF
   -1
   STO 88
   FS?C 04
   XEQ 70
   ^TW = ?
   PROMPT
   STO 28

100 ^TL = ?
    PROMPT
    STO 29
    /
    STO 20
    ^TVD = ?
    PROMPT
    STO 11
    ^TVG = ?
    PROMPT
110 STO 10
    ^TVS = ?
    PROMPT
    STO 12
    ^TVB = ?
    PROMPT
    STO 13
    RCL 10
    RCL 12
    -
120 RCL 88
    *
    STO 10
    RCL 11
    RCL 12
    -
    RCL 88
    *
    STO 11
    RCL 12
    RCL 13
130 -
    RCL 88
    *
    STO 12
    FS? 03
    GTO 55
    RCL 19
    +
    SQRT
140 RCL 19
    SQRT
    -
    RCL 18
    *
    RCL 17
    RCL 88
    *
    +
    STO 17
```

679

```
150  RCL 10              200  STOP                 250  ARCL 16
     RCL 17                   TVSB =                     AVIEW
     —                        ARCL 12                    STOP
     STO 00                   AVIEW                      XEQTSPICE
     X< = 0?                  STOP                       LBL 97
     GTO 04                   FS? 03                     RCL 01
     RCL 11                   GTO 97                     RCL 07
     X< = ?                   RCL 01                     *
     GTO 61                   RCL 14                      2
     GTO 62                   *                           *
160  LBL 61              210  RCL 20              260  SQRT
     STO 38                   *                          STO 14
     GTO 05                   RCL 15                     RCL 12
     LBL 62                   *                          RCL 16
     RCL 00                   2                          2
     STO 38                   *                          *
     LBL 05                   SQRT                       +
     RCL 14                   STO 14                     SQRT
     RCL 15                   RCL 19                     2
     *                        RCL 12                     *
170  RCL 20              220  +                   270  RCL 35
     *                        SQRT                       X<>Y
     RCL 38                   2                          /
     *                        *                          RCL 14
     RCL 00                   RCL 18                     *
     RCL 38                   X <> Y                     STO 15
     2                        /                          RCL 01
     /                        RCL 14                     RCL 08
     —                        *                          *
     *                        STO 15                     RCL 08
180  RCL 16              230  RCL 01              280  RCL 11
     RCL 11                   RCL 16                     *
     *                        *                          1
     1                        RCL 16                     +
     +                        RCL 11                     /
     *                        *                          STO 16
     STO 01                   1                          GTO 98
     GTO 20                   +                          LBL 08
     LBL 20                   /                          TFILE 1/2/3?
     TID =                    STO 16                     PROMPT
190  ARCL 01             240  LBL 98              290  1
     AVIEW                    TGM =                      X=Y?
     STOP                     ARCL 14                    GTO 10
     TVGS =                   AVIEW                      RDN
     TARCL 10                 STOP                       2
     AVIEW                    TGMBS =                    X = Y?
     STOP                     ARCL 15                    GOT 11
     TVDS =                   AVIEW                      RDN
     TARCL 11                 STOP                       3
     AVIEW                    TGDS =                     X = Y?
```

680

```
300  GTO 12        350  FS?C 01       400  STO 01
     LBL 10             GTO 07              76
     FS? 03             XEQ 15              STO 02
     GTO 30             GTO 01              77
     45                 LBL 07              STO 03
     STO 01             RCL IND 01          78
     46                 STO 14              STO 04
     STO 02             RCL IND 02          79
     47                 STO 15              STO 05
     STO 13             RCL IND 03          80
310  48            360  STO 16         410  STO 06
     STO 14             RCL IND 04          81
     49                 STO 17              STO 39
     STO 05             RCL IND 05          82
     50                 STO 18              STO 40
     STO 06             RCL IND 06          83
     GOT 09             STO 19              STO 41
     LBL 11             GTO 06              84
     FS? 03             LBL 04              STO 42
     GTO 31             0                   84
320  51            370  STO 01         420  STO 43
     STO 01             GTO 20              86
     52                 LBL 30              STO 44
     STO 02             63                  GTO 40
     53                 STO 01              LBL 32
     STO 03             64                  TERROR
     54                 STO 02              AVIEW
     STO 04             65                  STOP
     55                 STO 03              LBL 40
     STO 05             66                  FS?C 01
330  56            380  STO 04         430  GTO 41
     STO 06             67                  XEQ 42
     GTO 09             STO 05              GTO 01
     LBL 12             68                  LBL 41
     FS?03              STO 06              RCL IND 01
     GTO 32             69                  STO 14
     57                 STO 39              RCL IND 02
     STO 01             70                  STO 15
     58                 STO 40              RCL IND 03
     STO 02             71                  STO 16
340  59            390  STO 41         440  RCL IND 04
     STO 03             72                  STO 17
     60                 STO 42              RCL IND 05
     STO 04             73                  STO 18
     61                 SOT 43              RCL IND 06
     STO 05             74                  STO 19
     62                 STO 44              RCL IND 39
     STO 06             GTO 40              STO 21
     GTO 09             LBL 31              RCL IND 40
     LBL 09             75                  STO 22
```

```
450 RCL IND 41          500 T_NSUB = ?          550 1
    STO 23                  PROMPT                  +
    RCL IND 42              STO IND 43              STO 31
    STO 24                  STO 25                  RCL 16
    RCL IND 43              T_LD = ?                2
    STO 25                  PROMPT                  *
    RCL IND 44              STO IND 44              RCL 12
    STO 26                  STO 26                  +
    GTO 06                  RTN                     STO 32
    LBL 42                  LBL 55                  RCL 30
460 T_MOB. = ?          510 RCL 22              560 *
    PROMPT                  1.45 E10                RCL 16
    STO IND 01             /                        2
    STO 14                 LN                       *
    T_COX = ?              0.026                     +
    PROMPT                 *                        RCL 89
    STO IND 02             RCL 88                    +
    STO 15                 *                        STO 30
    T_PSI = ?              STO 89                   RCL 32
    PROMPT                 1.45 E10                 SQRT
470 STO IND 03         520 RCL 25              570 RCL 27
    STO 16                 /                        2
    T_UCRIT=?              LN                       *
    PROMPT                 0.026                     1.6 E-19
    STO IND 04             *                        /
    STO 17                 RCL 88                    RCL 25
    T_UTRA = ?             *                        /
    PROMPT                 RCL 89                    SQRT
    STO IND 05             −                        STO 34
    STO 18                 1.6 E-19                 *
480 T_UEXP =?          530 RCL 21              580 STO 33
    PROMPT                 *                        RCL 32
    STO IND 06             RCL 15                   RCL 11
    STO 19                 /                        +
    T_NSS = ?              CHS                      SQRT
    PROMPT                 +                        RCL 34
    STO IND 39             STO 89                   *
    STO 21                 1.0359 E-12              STO 34
    T_NPOLY = ?            STO 27                   RCL 33
    PROMPT                 PI                       2
490 STO IND 40         540 *                   590 *
    STO 22                 4                        RCL 24
    GAMMA = ?              /                        1 E-4
    PROMPT                 RCL 15                   *
    STO IND 41             /                        /
    STO 23                 RCL 28                   1
    T_XJ = ?               1 E-4                    +
    PROMPT                 *                        SQRT
    STO IND 42             /                        1
    STO 24                 STO 30                   −
```

```
600  RCL 24          650  RCL 89          700  —
     *                    +                    RCL 31
     2                    STO 00               /
     /                    RCL 16               RCL 16
     RCL 29               2                    2
     /                    *                    *
     STO 33               RCL 12               +
     RCL 34               +                    RCL 12
     2                    SQRT                 +
     *                    RCL 16               RCL 31
610  RCL 24         660  2               710  RCL 35
     1 E-4                *                    /
     *                    SQRT                 X↑2
     /                    —                    *
     1                    RCL 23               1
     +                    *                    +
     SQRT                 RCL 00               SQRT
     1                    +                    1
     —                    STO 37               X<>Y
     RCL 24               STO 99               —
620  *              670  RCL 17         720  RCL 35
     2                    RCL 27               RCL 31
     /                    *                    /
     RCL 29               RCL 15               X↑2
     /                    /                    *
     STO 34               STO 01               2
     1                    RCL 10               /
     RCL 33               RCL 37               RCL 10
     —                    —                    RCL 30
     RCL 34               RCL 18               —
630  —              680  RCL 11         730  RCL 31
     RCL 23               *                    /
     *                    —                    +
     STO 35               RCL 01               STO 33
     1.6 E-19             X <> Y               X < = 0?
     RCL 27               X < = Y?             GTO 04
     *                    GTO 64               RCL 26
     RCL 25               /                    1E —4
     *                    RCL 19               *
     RCL 16               Y↑X                  2
640  *              690  RCL 14         740  *
     2                    *                    RCL 29
     *                    STO 36               1 E-4
     SQRT                 GTO 66               *
     RCL 15               LBL 64               X<>Y
     /                    RCL 14               —
     2                    STO 36               STO 34
     RCL 16               LBL 66               RCL 11
     *                    RCL 10               RCL 33
     +                    RCL 30               —
```

683

```
750  4
     /
     X↑2
     1
     +
     SQRT
     RCL 11
     RCL 33
     —
     4
760  /
     +
     SQRT
     2
     *
     RCL 27
     *
     1.6 E-19
     /
     RCL 25
770  /
     RCL 34
     /
     RCL 11
     /
     STO 37
     STO 08
     RCL 11
     *
     1
780  X<>Y
     —
     RCL 34
     *
     STO 37
     RCL 33
     RCL 11
     X>Y?
     GTO 91
     GTO 92
790  LBL 91
     RCL 33
     STO 38
     GTO 93
     LBL 92
     RCL 11
     STO 38
     LBL 93
     RCL 16
     2

800  *
     RCL 12
     +
     STO 00
     RCL 38
     +
     1.5
     Y↑X
     RCL 00
     1.5
810  Y↑X
     —
     RCL 35
     *
     2
     *
     3
     /
     STO 00
     RCL 10
820  RCL 30
     —
     RCL 31
     RCL 38
     *
     2
     /
     —
     RCL 38
     *
830  RCL 00
     —
     STO 00
     RCL 36
     RCL 15
     *
     RCL 28
     1 E-4
     *
     *
840  RCL 37
     /
     STO 07
     RCL 00
     *
     STO 01
     GTO 20
     END
```

Appendix

TIME AND FREQUENCY DOMAIN RELATIONSHIPS FOR SECOND-ORDER SYSTEMS

There are many reasons for considering the time and frequency domain relationships of a second-order system in the study of operational amplifiers. One is that many operational amplifier configurations can be modeled with reasonable accuracy assuming just a second-order system. Such a procedure represents a reasonable compromise between complexity and accuracy of the model. Another reason is that these relationships allow us to predict frequency-domain performance from the simpler to measure time-domain performance.

General Second-Order Sysem in the Frequency Domain

The general transfer function of a low-pass, second-order system in the frequency domain using voltage variables is

$$A(s) = \frac{V_o(s)}{V_{in}(s)} = \pm \frac{A_o \omega_n^2}{s^2 + 2\zeta\omega_n s + \omega_n^2} = \pm \frac{A_o \omega_o^2}{s^2 + (\omega_o/Q)s + \omega_o^2} \quad \textbf{(1)}$$

where

A_o = the low-frequency gain of $V_o(s)/V_{in}(s)$

$\omega_o = \omega_n$ = the pole frequency in radians per second

685

ζ = the damping factor ($= 1/2Q$)

Q = the pole Q ($= 1/2\zeta$)

The roots of equation (1) are illustrated in Fig. C-1.

The magnitude of the frequency response can be found from Eq. (1) as

$$|A(j\omega)| = \frac{A_o\omega_n^2}{\sqrt{(\omega_n^2 - \omega^2)^2 + 4\zeta^2\omega_n^2\omega^2}} \tag{2}$$

However, Eq. (2) may be generalized by normalizing the amplitude with respect to A_o and the radian frequency by ω_n to give

$$\frac{|A(j\omega/\omega_n)|}{A_o} = \frac{1}{\sqrt{[1 - (\omega/\omega_n)^2]^2 + 4\zeta^2(\omega/\omega_n)^2}} \tag{3}$$

A plot of Eq. (3) in dB versus log ω/ω_n is shown in Fig. C-2 where ζ or $1/2Q$ is used as a parameter. By taking the derivative of Eq. (3) with respect to ω/ω_n and setting it to zero, the peak of $|A(j\omega/\omega_n)/A_o|$ can be found as

$$M_p = \frac{1}{2\zeta\sqrt{1-\zeta^2}} \tag{4}$$

when $\zeta < 0.707$.

The second order function of Eq. (1) is found in the analysis of many practical systems. Consider the single-loop, feedback block diagram of Fig. C-3. The closed loop gain $A(s)$ can be expressed as

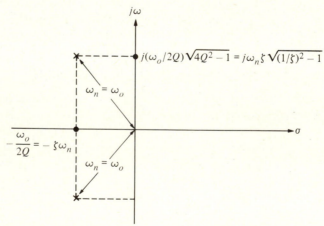

Figure C-1   Pole locations of a general second-order system.

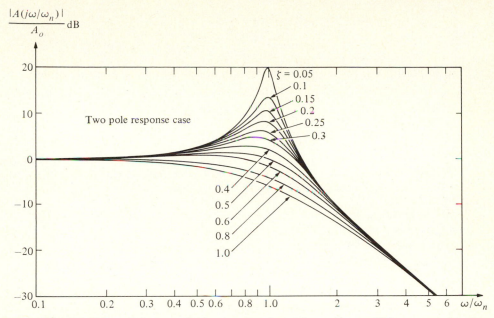

Figure C-2   Gain magnitude response for various values of ζ for a second-order, low-pass system.

$$A(s) = \frac{V_o(s)}{V_i(s)} = \frac{\alpha a \beta}{1 + a\beta} \qquad \textbf{(5)}$$

Let us assume that α and β are real and that a is the amplifier's gain and can be approximated as

$$a(s) \simeq \frac{a_o \omega_1 \omega_2}{(s + \omega_1)(s + \omega_2)} \qquad \textbf{(6)}$$

where a_o is the dc gain of the amplifier and ω_1 and ω_2 are real axis poles. Substitution of Eq. (6) into Eq. (5) gives

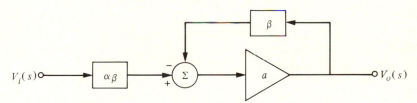

Figure C-3   Single-loop, feedback block diagram.

$$A(s) = (\alpha\beta) \frac{a_o\omega_1\omega_2}{s^2 + (\omega_1 + \omega_2)s + \omega_1\omega_2(1 + a_o\beta)} \tag{7}$$

Comparing Eq. (7) with Eq. (1) results in the following identifications.

$$A_o = \alpha a_o\beta/(1 + a_o\beta) \tag{8}$$

$$\omega_n = \omega_o = \sqrt{\omega_1\omega_2(1 + a_o\beta)} \tag{9}$$

$$2\zeta = 1/Q = \frac{\omega_1 + \omega_2}{\sqrt{\omega_1\omega_2(1 + a_o\beta)}} \tag{10}$$

The same principles can be applied to a second-order bandpass or high-pass system, but the low-pass case is of more practical interest to us and will be the only one considered. It is also possible for β (and thus α) to become frequency dependent which further complicates the analysis.

Low-Pass Second-Order System in the Time Domain

Unfortunately, it is time-consuming to make measurements in the frequency domain. Therefore we are interested in determining the frequency-domain performance from the time domain performance. This information is developed as follows. The general response of Eq. (1) to a unit step can be written as

$$v_o(t) = A_o\left[1 - \frac{1}{\sqrt{1 - \zeta^2}} e^{-\zeta\omega_n t} \sin(\sqrt{1 - \zeta^2}\,\omega_n t + \phi)\right] \tag{11}$$

where

$$\phi = \tan^{-1}\left[\frac{\sqrt{1 - \zeta^2}}{\zeta}\right] \tag{12}$$

The step response plotted in normalized amplitude versus radians is shown in Fig. C-4.

Let us consider first the underdamped case where $\zeta < 1$. For the underdamped case there will always be an overshoot as defined by Fig. C-5. The overshoot can be expressed as

$$\text{Overshoot} = \frac{\text{Peak value} - \text{Final vlaue}}{\text{Final value}} = \exp\left[\frac{-\pi\zeta}{\sqrt{1 - \zeta^2}}\right] \tag{13}$$

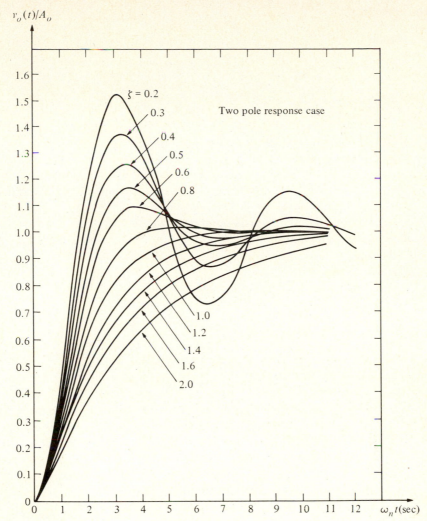

$v_o(t)/A_o$

1.6

1.5 $\zeta = 0.2$

1.4   0.3   Two pole response case

1.3   0.4

1.2   0.5

1.1   0.6

1.0   0.8

0.9

0.8

0.7   1.0

0.6   1.2

0.5   1.4

0.4   1.6

0.3   2.0

0.2

0.1

0

0 1 2 3 4 5 6 7 8 9 10 11 12 $\omega_n t$(sec)

Figure C-4   Step response as a function of ζ for a low-pass, second-order system.

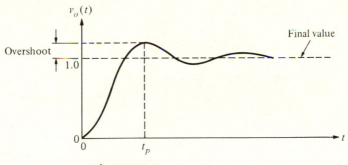

$v_o(t)$

Final value

Overshoot

1.0

0

0   t_p t

Figure C-5   Overshoot and t_p.

689

The time t_p at which the overshoot occurs is shown in Fig. C-5 and can be found as

$$t_p = \frac{\pi}{\omega_n \sqrt{1 - \zeta^2}}$$ (14)

Thus the measurement of the overshoot permits the calculation of ζ (or $1/2Q$). With this information and the measurement of t_p, one can calculate ω_n from Eq. (14). Therefore the frequency response of a second-order, low-pass system with $\zeta < 1$ can be determined by measuring the overshoot and t_p of the step response.

Next consider the overdamped case where $\zeta \geq 1$. In this case there is no overshoot. The unit step response can be simplified from Eq. (10) to yield

$$v_o(t) = A_o \left[1 - \frac{1}{2\sqrt{\zeta^2 - 1}} \left(\frac{e^{-\omega_n t}(\zeta - \sqrt{\zeta^2 - 1})}{\zeta - \sqrt{\zeta^2 - 1}} \right. \right.$$
$$\left. \left. - \frac{e^{-\omega_n t}(\zeta + \sqrt{\zeta^2 - 1})}{\zeta + \sqrt{\zeta^2 - 1}} \right) \right]$$ (15)

It is difficult to measure various aspects of this response and thus determine ζ and ω_n. Fortunately there are very few occasions where we have $\zeta > 1$. If $\zeta > 1$, then the best result is probably obtained by matching the step response to one of the curves for $\zeta > 1$ of Fig. C-4. More accuracy could be achieved by evaluating $v_o(t)$ (for say $\omega_n t = 4$) and selecting values of ζ until $v_o(4/\omega_n)$ matches with the experimental data at this point.

Determination of Phase Margin and Cross-over Frequency from ζ and ω_n

In the previous discussion we have seen how ζ and ω_n of a second-order system can be determined by the time-domain step response. It is the objective of this section to show how to find the phase margin ϕ_m and the crossover frequency ω_c from ζ and ω_n. Fig. C-6 shows the meaning of ϕ_m and ω_c.

In order to assist in developing the desired relationships, it will be convenient to assume that β (thus α) is real. From Eq. (5) we may solve for a to get

$$a\beta = \frac{1/\alpha}{(1/A) - (1/\alpha)}$$ (16)

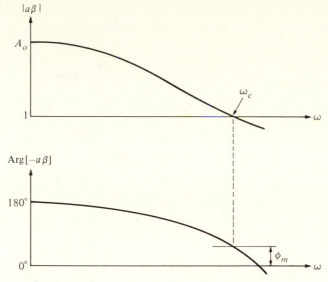

Figure C-6   Crossover frequency and phase margin.

Substituting Eq. (1) into Eq. (16) gives the loop gain as

$$\alpha\beta = \frac{A_o\omega_n^2/\alpha}{(s^2 + 2\zeta\omega_n s + \omega_n^2) - (A_o\omega_n^2/\alpha)}$$

$$= \frac{A_o/\alpha}{\left(\dfrac{s}{\omega_n}\right)^2 + 2\zeta\left(\dfrac{s}{\omega_n}\right) + 1 - (A_o/\alpha)} \qquad \textbf{(17)}$$

when $|\alpha\beta| = 1$, then $\omega = \omega_c$ so that Eq. (16) becomes

$$|\alpha\beta| = \frac{A_o/\alpha}{\sqrt{[1 - A_o/\alpha) - (\omega_c/\omega_n)^2]^2 + [2\zeta(\omega_c/\omega_n)]^2}} \qquad \textbf{(18)}$$

Since $|\alpha\beta| = 1$, we may solve Eq. (18) for ω_c to get

$$\omega_c = \omega_n[\sqrt{[2\zeta^2 - (1 - A_o/\alpha]^2 - (1 - 2A_o/\alpha)} \\ - 2\zeta^2 + (1 - A_o/\alpha)]^{1/2} \qquad \textbf{(19)}$$

Knowing A_o, and α, ω_n, and ζ, we may calculate the cutoff frequency of a second order system. In an operational amplifier circuit, $\alpha = A_o$ so that Eq. (19) becomes

$$\omega_c = \omega_n[\sqrt{4\zeta^4 + 1} - 2\zeta^2]^{1/2} \tag{20}$$

Fig. C-7 gives a plot of this useful function.

The phase of $a\beta$ can be found from Eq. (17). However we must add $\pm \pi$ to this value to account for the minus sign of the summing junction of Fig. C-3. Thus

$$\phi_m = -\tan^{-1}\left[\frac{2\zeta\omega_c/\omega_n}{(1 - A_o/\alpha) - (\omega_c/\omega_n)^2}\right] \tag{21}$$

Since $A_o = \alpha$, we may write Eq. (21) as

$$\phi_m = \tan^{-1}\left[\frac{2\zeta}{\omega_c/\omega_n}\right] \tag{22}$$

Substituting Eq. (20) into Eq. (22) yields

$$\phi_m = \tan^{-1}\left[\frac{2\zeta}{(\sqrt{4\zeta^4 + 1} - 2\zeta^2)^{1/2}}\right] \tag{23}$$

An equivalent form of Eq. (23) is

$$\phi_m = \cos^{-1}[\sqrt{4\zeta^4 + 1} - 2\zeta^2] \tag{24}$$

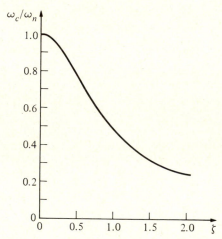

Figure C-7   Plot pf ω_c/ω_n versus ζ for a low-pass, second-order system.

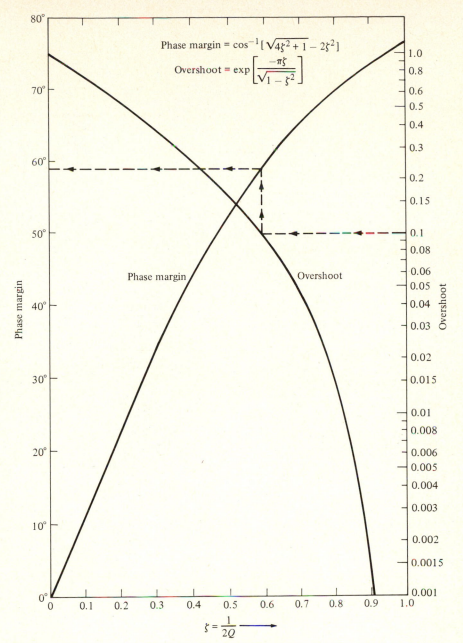

Phase margin = $\cos^{-1}[\sqrt{4\zeta^2 + 1} - 2\zeta^2]$

Overshoot = $\exp\left[\dfrac{-\pi\zeta}{\sqrt{1 - \zeta^2}}\right]$

$\zeta = \dfrac{1}{2Q}$

Figure C-8 Relationship between the phase margin and the overshoot for a second-order system as a function of the damping factor.

Fig. C-8 gives a plot of ϕ_m of Eq. (23) or Eq. (24) and of Eq. (13) as a function of the damping factor ζ. Therefore, the time-domain performance characterized by ζ permits the designer to estimate a value of phase margin using Eq. (24) or Fig. C-8.

*I*ndex

FOX *see* field oxide
fractional temperature coefficient *see*
    temperature coefficient, fractional
frequency doubler, 607
full scale error, 525. (*See also* gain error)
full-wave rectifier *see* rectifier, full-wave

gain error, 525–526, 539
gainbandwidth (GB), 379, 381, 383, 385, 411,
    413, 424, 499–500
GAMMA (bulk threshold parameter), 134,
    160–161
gate, 48
GB *see* gainbandwidth, and/or unity-gain
    bandwidth
gds *see* conductance drain/source
geometry, 154
gmbs *see* transconductance bulk-channel
grading coefficient, 42, 105. (*See also* MJ,
    MJSW)
guard ring, 70
guidelines for simulation *see* simulator usage

half-wave rectifier *see* rectifier, half-wave
harmonic distortion, 302
hierarchy, 8, 198
high speed op amps, 479–486
hysteresis, 349–357

integral linearity *see* linearity, integral
intrinsic carrier concentration (n_i), 40, 98, 592
inverter, 258–273
    active resistor load inverter, 258, 260–
        262
    current source load inverter, 263–267,
        300
    push-pull, 269, 305
inversion *see* moderate inversion, strong
    inversion, weak inversion
ion implantation, 31, 33
ion implanted resistor, 63
IS (reverse current of the bulk-drain/source
    junctions), 47, 135, 243

JS (reverse current density of the bulk-drain/
    source junctions), 135, 594
junction capacitance *see* capacitance, junction
junction depth, 33

KAPPA (field correlation factor), 136
K_L', 156, 160
K_S', 156

KF (flicker noise coefficient), 112, 135, 172–
    174
KP (intrinsic transconductance parameter),
    134

LAMBDA (channel length modulation
    parameter), 97, 134
latch up, 69
lateral BJT *see* BJT, lateral
lateral diffusion (LD), 108–109, 120, 135.
    (*See also* overlap)
layout rules, 76, 78, 81–82. (*See also* design
    rules)
length, 48
length, effective; 108, 120, 157, 165
Level 1 model 102, 135. (*See also* simple MOS
    model)
Level 2 model *see* extended MOS model
Level 3 model, 136
LHP, 379, 383, 471
linear regression, 158–159
linearity
    best straight-line, 527–528
    differential, 527, 542–543
    endpoint, 527–528
    integral, 528, 543
low noise amplifiers, 487–490

mask, 76, 232
matching, 229, 232, 234
mesh analysis, 665
micropower op amps, 497–504, 512
Miller
    capacitance, 296, 421
    compensation, 377, 380, 408, 460, 467
    simplification, 673–675
minimum dimension resolution, (lambda) 78
minority carrier concentration, 46
MJ, 105, 184, 187–189
MJSW, 105–107, 135, 184, 187, 189
moat, 55, 78, 81–82, 186. (*See also* active
    area)
mobility, 72, 160
    degradation, 168, 171
    surface, 97
    temperature dependence, 72, 122
model, modeling, 95, 662
    extended MOS, 119–122
    large signal, 104
    parameters
        for a typical CMOS bulk process, 99
        for the extended model, 121